Determination of Anions

Springer
*Berlin
Heidelberg
New York
Barcelona
Budapest
Hong Kong
London
Milan
Paris
Santa Clara
Singapore
Tokyo*

T.R. Crompton

Determination of Anions

A Guide for the Analytical Chemist

With 97 Figures and 202 Tables

Thomas Roy Crompton
Hill Cottage (Bwthyn Yr Allt)
LL58 8SUM Beaumaris
Anglesey Gwynedd
Great Britain

Coventry University

ISBN 3-540-60162-7 Springer-Verlag Berlin Heidelberg New York

Die Deutsche Bibliothek – CIP-Einheitsaufnahme
Crompton, Thomas R.:
Determination of anions; a guide for the analytical chemist; with 202 tables
/T.R. Crompton - Berlin; Heidelberg; New York; Barcelona; Budapest;
Hong Kong; London; Milan; Paris; Santa Clara; Singapore; Tokyo:
Springer, 1996
  ISBN 3-540-60162-7

This work is subject to copyright. All rights are reserved, whether the whole or part of the material is concerned, specifically the rights of translation, reprinting, reuse of illustrations, recitation, broadcasting, reproduction on microfilm or in other way, and storage in data banks. Duplication of this publication or parts thereof is permitted only under the provisions of the German Copyright Law of September 9, 1965, in its current version, and permission for use must always be obtained from Springer-Verlag. Violations are liable for prosecution under the German Copyright Law.

© Springer-Verlag Berlin Heidelberg 1996
Printed in Germany

The use of general descriptive names, registered names, trademarks, etc. in this publication does not imply, even in the absence of a specific statement, that such names are exempt from the relevant protective laws and regulations and therefore free for general use.

Cover design: LEWIS + LEINS GmbH, Berlin

Typesetting: Thomson Press (India) Ltd., Madras

SPIN: 10493572   52/3020/SPS – 5 4 3 2 1 0 – Printed on acid-free paper

Preface

Many books have been written on the determination of metals (cations) in water and in other materials, including several by the author.

No up to date comprehensive book exists, however, on the determination of anions which is a subject of equal importance to the determination of cations.

The author has drawn together in one volume all material available in the world literature published since 1975 on the determination of anions, and has presented it in a logical manner so that the reader can quickly gain access to methods for the determination of anions, and to the types of instrumentation available for carrying out these analyses. There is a general discussion of preferred methodologies, and the contents pages indicate the coverage of the book and demonstrate the wide range of anions and types of sample covered.

The contents pages list under each anion the various types of sample. In Chap. 1, types of instruments and determinands are cross referenced.

Chapter 1 cross references with the text the types of anions determined, the types of sample, and the types of instrumentation employed. It also gives a brief summary of the theory of the types of instrumentation used in anion analyses.

Chapters 2–10 cover systematically the various types of anions, including simple anions, complex anions, metal-containing anions and organic anions.

Chapter 11 discusses in detail the rapidly developing subject of ion chromatography and its applications to the analysis of mixtures of anions.

Chapter 12 covers rapid spot tests for use in the field, e.g. by water chemists, whilst Chap. 13 discusses automated on-line instrumentation for the determination of anions in industrial process liquids.

Gwynedd, Great Britain
June 1996

T.R. Crompton

Contents

1	**Introduction**	1
1.1	Rationale: Water Samples	1
1.2	Rationale: Solid and Gaseous Samples	16
1.3	Brief Summary of Methodologies	16
1.3.1	Spectroscopic Methods	16
1.3.2	Segmented Flow Analysis	24
1.3.3	Flow Injection Analysis	26
1.3.4	Spectrometric Methods	30
1.3.5	Titration Procedures	35
1.3.6	Ion-Selective Electrodes	38
1.3.7	Polarographic Methods	41
1.3.8	Chromatographic Methods	44
1.4	References	57
2	**Halogen-Containing Anions**	58
2.1	Chloride	58
2.1.1	Natural Waters	58
2.1.2	Rainwater and Snow	69
2.1.3	Potable Waters	69
2.1.4	Industrial Effluents	70
2.1.5	Waste Waters	71
2.1.6	Sewage	72
2.1.7	Boiler Feed and High Purity Waters	72
2.1.8	Soil and Plant Extracts	77
2.1.9	Soils	77
2.1.10	Foodstuffs	78
2.2	Bromide	78
2.2.1	Natural Waters	78
2.2.2	Rainwater and Snow	89
2.2.3	Potable Water	89
2.2.4	Industrial Effluents	91

2.2.5	Soil and Foodstuffs	91
2.3	Iodide	92
2.3.1	Natural Waters	92
2.3.2	Rainwater	96
2.3.3	Potable Water	96
2.3.4	Industrial Effluents	102
2.3.5	Cooling Waters	102
2.3.6	Foodstuffs	103
2.3.7	Seaweed	103
2.3.8	Animal Feeds	103
2.3.9	Table Salt	103
2.3.10	Milk	104
2.3.11	Pharmaceuticals	104
2.4	Fluoride	104
2.4.1	Natural Waters	104
2.4.2	Rainwater	107
2.4.3	Potable Water	107
2.4.4	Industrial Effluents	113
2.4.5	Waste Waters	113
2.4.6	Sewage	114
2.4.7	Biological Materials	114
2.4.8	Plant Extracts and Vegetative Matter	115
2.4.9	Milk	115
2.4.10	Coal	115
2.5	Mixed Halides	116
2.5.1	Natural Waters	116
2.5.2	Rainwater	119
2.5.3	Potable Waters	124
2.5.4	Industrial Effluents	125
2.5.5	Waste Waters	125
2.6	Chlorate, Chlorite, Perchlorate and Hypochlorite	126
2.6.1	Natural Waters	126
2.6.2	Waste Waters	128
2.6.3	Potable Waters	128
2.6.4	Soil Extracts	128
2.7	Bromate	128
2.7.1	Natural Waters	128
2.7.2	Rainwater	129
2.7.3	Potable Waters	129
2.7.4	Foodstuffs	129
2.8	Iodate	130
2.8.1	Natural Waters	130
2.8.2	Potable Waters	133
2.9	References	133

3	**Nitrogen-Containing Anions**	**140**
3.1	Nitrate	140
3.1.1	Natural Waters	140
3.1.2	Rainwater and Snow	168
3.1.3	Potable Water	172
3.1.4	Industrial Effluents	183
3.1.5	Waste Waters	183
3.1.6	Sewage	184
3.1.7	Plant and Soil Extracts	185
3.1.8	Soils	185
3.1.9	Plants	186
3.2	Nitrite	187
3.2.1	Natural Water	187
3.2.2	Rainwater	209
3.2.3	Industrial Effluents	210
3.2.4	Waste Waters	211
3.2.5	High Purity Water	216
3.2.6	Soil	216
3.2.7	Milk	217
3.3	Nitrate and Nitrite	217
3.3.1	Natural Waters	217
3.3.2	Rainwater	234
3.3.3	Potable Water	234
3.3.4	Industrial Effluents	239
3.3.5	Sewage	239
3.3.6	Soil	240
3.3.7	Biological Fluids	240
3.3.8	Meat	240
3.3.9	Vegetables	241
3.4	Free Cyanide	241
3.4.1	Natural Waters	241
3.4.2	Potable Waters	247
3.4.3	Industrial Effluents	247
3.4.4	Waste Waters	251
3.4.5	Plant Materials	261
3.4.6	Feeds, Fertilizers and Plant Tissues	261
3.4.7	Soil	261
3.4.8	Biological Fluids	262
3.4.9	Air	262
3.4.10	Sewage Sludge	262
3.5	Total Cyanide	262
3.5.1	Natural Waters	263
3.5.2	Waste Waters	263
3.6	Cyanate	271

3.6.1	Natural Waters	271
3.7	References	271

4	**Sulphur Containing Anions**	278
4.1	Sulphate	278
4.1.1	Natural Waters	278
4.1.2	Rainwater	300
4.1.3	Potable Water	305
4.1.4	Industrial Effluents	307
4.1.5	Wastewaters	308
4.1.6	Boiler Feed Water	309
4.1.7	Soil	309
4.1.8	Plant and Soil Extracts	310
4.1.9	Grain	311
4.2	Sulphite	311
4.2.1	Natural Waters	311
4.2.2	Rainwater	315
4.2.3	Industrial Effluents	315
4.2.4	Waste Waters	315
4.2.5	Boiler Feed Water	316
4.2.6	Wine	316
4.2.7	Foodstuffs	317
4.2.8	Pharmaceuticals	318
4.3	Thiosulphate and Polythionates	318
4.3.1	Natural Waters	318
4.3.2	Industrial Effluents	319
4.4	Thiocyanate	320
4.4.1	Natural Waters	320
4.4.2	Industrial Effluents	321
4.4.3	Waste Waters	323
4.4.4	Urine	323
4.5	Sulphide	324
4.5.1	Natural Waters	324
4.5.2	Potable Water	329
4.5.3	Industrial Effluents	329
4.5.4	Waste Waters	330
4.5.5	Sewage	331
4.5.6	Soil	334
4.5.7	Environmental Samples	335
4.5.8	Blood	335
4.5.9	Air	336
4.6	Polysulphide	336
4.6.1	Natural Waters	336

4.6.2	Industrial Effluents	336
4.6.3	Petroleum Fractions	336
4.7	References	336

5	**Phosphorus Containing Anions**	342
5.1	Phosphate	342
5.1.1	Natural Waters	342
5.1.2	Sea and Estuary Water	366
5.1.3	Rainwater	367
5.1.4	Potable Water	368
5.1.5	Industrial Effluents	368
5.1.6	Waste Waters	368
5.1.7	Sewage	372
5.1.8	Boiler Feed Water	373
5.1.9	Soil and Plant Extracts	373
5.1.10	Blood and Serum	374
5.2	Condensed Phosphates	374
5.2.1	Ion Exclusion Chromatography	375
5.2.2	High Performance Liquid Chromatography . .	375
5.3	References	379

6	**Silicon-Containing Anions**	381
6.1	Silicate	381
6.1.1	Natural Waters	381
6.1.2	Potable Water	383
6.1.3	Industrial Effluents	385
6.1.4	Waste Waters	386
6.1.5	Boiler Feed Water	386
6.2	References	387

7	**Boron-Containing Anions**	388
7.1	Borate	388
7.1.1	Natural Waters	388
7.1.2	Industrial Effluents	396
7.1.3	Sewage	397
7.1.4	Soil	397
7.1.5	Plants	397

7.1.6	Plant Extracts		398
7.2	References		398

8 Carbonate, Bicarbonate and Total Alkalinity . 400

8.1	Total Alkalinity	400
8.1.1	Natural Waters	400
8.1.2	Potable Waters	411
8.2	Carbonate	414
8.2.1	Natural Waters	414
8.2.2	Waste Waters	414
8.2.3	High Purity Waters	414
8.2.4	Soil	415
8.3	Bicarbonate	415
8.3.1	Natural Waters	415
8.3.2	Potable Waters	416
8.4	References	418

9 Metal-Containing Anions 419

9.1	Arsenite	419
9.1.1	Natural Waters	419
9.2	Arsenate	419
9.2.1	Natural Waters	419
9.2.2	Potable Water	420
9.3	Arsenite/Arsenate	420
9.3.1	Natural Waters	420
9.4	Selenite	421
9.4.1	Natural Waters	421
9.4.2	Potable Water	423
9.4.3	Soil Extracts	423
9.4.4	Milk	423
9.4.5	Urine	424
9.4.6	Biological Samples	424
9.5	Selenate	425
9.5.1	Natural Waters	425
9.5.2	Potable Water	425
9.6	Selenate/Selenite	425
9.6.1	Sea Water	425
9.6.2	Natural Waters	429
9.6.3	Potable Waters	433
9.6.4	Industrial Effluent	433
9.7	Tellurate	434

9.7.1	Natural Waters	434
9.8	Chromate/Dichromate	434
9.8.1	Natural Waters	434
9.8.2	Waste Waters	442
9.8.3	Potable Water	445
9.8.4	Oxides, Silicate Glasses and Cements	445
9.9	Germanate	445
9.9.1	Natural Waters	445
9.10	Molybdate	445
9.10.1	Natural Waters	445
9.10.2	Sea Water	446
9.11	Tungstate	446
9.11.1	Natural Waters	446
9.11.2	Tissues	447
9.12	Uranate	447
9.12.1	Natural Waters	447
9.13	Vanadate	447
9.13.1	Natural Waters	447
9.13.2	Soil and Plant Extracts	447
9.14	Titanate	448
9.14.1	Natural Waters	448
9.15	Ferrocyanide	448
9.15.1	Industrial Effluents	448
9.16	Metal Cyanide Complexes	449
9.16.1	Natural Waters	449
9.17	References	450
10	**Organic Anions**	**453**
10.1	Formate and Acetate	453
10.1.1	Natural Waters	453
10.2	Citrate	454
10.2.1	Foodstuffs and Beverages	454
10.3	Isocitrate	454
10.3.1	Foodstuffs	454
10.4	Oxalate	455
10.4.1	Foodstuffs and Beverages	455
10.5	Malate	456
10.5.1	Foodstuffs and Beverages	456
10.6	Lactate	456
10.6.1	Silage	456
10.6.2	Foodstuffs and Beverages	456
10.7	Salicylate	457
10.7.1	Urine	457

10.7.2	Pharmaceuticals	457
10.8	Ascorbate	457
10.8.1	Natural Waters	457
10.8.2	Foodstuffs	458
10.8.3	Pharmaceuticals	458
10.8.4	Crustacea	459
10.9	Dehydroascorbate	459
10.9.1	Foodstuffs	459
10.10	Pyruvate	460
10.10.1	Milk	460
10.11	Glutamate	460
10.11.1	Foodstuffs	460
10.12	Amino Acids	461
10.12.1	Natural Waters	461
10.13	Nitroloacetate and Ethylene Diamine Tetraacetates	462
10.13.1	Natural Waters	462
10.13.2	Waste Waters	462
10.14	Carboxylates	463
10.14.1	Natural Waters	463
10.15	Sulphonates and Chlorolignosulphonates	464
10.15.1	Natural Waters	464
10.15.2	Potable Waters	464
10.16	Acetoacetate	466
10.16.1	In Body Fluids	466
10.17	Mixed Organic Anions	466
10.17.1	Natural Waters	466
10.17.2	Sea Water	467
10.18	References	467
11	**Applications of Ion Chromatography**	**469**
11.1	Natural Waters	469
11.1.1	Mixtures of Chloride, Bromide, Fluoride, Nitrite, Nitrate, Sulphate, Phosphate and Bicarbonate	469
11.1.2	Mixtures of Arsenate, Arsenite, Selenate and Selenite	486
11.1.3	Borate	490
11.1.4	Miscellaneous Anions	495
11.2	Rainwater	495
11.2.1	Mixtures of Chloride, Bromide, Fluoride, Nitrite, Nitrate, Sulphite, Sulphate and Phosphate	495
11.3	Potable Water	516

11.3.1	Mixtures of Chloride, Bromide, Nitrate and Sulphate	516
11.3.2	Mixtures of Sulphide and Cyanide	519
11.3.3	Mixtures of Arsenate, Selenate and Selenite . .	520
11.4	Industrial Effluents and Waste Waters	521
11.4.1	Mixtures of Chloride, Bromide, Fluoride, Nitrite, Nitrate, Sulphate and Phosphate	521
11.4.2	Mixtures of Cyanide, Sulphide, Iodide and Bromide	524
11.4.3	Sulphite and Dithionate	530
11.4.4	Borate	530
11.5	Boiler Feed Water	531
11.5.1	Mixtures of Chloride and Sulphate	531
11.5.2	Miscellaneous Anions	533
11.6	Plant and Soil Extracts	533
11.6.1	Mixtures of Chloride, Nitrate, Sulphate and Phosphate	533
11.7	Foodstuffs	534
11.8	Column Coupling Isotachophoresis of Natural Waters	536
11.8.1	Mixtures of Chloride, Fluoride, Nitrite, Nitrate Sulphate and Phosphate	536
11.8.2	Miscellaneous Anions	539
11.9	References	542
12	**On-Site Measurement of Anions**	544
13	**On-Line Measurement of Anions**	550
14	**Preconcentration of Anions**	555
14.1	Iodide	556
14.2	Phosphate	556
14.3	Sulphide	557
14.4	Borate	558
14.5	Arsenite	558
14.6	Selenate/Selenite	558
14.7	Chromate	559
14.8	Molybdate	563
14.9	Rhenate	564
14.10	Complex Anions	564
14.11	References	569
Subject Index		571

1 Introduction

1.1 Rationale: Water Samples

The contents pages of this book list the types of anions discussed, together with subsections dealing with each type of sample. For some readers it may be of interest to list the analytical techniques used for the determination of each type of anion. This is discussed below.

Aqueous sample types discussed include natural waters, rainwater and snow, potable waters, effluents of various types including sewage works and industrial effluents and wastewaters, and high purity and boiler feed waters. Rainwater and potable waters are separated from natural waters and the concentrations of anions in these, particularly in rainwater, are usually extremely low, and may require different methods from those adopted for natural waters.

Table 1.1 reviews the major methods used for the determination of anions in water samples. In Table 1.2a–c, analytical techniques are cross referenced with the anion determined and the section number in the book. It is seen that by far the most widely applicable methods are ion chromatography (29 anions), spectrometry (28 anions), gas chromatography (17 anions), atomic absorption spectrometry (15 anions increasing to 21 anions if inductively coupled plasma atomic emission spectrometry is included), high performance liquid chromatography (15 anions), flow injection analysis and polarography (14 anions each), titration methods (13 anions), ion selective electrodes (12 anions) and segmented flow analysis (11 anions).

Less commonly used methods include spectrofluorimetry and ion exchange chromatography (8 anions each), column coupling isotachophoresis (6 anions), micelle chromatography and ultraviolet spectroscopy (5 anions each), continuous flow analysis (4 anions), neutron activation analysis, isotope dilution analysis, and radiochemical methods (3 anions each), and potentiometric methods, enzyme methods, Raman spectroscopy, flame emission spectrometry, coulometry, and turbidimetric and nepthelometric methods (1 anion each).

Table 1.1. Major methods used in determination of anions in liquid samples

	Cl	Br	I	ClO₃ BrO₃ ClO₂ OCl	IO₃	NO₃	NO₂	Free CN	Total CN	CNO	SO₄	SO₃	S₂O₃	CNS	S	Poly PO₄ S	SiO₃	BO₄	CO₃	HCO₃	AsO₃
Spectrophotometry	✓	✓	✓	✓	✓	✓	✓	✓	✓		✓	✓	✓	✓	✓	✓	✓	✓	✓		✓
Ion Chromatography	✓	✓	✓	✓	✓	✓	✓	✓	✓	✓	✓	✓	✓	✓	✓	✓	✓	✓	✓		
Gas Chromatography	✓	✓	✓	✓		✓	✓	✓	✓												
Titration	✓	✓	✓	✓							✓				✓						
Flow injection analysis	✓	✓	✓	✓	✓	✓	✓	✓	✓		✓	✓	✓	✓	✓		✓				
Polarography		✓	✓	✓		✓	✓				✓	✓	✓	✓	✓						
HPLC		✓	✓			✓	✓					✓	✓	✓	✓						
Ion Selective Electrodes	✓	✓	✓	✓		✓	✓	✓			✓			✓	✓						
Segmented flow analysis						✓	✓		✓		✓	✓			✓		✓	✓			
Atomic absorption spectrometry	✓										✓										
Spectro fluorometry						✓	✓				✓				✓						
Icpaes		✓	✓								✓										
Column Coupling Isotachophoresis	✓	✓				✓	✓				✓										
Ion exchange chromatography	✓		✓			✓	✓				✓										
Ion exclusion chromatography																✓			✓	✓	
Micelle chromatography	✓	✓	✓		✓	✓	✓														

Rationale: Water Samples

Method										
Spectrophotometry	✓	✓								
Ion Chromatography	✓	✓	✓	✓	✓	✓	✓	✓		
Gas Chromatography	✓	✓					✓			
Titration										
Flow injection analysis	✓	✓	✓	✓						
Polarography	✓	✓	✓	✓						
HPLC		✓	✓							
Ion Selective Electrodes			✓							
Segmented flow analysis			✓							
Atomic absorption spectrometry	✓	✓	✓	✓						
Spectrofluorometry		✓	✓							
Icpaes		✓	✓							
Column Coupling Isotachophoresis										
Ion exchange chromatography										
Ion exclusion chromatography										
Micelle chromatography										

4 Introduction

Table 1.2a. Methods used in determination of anions in liquid samples. Cross referenced with Section number

Anion	Technique									
	Titration	Spectro-photometry	Turbid-metry	Nephelo-metry	Continuous Flow Analysis	Segmented Flow Analysis	Spectro-fluorimetry	Ultra Violet Spectro-scopy	Flow Injection Analysis	Atomic Absorption Spectro-metry
Chloride	2.1.1 2.1.3 2.1.4 2.1.5 2.1.7	2.1.1 2.1.7			2.1.2				2.1.1 2.1.2 2.1.7	2.1.1
Bromide	2.2.1	2.2.1 2.2.3							2.2.1	
Iodide	2.3.1 2.3.5	2.3.1, 2.3.3 2.3.2								
Fluoride		2.4.1 2.4.3 2.4.6				2.4.1				
Mixed halide										
Chlorate, Chlorite, Hypochlorite	2.6.1	2.6.2							2.6.1	
Bromate										
Iodate		2.8.1								
Nitrate		3.1.1 – 3.1.5			3.1.2 3.1.3	3.3.1 3.3.5	3.1.1	3.1.1 3.1.3	3.1.1 3.1.2 3.1.3	3.1.1
		3.3.1 3.3.3 3.3.5								

Rationale: Water Samples   5

Determinand								
Nitrite		3.2.1 3.2.3 3.2.4 3.3.1 3.3.3 3.3.5		3.2.2	3.2.1	3.3.1 3.2.1	3.2.1 3.2.2	
Free cyanide	3.4.3	3.4.1 3.4.3 3.4.4		3.4.3	3.4.3			3.4.1
Total cyanide	3.5.2	3.5.2			3.3.5			3.5.1
Cyanate								
Sulphate	4.1.1 4.1.4	4.1.1 4.1.2 4.1.3 4.1.4	4.1.1	4.1.2		4.1.1	4.1.1 4.1.2 4.1.3	4.1.1 4.1.3
Sulphite	4.2.1	4.2.1 4.2.4			4.2.1	4.2.1		
Thiosulphate	4.3.1 4.3.2	4.3.1					4.3.1	4.3.2
Thiocyanate	4.4.2	4.4.1 4.4.3						
Sulphide	4.5.1 4.5.3 4.5.4	4.5.1 4.5.4			4.5.1	4.5.1	4.5.1	4.5.1

Table 1.2a (*Contd.*)

Anion	Titration	Spectro-photometry	Turbid-metry	Nephelo-metry	Continuous Flow Analysis	Segmented Flow Analysis	Spectro-fluorimetry	Ultra Violet Spectro-scopy	Flow Injection Analysis	Atomic Absorption Spectro-metry
Polysulphide	4.6.1									
Phosphate		5.1.1, 5.1.6 5.1.2, 5.1.7			5.1.2 5.1.3	5.1.1		5.1.6	5.1.1 5.1.2	5.1.1
Silicate		6.1.1 6.1.3 6.1.5				6.1.1			6.1.1	6.1.4 6.1.3
Borate		7.1.1 7.1.3				7.1.1			7.1.1	
Total alkalinity Carbonate	8.1.1	8.1.1 8.1.2				8.1.1			8.1.1 8.1.2	
Bicarbonate	8.3.1					8.3.1				
Arsenate		9.3.1							9.2.1	9.1.1 9.2.1
Arsenite		9.2.1 9.3.1							9.2.1	
Selenite		9.4.1 9.6.1					9.4.1, 9.6.1 9.6.4			9.6.1
Selenate							9.5.2 9.6.1 9.6.4			9.6.1
Tellurate		9.7.1								
Chromate		9.8.1 9.8.2				9.8.1			9.8.1	9.8.1 9.8.2

Germanate		9.9.1
Molybdate		9.10.1
		9.10.2
Tungstate		9.11.1
Uranate	9.12.1	
Vanadate	9.13.1	
Ferrocyanide	9.15.1	
Titanate	9.14.1	
Formate		
Ascorbate	10.8.1	
Dehydro-ascorbate		10.9.1
Aminoacids NTA, EDTA		10.13.1
Sulphonic and carboxylic acids		

Table 1.2b. Methods used in determination of anions in liquid samples. Cross referenced with Section number

Anion	Technique									
	ICPAES	Flame Emission Spectrometry	Ion Selective Electrodes	Polarography	Coulometry	Potentiometric Methods	Neutron Activation Analysis	Isotope Dilution Methods	Radio-Chemical Methods	Raman spectroscopy
Chloride	2.1.1		2.1.2 2.1.6 2.1.7		2.1.1			2.1.1 2.1.2	2.1.1	
Bromide	2.2.1		2.2.1	2.2.1			2.2.1	2.2.2	2.2.1	
Iodide	2.3.1		2.3.1	2.3.1		2.3.1	2.3.1	2.3.1	2.3.1	
Fluoride	2.4.1		2.4.1–2.4.3 2.4.6							
Mixed halide			2.5.1, 2.5.2							
Chlorate, Chlorite, Hypochlorite										
Bromate										
Iodate			2.8.1							
Nitrate			3.1.1 3.1.6	3.1.2 3.1.3 3.3.1						3.1.4 3.3.1
Nitrite				3.2.1 3.3.1						3.2.4 3.3.1
Free cyanide			3.4.1 3.4.3 3.4.4	3.4.3						

Rationale: Water Samples

Total cyanide			3.5.1 3.5.2				
Cyanate	4.1.1	4.1.2					
Sulphate			4.1.1 4.1.2 4.1.5				
Sulphite			4.2.1				
Thiosulphate							
Thiocyanate			4.4.1 4.4.3				
Sulphide			4.5.1 4.5.4	4.5.1 4.5.3			
Polysulphide					4.6.3		
Phosphate				5.1.1			5.1.1
Silicate				6.1.2			
Borate			7.1.1				
Total alkalinity Carbonate Bicarbonate							
Arsenite				9.3.1			
Arsenate				9.2.1 9.3.1			

Table 1.2b (*Contd.*)

Anion	Technique										
	ICPAES	Flame Emission Spectrometry	Ion Selective Electrodes	Polarography	Coulometry	Potentiometric Methods	Neutron Activation Analysis	Isotope Dilution Methods	Radio-Chemical Methods	Raman spectroscopy
Selenite				9.4.1, 9.6.2 9.6.4						
Selenate				9.6.2 9.6.4						
Tellurate										
Chromate				9.8.1 9.8.2						
Germanate										
Molybdate										
Tungstate	9.11.1									
Uranate										
Vanadate										
Ferrocyanide										
Metal cyanide complex										
Acetate			10.1.1							
Isoascorbate										
Dehydroascorbate										
Aminoacids										
NTA, EDTA				10.13.1						
Sulphonic and carboxylic acids										

Table 1.2c. Methods used in determination of anions in waters, cross referenced with section

Anion	Mass Spectro-metry	Enzymic Methods	HPLC	Gas Chromato-graphy	Ion Exchange Chromato-graphy	Micelle Chromato-graphy	Column Coupling Isotacho-phoresis	Ion Exclusion Chroma-tography	Ion Chromato-graphy
Chloride				2.1.1	2.1.1		2.1.1 11.7.1	2.1.5	2.1.1–2.1.5 2.1.7 11.1.1 11.5.1 11.4.1 11.3.1 11.2.1
Bromide			2.2.1 2.2.3	2.2.1 2.2.3	2.2.1	2.2.1			2.2.1–2.2.4 11.4.2 11.4.1 11.3.1 11.1.1
Iodide			2.3.1 2.3.3	2.3.1 2.3.3		2.3.1			2.3.1 2.3.4 11.4.2
Fluoride			2.4.4	2.4.1	2.4.1		2.4.1 11.7.1		2.4.1–2.4.5 11.4.1 11.3.1 11.2.1 11.1.1
Mixed halides			2.5.1 2.5.2	2.5.3 2.5.5	2.5.2	2.5.2	2.5.2		2.5.1–2.5.4

Table 1.2c (Contd.)

Anion	Mass Spectrometry	Enzymic Methods	HPLC	Gas Chromatography	Ion Exchange Chromatography	Micelle Chromatography	Column Coupling Isotachophoresis	Ion Exclusion Chromatography	Ion Chromatography
Chlorate, Chlorite, Hypochlorite			2.6.1						2.6.3
Bromite			2.7.1						2.6.1
Iodate			2.8.1			2.8.1			2.7.2
Nitrate	3.1.1	3.1.1	3.3.1 3.3.2	3.1.2 3.3.3	3.1.1 3.3.3	3.3.1	3.3.1 11.7.1		3.1.1–3.1.4 3.3.1, 3.1.6 3.3.2–3.3.4 11.4.1 11.3.1 11.2.1–11.2.3 11.1.1
Nitrite			3.3.1 3.3.2	3.2.1 3.3.3 3.3.1	3.3.3	3.3.1	3.3.1 11.7.1		3.2.1–3.2.3 3.2.6 3.3.4
Free Cyanide				3.4.1 3.4.4					11.5.1 11.4.1 11.2.1 11.1.1 3.4.1 3.4.2 3.4.3 11.3.2 11.4.2

Rationale: Water Samples 13

Total cyanide			3.5.1				
Cyanate			3.6.1				
Sulphate		4.1.5	4.1.2	4.1.1	11.7.1	4.1.5	4.1.1–4.1.6 11.5.1 11.4.1, 11.3.1 11.2.1–11.2.3 11.1.1
Sulphite							4.2.1–4.2.5 11.2.1 11.4.3 11.5.1
Thiosulphate		4.3.2					4.3.1 11.4.3
Thiocyanate		4.4.1	4.4.1 4.4.2				
Sulphide			4.5.3		11.7.1		4.5.1–4.5.3 11.3.2 11.4.2
Polysulphide	5.1.1		4.6.2				
Phosphate		5.1.6 5.2	5.1.2 5.1.3			5.2 5.1.6	5.1.1–5.1.5 5.1.7–5.1.6 11.1.1, 5.1.8 11.2.1 11.3.1 11.4.1 11.5.1

Table 1.2c (Contd.)

Anion	Mass Spectro-metry	Enzymic Methods	HPLC	Gas Chromato-graphy	Ion Exchange Chromato-graphy	Micelle Chromato-graphy	Column Coupling Isotacho-phoresis	Ion Exclusion Chromato-graphy	Ion Chromato-graphy
Silicate									6.1.1
Borate									7.1.1
									7.1.2
									11.1.3, 11.4.4
Total alkalinity									
Carbonate					8.2.2				8.2.1
									8.2.3
									11.5.1
Bicarbonate								8.3.2	8.3.2
									11.1.1
Arsenite									9.1.1, 9.3.1
									11.1.2
Arsenate									9.2.1, 9.2.2
									9.3.1
									11.1.2, 11.3.3
Selenite			9.6.2	9.6.2					9.4.1, 9.4.2
									9.6.3
									11.1.2, 11.3.3
Selenate			9.6.2	9.6.2					9.5.1, 11.1.2
									9.6.3
									11.3.3

Tellurate			
Chromate	9.8.1		
Germanate			
Molybdate			
Tungstate			
Uranate			
Vanadate			
Ferrocyanide			9.16.1
Metal cyanide complexes			11.4.2
Formate	10.1.1		10.1.1
Isoascorbate			
Dehydroascorbate			
Aminoacids		10.12.1	10.12.1
NTA, EDTA	10.13.1		
Sulphonic and carboxylic acids	10.14.1	10.14.1	10.14.1
	10.15.1		10.16.1
Ascorbate		10.6.1	10.5.1

1.2 Rationale: Solid and Gaseous Samples

As well as aqueous samples, various other types of samples are discussed. These are usually solids (foodstuffs, grains, meat, milk, seaweed, animal feeds, table salt, coal, soil, pharmaceuticals, plants, vegetative matter, silage, grass, biological materials such as blood, serum, urine and tissues, and plant and soil extracts. Again, as shown in Table 1.3a, b, spectrophotometric methods predominate (13 anions), followed by ion chromatography (9 anions), ion selective electrodes (7 anions), titration and ultraviolet spectroscopy (6 anions each), flow injection analysis (5 anions), and atomic absorption spectrometry and polarography (3 anions each). Less commonly used techniques include gas chromatography (2 anions), spectrofluorimetry, emission cavity analysis, gasometric methods, and high performance liquid chromatography (1 anion each).

1.3 Brief Summary of Methodologies

The principles of the more important methodologies are discussed below together with listings of the anions that can be determined by these procedures.

1.3.1 Spectroscopic Methods

1.3.1.1 Spectrometric Methods

Spectrometric methods are applicable to chloride, bromide, iodide, fluoride, chlorate, chlorite, hypochlorite, iodate, nitrite, nitrate, free and total cyanide, sulphite, sulphate, thiosulphate, thiocyanate, sulphide, phosphate, total alkalinity, silicate, borate, arsenite, arsenate, selenite, tellurate, chromate, uranate, vanadate, titanate, ferrocyanide, ascorbate, glutamate, isocitrate, citrate, malate, lactate, oxalate and isoascorbate.

Visible spectrophotometers are commonly used in the water industry for the estimation of colour in a sample or for the estimation of coloured products produced by reacting a colourless component of the sample with a reagent which produces a colour, which can be evaluated spectrophotometrically.

Formerly, visible spectrophotometry was used extensively for the determination of metals. Thus, lead was determined by reaction of the lead ions with dithizone, with which it produces a red coloured complex. However, following the introduction of techniques such as atomic absorption spectrometry, these applications are becoming fewer.

Visible spectrophotometry still finds extensive use in the determination of some anions such as phosphate, silicate, chloride and nitrate, although these

Table 1.3a. Methods used in determination of anions in solid gaseous samples, cross referenced with section

Type of sample	Anion	Technique							
		Titration	Spectro-photo-metric	Spectro-fluorimetry	Ultra violet spectroscopy	Flow Injection Analysis	Atomic Absorption Spectrometry	Ion Selective Electrode	Polarography
Foods	Bromide								
	Iodide	2.3.5	2.3.6					2.3.8	
	Bromate								
	Sulphite	4.2.7	4.2.7			2.7.4			
	Ascorbate		10.8.2			4.2.7			
	Glutamate		10.11.1						
	Isocitrate		10.3.1		10.3.1				
	Malate		10.5.1		10.5.1				
	Citrate		10.2.1		10.2.1				
	Lactate		10.6.1		10.6.2				
	Oxalate		10.4.1		10.4.1				
	Nitrite		3.3.9						
	Nitrate		3.3.9					2.1.10	
	Chloride							2.4.9	
Milk	Fluoride								
	Nitrite		3.2.7						
	Pyruvate								
	Selenite								
Meat	Nitrate				10.10.1	10.10.1		3.3.8	
	Nitrite							3.3.8	
Grain	Sulphate		4.1.9						
Wine	Sulphate		4.2.6		4.2.6	4.2.6			
Plants	Chloride								
	Borate		7.1.5						
	Free cyanide		3.4.5					3.4.6	
	Nitrate		3.1.9						3.1.9

Table 1.3a (Contd.)

Type of sample	Anion	Technique							
		Titration	Spectro-photo-metric	Spectro-fluorimetry	Ultra violet spectroscopy	Flow Injection Analysis	Atomic Absorption Spectrometry	Ion Selective Electrode	Polarography
Plant extracts	Fluoride							2.4.8	
	Sulphate		7.1.6				4.1.8		
	Borate		9.13.2				2.3.7		
	Vanadate								
Soil	Chloride	2.1.9, 2.1.8							
	Bromide					2.2.5			
	Nitrate		3.1.8, 3.3.6					3.1.8	
	Nitrite		3.2.6, 3.3.6						
	Free cyanide	3.4.7	3.4.7						
	Sulphate	4.1.7					4.5.6	4.5.7	
	Sulphide								
	Carbonate								
	Borate		7.1.4						
Soil extracts	Chlorate, Chlorite, Hypochlorite		2.6.4						
	Sulphate						4.1.8		
	Phosphate		5.1.9						
	Selenite								
	Vanadate		9.13.2						
Biological Solids and fluids	Fluoride		2.4.7						
	Nitrite		3.3.7						
	Nitrate		3.3.7						
	Free cyanide		3.4.8						
	Selenite								
Urine	Salicylate			10.7.1					
Tissue	Tungstate						9.11.2		9.4.6

Brief Summary of Methodologies

Matrix	Analyte				
Blood/Serum	Sulphide				4.5.8
Blood/Serum	Phosphate		5.1.10		
Air	Free cyanide	3.4.9			
Animal	Sulphide	4.5.9			3.4.6
Animal Feed	Free cyanide				3.4.6
Fertilizers	Free cyanide				
Sewage sludge	Free cyanide			3.4.10	
Pharmaceuticals	Sulphite	4.2.8		3.4.10	
	Ascorbate		10.8.3		
Silage	Lactate	10.6.1			
Table Salt	Iodide			2.3.9	2.3.9

Table 1.3b. Matrials used in determinations of anions in solid samples, Cross referenced with section

Type of sample	Anion	Gas Chromatography	Ion Exchange Chromatography	Ion Chromatography	Emission cavity analysis	Potentio-metric methods	Gasometric methods	High Performance bound chromatography
Food	Bromide	2.2.5						
	Glutamate					10.11.1		
Milk	Iodide							2.3.10
	Selenite	9.4.4						
Plants	Chloride			2.1.8				
	Nitrate			3.1.9				
	Free cyanide			3.4.5				
Plant extrats	Phosphate			5.1.8, 11.6.1				
	Chloride			11.6.1				
	Bromide			11.6.1				
	Fluoride			11.6.1				
	Nitrate			11.6.1				
	Sulphate			11.6.1				
Coal	Fluoride			2.4.10				
Soil	Chloride			2.1.8				
	Nitrate			3.1.8				
	Sulphate			4.1.7	4.1.7			
	Carbonate						8.2.4	
Soil extracts	Sulphate			4.1.8				
	Phosphate			5.1.9				
	Selenite			9.4.3				
Urine	Selenite	9.4.6	9.4.5					
Pharmaceticals	Iodide							2.3.11
	Salicylate		10.7.2					

analyses are now normally accomplished using an autoanalyser in which the spectrophotometer is embodied (discussed elsewhere).

An extensive modern application of visible spectrophotometry is in the determination of organic substances in water, including non-ionic detergents, alcohols (by estimation of the orange-red product produced upon reaction with ceric ammonium nitrate), aldehydes (by formation of the red coloured 2:4 dinitrophenyl-hydrazones) to name but a few determinations.

Some commercially available instruments, in addition to visible spectrophotometers, can also perform measurements in the UV and near IR regions of the spectrum. These have not yet found extensive applications in the field of water analysis.

Suppliers of visible spectrophotometers are reviewed in Table 1.4.

Table 1.4. Visible-ultraviolet-near infrared spectrophotometers

Spectral region	Range (nm)	Manufacturer	Model	Single or double beam	Cost range
UV/visible	–	Philips	PU 8620 (optional PU 8620 scanner)	Single	Low
Visible	325–900	Celcil Instruments	CE 2343 Optical Flowcell	Single	Low
Visible	280–900	Celcil Instruments	CE 2393 (grating, digital)	Single	High
Visible	280–900	Celcil Instruments	CE 2303 (grating, non-digital)	Single	Low
Visible	280–900	Celcil Instruments	CE 2373 (grating, linear)	Single	High
UV/visible	190–900	Celcil Instruments	CE 2292 (digital)	Single	High
UV/visible	190–900	Celcil Instruments	CE 2202 (non-digital)	Single	Low
UV/visible	190–900	Celcil Instruments	CE 2272 (linear)	Single	High
UV/visible	200–750	Celcil Instruments	CE 594 (microcomputer controlled)	Double	High
UV/visible	190–800	Celcil Instruments	CE 6000 (with CE 6606 graphic plotter option)	Double	High
UV/visible	190–800	Celcil Instruments	5000 series (computerized and data station)	Double	High
UV/visible	–	Philips	PU 8800		High
UV/visible	–	Kontron	Unikon 860 (computerized with screen)	Double	High
UV/visible	–	Kontron	Unikon 930 (computerized with screen)	Double	High
UV/visible	190–1100	Perkin-Elmer	Lambda 2 (microcomputer electronics screen)	Double	High
UV/visible	190–750 or 190–900	Perkin-Elmer	Lambda 3 (microcomputer electronics)	Double	Low to High
UV/visible	190–900	Perkin-Elmer	Lambda 5 and Lambda 7 (computerized with screen)	Double Double	High
UV/visible	185–900 & 400–3200	Perkin-Elmer	Lambda 9 (computerized with screen)	UV/vis/ NIR	High
UV/visible	190–900	Perkin-Elmer	Lambda Array 3840 (computerized with screen)	Photodiode	High

For many water industry applications a simple, basic UV/visible single-beam instrument such as the Phillips PU 8620 or the Cecil Instrument CE 2343, CE2303 or CE2202 will suffice. If better instrument stability is required, then a double-beam instrument is preferable and can be purchased at little extra cost.

Moving up-market, spectrophotometers are available which have computer interfaces which enable reaction kinetics studies to be carried out (e.g. the Cecil CE 2202, CE 2272, CE 2292, CE 2303, CE 2373 and CE 2393 single-beam instruments and the Cecil CE 594 double-beam instrument). These instruments also have an autosampler facility capable of handling up to 40 samples. Even more sophisticated is the Cecil 6000 double-beam instrument equipped with a real-time graphics plotter. In this instrument, sets of parameters, i.e. methods, can be stored and curve fitting carried out. It also enables multi-component analysis to be performed. Analyses of mixtures of up to nine different materials may be carried out.

1.3.1.2 Spectrofluorimetric Methods

Chloride, nitrite, nitrate, free cyanide, sulphate, sulphide, selenite, selenate, salicylate and isoascorbate may be determined using spectrofluorimetric methods. The anions can be converted to fluorescent products by reaction with suitable reagents and these products can be evaluated spectrofluorimetrically. Thus both resorcinol and 2,2-hydroxy-4,4^1-dimethoxybenzophenone form fluorescent products upon reaction with nitrate ions whilst nitrite oxidizes pyridoxal-5-phosphate-2-pyridyl hydrozone to a fluorescent product. Chloride can be estimated by the quenching effect of silver chloride on sodium fluorescein. These methods have very high sensitivity.

Luminescence is the generic name used to cover all forms of light emission other than that arising from elevated temperature (thermoluminescence). The emission of light through the absorption of UV or visible energy is called photo-luminescence, and that caused by chemical reactions is called chemiluminescence. Light emission through the use of enzymes in living systems is called bio-luminescence. Photoluminescence may be further subdivided into fluorescence, which is the immediate release (10^{-8} s) of absorbed light energy as opposed to phosphorescence which is the delayed release ($10^{-6} - 10^2$ s) of absorbed light energy.

The excitation spectrum of a molecule is similar to its absorption spectrum while the fluorescence and phosphorescence emission occur at longer wavelengths than the absorbed light. The intensity of the emitted light allows quantitative measurement since, for dilute solutions, the emitted intensity is proportional to concentration. The excitation and emission spectra are characteristic of the molecule and allow qualitative measurements to be made. The inherent advantages of the techniques, particularly fluorescence, are

1 sensitivity, picogram quantities of luminescent materials are frequently studied

2 selectivity, derived from the two characteristic wavelengths and
3 the variety of sampling methods that are available, i.e. dilute and concentrated samples, suspensions, solids, surfaces and combination with chromatographic methods, such as, for example is used in the HPLC separation of o-phthalyl dialdehyde derivatized amino acids in natural and sea water samples.

Fluorescence spectroscopy forms the majority of luminescence analyses. However, the recent developments in instrumentation and room-temperature phosphorescence techniques have given rise to practical and fundamental advances which should increase the use of phosphorescence spectroscopy. The sensitivity of phosphorescence is comparable to that of fluorescence and complements the latter by offering a wider range of molecules for study.

The pulsed xenon lamp forms the basis for both fluorescence and phosphorescence measurement. The lamp has a pulse duration at half peak height of 10 µs. Fluorescence is measured at the instant of the flash. Phosphorescence is measured by delaying the time of measurement until the pulse has decayed to zero.

Luminescence Spectrometers

Perkin-Elmer and Hamilton both supply luminescence instruments (see Appendix).

Perkin-Elmer LS-3B and LS-5B Luminescence Spectrometers

The LS-3B is a fluorescence spectrometer with separate scanning monochromators for excitation and emission, and digital displays of both monochromator wavelengths and signal intensity. The LS-5B is a ratioing luminescence spectrometer with the capability of measuring fluorescence, phosphorescence and bio- and chemiluminescence. Delay time (t_d) and gate width (t_g) are variable via the keypad in 10 µs intervals. It corrects excitation and emission spectra.

Both instruments are equipped with a xenon discharge lamp source and have an excitation wavelength range of 230–720 nm and an emission wavelength range of 250–800 nm.

These instruments feature keyboard entry of instrument parameters which combined with digital displays, simplify instrument operation. A high-output pulsed xenon lamp, having low power consumption and minimal ozone production, is incorporated within the optical module.

Through the use of an RS 232C interface, both instruments may be connected to Perkin-Elmer computers for instrument control and external data manipulation.

With the LS-5B instrument, the printing of the sample photomultiplier can be delayed so that it no longer coincides with the flash. When used in this mode, the instrument measures phosphorescence signals. Both the delay of the start of the gate (t_d) and the duration of the gate (t_g) can be selected in multiples of 10 μs from the keyboard. Delay times may be accurately measured, by varying the delay time and noting the intensity at each value.

Specificity in luminescence spectroscopy is achieved because each compound is characterized by an excitation and emission wavelength. The identification of individual compounds is made difficult in complex mixtures because of the lack of structure from conventional excitation or emission, spectra. However, by collecting emission or excitation spectra for each increment of the other, a fingerprint of the mixture can be obtained. This is visualized in the form of a time-dimensional contour plot on a three-dimensional isometric plot.

Fluorescence spectrometers are equivalent in their performance to single-beam UV-visible spectrometers in that the spectra they produce are affected by solvent background and the optical characteristics of the instrument. These effects can be overcome by using software built into the Perkin-Elmer LS-5B instrument or by using application software for use with the Perkin-Elmer models 3700 and 7700 computers.

The Perkin-Elmer LS-2B microfilter fluorimeter is a low-cost, easy-to-operate, filter fluorimeter that scans emission spectra over the wavelength range 390–700 nm (scanning) or 220–650 nm (individual interference filters).

The essentials of a filter fluorimeter are:

a source of UV/visible energy (pulsed Xenon)
a method of isolating the excitation wavelength
a means of discriminating between fluorescence emission and excitation energy
a sensitive detector and a display of the fluorescence intensity.

The model LS-2B has all of these features arranged to optimize sensitivity for microsamples. It can also be connected to a highly sensitive 7 μl liquid chromatographic detector for detecting the constituents in the column effluent. It has the capability to measure fluorescence, time-resolved fluorescence, and bio- and chemiluminescent signals. A 40-portion autosampler is provided. An excitation filter kit containing six filters – 310, 340, 375, 400, 450 and 480 nm – is available to enable a range of analyses to be performed.

1.3.1.3 Ultraviolet Spectroscopy

Sulphite, nitrite, nitrate, phosphate, pyruvate, lactate, citrate, isocitrate, malate, dehydroascorbate, NTA and EDTA can be determined by this technique. See Sect. 1.1.

1.3.2 Segmented Flow Analysis

This technique is applicable to fluoride, nitrite, nitrate, free cyanide, sulphite, phosphate, silicate, bicarbonate, borate, and total alkalinity.

Segmented flow analysers, alternatively known as autoanalysers, are extensively used for the routine batch determination (up to 80 samples per hour) of a wide range of determinants including those listed above. This data is applicable to equipment supplied by Skalar BV, Holland.

Segmented flow analysis is based on the principle of pumping a liquid through a system of tubing, dividing it by air bubbles into equal parts or segments and then measuring it. In practice, the liquid is a reagent to which the sample to be analysed is added; the resulting colour change is then measured by various methods. Samples are analysed in batches and the system is rinsed between each sample to prevent 'carry-over'. Either a single measurement can be made repeatedly on the sample or it can be divided between a number of analytical modules, so that various measurements are made simultaneously on the same sample.

The advantages of segmented flow analysis are:

many different analytical measurements can be made on a single sample at the same time
all analytical results are produced in analogue forms if necessary, reducing sample administration work and transcription errors
small sample and reagent volumes are required, effectively reducing costs.

The Skalar analyser unit consists of five components, a sampler, a pump unit, an analytical section, a detection unit and a calculator/computer (Fig. 1.1) Detections based on spectrophotometry and ion selective electrodes (chloride, fluoride) are possible.

Other companies, e.g. Chemlab in UK and Technicon Corporation in USA produce segmented flow analysers. The continuous flow system supplied by Chemlab has been in use in water laboratories for several years.

The original Chemlab system (CAA I) utilizes separate dialysis and high-temperature baths and requires larger volumes of samples and reagents, whereas in the second series system (CAA II) the amounts of samples and reagents required are smaller, due to the heating and dialysis baths being miniaturized and fitted, with all the requisite glassware, into a neat, compact, analytical cartridge. A separate cartridge is constructed for each constituent in the sample to be analysed. Both systems are in current use although the CAA II system is the one most frequently used because it uses smaller volumes of samples and reagents and is more compact.

The ChemLab Flow Analysis Systems give a graphical output on the recorders. If a number of channels are being used, the results can be automatically computed and printed out by a data processor. This instrument can be connected to up to eight separate analytical channels. If further statistical analysis of these results is required, the data processor can be connected to a mainframe computer by various interfaces.

ChemLab System 4 is a new analytical system for the user who wants to perform single or multiple analyses of a sample with maximum efficiency. The analyser is designed in modular form which allows the user to select the system

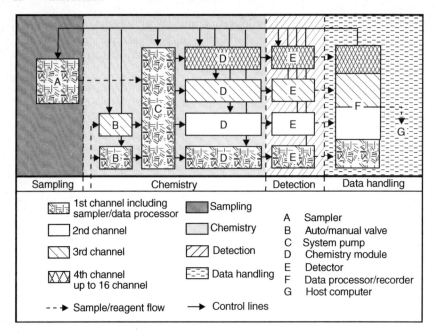

Fig. 1.1. Segmented flow analysis system employed by Skalar BV

that best suits his needs, and yet can be easily changed or upgraded as the workload alters. System 4 is fast, accurate and precise with a wide range of chemistries available.

1.3.3 Flow Injection Analysis

Chloride, bromide, chlorate, chlorite, hypochlorite, bromate, nitrite, nitrate, sulphite, sulphate, phosphate, silicate, arsenate, arsenite, chromate, germanate, thiosulphate, sulphide, borate, total alkalinity and pyruvate may be measured by this method.

Flow-injection analysis (FIA) is a rapidly growing analytical technique. Since the introduction of the original concept by Ruzicka and Hansen [1.16] in 1975, about 1000 papers have been published.

Flow-injection analysis is based on the introduction of a defined volume of sample into a carrier (or reagent) stream. This results in a sample plug bracketed by carrier (Fig. 1.2a)

The carrier stream is merged with a reagent stream to obtain a chemical reaction between the sample and the reagent. The total stream then flows

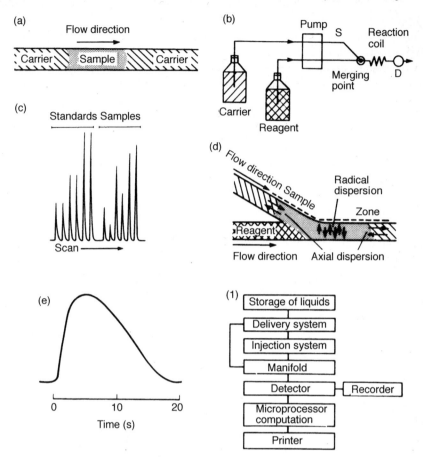

Fig. 1.2 a Schematic diagram of the flow pattern in an FIA system directly after injection of sample. **b** Simple FIA system for one reagent, S denotes the sample injection site and D is the flow through detector. **c** Typical FIA peaks (detector output signals). **d** Radial and axial dispersion in an injected sample plug. **e** Rapid scan of an FIA curve. **f** Configuration of an FIA system

through a detector (Fig. 1.2b). Although spectrophotometry is the commonly used detector system in this application, other types of detectors have been used, namely fluorimetric, atomic absorption emission spectroscopy and electrochemical (e.g. ion selective electrodes).

The pump provides constant flow and no compressible air segments are present in the system. As a result the residence time of the sample in the system is absolutely constant. As it moves towards the detector the sample is mixed with both carrier and reagent. The degree of dispersion (or dilution) of the sample can be controlled by varying a number of factors, such as sample volume, length and diameter of mixing coils and flow rates.

When the dispersed sample zone reaches the detector, neither the chemical reaction nor the dispersion process has reached a steady state. However, experimental conditions are held identical for both samples and standards in terms of constant residence time, constant temperature and constant dispersion. The sample concentration can thus be evaluated against appropriate standards injected in the same manner as samples (Fig. 1.2c).

The short distance between the injection site and the merging point ensures negligible dispersion of the sample in this part of the system. This means that sample and reagent are mixed in equal proportions at the merging point.

The mixing technique can be best understood by having a closer look at the hydrodynamic conditions in and around the merging point (Fig. 1.2d).

In Fig. 1.2d the hydrodynamic behaviour is simplified in order to explain the mixing process. Let us assume that there is no axial dispersion and that radial dispersion is complete when the sampler reaches the detector. The volume of the sample zone is thus 200 μl after the merging point (100 μl sample + 100 μl-reagent as flow rates are equal). The total flow rate is 2.0 ml min^{-1}. Simple mathematics then gives a residence time of 6s for the sample in the detector flow cell. In reality, response curves reflect some axial dispersion. A rapid scan curve is shown in Fig. 1.2e. The baseline is reached within 20 s. This makes it possible to run three samples per minute and obtain baseline readings between each sample (no carryover) i.e. 180 samples per hour.

The configuration of an FIA system is shown schematically in Fig. 1.2f. The (degassed) carrier and reagent solution(s) must be transported in a pulse-free transport system and at constant rate through narrow Teflon (Du Pont) tubing.

In a practical FIA system, peristaltic pumps are usually used since they have several channels, and different flow rates may be achieved by selection of a pump tube with a suitable inner diameter.

A manifold provides the means of bringing together the fluid lines and allowing rinsing and chemical reaction to take place in a controlled way. Manifolds with several lines can be assembled as required. These manifolds are mounted on plastic trays and allow the use of different reaction coils.

Flow-injection analysers available range from relatively low-cost unsophisticated instruments such as those supplied by Advanced Medical Supplies, Skalar and ChemLab to the very sophisticated instruments such as the FIA star 5010 and 5020 supplied by Tecator (Table 1.5).

One of the advantages claimed for flow-injection analysis is a study of speciation of anions and cations in water samples. As an example of this, Ruz et al. [1.1] speciated the different oxidation states of chromium. They were able to obtain a concentration profile for the chromium III and chromium VI species, $HCrO_4^-$, $Cr_2O_4^{2-}$ and CrO_4^{2-}.

They also developed on-line flow-injection analysis preconcentration methods using a microcolumn of Chelex 100 resin which enabled them to determine lead in sea water at concentrations down to 10 μgl^{-1} and cadmium and zinc down to 1 μgl^{-1} at a sampling rate of 30–60 samples per hour. Fang et al. [1.2] also used a flow-injection system comprising on-line preconcentration on an ion-exchange resin for the determination of heavy metals (copper, zinc, lead and cadmium) in sea water by atomic absorption spectrometry. The respective detection limits were 0.07, 0.03, 0.5 and 0.05 µgl^{-1}.

Marshall and Mottola [1.3] evaluated a method for determining the ionic forms of chromium in water:

Table 1.5. Equipment for flow-injection analysis[a]

Supplier	Model	Features	Detectors available
Advanced Medical Supplies	LGC 1	Relatively low-cost instrument, recorder output. No computerization on data processing (8 channels)	Colorimeter (other detectors can be used but are not linked in, e.g. atomic absorption, fluorimeter ion-selective electrodes)
Chemlab	–	Relatively low-cost, recorder output or data analysis by microprocessor (3 channels)	Colorimeter
Skalar	–	Relatively low-cost, recorder output on data analysis by microprocessor also carries out segmented flow analysis	Colorimeter, flow cells for fluorimeter and ion-selective electrodes available
Fiatron	Finlite 600	Laboratory process control and pilot plant instrument computerized	Colorimeter
	Fiatrode 400 Fiatrode 410 Fiatrode 430	Flow through analyser/controller, process control analyser	pH and ion-selective electrode
Tecator	FIA star 5025	Relatively low cost manual instrument specifically designed for fluoride, cyanide, potassium, iodide, etc.	Specially designed for use with ion-selective electrodes
	FIA star 5032	Relatively low-cost manual instrument (400–700 nm)	Spectrophotometer and/or photometer detectors
	Aquatec	Modular, semi- or fully automatic operation. Microprocessor controlled. A dedicated instrument designed for water analysis, i.e. dedicated method cassettes for phosphate and chloride, 60–100 samples h^{-1}	Flow through spectrophotometer (400–700 nm)
	FIA star 5010	Modular, semi- or fully automatic operation. May be operated with process controller microprocessor. Can be set up in various combinations with 5017 sampler and superflow software which is designed to run on IBM PC/XT computer; 60–180 samples h^{-1}. Dialysis for in-line sample preparation and in-line solvent extraction. Thermostat to speed up reactions	Spectrophotometer (400–700 nm) or photometer can be connected to any flow through detector, e.g. UV/visible, inductively coupled plasma, atomic absorption spectrometer and ion-selective electrodes

Table 1.5 (*Contd.*)

Supplier	Model	Features	Detectors available
	FIA star 5020	As above, top of the market, higher sample throughput (up to 300 samples h^{-1}). Microprocessor controlled functions, automatic calculation of results, digital presentation of results, automatic recalibration, stopped flow and intermittent pumping for slow reactions, 100-sample sampler, 5 chemifolds, gas diffusion measurements dialysis and solvent extraction. Non-aqueous and corrosive reagents	As FIA star 5010

ᵃThe following ions can be determined by flow injection analysis: fluoride, cyanide, potassium, iodide, calcium, phosphate, carbonate, total alkalinity, nitrate, nitrite; also alkalinity, boron, calcium, color, cyanide, hardness, total nitrogen, phenol, phosporus, FIA star surfactants, Kjeldahl nitrogen, urea (Tecator 5010 analyser).

Chromium VI

$$H_2CrO_4 \rightleftharpoons HCrO_4-$$
$$HCrO_4^- \rightleftharpoons CrO_4^{2-} + H^+$$
$$Cr_2O_7^{2-} + H_2O - \rightleftharpoons 2HCrO_4^-$$

Chromium III

$$Cr^{3+} + H_2O \rightleftharpoons Cr(OH)^{2+} + H^+$$
$$Cr(OH)^{2+} + H_2O \rightleftharpoons Cr(OH)_2^+ + H^+$$
$$Cr^{3+} + 3OH^- = \rightleftharpoons Cr(OH)_3$$
$$Cr^{3+} + 4OH^- \rightleftharpoons Cr(OH)_4^-.$$

1.3.4 Spectrometric Methods

1.3.4.1 Atomic Absorption Spectrometry

Iodide, chloride free cyanide, nitrate, sulphate, thiosulphate, sulphide, phosphate, silicate, arsenite, selenite, selenate, chromate, molybdate and tungstate, borate and total cyanide are measurable using this technique.

Since shortly after its inception in 1955, atomic absorption spectrometry has been the standard tool employed by analysts for the determination of trace levels of metals in water samples. In this technique a fine spray of the analyte is passed into a suitable flame, frequently oxygen acetylene or nitrous oxide acetylene, which converts the elements to an atomic vapour. Through this vapour radiation is passed at the right wavelength to excite the ground state atoms to the first excited electronic level. The amount of radiation absorbed can then be measured and directly related to the atom concentration: a hollow cathode lamp is used to emit light with the characteristic narrow line spectrum of the analyte element. The detection system consists of a monochromator (to reject other lines produced by the lamp and background flame radiation) and a photomultiplier. Another key feature of the technique involves modulation of the source radiation so that it can be detected against the strong flame and sample emission radiation.

The technique can determine a particular element with little interference from other elements. It does, however have two major limitations. One of these is that the technique does not have the highest sensitivity, and the other is that only one element at a time can be determined. This has reduced the extent to which it is currently used.

Basically, atomic absorption spectrometry was designed for the determination of cations. However, the technique can be adapted to the indirect determination of anions. For example, to determine sulphate, the sample is reacted with barium chloride to produce barium sulphate, and this is filtered off and washed to remove soluble barium ions. The barium sulphate is then dissolved and the barium equivalent to the sulphate content of the original sample is determined. Suitable instrumentation is listed in Table 1.6.

1.3.4.2   Inductively Coupled Plasma Atomic Emission Spectroscopy

Chloride, bromide, iodide, fluoride, sulphate and tungstate may be determined by means of this technique. It was originally developed for the determination of cations and can be adapted to the indirect determination of anions.

An inductively coupled plasma is formed by coupling the energy from a radiofrequency (1–3 kW or 27–50 MHz) magnetic field to free electrons in a suitable gas. The magnetic field is produced by a two- or three-turn water-cooled coil and the electrons are accelerated in circular paths around the magnetic field lines that run axially through the coil. The initial electron 'seeding' is produced by a spark discharge but, once the electrons reach the ionization potential of the support gas, further ionization occurs and a stable plasma is formed.

The neutral particles are heated indirectly by collisions with the charged particles upon which the field acts. Macroscopically the process is equivalent to heating a conductor by a radio-frequency field, the resistance to eddy-current flow producing Joule heating. The field does not penetrate the conductor uniformly and therefore the largest current flow is at the periphery of the plasma. This is the so-called 'skin' effect and, coupled with a suitable gas-flow geometry, it produces an annular or doughnut-shaped plasma. Electrically, the coil and plasma form a transformer with the plasma acting as a one-turn coil of finite resistance.

Table 1.6. Available flame and graphite furnace atomic absorption spectrometers

Type instrument	Supplier	Model no. and type	Microprocessor	Hydride and mercury attachment	Autosampler	Wavelength range
Flame (direct injection)	Thermo-electron	1L 157 single channel 1L 357 single beam	—	Yes	Yes	
	Perkin-Elmer	1L 457 single channel double beam	with graphics			
		Video 11 single channel single beam	with graphics			
		Video 12 single channel double beam	with graphics			
		Video 22 two double channels	computer interference			
Graphite furnace Direct injection	Thermo-electron Perkin-Elmer	1L 655 CTF 2280 single beam 2380 double beam	— Yes Yes (with automatic background correction)	— — —	Yes — —	190–870 190–870 190–870
Graphite furnace	Perkin-Elmer	100 single beam 2100 single path double beam	Yes Yes	— Yes		
Graphite furnace	Varian Associates	Spectr A30 + 40 multielement analysis. Method storage	Yes	Yes	Yes	190–900
		SpectrA A10 (low cost, single beam)	Yes (built in VDU)	Yes	Yes	190–900
		SpectrA A20 (medium cost, double beam)	Yes (built in VDU)	Yes	Yes	190–900

Flame graphite furnace		SpectrA A 300/400 multi-element analysis, centralized instrument control	Yes (with colour graphics and 90 elements ondisk)	Yes	Yes	190–900
		STA 9S and GTA 96 graphite tube atomizer units – compatible with all SpectrA A instruments	Furnace and programmable sample dispenser operated from SpectrA A keyboard. Rapid interchange between flame and furnace operation			
Graphite furnace	GBC Scientific Pty Ltd.	903 single beam 902 double beam (both with impact head option)	Yes Yes	Yes Yes	Yes Yes	176–900 170–900
Flame (direct injection) graphite furnace	Shimadzu	AA670 double beam AA670 G Double beam	Yes Yes	Yes Yes	Yes Yes	190–900 190–900

The properties of an inductively coupled plasma closely approach those of an ideal source for the following reasons:

the source must be able to accept a reasonable input flux of the sample and it should be able to accommodate samples in the gas, liquid or solid phases;

the introduction of the sample should not radically alter the internal energy generation process or affect the coupling of energy to the source from external supplies;

the source should be operable on commonly available gases and should be available at a price that will give cost-effective analysis;

the temperature and residence time of the sample within the source should be such that all the sample material is converted to free atoms irrespective of its initial phase or chemical composition; such a source should be suitable for atomic absorption or atomic fluorescence spectrometry;

if the source is to be used for emission spectrometry, then the temperature should be sufficient to provide efficient excitation of a majority of elements in the periodic table;

the continuum emission from the source should be of a low intensity to enable the detection and measurement of weak spectral lines superimposed upon it,

the sample should experience a uniform temperature field and the optical density of the source should be low so that a linear relationship between the spectral line intensity and the analyte concentration can be obtained over a wide concentration range.

Greenfield et al. [1.4] were the first to recognize the analytical potential of the annular inductively coupled plasma.

Wendt and Fassel [1.5] reported early experiments with a 'tear-drop' shaped inductively coupled plasma but later described the medium power 1–3 kW, 18 mm annular plasma now favoured in modern analytical instruments [1.6].

The current generation of inductively coupled plasma emission spectrometers provide limits of detection in the range of $0.1-500\,\mu g\,l^{-1}$ in solution, a substantial degree of freedom from interferences and a capability for simultaneous multi-element determination facilitated by a directly proportional response between the signal and the concentration of the analyte over a range of about five orders of magnitude.

The most common method of introducing liquid samples into the inductively coupled plasma is by using pneumatic nebulization (Thompson and Walsh [1.14] in which the liquid is dispensed into a fine aerosol by the action of a high-velocity gas stream. To allow the correct penetration of the central channel of the inductively coupled plasma by the sample aerosol, an injection velocity of about $7\,ms^{-1}$ is required. This is achieved using a gas injection with a flow rate of about $0.5-1\,l\,min^{-1}$ through an injector tube of 1.5–2.0 mm internal diameter. Given that the normal sample uptake is $1-2\,ml\,min^{-1}$ this is an insufficient quantity of gas to produce efficient nebulization and aerosol transport. Indeed, only about 2% of the sample reaches the plasma. The fine gas jets and liquid capillaries used in inductively coupled plasma nebulizers may cause inconsistent operation and even blockage when solutions containing high levels of dissolved solids, such as sea water or particulate matter, are used. Such problems have led to the development of a new type of nebulizer, the most successful being based on a principle originally described by Babington (US Patents). In these, the liquid is pumped from a wide bore tube and thence conducted to the nebulizing orifice by a V-shaped groove (Suddendorf and Boyer [1.7] or by the divergent wall of an over-expanded nozzle (Sharp [1.8]). Such devices handle most liquids and even slurries without difficulty.

Nebulization is inefficient and therefore not appropriate for very small liquid samples. Introducing samples into the plasma in liquid form reduces the potential sensitivity because the analyte flux is limited by the amount of solvent that the plasma will tolerate. To circumvent these problems a variety of thermal and electrothermal vaporization devices have been investigated. Two basic approaches are in use. The first involves indirect vaporization of the sample in an electrothermal vaporizer, e.g. a carbon rod or tube furnace or heated metal filament as commonly used in atomic absorption spectrometry (Gunn et al. [1.9], Matusiewicz and Barnes [1.10], Tikkanen and Niemczyk [1.11]). The second involves inserting the sample into the base of the inductively coupled plasma on a carbon rod or metal filament support (Salin and Harlick [1.12], Salin and Szung [1.13]). Available instrumentation is reviewed in Table 1.7.

1.3.5 Titration Procedures

Chloride, bromide, iodide, chlorate, chlorite, hypochlorite, free and total cyanide, sulphite, sulphate, thiosulphate, thiocyanate, sulphide, polysulphide, bicarbonate and total alkalinity are of relevance to these procedures:

The titration process has been automated so that batches of samples can be titrated non-manually and the data processed and reported via printouts and screens. One such instrument is the Metrohm 670 titroprocessor. This incorporates a built-in control unit and sample changer so that up to nine samples can be automatically titrated. The 670 titroprocessor offers incremental titrations with variable or constant-volume steps (dynamic or monotonic titration). The measured value transfer in these titrations is either drift controlled (equilibrium titration) or effected after a fixed waiting time; pK determinations and fixed end points (e.g. for specified standard procedures) are naturally included. End-point titrations can also be carried out.

Sixteen freely programmable computational formulae with assignment of the calculation parameters and units, mean-value calculations and arithmetic of one titration to another (via common variables) are available. Results can be calculated without any limitations.

The 670 titroprocessor can also be used to solve complex analytical tasks. In addition to various auxiliary functions which can be freely programmed, up to four different titrations can be peformed on a single sample.

Table 1.7. Inductively coupled plasma optical emission spectrometers available on the market

Supplier	Model	System	Number of elements claimed	Maximum analysis rate (elements min^{-1})	Microprocessor	Autosampler	Range (nm)
Perkin-Elmer	Plasma II	Optimized sequential system	70	Up to 50	Yes	Yes	160–800
Perkin-Elmer	ICP 5500	Sequential		15	Yes	Yes	170–900
Perkin-Elmer	ICP 5500B	Sequential		20	Yes	Yes	170–900
Perkin-Elmer	ICP 6500	Sequential		20	Yes	–	170–900
Perkin-Elmer	ICP 5000	Can be used for flame and graphite furnace ASS and inductively coupled plasma atomic emission spectrometry (sequential)		–	Yes	Yes	175–900
Perkin-Elmer	Plasma 40	Lower-cost sequential			Personal computer control	Yes	160–800
Labtam	Plasma Scan 8440	Simultaneous (polychromator or with optional monochromator for sequential)	60–70	Up to 64	Yes	Yes	170–820
Labtam	Plasma 8410	Sequential	More than 70	–	Yes	Yes	170–820
Thermo-electron	Plasma 300 (replacing the Plasma 200) (single (air) or double (air/vacuum) available	Sequential	Up to 63	Up to 18 (single channel air); up to 24 (double channel air/vacuum)	Yes	Yes	160–900

Philips	PV 8050 series PV 8055 PV 8060 PV8065	Simultaneous	56	–	Yes	165–485 and 530–860
Philips	PU 7450	Sequential	70	–	Yes	190–800
Baird	Spectrovac PS3/4 plasma hydride device option	Simultaneous and sequential	Up to 60	Up to 80 samples h^{-1} each up to 60 elements	Yes	162–766 and 162–800
Baird	Plasmatest system 75	Simultaneous and sequential	Up to 64	–	Yes	175–768 and 168–800
Spectro Analytical Ltd.	Spectroflame plasmahydride device option	Simultaneous and sequential	Up to 64	–	Yes	165–800

In addition to the fully automated 670 system, Metrohm also supply simpler units with more limited facilities which nevertheless are suitable for more simple titrations. Thus the model 682 titroprocessor is recommended for routine titrations with automatic equivalence pointer cognition or to preset end points. The 686 titroprocessor is a lower-cost version of the above instrument, again with automatic equivalence point recognition and titration to preset end points.

Mettler produce two automatic titrimeters the DL 40 GP memotitrator and the lower-cost DL 20 compact titrator. Features available on the DL 40GP include absolute and relative end-point titrations, equivalence point titrations, back-titration techniques, multi-method applications, dual titration, pH stating, automatic learn titrations, automatic determination of standard deviation and means, series titrations, correction to printer, acid balance analogue output for recorder and correction to the laboratory information system. Up to 40 freely definable methods can be handled and up to 20 reagents held on store. Six control principles can be invoked. The DL 20 can carry out absolute (not relative) end-point titrations and equivalence point titrations, back-titration, series titrations, and correction to printer and balance and the laboratory information system. Only one freely definable method is available. Four control principles can be invoked.

The DL 40GP can handle potentiometric, voltammetric or photometric titrations.

1.3.6 Ion-Selective Electrodes

Chloride, bromide, iodide, fluoride, iodate, nitrite, nitrate, free and total cyanide, sulphite, sulphate, thiocyanate, sulphide, borate and acetate are relevant to this method.

Ion-selective electrode technology is based on the simple measuring principle consisting of a reference electrode and a suitable sensing or indicator electrode sample solution (for the ion being dipped) dipped in the sample solution and connected by a sensitive voltameter. The sensing electrode responds to a difference between the composition of the solution inside and outside the electrode and requires a reference electrode to complete the circuit.

The Nernst equation, $E = E_0 + S \log C$, which gives the relationship between the activity and concentration (C) contains two terms which are constant for a particular electrode. These are E_0 (a term based on the potentials which remain constant for a particular sensing/reference electrode pair) and the slope S (which is a function of the sign and valency of the ion being sensed and the temperature). In direct potentiometry, it has to be assumed that the electrode response follows the Nernst equation in the sample matrix and in the range of measurement. E_0 and slope are determined by measuring the electrode potential in two standard solutions of known composition and the activity of the ion in the unknown sample is then calculated from the electrode potential measured in the sample.

Reference electrodes are of two types – single function and double function. Indicating or sensing electrodes for the determination of anions are of two types:

solid state   Determination of Br^{1-}, Cl^{1-}, $F^{1-}.I^{1-}$., Redox, silver/sulphide^{1-} CNS^{1-}

liquid membrane fluoroborate, NO_3^{1-}, ClO_4^{1-}, HF,

Variables which effect precise measurement by ion selective electrodes are the following:

concentration range;
ionic strength – an ionic strength adjuster is added to the samples and standards to minimize differences in ionic strength;
temperature;
pH;
stirring;
interferences;
complexation

Traditionally, electrodes have been used in two basic ways, direct potentiometry and potentiometric titration. Direct potentiometry is usually used for pH measurement and for measurement of ions like sodium, fluoride, nitrate and ammonium, for which good selective electrodes exist.

Direct potentiometry is usually done by manually preparing ionic activity standards and recording electrode potential in millivolts, using a high-impedance millivoltmeter, and plotting a calibration graph on semilogarithmic graph paper (or using a direct reading pH/ion meter which plots the calibration graph internally).

In potentiometric titration techniques, the electrode is simply used to determine the end-point of a titration, much as a coloured indicator would be used.

Direct potentiometry is an accurate technique but the precision and repeatability are limited because there is only one data point. Electrodes drift, and potential can rarely be reproduced to better than ± 0.5 mV so that the best possible repeatability in direct measurement is usually considered to be $\pm 2\%$.

Orion, the leading manufacturers of ion-selective electrodes, supply equipment for both direct potentiometry (EA 940, EA 920, SA 720 and SA 270 meters) and potentiometric titration (Orion 90 autochemistry system).

A review of these four meters in Table 1.8 shows that only the EA 940 has a facility for multiple point calibration and this places it at the top of the Orion range of direct-potentiometry meters. This instrument automatically prints out results. It has a memory for storing calibration information for all the electrodes.

The EA 920 is a lower-priced instrument for two-step calibration. It also has a memory for storing calibration information. The SA 720 and the portable SA 270 are relatively inexpensive bottom-of-the-range instruments with more limited capabilities.

Table 1.8. Orion pH/ISE meter features chart

Feature	Orion pH/SE meters			
	EA 940	EA 920	SA 720	SA 270
pH	✓	✓	✓	
Direct concentration readout in any unit	✓	✓	✓	✓
mV	✓	✓	✓	✓
Rel mV	✓	✓	✓	
Temperature	✓	✓	✓	✓
Oxygen	✓	✓		✓
Redox	✓	✓	✓	✓
Dual electrode inputs	✓	✓		
Expandable/upgradable	✓	✓		
Automatic anion/cation electrode recognition	✓	✓	✓	✓
Multiple point calibration	✓			
Incremental analytical techniques	✓	✓[a]		
Multiple electrode memory	✓	✓		
Prompting	✓	✓	✓	✓
Ready indicator	✓	✓	✓	
Resolution and significant digit selection	✓	✓	✓	✓
pH autocal	✓	✓	✓	
Blank correction	✓	✓		
Multiple print option	✓	✓		
Recorder output	✓	✓	✓	
RS-232C output	✓	✓	✓	
Adjustable ISO	✓	✓	✓	✓
Automatic temperature compensation-line and battery operation	✓	✓	✓	✓

[a] With PROM upgrade.

The Orion 960 autochemistry system (direct potentiometry-potentiometric titration) is a top-of-the-range instrument. In addition to direct potentiometry and potentiometric titrations it has other features not previously incorporated in potentiometric analysers.

The 960 uses 12 basic analytical techniques. To do an analysis, one of these techniques is chosen and modified to suit the requirements of the paricular sample. The memory will accommodate up to 20 methods.

KAP analysis is a time-saving technique that eliminates sample preparation and calibration. Simply weigh sample into a beaker, add water and measure. Aliquots of one standardizing solution or reagent are added automatically to the sample and sample concentration is determined from the changes in potential observed after each addition. Every step is performed in one beaker.

Results from KAP analysis are automatically verified in two ways. First, a check for electrode drift and noise is performed at the beginning of each analysis. Second, each sample is spiked with standard as part of the analysis and recovery of the spike is calculated.

GAP analysis is a faster way to perform many titrations. It actually predicts the location of the end point so there is no need to titrate all the way. And GAP analysis allows titrations to be performed at low levels where conventional techniques yield very poor end-point breaks or asymmetrical curves.

HELP analysis is a diagnostic technique in which the instrument studies the data collected and recommends the optimum procedure for repetitive analysis of similar samples.

The heart of the Orion 960 autochemistry system is the EA 940 expandable analyser – an advanced pH/ISE meter.

Orion supply both electrodes and measuring equipment. Ingold, on the other hand, supply only electrodes. EDT Analytical (UK) also manufacture ion-selective electrodes.

1.3.7 Polarographic Methods

Bromide, iodide, nitrite, nitrate, free cyanide, sulphide, phosphate, silicate, arsenite, arsenate, selenate, chromate, NTA and EDTA can be determined by polarogaphic technique.

Three basic techniques of polarography are of interest and the basic principles of these are outlined below:

Universal: Differential Pulse (DPN, DPI, DPR)
In this technique a voltage pulse is superimposed on the voltage ramp during the last 40 ms of controlled drop growth with the standard dropping mercury electrode; the drop surface is then constant. The pulse amplitude can be preselected. The current is measured by integration over a 20 ms period immediately before the start of the pulse and again for 20 ms as the pulse nears completion. The difference between the two current integrals $(l_2 - l_1)$ is recorded and this gives a peak-shaped curve. If the pulse amplitude is increased, the peak current value is raised but the peak is broadened at the same time.

Classical: Direct Current (DCT)
In this direct current method, integration is performed over the last 20 ms of the controlled drop growth (Tast procedure): during this time, the drop surface is constant in the case of the dropping mercury electrode. The resulting polarogram is step-shaped. Compared with classical DC polarography according to Heyrovsky, i.e. with the free-dropping mercury electrode, the DCT method offers great advantages; considerably shorter analysis times, no disturbance due to current oscillations; simpler evaluation and larger diffusion-controlled limiting current.

Rapid: Square Wave (SQW)
Five square-wave oscillations of frequency around 125 Hx are superimposed on the voltage ramp during the last 40 ms of controlled drop growth – with the

dropping mercury electrode the drop surface is then constant. The oscillation amplitude can be preselected. Measurements are performed in the second, third and fourth square-wave oscillation; the current is integrated over 2 ms at the end of the first and the end of the second half of each oscillation. The three differences of the six integrals $(l_1 - l_2, l_3 - l_4, l_5 - l_6)$ are averaged arithmetically and recorded as one current value. The resulting polarogram is peak shaped.

Metrohm are leading suppliers of polarographic equipment. They supply three main pieces of equipment: the Metrohm 646 VA Processor, the 647 VA Stand (for single determinations) and the 675 VA sample changer for a series of determinations). Some features of the 646 VA processor are listed below:

Optimized data acquisition and data processing
High-grade electronics for a better signal-to-noise ratio
Automatic curve evaluation as well as automated standard addition for greater accuracy and smaller standard deviation
Large, non-volatile methods memory for the library of fully developed analytical procedures
Connection of the 675 VA sample changer for greater sample throughout
Connection of an electronic balance
Simple, perfectly clear operation principle via guidance in the dialogue mode yet at the same time high application flexibility thanks to the visual display and alphanumeric keyboard
Complete and convenient result recording with built-in thermal recorder/printer

The 675 VA sample changer is controlled by the 646 VA Processor on which the user enters the few control commands necessary. The 646 VA processor also controls the 677 drive unit and the 683 pumps. With these auxiliary units, the instrument combination becomes a polarographic analysis station which can be used to carry out on-line measurements.

The 646 VA processor is conceived as a central, compact component for automated polarographic and voltammetric systems. Thus, two independent 647 VA stands or a 675 VA sample changer can be added. Up to 4 multi-dosimats of the 665 type for automated standard additions and/or addition of auxiliary solutions can be connected to each of these wet-chemical workstations. Connection of an electronic balance for direct transfer of data is also possible.

Program-controlled automatic switching and mixing of these three electrode configurations during a single analysis via software commands occur. The complete electrode is pneumatically controlled. A hermetically sealed mercury reservoir of only a few millilitres suffices for approximately 200 000 drops. The mercury drops are small and stable, consequently there is a good signal-to-noise ratio. Mercury comes into contact only with the purest inert gas and plastic free of metal traces. Filling is seldom required and very simple to carry out. The system uses glass capillaries which can be exchanged simply and rapidly.

Up to 30 complete analytical methods (including all detailed information and instructions) can be filed in a non-volatile memory and called up. Consequently, a large, extensive and correspondingly efficient library of analytical methods can be built up, comprehensive enough to carry out all routine determinations conveniently via call-up of a stored method.

The standard addition method (SAM) is the procedure generally employed to calculate the analyte content from the signal of the sample solution; electric current SAM amount of substance/mass concentration. The SAM is coupled directly to the determination of the sample solution so that all factors which influence the measurement remain constant. There is no doubt that the SAM provides results that have proved to be accurate and precise in virtually every case.

The addition of standard solutions can be performed several times if need be (multiple standard addition) to raise the level of quality of the results still further.

Normally, a real sample solution contains the substances to be analysed in widely different concentrations. In a single multi-element analysis, however, all components must be determined simultaneously. The superiority of the facilities offered by segmented data acquisition in this respect is clear when a comparison is made with previous solutions. The analytical conditions were inevitably a compromise: no matter what type of analytical conditions were selected, such large differences could rarely be reconciled. In the recording, either the peaks of some of the components were shown meaningfully – each of the other two were either no longer recognizable or led to gigantic signals with cut-off peak tips. And all too often the differences were still too large even within the two concentration ranges. Since the recorder sensitivity and also all other instrument and electrode functions could only be set and adjusted for a single substance, even automatic range switching of the recorder was of very little use.

The dilemma is solved with the 646 VA processor: the freedom to divide the voltage sweep into substance-specific segments and to adjust all conditions individually and independently of one another within these segments opens up quite new and to date unknown analytical possibilities. Furthermore, it allows optimum evaluation of the experimental data.

Various suppliers of polarographs are summarized below.

Supplier	Type	Model No.	Detection limits
Metrohm	Differential Pulse Direct current Square wave	646 VA processor 647 VA stand 675 Sample changer 665 Dosimat (motor driven piston burettes for standard additions)	$2-10\,\mu g\,l^{-1}$ quoted for nitriloacetate
	Direct current normal pulse differential pulse 1st harmonic ac. 2nd harmonic ac. Kalousek	506 Polarecord	
	Direct current sampled differential pulse	DC 626 Polarecord	

Supplier	Type	Model No.	Detection limits
Chemtronics Ltd	On-line voltammetric analyser for metals in effluents and field work	PDV 2000	$\sim 0.1\ \mu g\,l^{-1}$
RDT Analytical Ltd	Differential pulse anodic stripping on-line voltammetric analyser for metals in effluents and field work	ECP 100 plus ECP 104 programmes ECP 140 PDV 200	-
	On-line voltametric analyser for continuous measurement of metals in effluents and water	OVA 2000	-
EDT Analytical Ltd	Cyclic voltammetry differential pulse voltammetry linear scam voltammetry, square-wave voltammetry, single- and double-step chronopotentiometry and chronocoulometry	Cipress Model CYSY-1B (basic system) CY57-1H-(high-sensitivity system)	

1.3.8 Chromatographic Methods

1.3.8.1 Ion Chromatography

Chloride, bromide, iodide, fluoride, chlorate, and total cyanide free chlorite, hypochlorite, bromite, nitrite, nitrate, cyanate, sulphite, sulphate, thiocyanate, thiosulphate, sulphide, phosphate, silicate, borate, carbonate, bicarbonate, arsenite, arsenate, selenite, selenate, metal cyanide complexes and formate, ascorbate, amino acetates and sulphonates may be determined by ion chromatography which was originally developed by Small et al. [1.15] in 1975 for rapid and sensitive analysis of inorganic anions using specialized ion-exchange columns and chemically suppressed conductivity detection. Advances in column and detection technologies have expanded this capability to include a wider range of anions as well as organic ions. These recent developments, discussed below, provide the chemist with a means of solving many problems that are difficult, if not impossible, using other instrumental methods. Ion chromatography can analyse a wide variety of anions more easily than either atomic absorption spectrometry or inductively coupled plasma techniques. These include halides, ammonia, nitrate, phosphate, sulphate, borate and silicate.

Metals determination is an excellent example of the problem-solving power of ion chromatography. As a stand-alone instrument, an ion chromatograph offers several advantages over atomic absorption spectrometry and inductively coupled plasma spectroscopy.

Ion chromatography can complement atomic absorption and plasma techniques as a back-up technique and as an alternative to wet chemistry for cross checking results.

At the heart of the ion chromatography system is an analytical column containing an ion-exchange resin on which various anions and/or cations are separated before being detected and quantified by various detection techniques such as spectrophotometry, atomic absorption spectrometry (metals) or conductivity (anions).

Ion chromatography is not restricted to the separate analysis of only anions or cations, and, with the proper selection of the eluent and separator columns, the technique can be used for the simultaneous analysis of both anions and cations.

This original method for the analysis of mixtures of anions used two columns attached in series packed with ion-exchange resins to separate the ions of interest and suppress the conductance of the eluent, leaving only the species of interest as the major conducting species in the solution. Once the ions were separated and the eluent suppressed, the solution entered a conductivity cell, where the species of interest were detected.

The analytical column is used in conjuction with two other columns, a guard column which protects the analytical column from troublesome contaminants, and a pre-concentration column.

The intended function of the pre-concentration column is twofold. First, it concentrates the ions present in the sample, enabling very low levels of contaminants to be detected. Second, it retains non-complexed ions on the resin, while allowing complexed species to pass through.

Dionex Series 4000i Ion Chromatographs. Some of the features of this instrument are:

> chromatography module;
> up to six automated valves made of chemically inert, metal-free material eliminate corrosion and metal contamination;
> liquid flow path is completely compatible with all HPLC solvents;
> electronic valve switching, multidimensional, coupled chromatography or multi-mode operation;
> automated sample clean up or pre-concentration;
> environmentally isolates up to four separator columns and two suppressors for optimal results;
> manual or remote control with Dionex Autoion 300 or Autoion 100 automation accessories;
> individual column temperature control from ambient to 100 °C (optional).

Dionex Ion-Pac Columns. Features are:

> polymer ion exchange columns are packed with new pellicular resins for anion or cation exchange applications;

new 4 µ polymer ion exchange columns have maximum efficiency and minimum operating pressure for high-performance ion and liquid chromatography applications;

new ion exclusion columns with bifunctional cation exchange sites offer more selectivity for organic acid separations;

neutral polymer resins have high surface area for reversed phase ion-pair and ion-suppression applications without pH restriction;

5 and 10 µ silica columns are optimized for ion-pair, ion suppression and reversed phase applications.

Micromembrane Suppressor. The micromembrane suppressor makes possible detection of non-UV-absorbing compounds such as inorganic anions and cations, surfactants, antibiotics, fatty acids and amines in ion-exchange and ion-pair chromatography.

Two variants of this exist: the anionic (AMMS) and the cationic (CMMS) suppressor. The micromembrane suppressor consists of a low dead volume eluent flow path through alternating layers of high-capacity ion-exchange screens and ultra-thin ion exchange membranes. Ion-exchange sites in each screen provide a site-to-site pathway for eluent ions to transfer to the membrane for maximum chemical suppression.

Dionex anion and cation micromembrane suppressors transform eluent ions into less conducting species without affecting sample ions under analysis. This improves conductivity detection, sensitivity, specificity, and baseline stability. It also dramatically increases the dynamic range of the system for inorganic and organic ion chromatography. The high ion-exchange capacity of the MMS permits changes in eluent composition by orders of magnitude, making gradient ion chromatography possible.

In addition, because of the increased detection specificity provided by the MMS sample, preparation is dramatically reduced, making it possible to analyse most samples after simple filtering and dilution.

Conductivity Detector. Features include:

high-sensitivity detection of inorganic anions, amines, surfactants, organic acids, Group I and II metals, oxy-metal ions and metal cyanide complexes (used in combination with MMS);

bipolar-pulsed excitation eliminates the non-linear response with concentration found in analogue detectors;

microcomputer-controlled temperature compensation minimizes the baseline drift with changes in room temperature.

UV Vis Detector. Important factors are:

high-sensitivity detection of amino acids, metals, silica, chelating agents, and other UV absorbing compounds using either post-column reagent addition or direct detection;

non-metallic cell design eliminates corrosion problems;
filter-based detection with selectable filters from 214 to 800 nm;
proprietary dual wavelength detection for ninhydrin-detectable amino acids and PAR-detectable transition metals.

Optional Detectors. In addition to the detectors shown, Dionex also offer visible, fluorescence, and pulsed amperometric detectors for use with the series 4000i.

Dionex also supply a wide range of alternative instruments, e.g. single channel (2010i) and dual channel (2020i). The latter can be upgraded to an automated system by adding the Autoion 100 or Autoion 300 controller to control two independent ion chromatograph systems. They also supply a 2000i series equipped with conductivity pulsed amperometric, UV-visible, visible and fluorescence detectors.

1.3.8.2   High Performance Liquid Chromatography

Bromide, iodide, fluoride, nitrite, nitrate, sulphate, thiocyanate, phosphate, selenate and chromate, chlorate, chlorite, hypochlorite, bromite, iodate, thiosulphate, formate, N.T.A, EDTA and sulphonates can be determined using this technique.

Modern high-performance liquid chromatography has been developed to a very high level of performance by the introduction of selective stationary phases of small particle sizes, resulting in efficient columns with large plate numbers per litre.

There are several types of chromatographic columns used in high-performance liquid chromatography.

Reversed-Phase Chromatography. The most commonly used chromatographic mode in HPLC is reversed-phase chromatography. Reversed-phase chromatography is used for the analysis of a wide range of neutral and polar organic compounds. Most common reversed phase chromatography is performed using bonded silica-based columns, thus inherently limiting the operating pH range to 2.0–7.5. The wide pH range (0–14) of some columns (e.g. Dionex Ion Pac NSI and NS 1–5 μ columns) removes this limitation, and consequently they are ideally suited for ion-pairing and ion-suppression reversed-phase chromatography, the two techniques which have helped extend reverse-phase chromatography to ionizable compounds.

High-sensitivity detection of non-chromophoric organic ions can be achieved by combining the power of suppressed conductivity detection with these columns. Suppressed conductivity is usually a superior approach to using refractive index or low UV wavelength detection.

Reversed-Phase Ion-Pairing Chromatography. Typically, reversed-phase ion-pairing chromatography is carried out using the same stationary phase as reversed-phase chromatography. A hydrophobic ion of opposite charge to the solute of interest is added to the mobile phase. Samples which are determined by reversed-phase ion-packing chromatography are ionic and thus capable of forming an ion pair with the added counter ion. This form of reversed-phase chromato-graphy can be used for anion and cation separations and for the separation of surfactants and other ionic types of organic molecules. An unfortunate drawback to using silica-based columns is that ion-pairing reagents increase the solubility of silica in water, leading to loss of bead integrity and drastically reducing column life. Some manufacturers (e.g. Dionex) employ neutral macroporous resins, instead of silica, in an attempt to widen the usable pH range and eliminate the effect of ion-pairing reagents.

Ion-Suppression Chromatography. Ion suppression is a technique used to suppress the ionization of compounds (such as carboxylic acids) so they will be retained exclusively by the reversed-phase retention mechanism and chromatographed as the neutral species.

Column packings with an extended pH range are needed for this application as strong acids or alkalies are used to suppress ionization. In addition to carboxylic acids, the ionization of amines can be suppressed by the addition of a base to the mobile phase, thus allowing chromatography of the neutral amine.

Ion-Exclusion Chromatography. Chloride, sulphate, phosphate and bicarbonate are detected by this means. Unlike the pellicular packings used for ion exchange, the packings used in ion exclusion are derived from totally sulphonated polymeric materials. Separation is dependent upon three different mechanisms Donnan exclusion, steric exclusion, and adsorption/partitioning.

Donnan exclusion causes strong acids to elute in the void volumes of the column. Weak acids which are partially ionized in the eluent are not subject to Donnan exclusion and can penetrate into the pores of the packing. Separation is accomplished by differences in acid strength, size and hydrophobicity. The major advantage of ion exclusion lies in the ability to handle samples that contain both weak and strong acids. A good example of the power of ion exclusion is the routine determination of organic acids in sea water. Without ion exclusion, the high chloride ion concentration would present a serious interference.

Four basic types of elution system are used in HPLC. This is illustrated below by the systems offered by LKB, Sweden.

The Isocratic System. This consists of a solvent delivery for isocratic reversed phase and gel filtration chromatography.

The isocratic system (Fig. 1.3a) provides an economic first step into high-performance liquid chromatography techniques. The system is built around a high-performance, dual-piston, pulse-free pump providing precision flow from 0.01 to 5 ml min^{-1}.

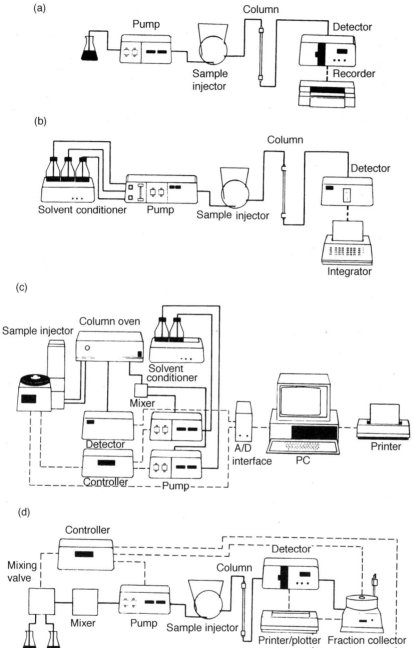

Fig. 1.3a–d. Elution systems supplied by LKB, Sweden: **a** isocratic bioseparatio system; **b** basic system; **c** advanced chromatography system; **d** inert system

Any of the following detectors can be used with this system:

fixed wavelength ultraviolet detector (LKB Unicord 2510);
variable UV visible (190–600 nm);
wavelength monitor (LKB 2151);
rapid diode array spectral detector (LKB 2140) (discussed later);
refractive index detector (LKB 2142);
electrochemical detector (LKB 2143);
wavescan EG software (LKB 2146).

Basic Gradient System. This is a simple upgrade of the isocratic system with the facility for gradient elution techniques and greater functionality (Fig. 1.3b). The basic system provides for manual operating gradient techniques such as reversed phase, ion-exchange and hydrophobic interaction chromatography. Any of the detectors listed above under the isocratic system can be used.

Advanced Gradient System. For optimum functionality in automated systems designed primarily for reversed-phase chromatography and other gradient techniques, the LKB advanced-gradient system is recommended (Fig. 1.3c). Key features include the following:

a configuration that provides the highest possible reproducibility of results;
a two-pump system for highly precise and accurate gradient formation for separation of complex samples;
full system control and advanced method development provided from a liquid chromatography controller;
precise and accurate flows ranging from 0.01 to 5 ml min^{-1}.

This system is ideal for automatic method for development and gradient optimization.

The Inert System. By a combination of the use of inert materials (glass, titanium, and inert polymers) this system offers totally inert fluidics. Primary features of the system include: (Fig. 1.3d)

the ability to perform isocratic or gradient elution by manual means;
full system control from a liquid chromatography controller;
precise and accurate flows from 0.01–5 ml min^{-1}

This is the method of choice when corrosive buffers, e.g. those containing chloride or aggressive solvents, are used.

Chromatographic Detectors. Details concerning the types of detectors used in high-performance liquid chromatography are given in Table 1.9. The most commonly used detectors are those based on spectrophotometry in the region 185–400 nm, visible ultraviolet spectroscopy in the region 185–900 nm, post-column derivatization with fluorescence detection (see below), conductivity and those based

Table 1.9. Detectors used in HPLC

Type of detector		Supplier	Detection part no.	HPLC instrument part no.
Spectrophotometric (variable wavelength)	190–390 nm	Perkin-Elmer	LC-90	–
	195–350 nm	Kontron	735 LC	Series 400
	195–350 nm	Shimadzu	SPD-7A	LC-7A
	195–350 nm	Shimadzu	SPD-6A	LC-8A
	195–350 nm	Shimadzu	SPD-6A	LC-6A
	206–405 nm (fixed wavelength choice of seven wavelengths between 206 and 405 nm)	LKB	2510 Uvicord SD	–
	190–370 nm } 190–400 nm }	Cecil Instruments	Model 1937 CE 1220	Chrom-A-Scope Series 100
Variable wavelength UV-visible	190–600 nm	Varian	2550	2500
	190–600 nm	LKB	2151	Uvicord SD
	190–700 nm	Kontron	432	Series 400
	190–800 nm	Kontron	430	Series 400
	185–900 nm	Kontron	720 LC	Series 400
	200–570 nm	Kontron	740 LC	Series 400
	190–800 nm	Dionex	VDM II	Series 400
	190–750 nm	Isco	V4	
	214–660 nm (18 preset wavelengths)	Isco	USA and 228 }	Microbo system
	195–700 nm	Shimadzu	SPD 7A	LC-7A
	195–700 nm	Shimadzu	SPD-6A V	LC-8A
	195–700 nm	Shimadzu	SPD-6AV	LC-6A
	195–700 nm	Hewlett Packard	Programmably variable wavelength detector	9050 series
	380–600 nm	Cecil Instruments	CE 1200	Series 1000
	190–800 nm	Applied Chromatography systems	750/16 and 5750/11	–
Conductivity	–	Dionex	CDM 11	4500 i
		Roth Scientific	–	Chrom-A-Scope
Electrochemical detector	–	Dionex	PAD-11	4500i
		LKB	2143	Wave-scan EG
		Roth Scientific	–	Chrom-A-Scope
		Cecil Instruments	CE 1500	–
		PSA Inc.	5100A	–
		Applied Chromatography systems	650*350*06	–
Refractive index detector		LKB	2142	Wavescan E.G.
		Roth Scientific	–	Chrom-A-Scope
		Cecil Instruments	CE 1400	Series 1000

Table 1.9 (*Contd.*)

Type of detector	Supplier	Detection part no.	HPLC instrument part no.
Differential viscosity	Roth Scientific	–	Chrom-A-Scope
mass detection (evaporative)	Applied Chromatography systems	750/14	–
Diode array	Varian	9060	2000L and 5001 5500 series
	Perkin-Elmer	LC135, LC 235 and LC 480	–
	LKB	2140	–
	Hewlett Packard	Multiple wavelength detector	1050 series

on the relatively new technique of multiple wavelength ultraviolet detectors using a diode array system detector (see below). Other types of detectors available are those based on electrochemical principles, refractive index, differential viscosity and mass detection.

Electrochemical Detectors. These are available from several suppliers (Table 1.9). ESA supply the model PS 100A coulochem multi-electrode electrochemical detector. Organics, anions and cations can be detected by electrochemical means.

The Gilson Aspec automatic sample preparation system is a fully automated system for solid-phase extraction on disposable columns and on-line HPLC analysis. The Aspec system offers total automation and total control of the entire sample preparation process including clean-up and concentration. In addition, Aspec can automatically inject prepared samples into on-line HPLC systems.

Aspec is designed to receive up to 108 samples. The system is compatible with most standard disposable extraction columns. Analytichem Bond-Elut, Baker SPE, Supelco Supelclean, Alltech Extract Clean, etc. There is a choice of more than 20 different stationary phases.

1.3.8.3 Ion-Exchange Chromatography

Chloride, bromide, fluoride, nitrite, nitrate, sulphate, carbonate and salicylate are detectable. See also previous section.

Ion-exchange chromatography is based upon the differential affinity of ions for the stationary phase. The rate of migration of the ion through the column is directly dependent upon the type and concentration of ions that comprise the eluent. Ions with low or moderate affinities for the packing generally prove to be the best eluents. Examples are hydroxide and carbonate eluents for anion separations.

The stationary phases commonly used in HPLC are typically derived from silica substrates. The instability of silica outside the pH range 2–7.5 represents one of the main reasons why ion exchange separations have not been extensively used in HPLC. To overcome this, some manufacturers (e.g. Dionex Ion-Pac) columns supply a packing which is derived from crosslinked polystyrene which is stable throughout the entire pH range. This pH stability allows eluents of extreme pH values to be used so that weak acids such as carbohydrates (and bases) can be ionized.

1.3.8.4 Column Coupling Capillary Isotachophoresis

Chloride, fluoride, nitrite, nitrate, sulphate and phosphate are detectable. This technique offers many similar advantages to ion chromatography in the determination of anions in water, namely multiple ion analysis, little or no sample pretreatment, speed, sensitivity and automation. However, to date, very little attention has been paid to this technique.

Separation capillary columns are made in fluorinated ethylene-propylene copolymer. Detection is achieved by conductivity cells and an a.c. conductivity mode of detection is used for making the separations visible. The driving current is supplied by a unit enabling independent currents to be preselected for the preseparation and final analytical stages. The run of the analyser is controlled by a programmable timing and control unit. The zone lengths from the conductivity detector, evaluated electronically, can be printed on a line printer.

The use of column coupling configuration of the separation unit provides the possibility of applying a sequence of two leading electrolytes to the analytical run. Therefore, the choice of optimum separation conditions can be advantageously divided into two steps, first, the choice of a leading electrolyte suitable for the separation and quantitation of the micro-constituents in the first stage (pre-separation column) simultaneously having a retarding effect on the effective mobilities of micro-constituents (nitrite, fluoride, phosphate), and, second, the choice of the leading electrolyte for the second stage in which only micro constituents are separated and quantified (macro constituents are removed from the analytical system after their evaluation in the first stage).

To satisfy the requirements for the properties of the leading electrolyte applied in the first stage and, consequently, to decide its composition, two facts had to be taken into account, i.e. the pH value of the leading electrolyte needs to be around 4 or less (retardation of nitrite relative to the macro constituents in this stage) and at the same time the separations of the macro constituents need to be optimised by means other than adjusting the pH of the leading electrolyte (anions of strong acids). For the latter reason, complex equilibria and differentiation of anions through the charge number of the counter ions were tested at a low pH as a means of optimising the separation conditions for chloride, nitrate and sulphate in the presence of nitrite, fluoride and phosphate. The retardation of chloride and sulphate through complex formation with cadmium, enabling the separation of anions of interest to be carried out, was found to be unsuitable. This was due to the high concentrations of cadmium ions necessary to achieve the desired effect (loss of fluoride and phosphate probably owing to precipitation).

Similarly, the use of calcium and magnesium as complexing co-counter ions to decrease the effective mobility of sulphate was found to be ineffective, as a very strong retardation of fluoride occurred.

Better results were achieved when a divalent organic cation was used as a co-counter ion in the leading electrolyte employed in the first separation stage. When, simultaneously, the pH of the leading electrolyte was 4 or less, the steady state configuration of the constituents to be separated was chloride, sulphate, nitrite, fluoride and phosphate.

The choice of the leading electrolyte for the second stage, in which the microconstituents were finally separated and quantitatively evaluated, was straightforward, involving a low concentration of the leading constituent (low detection limit) and a low pH of the leading electrolyte (separation according to pK values).

1.3.8.5 Gas Chromatography

Chloride, bromide, iodide, fluoride, nitrite, nitrate, free and total cyanide, thiocyanide, sulphide, sulphate, polysulphide, phosphate, selenite, selenate, amino carboxylates, sulphonates and carboxylates may be determined.

This technique has been applied to the determination of anions. The anions is first converted to a volatile derivative which is amenable to gas chromatography. In one method for the determination of chloride, for example, the sample is reacted with phenyl mercuric nitrate to produce volatile phenyl mercuric chloride. A carbon tetrachloride extraction of the reaction mixture separates phenyl mercuric chloride which is determined by gas chromatography:

$$PhHgNO_3 + MCl = PhHgCl + MNO_3.$$

In another method, the determining selenium IV reaction with 1,2-diamino-3,5-dibromobenzene produces 4,6-dibromo-piazaselenol. This is extracted with toluene and determined by gas chromatography.

The basic requirements required of a high-performance gas chromatograph are as follows:

> sample is introduced to the column in an ideal state, i.e. uncontaminated by septum bleed or previous sample components, without modification due to distillation effects in the needle and quantitatively, i.e. without hold-up or adsorption prior to the column;
> the instrument parameters that influence the chromatographic separation are precisely controlled;
> sample components do not escape detection, i.e. highly sensitive, reproducible detection and subsequent data processing are essential.

There are two types of separation column used in gas chromatography – capillary columns and packed columns.

Packed columns are still used extensively, especially in routine analysis. They are essential when sample components have high partition coefficients and/or high concentrations. Capillary columns provide a high number of theoretical plates, hence a very high resolution, but they cannot be used in all applications because there are not many types of chemically bonded capillary columns. Combined use of packed columns of different polarities often provides better separation than with a capillary column. It sometimes happens that a capillary column is used as a supplement in the packed-column gas chromatograph. It is best, therefore, to house the capillary and packed columns in the same column oven and use them selectively. In the screening of some types of samples, the packed column is used routinely and the capillary column is used when more detailed information is required.

Conventionally, it is necessary to use a dual column flow line in packed-column gas chromatography to provide sample and reference gas flows. The recently developed electronic base-line drift compensation system allows a simple column flow line to be used reliably.

Recent advances in capillary column technology presume stringent performance levels for the other components of a gas chromatograph as column performance is only as good as that of the rest of the system. One of the most important factors in capillary column gas chromatography is that a high repeatability of retention times be ensured even under adverse ambient conditions.

These features combine to provide ± 0.01 min repeatability for peaks having retention times as long as 2 h (other factors being equal).

Another important factor for reliable capillary column gas chromatography is the sample injection method. Various types of sample injection ports are available. The split/splitless sample injection port unit series is designed so that the glass insert is easily replaced and the septum is continuously purged during operation. This type of sample injection unit is quite effective for the analysis of samples having high boiling point compounds as the major components.

In capillary column gas chromatography, it is often required to raise and lower the column temperature very rapidly and to raise the sample injection port temperature. In one design of gas chromatograph, the Shimadzu GC 14-A, the computer-controlled flap operates to bring in the external air to cool the column oven rapidly – only 6 min from 500 °C to 100 °C. This computer-controlled flap also ensures highly stable column temperature when it is set to a near-ambient point. The lowest controllable column temperature is about 26 °C when the ambient temperature is 20 °C.

Some suppliers of gas chromatography are listed in Table 1.10.

Table 1.10. Commercial gas chromatographs

Manufacturer	Model	Packed colmn	Capillary column	Detectors	Sample injection point system	Keyboard control	Link to computer	Visual display	Printer	Core instrument amenable to tap automation	Temperature programming/ isothermal	Cryogenic unit (sub-ambient chromatography)
Shimadzu	GC-14A	Yes	Yes	FID ECD FTD FPD TCD (all supplied)	1. Split-splitters 2. Glass insert for single column 3. Glass insert to dual column 4. Cool on column system unit 5. Moving needle system 6. Rapidly ascending temperature vaporizer	Yes	Yes	No	No	No	Yes/Yes	No
Shimadzu	GC 15A	Yes	Yes	FID ECD FTD FPD TCD	1. Split-splitters 2. Direct sample injection (capillary column) 3. Standard sample injector (packed column) 4. Moving precolumn system (capillary columns) 5. On column (capillary columns)	No	Yes	Yes	Yes	No	Yes/Yes	No
Shimadzu	GC 16A		Yes	FID FCD FTD FPD TCD (all supplied)	Split-splitters	Yes	Yes	Yes	Yes	No	Yes/Yes	No
Shimadzu	GC 8A	Yes	Yes	FID FCD FPD TCD single detector instruments (detector chosen on purchase)	1. Point for packed columns 2. Point for capillary columns 3. Split-splitters	Yes Not built in	Optional	No	Optional	No	Yes temp prgramming: GC 8APT (TCD detector) GC 8 APF (FID detector) GV 8APFD (FID detector) isothermal: GC 8AIF (FID detector) GC 8AIE (ECD detector)	No
Perkin-Elmer	8420	NO	Yes	Single detector instrument (chosen on purchase) FID ECD FTD FPD TCD	1. Programmable temperature vaporizer 2. Split-splitless injector 3. Direct on column injector	No	No	No	No	No	Yes/Yes	Yes, down to −80 °C

1.4 References

1.1 Ruz J, Torres A, Rios A, Luque de castro MD, Valcarcel MJ (1986) Automatic Chem 8: 70
1.2 Fang Z, Ruzicka J, Hansen EH (1984) Anal Chem Acta 164: 23
1.3 Marshall MA, Mottola HA (1985) Anal Chem 57: 729
1.4 Greenfield S, Jones IL, Berry CT (1964) Analyst (London) 89: 713
1.5 Wendt RH, Fassel UA (1965) Anal Chem 37: 920
1.6 Scott RH (1974) Anal Chem 46: 75
1.7 Suddendorf RF, Boyer KW (1978) Anal Chem 50: 1769
1.8 Sharp BL (1984) The conespray Nebulizer, British Technology Group, Patent assignment No.8, 432, 338
1.9 Gunn AM, Millard DL, Kirkbright GF (1978) Analyst (London) 103: 1066
1.10 Matusiewicz H, Barnes RM (1984) Applied Spectroscopy 38: 745
1.11 Tikkanen MW, Niemczyk TM (1984) Anal Chem 56: 1997
1.12 Salin ED, Harlick G (1979) Anal Chem 51: 2284
1.13 Salin ED, Szung RLA (1984) Anal Chem 56: 2596
1.14 Thompson M, Walsh JN, In "A Handbook of Reductively Coupled Plasma Spectrometry", Blackie, London and Glasgow, p 55, (1983)
1.15 Small H, Stevens TS, Bauman WC (1975) Anal Chem 47: 1801
1.16 Ruzicka J, Hansen EA (1975) Anal Chem Acta 78: 145

2 Halogen-Containing Anions

Titration methods for the determination of halides are convenient but limited in sensitivity or ability to automate. When these characteristics are required then spectrophotometric methods are recommended. Methods based on ion-selective electrodes are also useful. When mixtures of anions are to be analysed and high sensitivity and speed are essential, then ion chromatography becomes the method of choice. For the determination of fluoride, ion selective electrodes are the method of choice.

2.1 Chloride

2.1.1 Natural Waters

2.1.1.1 Titration Methods

Three basic approaches have been made to the determination of chloride in water:

Mohr titration;
Mercuric nitrate and diphenyl carbazone titration;
Technicon autoanalyser method (mercuric thiocyanate-ferric salt spectrophotometric method).

The precision and bias of these three standard approaches to the determination of chloride in river water has been evaluated. [2.1, 2.2] in trials carried out in a number of laboratories in the U.K. The following targets were set in this exercise:

maximum tolerable bias −10% of the chloride concentration or 0.5% mg l^{-1} chloride, whichever is the greater;
maximum tolerable standard deviation −5% of the chloride concentration, or 0.25% mg l^{-1} chloride, whichever is the greater.

The results were analysed statistically to provide estimates of the standard deviation of individual analytical results in any batch of analyses. These estimates are summarized in Table 2.1, which shows that all laboratories and methods achieved the target standard deviation.

Portions of a standard chloride solution and two river water samples were analysed once on each of five days. The results, summarized in Tables 2.1 and 2.2, show that the confidence intervals of the mean results for a given solution do not all overlap, that is, some inter-laboratory bias is present. To assess whether or not the bias of any laboratory exceeded the target value, the following procedure was adopted. The mean result and its 90% confidence interval from laboratory i was denoted by $X_i \pm L_i$ and the true value of the standard chloride solution or, for the two river waters, the overall mean of all laboratories, was denoted by $\bar{\bar{x}}$. The value of the maximum possible bias (95% confidence level) was then calculated as

$$\frac{100(X_2 + L_i - \bar{\bar{x}})}{\bar{\bar{x}}} \quad \text{if } \bar{X}_2 > \bar{\bar{x}}$$

or

$$\frac{100(\bar{X}_2 + L_i - \bar{\bar{x}})}{\bar{\bar{x}}} \quad \text{if } \bar{X}_2 < \bar{\bar{x}}$$

The values for the maximum bias shown in Table 2.1 are almost invariably less than 10% and often much less.

It is concluded that precision and bias are satisfactory in all three methods for the determination of chloride mentioned above.

Chloride can be automatically titrated with standard silver nitrate solution using a chloride selective electrode [2.3], and non-ionic detergent must be added to prevent fouling of the electrodes by precipitated silver chloride. A modification to the titrant-delivery tube on the commercial apparatus used is required to deliver the titrant against the indicating surface of the chloride-selective electrode. Reproducibility is good for all concentrations in the range 1 to 100 mg l^{-1} chloride; the standard deviation was ± 0.28 to 0.47 mg.

Kuttel et al. [2.4] have described a potentiometric titration procedure for the simultaneous determination of chloride (and bromide and iodide) in natural water. Errors caused by adsorption of the common ion on to the surface of the precipitate in the potentiometric titration of mixtures of halides with use of a silver electrode or a AgI, AgBr or AgCl membrane-electrode can be overcome by the addition of sodium nitrate at a concentration exceeding 1 N. Because of the small difference in the solubilities of silver bromide and silver chloride, the determination of bromide in the presence of chloride is not as accurate as the determination of other pairs of halides.

2.1.1.2 Spectrophotometric Methods

Dojlido and Bierwagen [2.5] have described an automated procedure utilizing an autoanalyser for the determination of chloride in natural water. The method is

Table 2.1. Results of precision tests. (From [2.1]; Copyright 1979, Royal Society of Chemistry)

Laboratory	Analytical method	Standard solution 1			Standard solution 2			River Water			'Spiked' river water		
		Concn (mg Cl⁻¹)	Standard[a] deviation (mg Cl⁻¹)	Relative[b] SD (%)	Concn (mg Cl⁻¹)	Standard[a] deviation (mg Cl⁻¹)	Relative[b] SD (%)	Concn found (mg Cl⁻¹)	Standard[a] deviation (mg Cl⁻¹)	Relative[b] SD (%)	Concn found (mg Cl⁻¹)	Standard[a] deviation (mg Cl⁻¹)	Relative[b] SD (%)
1	Mohr titration	25	0.065	2.6	200	0.52	0.26	6.1	0.31	5.1/	106	0.38	0.36
2	Mohr titration	50	(0.0)	(0.0)	250	0.49	0.20	54	0.68	1.3	99	0.58	0.59
3	Mercuric nitrate titration	20	0.29	1.5	180	0.42	0.23	47	0.55	1.2	139	0.70	0.50
4	Mercuric nitrate titration	8	0.12	1.5	80	0.14	0.18	16	0.091	0.57	65	0.073	0.11
5	AutoAnalyzer	20	0.71	3.6	180	1.6	0.89	42	0.94	2.2	124	2.4	1.9
6	AutoAnalyzer	20	1.2	6.0	180	2.1	1.2	40	0.89	2.2	137	1.6	1.2
7	AutoAnalyzer	30	1.1	3.7	270	2.7	1.0	18	0.97	5.4/	111	1.5	1.4
8	AutoAnalyzer	5	0.19	3.8	45	0.15	0.33	26	0.24	0.92	45	0.19	0.42
9	AutoAnalyzer	5	0.17	3.4	45	0.42	0.93	23	0.71	3.1	36	1.0	2.8
10	SMA AutoAnalyzer	20	0.84	4.2	180	1.6	0.89	53	1.52	2.9	144	1.38	0.96

[a] The estimate of the standard deviation of any one result in any one batch of analyses; the estimates have between 9 and 19 effective degrees of freedom.
[b] The standard deviation expressed as a percentage of the concentration of the solution.
* Not statistically significantly greater than 5.0% (0.05 probability level).

Chloride 61

Table 2.2. Results of inter-laboratory bias tests. (From [2.1]; Copyright 1979, Royal Society of Chemistry)

Laboratory	Analytical method	Std soln 46.0 mg Cl l^{-1}			River Water A				River Water B			
		Mean concn found[a] (mg Cl l^{-1})	Deviation from true value (%)	Maximum possible bias (%)	Mean concn found[a] (mg Cl l^{-1})	Deviation from mean value (%)	Maximum possible bias (%)		Mean concn found[a] (mg Cl l^{-1})	Deviation from mean value (%)	Maximum possible bias (%)	Average deviation (%)
1	Mohr titration	46.0 ±0.09	0.0	±0.2	97.9 ±0.17	+1.4	+1.6		8.52 ±0.17	−0.7	−2.7	+0.2
2	Mohr titration	45.0 ±0.26	−2.2	−2.7	97.4 ±0.49	+0.9	+1.4		8.28 ±0.13	−3.3	−5.0	−1.5
3	Mercuric nitrate titration	46.0 ±0.41	0.0	±0.9	97.0 ±0.25	+0.5	+0.8		8.80 ±0.23	+2.6	+5.2	+1.0
4	Mercuric nitrate titration	45.9 ±0.13	−0.2	−0.5	96.7 ±0.19	+0.2	+0.4		8.39 ±0.11	−2.2	−3.5	−0.7
5	AutoAnalyzer	48.0 ±0.89	+4.3	+6.3	94.6 ±1.03	−2.0	−3.0		8.90 ±0.40	+3.7	+8.4	+2.0
6	AutoAnalyzer	46.0 ±1.24	0.0	±2.7	95.7 ±1.47	−0.9	−2.4		8.24 ±0.60	−4.0	−11.0	−1.6
7	AutoAnalyzer	45.8 ±1.57	−0.4	−3.8	97.2 ±1.57	+0.7	+2.4		8.60 ±0.52	+0.2	+6.2	+0.2
8	AutoAnalyzer	45.0 ±0.20	−2.2	−2.6	95.2 ±0.27	−1.4	−1.6		9.02 ±0.43	+4.9	+10.1	+0.4
9	AutoAnalyzer	46.1 ±0.43	+0.2	+1.2	97.1 ±0.17	+0.6	+0.8		8.48 ±0.38	−1.2	−5.6	−0.1
10	SMA Auto-Analyzer	46.2 ±0.43	+0.4	+1.4	[b]	[b]	[b]		8.6 ±0.52	+0.2	+6.3	+3.9
Average of all laboratories		46.0	0.0	—	96.5	—	—		8.58	—	—	—
Concentration found by WRC		—	—	—	96.2	—	—		8.48	—	—	—

[a]The figures after ± sign are the 90% confidence limits of the means.
[b]This laboratory originally conducted the analytical quality control work on another AutoAnalyzer system and repeated the work on changing to the present system. Insufficient river water A sample remained for work on the new system, but the first of the regular 'follow-up' checks showed the bias target to be met.

based on the displacement by chloride ions of thiocyanate ions from mercuric thiocyanate and subsequent spectophotometric determination of thiocyanate as ferric thiocyanate. The standard deviations are ± 0.36 and ± 0.75 mg l^{-1} for 8.9 and 34.5 mg l^{-1} chloride ion concentration.

$$Hg(CNS)_2 + 2Cl^- = HgCl_2 + 2CNS.$$

A further method [2.6] is based on the oxidation of chloride to free chlorine by potassium permanganate solution. Chlorine is then swept out of the solution by a nitrogen purge and determined by the o-toluidine spectrophotometric method at a wavelength of 450 nm.

Rossner and Schwedt [2.7] have described a spectrophotometric method for the determination of chloride in natural water which is based on the FeII-Hg tripyridyl-s-triazine system. This method is applicable in the 10 µg l^{-1} to 10 mg l^{-1} concentration range. A modification of the technique based on continuous flow analysis has a detection limit of 5 µg l^{-1}.

2.1.1.3 Flow Injection Analysis

Van Staden [2.8] has described an automated prevalve dilution in-flow injection analysis method for the determination of chloride in ground water. This method can analyse samples containing up to 800 mg l^{-1} chloride. By adjusting the ratio of sample flow rate to dilution water flow rate, the concentration could be reduced to a level at which linearity of the calibration was preserved, (about 160 mg l^{-1} maximum). The method was based on reaction with mercury thiocyanate in the presence of ferric ions, giving the deeply coloured ferric thiocyanate. Absorption was measured at 480 nm.

Basson and Van Staden [2.9] have described an automatic method for the simultaneous determination of chloride and sulphate in natural water based on an adaptation of the flow injection concept. About 200 samples per hour can be analysed by this procedure.

Reagents. All reagents are of analytical reagent grade, unless otherwise specified.
(i) Chloride colour reagent.
  (a) Solution I. Dissolve 6.34 g mercuric thiocyanate with stirring in 950 ml distilled water at a temperature of 50–70 °C. After cooling, add 2 ml concentrated nitric acid (specific gravity 1.42) and 8.08 g Fe(NO$_3$)$_3$·9H$_2$O. Stir until dissolved. The reagent is stable for several months at room temperature if stored in a dark bottle.
  (b) Solution II. Dissolve 6.85 g mercuric nitrate in 100 ml distilled water. Add 0.50 ml of solution II to 100 ml of solution I to give the chloride colour reagent. Leave solution overnight to stabilise.
(ii) Barium chloride solution (2.5%).
    Dissolve 0.05 g thymol crystals with stirring in 500 ml distilled water at a temperature of about 80 °C. Cool to 40 °C and add 1500 ml distilled

water. Add 4 g gelatin and swirl until dissolved. When dissolved, add 50 g barium chloride and dissolve. Filter if necessary.

Standards.
(i) Stock solutions.
  (a) Dissolve 16.4846 g dried sodium chloride and dilute 1 l with distilled water. Then 1 ml = 10 mg Cl^-.
  (b) Dissolve 7.3932 g dried sodium sulphate and dilute to 500 ml with distilled water. Then 1 ml = 10 mg SO_4^{2-}.

Apparatus.
  (i) peristaltic pump;
  (ii) Manifold (see Fig. 2.1);
  (iii) Spectrophotometers: Bausch and Lomb Spectronic 21(Rochester, New York) fitted with 10 mm flowcells
  (iv) Recorders: Mettler Model GA12.

Procedure. A schematic diagram of the system employed is shown in Fig. 2.1. The manifold consists of Tygon tubing with tube lengths and inside diameters as indicated in the figure.

Inject the sample automatically into the system from a Breda Scientific flow-injection sampler containing a valve with a capacity of 200 µl. Using a sampling time of 18 s gives a capacity of 200 samples per hour. Actuate the valve

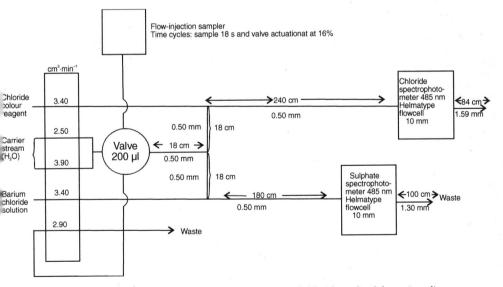

Fig. 2.1. Flow system for the simultaneous determination of chloride and sulphate. Sampling rate 200 samples h^{-1}. Tube length and i.d. are given in cm and mm respectively (From [2.9])

system on a time basis which is correlated with the sampler unit. The sampling valve actuates every 16 s after movement to the next sample.

The performance of the proposed flow-injection analysis procedure is shown in Table 2.3 for chloride. The coefficient of variation for chloride is less than 1.3%.

2.1.1.4 Atomic Absorption Spectrometry

Belcher et al. [2.10] have described a procedure in which chloride is reacted with phenyl mercuric nitrate. The phenyl mercuric chloride produced is extracted with chloroform and mercury (equivalent to original chloride content of sample) is determined by atomic absorption spectrometry in amounts down to $0\text{--}15\ \mu g\ l^{-1}$.

In a further atomic absorption method [2.11] the chloride in the sample is contacted with metallic copper in the flame of an atomic absorption spectrometer. Concentrations of copper (equivalent to chloride ions) in the flame are evaluated at 324.75 nm. Between 2 and 5 mg l^{-1} of chloride can be determined in samples by this method.

2.1.1.5 Inductively Coupled Plasma Atomic Emission Spectrometry

Michlewicz and Carnahan [2.12] have described a method for determination of chloride in a continuous flowing cell. Hydrogen chloride is then determined in amounts equivalent to $0\text{--}2$ mg l^{-1} in the sample by microwave techniques. These workers [2.13] have also described a procedure for determining chloride

Table 2.3. Precision of the proposed FIA method for chloride in surface, ground and domestic waters. (From [2.9], Copyright 1981, Pergamon Publishing Co., UK)

Sample	Number of tests	Automated segmented method [Cl$^-$] (mg l^{-1})	FIA method [Cl$^-$] (mg l^{-1})	Coefficient of variation (%)
1	15	8	11	1.21
2	15	66	74	0.87
3	15	7	10	1.05
4	15	239	251	0.21
5	15	26	26	0.99
6	15	107	110	0.74
7	15	149	150	0.66
8	15	6	4	1.30
9	15	239	245	0.15
10	15	7	5	1.14
11	15	26	30	0.91
12	15	302	291	0.17

Student's t-test at 5% probability level: calculated t-value = 1.298; critical t-value = 2.060.

down to 7 mg l^{-1} based on direct nebulization of aqueous sample into an atmospheric pressure helium microwave plasma.

Michlewicz and Carrnahan [2.14] give details of a method for the determination of chloride (and bromide and iodide) based on pneumatic nebulization and ultrasonic nebulization of the sample into a helium microwave induced plasma.

Detection limits in mg l^{-1} for chloride, bromide and iodide were 0.4, 3.0 and 0.8 respectively. Interference effects from added magnesium, potassium, sodium and lithium were moderate with pneumatic nebulization but significant with ultrasonic sample nebulization.

2.1.1.6 Ion-Selective Electrodes

Hara and Okazaki [2.15] have described a method for the two-point Gran titration of chloride in natural water using a silver sulphide ion-selective electrode. The analysis was performed without sample pre-treatment and results correlated well with those obtained by potentiometry and spectrophotometry or ion chromatography. Interferences and the detection limit were examined, together with the effects of temperature and number of titration points. The Monte Carlo simulation method was used to determine the precision of the method; two titration points were sufficient to obtain maximal precision of 0.2% provided that the potential difference between them exceeded 20 mV. The two-point Gran titration procedure was used to analyse stream and lake water with an error of less than 2% and compared with the usual seven-point Gran titration method.

Various workers have used ion-selective electrodes for the determination of chloride ions in water. End-points have been detected potentiometrically [2.16–2.18] and titrimetrically [2.19]. Zeinalova and Senyavin [2.20] examined conditions for the potentiometric determination of chloride in natural water using an ion-selective electrode. A pH range of 1–12 is suitable for the determination of chloride. Macro- and microcomponents of natural water did not interfere with determining chlorides.

Trojanowicz and Matuszewski [2.21] investigated the usefulness of potentiometric detection in the flow injection analysis of chloride in river waters. A direct, as opposed to logarimic, relation exists between electrode potential and chloride concentration.

Hara et al. [2.22] found that preliminary treatment of river or stream water samples with silver chloride was effective in removing interferences caused by bromide, iodide and sulphide prior to potentiometric chloride determination using a chloride ion-selective electrode. The effects of pH on interference by bromide were also studied. Complete removal of bromide interference was not possible by acidification of the sample alone.

Trojanowicz and Lewardowski [2.23] have described a multiple flow through potentiometric system for the continuous determination of chloride in admixture with fluoride and nitrate in natural waters. This method is discussed further under multi-halide analysis.

2.1.1.7 Gas Chromatography

Belcher et al. [2.24] determined chloride by a method based on reaction with phenyl mercuric nitrate to produce phenyl mercury chloride, followed by carbon tetrachloride extraction of the latter and gas chromatography. Down to 0.4 mg l^{-1} chloride in the sample could be determined.

2.1.1.8 High Performance Liquid Chromatography

Stetzenbach and Thompson [2.25] employed high performance liquid chromatography on anion exchange columns to determine chloride in admixture with bromide, iodide, nitrate and thiocyanate in natural waters. This method is discussed further under multihalide analysis.

Moss and Stephen [2.26] have described the determination of chloride (and bromide, iodide, and phosphate) in natural waters by high performance liquid chromatography.

Salov et al. [2.243] determined chloride (and bromide, iodide, iodate, bromate and chlorate) by high performance liquid chromatography using inductively coupled argon plasma mass spectrometry as a detector.

2.1.1.9 Ion-Chromatography

This technique has been applied to the determination of chloride and other anions in natural waters (Table 2.4) and in rainwater (Table 2.5) and is discussed in more detail in Sect. 11.2.

2.1.1.10 Isotope Dilution Analysis

Mass spectrometric isotope dilution analysis has been applied to the determination of chloride ion in rainfall and snow. Detection limits of 1 µg l^{-1} or lower can be achieved by this technique. For the isotope ratio measurement of chlorine and bromide, a single focusing mass spectrometer (Varian MAT, type CH5TH) with a double filament ion source and a Faraday cup as a detector was used. The filaments consist of rhenium. The best precision of the isotope ratio

Table 2.4. Application of Ion chromatography to determination of chloride in natural waters

Type of water	Codetermined anions	Reference
Natural	Br, NO_2, NO_3, SO_4	2.30
Natural	Br, F, NO_2, NO_3, SO_4, PO_4	2.31
	NO_3, SO_4, PO_4, F, Br	2.32

Table 2.5. Ion chromatography of chloride in rainwater

Type of water	Codetermined anions	Reference
Rain water	–	2.41–2.47
Rain water	F, NO_3, PO_4, SO_4	2.47
Rain water	NO_3, SO_4	2.42, 2.48, 2.49
Rain water	PO_4, NO_2, NO_3, SO_4	2.43
Rain water	F, Br, NO_2, NO_3, SO_4, SO_3, PO_4	2.50
Rain water	Br, F, NO_2, PO_4, BrO_3, NO_3, SO_4	2.51
Rain water	PO_4, NO_3, SO_4	2.45
Rain water	Br, SO_4	2.46

measurement is achieved when a solution of $(Ag(NH_3)_2)$ X ($X^- = Cl^-$, Br^-) is deposited on the evaporation filament and dried under a heat lamp. In the ion source the temperature of the ionization filament is increased stepwise up to a temperature of 1850 °C.

Using sample amounts of 1 μmol halide with natural isotope composition, ion intensities are achieved on the Faraday cup in the range of 10^{-11} to 10^{-10} for $^{35}Cl^-$ and in the range of 10^{-11} to 5×10^{-11} for $^{79}Br^-$.

Samples of 100–250 g rain and snow were collected in polyethylene funnels of diameter 35 cm connected by means of polyethylene tubes and Y-brackets with a 1-l polyethylene flask. Approximately 100 g of one sample was accurately weighed in a 250-ml polyethylene beaker and approximately 1 g ^{37}Cl spike solution was added. The spike solution contained 1.014×10^{19} chloride ions g^{-1}. Henmann employed isotope dilution analysis coupled with mass spectrometry to determine down to 13 μg l^{-1} chloride in water [2.36].

The spike solution had an isotope distribution of $^{35}Cl = 24.63\%$ and $^{37}Cl = 75.37\%$. After the solutions of the spike and the sample had been stirred up and thoroughly mixed, 0.5 ml of saturated silver nitrate solution was added and the mixture left to stand for about 30 min so that the silver chloride could precipitate. After this, the solution was filtered through a cellulose nitrate filter (Sartorius, Type 11301) and then, to purify it, the silver chloride was dissolved again in some drops of a semi-concentrated ammonia solution. The silver chloride was then precipitated again with concentrated nitric acid. After being filtered, the precipitate was dried at 60 °C and kept in a non-transparent flask until mass spectrometry began. The silver chloride was kept away from light until measurement because the chlorine isotopes fractionate slightly when silver chloride is decomposed by light of a natural wave-length distribution. Shortly before the mass spectrometric measurement was taken, 1 to 2 drops of a 15% ammonia solution were added to the sample and placed on the evaporator filament of the thermal ion source of the mass spectrometer using a micro-liter injector.

Henmann et al. [2.52] determined chloride and bromide in snow and rain and also measured their concentrations at different distances away from a highway (Table 2.6). Unlike the concentrations of bromide, the chloride concentration in snow continually decreases the further the snow is from the highway, a considerable proportion of the bromide originating from traffic exhausts.

Table 2.6. Cl^- and Br^- content in snow depending on its distance away from a highway. (From [2.52]; Copyright 1981, Gordon and Breach Science Publishers Ltd., OK)

Distance from highway [m]	Cl^- content ($\mu g\ g^{-1}$)		Br^- content ($\mu g\ g^{-1}$)	
	MS-IDA	Ion selective electrode	MS-IDA	Ion selective electrode
2	71	68	124	580
10	40	45	79	380
20	32	40	196	430
30	26	31	40	300
40	18	23	351	320
50	13	17	66	220
300	6	11		130

2.1.1.11 Ion Exchange Chromatography

Akaiwa et al. [2.27] have used ion exchange chromatography on hydrous zirconium oxide combined with a detection based on direct potentiometry with an ion selective electrode for the simultaneous determination of chloride and bromide in natural waters. This technique is discussed in further detail in the section on multihalide analysis.

2.1.1.12 Column Coupling Capillary Isotachophoresis

This technique offer very similar advantages to ion chromatography (Chap. 11) in the determination of anions in water, namely multiple analysis, little or no sample pretreatment, speed, sensitivity, and automation. Very similar advantages are offered by capillary isotachophoresis. Bocek et al. [2.28] determined chloride (and sulphate) in mineral waters by this technique and Zelinski et al. [2.29] applied the technique to the determination of 0.02–0.1 mg l^{-1} quantities of chloride, fluoride, nitrate, sulphate, nitrite, and phosphate in river water.

2.1.1.13 Coulometric Titration

Coulometric titration with dead stop end-point detection has been used for the determination of chloride in acetic acid and nitric acid media [2.33] and in methanol-nitric acid medium [2.34].

Interstitial chloride in very low volumes of pore waters from oil shales rocks has been determined by coulometric titration. [2.35]

2.1.1.14 Radiochemical Methods

Johannesson [2.37] treated the sample with $H^{35}Cl$ and evaporated it to dryness prior to counting. He found that 1 mg l^{-1} chloride could be determined.

2.1.2 Rainwater and Snow

2.1.2.1 Continuous Flow Analysis

Continuous flow analysis has been applied [2.38] to the determination of chloride in rainwater (together with nitrite, nitrate, sulphate and phosphate).

A computer controlled multichannel continuous flow analysis system [2.39] has been applied to the measurement of chloride (and nitrite, nitrate, sulphate and phosphate) in small samples of rainwater.

2.1.2.2 Flow Injection Analysis

Flow injection analysis has been applied to the determination of chloride in rainwater (and nitrite, nitrate, phosphate and sulphate [2.38]).

2.1.2.3 Ion-Selective Electrodes

Slanina [2.40] has discussed the application of ion selective electrodes to the determination of chloride in rainwater. Detection limits of 0.1 mg l^{-1} were achieved with a relative accuracy of 95% using a sample volume of 0.5 ml.

2.1.3 Potable Waters

2.1.3.1 Titration Methods

Levy [2.53] has described a titrimetric procedure for the determination of chloride ions in the presence of free chlorine. Previous workers have attempted to remove free chlorine by boiling the sample. This, however, leads to partial decomposition of hypochlorous acid to hydrochloric acid and oxygen:

$Cl_2 + H_2O \rightarrow HCl + HOCl$

$2HOCl = 2HCl + O_2$

Levy [2.53] overcame this problem by carrying out the reaction at a temperature close to $0\,°C$ and sweeping unhydrolysed chlorine out of solution with a purge of nitrogen before commencing the determination of chloride ions.

Reagents. All chemicals used were of analytical-reagent grade.
    Silver nitrate solution, 0.010 mol l^{-1}.
    Sodium thiosulphate solution, 0.010 mol l^{-1}.
    Sodium chloride solution, approximately 0.01 mol l^{-1}.

Soluble starch solution, 2%.
Concentrated hydrochloric acid.
Nitric acid, 5 mol l^{-1}.
Concentrated sodium hypochlorite solution.
Potassium iodide.

Keep all solutions, except those to be added from burettes, in an ice-bath throughout the analysis. Bubble nitrogen via a fine bleed or glass frit for 20 min through the cooled sample solution, which contains 0.5–10 mmol l^{-1} total chloride, to displace unhydrolysed chlorine. Analyse 5-ml aliquots for chloride by the addition of silver nitrate solution until one drop causes no further precipitation of silver chloride. Near the end-point, add three drops of nitric acid to ensure complete precipitation. Centrifugation enables the end-point to be clearly distinguished. Analyse separate 5-ml aliquots for hypochlorous acid by the addition of 10 mg of potassium iodide and titration with thiosulphate solution. Add one drop of starch solution just before the end-point to clarify when it has been reached. The difference between two concentrations thus determined gives the concentration of chloride exclusive of that produces by chlorine hydrolysis.

2.1.3.2 Ion Chromatography

Schwabe et al. [2.47] have described their determination of chloride (and fluoride, nitrate, phosphate and sulphate) in potable waters. This is discussed further in Sect. 11.3.

2.1.3.3 High Performance Liquid Chromatography

The procedure described in Sect. 2.1.1 [2.243] has also been applied to the determination of chloride in potable water.

2.1.3.4 Gas Chromatography

Backmann and Matusca [2.54] have described a gas chromatographic method involving the formation of halogenated derivatives of cyclohexanol for the determination of chloride, bromide and iodide in potable waters. This method is discussed further under multihalide analysis.

2.1.4 Industrial Effluents

2.1.4.1 Titration Methods

Official methods issued by the Department of the Environment UK [2.55] describe various methods for the determination of chloride in sewage, trade effluents and waters. These methods are silver nitrate titration with chromate indicator (Mohr's method), mercuric nitrate titration with diphenyl-carbazone indicator, potentiometric titration with silver nitrate, and an automated mercuric-ferric thiocyanate colorimetric method. With one or other of these methods, concentrations from 1 to 1000 mg l^{-1} chloride may be determined. Brief notes are given in respect of the chloride ion selective electrode, silver coulometry and ion chromatography, the latter being relevant at very low concentrations down to the µg l^{-1} level.

Ilyukhina et al. [2.56] have described a volumetric method for the determination of chloride in aqueous effluents produced in the manufacture of the fungicides captan and phaltan (tolpet). This effluent contained tetrahydrophthalic acid, tetrahydrophthalimide, phthalic acid and phthalimide. The sample was treated with hydrogen peroxide, acidified with nitric acid and then titrated with 0.1 N mercurous nitrate in the presence of diphenylcarbazide indicator.

2.1.4.2 Ion Chromatography

This technique has been applied to the determination of chloride (and bromide, fluoride, nitrite, nitrate and phosphate) in industrial effluents [2.57] the procedure is described in Sect. 11.4.

2.1.5 Waste Waters

2.1.5.1 Titration Methods

Barrera and Martinez [2.58] have compared two methods – the Mohr procedure based on precipitation of silver chloride with a chromate indicator, and the mercuric nitrate method based on the formation of soluble, slightly ionized, mercuric chloride – for the volumetric determination of chlorides in waste waters. Results are in agreement. The mercuric nitrate method is preferred because the end-point is sharper, a magnetic stirrer is not required, the range of concentrations is 0–700 mg l^{-1} (against 5–300 mg l^{-1}) and it is less expensive for both high and low chloride concentrations.

2.1.5.2 Ion Exclusion Chromatography

Ion exclusion chromatography has been applied to the determination of chloride [2.59], sulphate, phosphate and carbonate in waste waters. Thus technique is discussed in Sect. 5.1.6.

2.1.5.3 Ion Chromatography

Methods for the codetermination of chloride with other anions in treated waters are reviewed in Table 2.7 and discussed further in Sect. 11.4.

2.1.6 Sewage

2.1.6.1 Ion-Selective Electrodes

Hindin [2.62] has shown that chloride in concentrations in sewage can be determined by the ion-specific electrode method as an alternative to the standard mercuric nitrate method, providing two precautions are taken – the addition of an ionic strength adjusting solution to overcome any effect of the ionic strength of the sample or standard may have, and the removal of sulphide ions by a cadmium ion preciptating solution.

2.1.7 Boiler Feed and High Purity Waters

2.1.7.1 Titration Methods

It is generally accepted that the presence of chloride in the steam – water circuit of power stations can be associated with corrosive conditions within the boiler, and consequently chloride levels have to be controlled. The concentration of chloride treated in boiler waters varies with the design of the boiler and the chemical treatment accorded to the boiler water, but in every instance it is essential that any ingress of chloride is detected so that remedial action can be taken. In order to identify the source of the chloride, it is necessary to monitor plant streams, e.g. feed water or the effluent from a water treatment plant, where chloride levels may be as low as a few micrograms per litre.

A titrimetric method for the determination of milligram per litre quantities of chloride (and bromide) in boiler feed water involves titration of the sample at pH 1 to 4 with 0.1 N mercuric nitrate to the tetra(ethyl di-(1-sodiotetrazol-5-ylazo)acetate yellow to red end-point [2.63]. A 300-fold excess of sulphate,

Table 2.7. Application of ion chromatography to the determination of chloride in treated waters

Type of water waste	Codetermined anions	Reference
Waste	SO_3	2.60
	F, PO_4, SO_4	2.61

nitrate, carbonate or acetate can be tolerated, but cobalt, fluoride, iodide or nickel interfere.

Samples can be acidified with perchloric acid or nitric acid and ferric nitrate solution added followed by ethanolic [2.64] or methanolic [2.65] thiocyanate.

$$3Cl' + Fe(NO_3)_3 = FeCl_3 + 3NO_3'$$

$$2FeCl_3 + 3Hg(CNS)_2 = Fe(CNS)_3 + 3\ HgCl_2.$$

The extinction of the mercuric thiocyanate complex is measured at 460 nm [2.64] or 463 nm [2.65]. Chloride can be determined down to 1 µg l^{-1} by these procedures. Hydrogen peroxide or hydrazine used as boiler water treatment chemicals do not interfere. Interference arises from the presence of comparatively large amounts of cupric sulphate, potassium dichromate, lithium hydroxide or ammonium fluoride. The coefficient of variation at the 50 µg l^{-1} level was 12%.

2.1.7.2 Flow Injection Analysis

Two spectrophotometric procedures involving the formation of the red coloured ferric thiocyanate complex formed between chloride ion and ethanolic mercuric thiocyanate in the presence of ferric ions have been described for the determination of chloride in boiler feed water [2.66, 2.67].

2.1.7.3 Spectrofluorimetric Methods

Fluorescence spectroscopy has been used to determine down to 1 µg l^{-1} chloride in high purity water. The method is based on the quenching effect of silver chloride on the fluorescence of sodium fluorescein. The coefficient of variation is within 15% at the 10 µg l^{-1} chloride level [2.68].

2.1.7.4 Ion-Selective Electrodes

Silver-silver chloride electrodes [2.69–2.72] and solid state mercurous chloride electrodes [2.73, 2.74] have been used to determine traces of chloride in high purity power station waters. Detection limits claimed are 6–100. µg l^{-1} [2.72] for silver-silver chloride electrodes and 1–10 µg l^{-1} for solid-state mercurous chloride electrodes.

Silver-Silver Chloride Electrodes. Measurements have been made in two ways. In one method [2.72] two solid-state Orion 94-17A silver chloride membrane electrodes are used. The reference electrode is contained in a compartment connected to the sample compartment by a ground glass sleeve. The reference compartment contains a 100-µg l^{-1} solution of chloride in 0.1 mol l^{-1} nitric acid. A graph is constructed from cell potential readings after adding increments of standard chloride solution to the sample solution (made 0.1 mol l^{-1} in nitric acid) in the sample compartment. A similar graph is obtained by adding increments of the chloride solution to a blank solution containing 0.1 mol l^{-1} nitric acid. The chloride content of the sample is given by the distance along the concentration axis between the two lines.

Concentrations of copper II, iron II, nickel II and aluminium interfere when present at concentrations exceeding 10 mg l^{-1} and ammonia and hydrazine also interfere. It is claimed that better precision is obtained using these electrodes than by using the conventional silver-silver chloride electrodes discussed below.

In the alternative method, measurements are made using silver-silver chloride electrodes vs a mercurous sulphate reference electrode [2.69, 2.71].

A manual version of this procedure has been described having a detection limit of 100 µg l^{-1}. The continuous analyser version (discussed below) has been described with a detection limit nearer to 1 µg l^{-1} [2.69].

In this apparatus the electrodes are immersed in a buffered sample stream. The temperature of the electrode flow cell is controlled at 10 (± 0.1)°C at which level the sensitivity of 3.54 mV per 100 µg l^{-1} is approximately 50 % greater than that at 25 (± 0.1)°C, and a linear response of cell potential to changes in chloride concentration is observed up to 0.150 µg l^{-1}. The standard deviations at 10°C for nominal chloride concentrations of 50, 100 and 250 µg l^{-1} are ± 1.7, ± 2.8 and ± 2.7 µg l^{-1} respectively. Substances normally present in these waters do not interefere appreciably.

Table 2.8 showns the satisfactory recoveries obtained in some spiking experiments carried out using this system. In most instances there was no significant interference in the procedure by substances likely to occur in boiler feed water and in every instance except one, where an interference was detected, the effect was positive (Table 2.9). The small effects noted for sodium hydroxide, trisodium

Table 2.8. Recovery of chloride from samples of power station waters. (From [2.69]; Copyright 1976, Royal Society of Chemistry, UK)

Sample	Initial chloride concentration µg l^{-1}	Chloride spike added µg l^{-1}	Final chloride concentration µg l^{-1}	Recovery %
Condensed steam (A)	<1	5.0	5.3[a] (0.23)[b]	106.0
Condensed steam (B)	<1	19.7	19.5 (0.40)	98.9
Boiler water	108.0	47.2	156.9 (0.40)	103.6

[a] Each result is-the mean of six determinations.
[b] The figures in parentheses are the standard deviations for single determinations with five degrees of freedom.

Table 2.9. Effect of other substances at 10 °C. (From [2.69]; Copyright 1976, Royal Society of Chemistry, UK)

Substance	Concentration of substance mg l^{-1}	Interference effects (µg l^{-1}) of chloride at chloride concentrations of	
		0 µg l^{-1a}	50 µg l^{-1a}
Na$_2$SO$_4$	75	1.7	3.1
Na$_2$HPO$_4$	50	1.4	2.7
NaOH	4	<1	<1
NaOH	40	4.1	3.8
Na$_3$PO$_4$	50	2.7	<1
Na$_2$CO$_3$	50	2.7	<1
NH$_3$	10	<1	<1
Morpholine	10	<1	<1
Cyclohexylamine	10	<1	<1
Hydrazine	1	<1	<1
H$_2$SO$_4$	10	<1	<1
Na$_2$SO$_3$	5	4.4	4.5
Fe^{2+}, Cu^{2+}, Zn^{2+}	1+1+1	−2.9	<1
Ca^{2+}, Mg^{2+}, K$^+$	2+2+20	1.5	<1
O$_2$	~4	<1	—

[a] If the other substances had no effect the results would have been expected to fall within the limits of 0 ± 1.2 and 50 ± 3.4 µg l^{-1} at the 95% confidence level.

orthophosphate and sodium carbonate could be explained by a change in the pH of the system. The unexpected interference from sodium sulphate and sodium sulphite could arise from trace amounts of chloride in the reagent chemicals. However, the effects were so small, even at the excess levels of these substances added, that for most practical purposes they could be ignored.

Solid State Mercurous Chloride Electrodes. This is the more recent development in the determination of chloride in high purity water [2.73, 2.74]. Marshall and Midgley [2.73, 2.74] have described a procedure for the determination of 0–20 µg l^{-1} chloride in high purity water, using two types of ion selective membrane electrode incorporating mercurous chloride. At these low concentrations, more chloride will dissolve from the mercury I chloride in the electrode than is present in the sample itself. The extent of the dissolution is controlled, however, by the chloride in the sample. In these circumstances, the electrode potential is linearly related to the concentration of chloride in the sample. With the electrode housed in a flow cell with a thermostatically controlled water jacket, the correlation coefficient between EMF and concentration was always greater than 0.99. The sensitivity (0.18 mV µg l^{-1} of chloride at 25 °C and 0.4–0.5 mV µg l^{-1} of chloride at 4 °C) was about ten times greater than that of the silver-silver chloride electrode. Total standard deviations at 10, 5 and 2 µg l^{-1} of chloride were 0.4, 0.5 and 0.3 µg l^{-1} of chloride respectively.

Potentials were measured with a digital pH meter reading to 0.1 mV and displayed on a chart recorder. Two types of ion-selective electrodes were used, those made from Radiometer F3012 Universal electrodes, and the Ionel Model SL-01 (Ionel electrodes, Mount Hope, Ontario, Canada). Both have membranes made of a mixture of mercury II sulphide and mercury II chloride. In the electrode developed at CERL, the mixture is used to impregnate a graphite PTFE electrode, while in the Ionel electrode the mixture is hot pressed into a disc. The reference electrode was a mercury-mercury I sulphate electrode with a 0.5 mol l^{-1} sodium sulphate filling solution (instead of the usual 1 mol l^{-1} solution, which would have precipitated at the lowest operating temperature of 4 °C). The reservoir containing the filling solution and the reference element could be raised to 0.5 m above the remote liquid junction, which was of the ground glass sleeve type.

The chloride electrode and the remote junction were housed in a flow cell fitted with a water jacket through which thermostatically controlled water was circulated by means of a Techne C-100 thermocirculator operated in conjunction with a Techne Model 1000 chiller unit. The remote junction was in a separate compartment downsteam of the chloride electrode compartment, which contained a magnetic stirrer bar. The flow cell was painted black to exclude light.

A Technicon Model 1 proportioning pump delivered the solutions to the flow cell, pumping the sample or standard solutions at 3.9 ml min^{-1} and the acid reagent solution at 0.8 ml min^{-1}. Air was injected at 1 ml min^{-1} to improve mixing. The air-segmented mixture of sample and reagent solutions passed through a glass mixing coil to a T-piece, from which the air and a portion of the solution were extracted at 1.6 ml min^{-1}. The rate of delivery of solution to the flow cell was, therefore, 4.1 ml min^{-1}. The pump tubes were made of PVC, but the transmission lines were made of PTFE.

Before entering the cell, the solution passed through a short length of stainless steel tubing, which was connected to the chassis ground terminal on the pH meter. A stainless steel wire was immersed in the solution downstream of the electrode pair and also connected to the chassis ground point. Without these connections the signal was noisy.

The use of mercury I chloride electrodes for determining chloride concentrations can be extended from the levels attainable by manual analysis to very low levels (less than 20 µg l^{-1}) by housing the electrode in a flow cell at a controlled temperature. At a given temperature, the electrode is about ten times more sensitive than the silver chloride electrode.

One advantage of this greater sensitivity is that the mercury I chloride electrode does not need to be operated at sub-ambient temperatures in order to obtain adequate precision in the concentration range 1–20 µg l^{-1}. Two kinds of electrode can be used, that developed at CERL and the Ionel SL01. The two electrodes are almost equally sensitive and precise, but the Ionel has a faster response time and requires no preparation.

2.1.7.5 Ion Chromatography

This technique, discussed further in Chap. 11, has been applied to the determination of chloride in boiler feed water (2.75) and chloride (2.76) (and nitrite and sulphate) in high purity water.

2.1.8 Soil and Plant Extracts

2.1.8.1 Ion-Selective Electrodes

Davey and Bembrick (2.77) have described a method for the determination of chloride in water extracts of soils based on measurement of the EMF developed between two silver-silver chloride electrodes in a cell with a liquid junction and suitable electrolyte.

Mc Leod et al. (2.78) carried out simultaneous measurement of chloride, pH and electrical conductivity in soil suspensions using a triple electrode system mounted on a single unit. A glass electrode and a silver-silver chloride electrode with a common reference electrode and two pH meters were used for the determination of pH and chloride respectively. The coefficient of variation obtained in determinations of chloride at the 50 mg l^{-1} level was approximately 7.3%.

2.1.8.2 Ion Chromatography

Bradfield and Loake [2.79] have applied ion chromatography with indirect ultraviolet detection to the determination of chloride (and nitrate, phosphate and sulphate) in plants and soils.

Using a revised phase system with a dynamically generated ion chromatographic column, retention times, respectively, of 5.5, 7.9, 12.6 and 18 min were obtained from the above anions. This techinique is discussed in further detail in Chap. 11 (Applications of ion Chromatography).

2.1.9 Soils

2.1.9.1 Titration Methods

In a method (2.80) for determining chloride in a calcium sulphate extract of soil, the extract is acidified and the concentration of chloride determined by titration with 5 mmol l^{-1} mercuric chloride using diphenylcarbazone as indicator. Mercuric ion in the presence of chloride forms mercuric chloride which, although soluble, provides sufficient mercuric ion to form the mercuric diphenylcarbazone complex. When all the chloride has been so removed, addition of further mercuric ion produces a violet complex.

2.1.10 Foodstuffs

2.1.10.1 Ion-Selective Electrodes

A sodium responsive ion-selective electrode has been employed (2.81) for the determination of chloride in salted foodstuffs. This "dry sample addition" method of ion-selective potentiometry involved the addition of a known weight of solid samples directly to a standard solution containing chloride ions. Dissolution of the sample produces a change in electrode potential which is related to the sample composition. The method has been applied to foodstuffs, e.g. salad creams, soups, milk powder, crispbread, ketchup, peanut butter, containing up to 2.8% sodium chloride. The Phillips 15550-Cl solid state electrode was used for the determination of chloride. Coefficients of variation between 3.7 and 5.5% were observed in the determination of 16 g kg^{-1} chloride in salad cream and chicken soup.

2.2 Bromide

2.2.1 Natural Waters

2.2.1.1 Titration Methods

The National Water Council (U.K.) [2.82] has described a classical titration procedure for the determination of bromide in river and seawaters in which bromide is oxidized to bromate by sodium hypochlorite at about 100 °C in a medium buffered to pH 6.2 with phosphate. After reduction of the excess oxidant with formate, the bromate is determined iodometrically. The method is applicable up to 140 mg l^{-1} bromide but is not designed for trace amounts, for which an alternative method is available. Iodine and certain oxidizing or reducing agents was interfere.

The potentiometric titration procedure described by Kuttel [2.83] for the determination of chloride in natural water has also been applied to the determination of bromide.

2.2.1.2 Spectrophotometric Methods

Various spectrophotometric methods have been described for the determination of low concentrations of bromide in natural waters. These include methods based on the formation of coloured products with Chromotrope 2B (CI Acid Red) acid [2.84], roseaniline [2.85] (0.4 mg l^{-1}), phenol red [2.86, 2.87] (200 µg l^{-1}) and o-toluidine [2.88] (detection limits in parentheses). A further method is based on oxidation of bromide to bromate by hypochlorite and subsequent decoloration of methyl orange by bromine formed by interaction of the bromate with bromide [2.89]. This method will detect 8 µg l^{-1} bromide in natural water.

A more recent method is also based on the use of methyl orange [2.90]. This method involves conversion of bromide to bromate, catalysed decomposition of bromate in acetic acid solution, and spectrophotometric determination of the reduction of the optical density of methyl orange by the bromide.

In a further method [2.91], the sample is treated with hydrochloric acid and with chromotrope 2B (CI Acid Red) and 5 mmol l^{-1} potassium bromate. The time taken for the colour to change from red to yellow as evaluated spectrophotometrically is measured and compared with calibration data obtained to known bromide standards.

The indirect spectrophotometric method of Fishman and Skougstad [2.92] and Skougstad et al. [2.93] is based on the catalytic effect of bromide on the oxidation of iodine to iodate by potassium permanganate in sulphuric acid solution. When an aqueous solution of iodine and permanganate is mixed with carbon tetrachloride, the iodine is extracted into the carbon tetrachloride. If bromide is present with the aqueous iodine and permanganate for a certain time at a particular temperature prior to mixing with carbon tetrachloride, the amount of iodine remaining unreacted to iodate will decrease linearly with increasing bromide concentration. Iodine in a carbon tetrachloride extract can, therefore, serve as a sensitive measure of the bromide concentration in the aqueous solution. The reaction temperature and the time for each reaction step are very critical.

Pyen et al. [2.94] have described an automated version of the above manual kinetic method. This method is accurate and precise and quicker than the manual technique (20 samples per hour). The detection limit is 0.01 mg l^{-1} bromide.

Apparatus. The automated equipment consists of a Technicon sampler, proportioning pump, manifold, spectrophotometer, voltage stabiliser and recorder. The spectrophotometer is fitted with specially designed matched 50-mm flow cells and 529-nm filters. The sampler is used with a 20 h^{-1} (2/7) cam made by cutting off four sampling lobes from a 60 h^{-1} (2/1) cam. This results in a cycle time of 40 s for sampling and 140 s for rinsing. A 17-turn coil containing glass beads is used as the extractor. To separate the two phases, a five-turn mixing coil is placed perpendicular to the manifold; PTFE tubing is inserted inside the separating coil to facilitate rapid separation. All connections are made with glass tubing or silicone rubber sleeving. The final separator (Fig. 2.2) has a top portion that is enlarged to facilitate the clean separation of solvent from the aqueous phase. Three 1 l flasks are used for the displacement of carbon tetrachloride. Connections are made with 1.0 mm PTFE tubing. A circulating System-255 cold bath (GCA/Precision Scientific) maintains the reaction temperature of the condenser coil at 10 °C.

Reagents. All chemicals were of analytical-reagent grade.

Potassium bromide stock solution, 100 mg l^{-1} of bromide. Dissolve 0.149 g of potassium bromide in demineralised water and dilute to 1 l with water. Prepare a series of standards in the concentration range 0.01–0.2 mg l^{-1} of bromide by appropriate dilution of the stock solution.

Potassium permanganate solution, 0.04 mol l^{-1}. Dissolve 6.32 g of potassium permanganate in demineralised water and dilute to 1 l, with water. Store in an amber-glass bottle and refrigerate.

Potassium iodide-sulphuric acid solution, 1 mg ml^{-1} of iodide. Dissolve 1.31 g of potassium iodide, dried overnight over concentrated sulphuric acid, in about 600 ml of demineralised water. Add slowly, with stirring, 350 ml of concentrated sulphuric acid. Cool the solution and dilute to 1 l with demineralised water; Store in an amber-glass bottle and refrigerate. If iodine crystals form on standing, prepare a fresh solution.

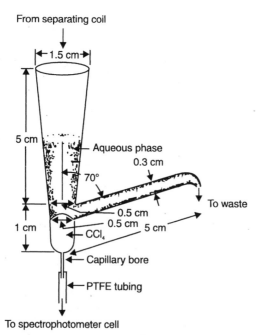

Fig. 2.2. Separator (From [2.94])

Oxalic acid solution, 50 g l^{-1}. Dissolve 50 g of oxalic acid (H$_2$C$_2$O$_4$·2H$_2$O) in demineralised water and dilute to 1 l in water.

Procedure. Set up the manifold system (Fig. 2.3) in a well ventilated hood to avoid contact with carbon tetrachloride vapour. Place the potassium iodide and potassium permanganate solutions in an ice-bath. Feed demineralised water into all reagent and sample lines, allowing sufficient time for good separation of carbon tetrachloride from the aqueous phase. Check both reference and sample flow cells to ensure that they are filled with carbon tetrachloride. If water droplets are present in the cells, the recorder pen will be deflected below the base line. Optically peak the spectrophotometer as instructed in the manufacturer's manual and then set the reversing switch under the spectrophotometer cover to position I (inverse). Set the CAL control to position 2.0 and lock, and the Display Rotary Switch to Damp I. Feed the reagents through the system. Reaction will occur, and iodine will be extracted into the carbon tetrachloride, resulting in deflection of the recorder from full scale to the base line. Adjust the base line to read approximately 5 chart divisions and allow it to stabilise (approximately 20 min). Beginning with the highest standard, place a complete set of standards covering a concentration range from 0.01 to 0.20 mg l^{-1} in the first positions of the sample tray followed by a blank. Place individual standards of different concentrations in several positions of the remainder of the tray. Place

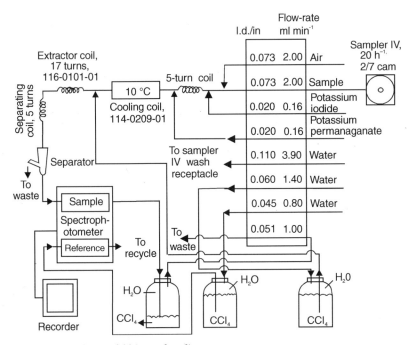

Fig. 2.3. Bromide manifold (From [2.94])

oxalic acid solution in the thirty-eighth position, followed by two blanks. Fill the remainder of the tray with unknown samples and begin the analysis.

Prepare a calibration graph by plotting the height of each standard peak against its respective bromide concentration. With the CAL control in position 2.0, the peak height reading for 0.20 mg l^{-1} of bromide using a Technicon recorder was about 30 chart units. Calculate the bromide concentration of each sample by comparing its peak height with the calibration graph. Any base-line drift that may occur must be taken into account when calculating the height of a sample or standard peak.

Spiking experiments carried out by this method gave recoveries in the range 94–110% (Table 2.10). No bias exists between the automated method and Manual methods [2.92, 2.93].

2.2.1.3 Flow Injection Analysis

Anfalt and Twengstrom [2.95] give details of equipment and a procedure for the determination of low concentrations of bromide in natural waters by modifying the phenol red method for use in a flow-injection system. The effect of various potential interfering substances was investigated. Ammonia, cyanide, and humic substances caused interference; ways of minimising the effects are indicated.

Table 2.10. Recovery of bromide added to surface-water samples as measured by the automated method. (From [2.94]; Copyright 1980, Royal Society of Chemistry, UK)

Sample Number	Present[a]	Bromide Added	Concentration/mg l^{-1}		Recovery %
			Total	Found[b]	
128–062	0.093	0.10	0.193	0.190	98
128–063	0.117	0.05	0.167	0.166	99
128–095	0.128	0.03	0.158	0.155	98
135–021	0.063	0.05	0.113	0.121	107
115–020	0.020	0.02	0.040	0.038	95
128–073	0.084	0.05	0.134	0.129	96
134–207	0.178	0.02	0.138	0.138	100
134–209	0.117	0.01	0.127	0.125	98
134–213	0.130	0.01	0.140	0.144	103
134–288	0.081	0.03	0.112	0.105	94
135–122	0.084	0.04	0.124	0.136	110
152–097	0.067	0.04	0.107	0.103	96
155–129	0.084	0.02	0.104	0.109	105
157–161	0.036	0.02	0.056	0.054	96
159–188	0.061	0.05	0.111	0.115	104

[a]Values based on ten replicate determinations.
[b]Values based on seven replicate determinations.

2.2.1.4 Inductively Coupled Plasma Atomic Emission Spectrometry

Michlewicz and Carnahan [2.96] give details of a method for the determination of bromide based on pneumatic nebulization of the sample into a helium microwave induced plasma. The detection limit was 0.4 mg l^{-1}. Potassium, magnesium, sodium and lithium interfered moderately with pneumatic nebulization and significantly with ultrasonic nebulization.

2.2.1.5 Ion-Selective Electrodes

Ion-selective electrodes have been used for the determination of bromide in water [2.97].

2.2.1.6 Gas Chromatography

Nota et al. [2.98] have devised a simple and sensitive method for the determination of amounts of bromide down to 0.05 mg l^{-1} in water. The procedure is in two stages. Cyanogen bromide is formed by the reaction between bromide, chlorine and cyanide. Cyanogen bromide is then separated by gas chromatography and selectively detected with an electron capture detector.

$$Br^- + 2Cl_2 + 2CN^- \rightarrow BrCN + ClCN + 3Cl^-.$$

Reagents. As follows.

Orthophosphoric acid, 85%.
Chlorine water, freshly prepared.
Potassium cyanide, 1 mol l^{-1}.
Nitromethane, aqueous, 5 wt%.

Equipment. Fractovap Model GI gas chromatograph (Carlo Erba) equipped with an electron capture detector (^{63}Ni (source) or equivalent. The column was made of borosilicate glass (1 m × 0.3 cm ID) and was packed with Porapak Q (80 to 100-mesh) (Waters Assoc., Milford, Mass., USA). Nitrogen was used as the carrier gas at a flow rate of 50 ml min^{-1}. The injector and detector temperatures were kept at 120 °C and 150 °C respectively. The retention times of cyanogen bromide and nitromethane (internal standard) were about 4 and 7 min respectively. The over temperature was 100 °C.

Preparation of Sample. Pipette a known volume of sample, not exceeding 5 ml and containing not less than 0.5 and not more than 10 µg of bromide in a 10-ml measuring flask. Add 0.5 ml of 85 wt% orthophosphoric acid, together with 0.5 ml of freshly prepared chlorine water. After standing for a few mins, add 0.5 ml of 1 mol l^{-1} potassium cyanide solution. Shake the solution thoroughly and then add 1 ml of 5 wt% aqueous nitromethane solution (internal standard). Finally dilute the mixture to 10 ml with water.

Inject a 1 µl volume of solution into the gas chromatograph.

Calibration Graphs. Obtain a calibration graph for bromide concentrations between 0.08 and 1.0 mg l^{-1} as described above using solutions prepared by diluting stock solutions of potassium bromide. As the peaks of cyanogen bromide and nitromethane are symmetrical, peak heights can conveniently be adopted in the calculations instead of the peak areas. Obtain the calibration graph by plotting the ratio of the peak height of cyanogen bromide to that of the internal standard (nitromethane) vs the amount of bromide.

It should be noted that the internal standard (nitromethane) must always be introduced after the addition of cyanide in order to prevent rapid darkening of the solutions, which causes non-reproducible results. No interferences are caused by oxidising or reducing substances or by mercury or cadmium at concentrations below 200 mg l^{-1}. Mercury and cadmium are, among the metals, the strongest complex-forming agents with bromide.

Even small amounts of aromatic compounds are likely to produce some interferences owing to their tendency to bind bromide, which is formed by reaction with chlorine water.

A gas chromatographic method for the microdetermination of bromide in water has been described [2.99]. It involves reaction of bromide with citric acid and potassium permanganate to form pentabromoacetone which is converted to bromoform, and measured by gas chromatography, with electron capture detection. Interference by chloride, humic acid, and phenols is eliminated by ion-exchange chromatography.

$$5(CH_2)_2C(OH)(COOH)_3 + 2MnO_4^- + 6H^+$$
$$= 5CO_2 + 2Mn^{2+} + 8H_2O + 5(CH_2)_2CO(COOH)_2$$
$$16Br^- + 2MnO_4^- = 6Br^- + 2Mn^{2+} + 8O^{2-} + 5Br_2$$
$$(CH_2)_2CO(COOH)_2 + 5Br_2 = 2CO_2 + 5HBr + C_2HBr_5CO$$
$$C_2HBr_5CO + NaOH = CHBr_2COONa + CHBr_3$$

(92% conversion)

Reagents. As follows.
  Sulfuric acid, 40 vol. %.
  Potassium permanganate, 5 vol. %
  Ferrous sulfate, 25% w/v.
  Ferrous sulfate heptahydrate in 1 vol.% sulfuric acid.
  Sodium Hydroxide, 1 mol l^{-1}.
  Citric acid, 0.05 wt%.
  Sodium nitrate, 0.1 mol l^{-1} and 0.5 mol l^{-1}.
  Amberlite CG. 400 (NO_3^- form).

Fill Amberlite CG-400 (Cl⁻ form, a strong basic anion exchange resin) into a column (10 mm i.d. x 10 mm high). Remove the fine particles from the resin by decantation. Treat the resin with 0.5 mol l^{-1} sodium nitrate until chloride is completely removed from the eluate.

Bromide stock solution: dissolve 0.1287 g of sodium bromide with 1 l of distilled water (0.1 g l^{-1} of bromide stock solution) and dilute it with distilled water to 0.005 mg of bromide l^{-1}.

Preparation. Wash the column with distilled water, pour 100 ml of sample solution carefully into the funnel of the column so that it will pass through the column at a rate of 2 ml min^{-1}. Drain the sample solution until its level reaches a few millilitres above the top of the column. Wash the column with a few millilitres of distilled water. Pass 70 ml of 0.1 mol l^{-1} sodium nitrate through the column at a flow rate of 2 ml min^{-1}. Pass 0.5 mol l^{-1} sodium nitrate through the column at the same flow rate to obtain 25 ml of eluate in a serum vial. Finally, run 50 ml of 0.5 mol l^{-1} sodium nitrate through the column to refresh the column for reuse.

Gas Chromatography. Shimazu 7A gas chromatograph with ^{63}Ni electron capture detector or equivalent, column: 3 mm i.d. × 3 m glass packed with 20% SF-96 + 20% silicone DC-550(8 + 2) Chromosorb W-AW DMCS, 80–100 mesh. Operation parameters: injection port and detector temperature 200 °C; column temperature 150 °C; carrier gas nitrogen 80 ml min injection volume 100–500 µl; head space, air; retention time of bromoform 2.7 min.

Procedure. Take the eluate of the sample from the column into a 50-ml serum vial. Add 0.1 ml of the citric acid solution and 1 ml of the 40% sulfuric acid solution to 25 ml of eluate or a standard bromide solution containing 1–150 µg of bromide, and stir. Add 2 ml of the potassium permanganate solution, and stir. Leave to stand for 10 min while stirring occasionally. After the 10 min oxidative bromination step, add 5 ml of the ferrous sulfate solution, and stir until the potassium permanganate solution is decoloured. Add 5 ml of the sodium hydroxide solution, and seal the serum vial with a teflon-lined rubber stopper and put on an aluminium cap. Shake, and leave to stand for 1 h at 20 °C. Sample 100–500 µl of vapour from the vial head space (13 ml) with a closable gas-tight syringe and inject it into the gas chromatograph. Identify the peaks by the retention time and quantify them by their height measurements.

The conversion rate of bromide ions to bromoform is reproducible at about 92%. However, it was found that the yield drops when chloride is present in amounts exceeding 500 µg. Humic acid and phenols have a similar effect. Ando and Sayato [2.99] found that the use of an ion-exchange gas chromatograph using strong anion exchange resins enabled them to separate and concentrate chloride and bromide. The bromide was eluted with 0.1 mol l^{-1} sodium nitrate through a 70-mm column, until 70 ml of eluate was obtained (total 130 ml). After elution of chloride with dilute sodium nitrate, the solution of bromide could be eluted with 0.5 mol l^{-1} sodium nitrate. Interfering effects of phenols and humic acids could also be overcome by these procedures.

Bromide concentrations down to 1–2 µg l^{-1} can be determined by this procedure. Calibration curves are linear up to 1.5 mg l^{-1} bromide.

2.2.1.7 High Performance Liquid Chromatography

Stetzenbach and Thompson [2.100] employed high performance liquid chromatography on anion exchange columns to determine bromide in admixture with chloride, iodide, nitrate and thiocyanate in natural waters. This method is discussed further under Multihalide Analysis, Sect. 2.5.1.

Moss and Stephen [2.101] determined bromide, chloride and iodide in natural waters by conversion to alkyl halide and measurement by high performance liquid chromatography.

2.2.1.8 Ion Exchange Chromatography

Akaiwa et al. [2.102] have used ion exchange chromatography on hydrous zirconium oxide, combined with detection based on direct potentiometry with an ion selective electrode, for the simultaneous determination of chloride and bromide in natural waters. This technique is discussed in further detail in Sect. 2.5.1.

Salov et al. [2.243] have described a procedure for determining bromide (and chloride, iodide, chlorate, bromate and iodate) in water employing high performance liquid chromatography with an indirectly coupled argon plasma mass spectrometric detector.

2.2.1.9 Micelle Chromatography

Okada [2.103] has used micelle exclusion chromatography to determine bromide in the presence of anions (iodide, iodate, nitrite and nitrate) in water. The method is based on partition of the anions to a cationic micelle phase and shows different selectivity from ion exchange chromatography.

Okada [2.244] was also used micelle chromatography to analyse mixtures of bromide, iodide, nitrite, nitrate and iodate.

2.2.1.10 Ion Chromatography

Applications of ion chromatography to the determination of bromide in natural waters are reviewed in Table 2.11 and discussed in further detail in Sect. 11.1.

Table 2.11. Determination of bromide in natural waters by ion chromotagraphy

Type of water	Codetermined anions	Reference
Natural	Cl, NO_2, NO_3, SO_4	2.207
Natural	Cl, NO_2, NO_3, SO_4, PO_4	2.208
River	PO_4, F,	2.106
River	Br	2.104

2.2.1.11 Polarography

In a specific method for determining down to 2 mg l^{-1} bromide in groundwaters [2.107] bromide is first converted to silver bromide which is then oxidized to silver bromate which is then determined polarographically.

Polarographic methods for bromide are based on the oxidation of bromide to bromate followed by electro-reduction. In this technique, iodide is partially oxidized to iodate.

The oxidation reaction is

$Br^- + 3ClO^- \rightarrow BrO_3 + 3Cl^-$

and the global electrode reaction

$BrO_3 + 6H^+ + 6e \rightarrow Br^- + 3H_2O$.

The half-wave potential for the reduction of iodate is less negative than is that of bromate.

Reagents. Merck unless otherwise stated.
Potassium chloride. Potassium bromate. Sodium acetate N.
Silver nitrate, 0.1 N.
Sodium hydrochlorite (BDH), low in bromine.
Activated carbon (p.a. Baker).

Apparatus. Polargraph (e.g. Bruker E 100) pH meter (Metrohm E, 632). Automatic titrator (Metrohm E 635) equipped with the electrode Metrohm EA. 0.4$^-$ µm membrane filtration equipment (Sartorius).

Sample. The method is directly applicable to samples containing 1–150 µg bromide. Adequate volumes are sampled and processed to fall within this range of applicability.

For surface waters or ground waters containing organic compounds shake the sample for 30 min after adding 1 g l^{-1} activated carbon powder, and subsequently filter on a standard bacteriological membrane filter with a porosity of 0.45 µm. Organobromated compounds are eliminated.

Sample Concentration Technique. Precipitate bromides simultaneously with chlorides and iodides by adding 0.1 N silver nitrate. Add 25 ml of silver nitrate solution to

1 l sample water and leave in the dark for 12 h. Bromate, if present, remains unprecipitated as silver bromate which is soluble.

Filter the precipitate under vacuum on a membrane with pore size of 0.45 μm. Then place the membrane with the deposit in a 100-ml vessel for about 30 min in a drying oven at 40–45 °C. Draw off the precipitate from the membrane and collect in the reaction vessel.

Conversion of Bromide to Bromate. Pour the isolated precipitate into 5 ml distilled water and treat with 4 ml 0.35 mol l^{-1} hypochlorous acid/sodium hypochloride at pH 7. (This reagent is obtained by partial neutralization with a normal solution of sulphuric acid AnalaR of a sodium hypochlorite solution of appropriate grade, e.g. BDH AnalaR, low in bromide.)

After mixing, evaporate the 9 ml of solution to dryness on an IR stand equipped with a 350 W lamp. At this stage, bromide is quantitatively oxidized to bromate. Take the residue remaining after evaporation up in 20 ml distilled water and transfer into a 25-ml volumetric flask. Adjust the pH to 7.5 with N sulphuric acid, then adjust the volume to 25 ml.

Method. Set the polarograph in the pulse mode under the following conditions: nitrogen stripping 10 min, mercury dropping frequency 0.5 s, scanning rate 5 mV s^{-1}. The half-wave potential for bromate in these experimental conditions is 0.64 V vs SCE.

The reaction medium is a self-supporting electrolyte. No further additions are necessary and may even hinder by their side effects such as formation of deposits and coprecipitation of bromate (with calcium chloride), superposition of the waves (with alkali-chlorides) and finally hindering convection currents (with lanthanum trichloride). A stable diffusion current is obtained at pH 1–2.

Using this method Masschelein and Denis [2.107] found levels of bromide in the range 20–280 mg l^{-1} in Belgian groundwater samples.

2.2.1.12   Neutron Activation Analysis

Slanina et al. [2.108] claim that a detection limit of 1 μg^{-1} can be achieved in the determination of bromide in rainwaters by neutron activation analysis. Typical accuracy was 5% relative using a 1-ml sample.

2.2.1.13   X-Ray Spectrometry

Direct X-ray spectrometry, utilising the bromine Kα line has now been used to determine mg l^{-1} quantities of bromide in natural waters. In one method the bromide was retained on an anion exchange column [2.109] and in another the liquid sample was contained in a lucite cup [2.110].

2.2.2 Rainwater and Snow

2.2.2.1 Ion Chromatography

Ion chromatography methods for the determination of traces of bromide in rainwater are reviewed in Table 2.12 and discussed further in Sect. 11.2.

Fishman et al. [2.111] compared an automated fluorescein method with ion chromatography for determination of bromide in rain water, and de-ionized water, at intervals over a period of 30 days. The fluorescein method involved buffering the sample to pH 5.6, oxidation of bromide to hypobromous acid with chloramine-T, and addition of fluorescein which reacted with hypobromous acid to form tetrabromofluorescein (eosin), the pink colour of which indicated bromide ion concentration. The results obtained by the two methods were not significantly different at a confidence level of 95% for samples containing 0.015 to 0.5 mg l^{-1} bromide and the correlation coefficient for the same sets of paired data was 0.9987. There appeared to be no loss of bromide from solution during storage in either polyethylene or polyproplene bottles. It was concluded that, where large numbers of samples were to be analysed, the automated fluorescein technique would be the method of choice, since it was more rapid than ion chromatography.

2.2.2.2 Isotope Dilution Analysis

The mass spectrometric isotope dilution method described by Henmann et al. [2.112, 2.113], for the determination of down to µg l^{-1} chloride in and snow has been applied to the determination of bromide (Sect. 2.1.2). Down to 0.4 µg l^{-1} bromide could be determined.

2.2.3 Potable Water

2.2.3.1 Spectrophotometric Methods

Moxon and Dixon [2.114] have described an automated method for the determination of µg l^{-1} amounts of bromide in potable water which is based on the

Table 2.12. Determination of bromide in rain water by ion chromatography

Type of water	Codetermined anions	Reference
Rain water	F, Cl, NO_2, NO_3, SO_4, SO_3, PO_4	2.41, 2.45, 2.51
Rain water	Cl, SO_4	2.41
Rain water	F, Cl, NO_2, NO_3, PO_4	2.46
Rain water	Br O_3, SO_4	2.51

catalytic effect of the bromide ion on the oxidation of iodine to iodate by permanganate and colorimetric measurement of residual iodine in carbon tetrachloride. A Technicon AutoAnalyser 1 is used. The system is kept at $0\pm0.2\,°C$ using an insulated ice bath. An automatic chloride method of similar sensitivity, accuracy and reproducibility is used to estimate chloride interference. The detection limit of the method is 4 $\mu g\,l^{-1}$ and the precision is of the order of 6%.

For most potable waters, the effects of interferences, except for that of chloride, will be negligible. Some contaminated drinking waters can contain elements such as iron and manganese in amounts which could give substantial interference, and these could be diluted to a level where the interference was removed.

Based on the bromine contents, bromide and bromate give the same response in this technique, indicating that bromide and bromate have a similar catalytic activity. Excellent precision data were obtained for this procedure. Coefficients of variation ranging from 2.2% at the 100 $\mu g\,l^{-1}$ bromide level to 7.0% at the 5 $\mu g\,l^{-1}$ bromide level. Spiking experiments gave recoveries between 98 and 103%, i.e. excellent accuracy.

2.2.3.2 Gas Chromatography

Backmann and Matusca [2.115] have described a gas chromatographic method involving the formation of halogenated derivatives of cyclohexanol for the determination of bromide, chloride and iodide in potable water. This method is discussed further under Multihalide Analysis (Sect. 2.5).

Grandet et al. [2.116] have described a gas chromatographic method for the determination of µg levels of bromide in potable water. This method is discussed further under Multihalide Analysis (Sect. 2.5).

2.2.3.3 Ion Chromatography

Morrow and Minear [2.117] have described an ion chromatographic method for the determination of bromide in potable water.

2.2.3.4 High Performance Liquid Chromatography

The procedure described in Sect. 2.1.1 [2.243] has been applied to the determination of bromide in potable water.

2.2.4 Industrial Effluents

2.2.4.1 Ion Chromatography

Applications of ion chromatography to the determination of bromide in treated waters are reviewed in Table 2.13 and discussed further in Sect. 11.4.

2.2.5 Soil and Foodstuffs

2.2.5.1 Flow Injection Analysis

Van Staden [2.119] employed flow injection analysis coupled with a coated tubular solid-state bromide selective electrode for the determination of bromide in soils. Soil extracted samples are injected into 10 mol l^{-1} potassium nitrate carrier solution containing 1000 mg l^{-1} chloride as an ionic strength adjustment buffer. The sample buffer zone formed is transported through the bromide selective electrode onto the reference electrode. The method is applicable in the range 10–50 000 mg l^{-1} bromide. The coefficient of variation of this method is better than 1.6%.

2.2.5.2 Gas Chromatography

The widespread use of methyl bromide as a soil fumigant has necessitated the development of convenient and specific analytical methods for the determination of bromide and total bromine in foodstuffs and soils subsequent to fumigation. Roughan et al. [2.120] have described a gas chromatographic method for carrying out this analysis in which the soil or foodstuff is mixed with sodium hydroxide solution and then treated with ethanol prior to evaporation to dryness. After muffling, the residue is digested with sulphuric acid and to this solution are added acetonitrile and ethylene oxide

$$H^+ + Br^- + C_2H_4O \rightarrow HOCH_2CH_2Br.$$

The 2- bromoethanol produced in this reaction is examined by gas chromatography using an electron capture detector. At the 10 mg kg^{-1} bromide level in soil a standard deviation of ± 0.34 mg kg^{-1} was obtained, i.e. coefficient of

Table 2.13. Determination of bromide in treated waters by ion chromatography

Type of Sample	Codetermined anions	Reference
Industrial effluents	S, CN, CNS, I	2.118
Industrial effluents	Cl, SO$_4$, NO$_2$, NO$_3$, PO$_4$	2.57

variation was ±3%. Recoveries from soil and vegetative matter were, respectively, 81–94% and 97–98%.

A method has been described [2.121] for the determination of bromide in grain. In this method, bromide ion is allowed to react selectively with ethylene oxide to form 2-bromoethanol which is separated and determined by gas chromatography using an electron capture detector. In calibration tests of this method carried out over seven laboratories on standard grains and maize containing 50 mg kg^{-1} bromide, inter-laboratory standard deviations between ±6.4 and ±6.2 mg kg^{-1} were obtained. Mean rates of recovery were in the range of 92–109%.

2.3 Iodide

2.3.1 Natural Waters

2.3.1.1   Titration Methods

The potentiometric titration proceedure described by Kuttel et al. [2.122] for the determination of chloride in natural waters has also been applied to the determination of iodide.

2.3.1.2   Spectrophotometric Methods

Kinetic spectrometric methods based on the catalytic effect of µg l^{-1} quantities of iodide ions in water on various chemical reactions have been described. These include reactions of cerium IV with arsenious acid [2.123], 3,3′ dimethyl naphthidine with hydrogen peroxide [2.124] and the oxidation of cadion IRFA by potassium persulphate [2.125]. The reaction of Malachite Green, Brilliant Green and Basic Turquoise Blue with iodide ions has been used as the basis of a method for the determination of microgram quantities of iodide [1.126].

Novak and Slama [2.127] determined iodide in the presence of bromide and chloride by adding potassium bromide to the sample to convert the iodide to iodine, followed by palladous chloride which is converted by iodine to palladous iodide. The latter was estimated spectrophotometrically at 465 nm. Using this method, 2–8 m mol iodide can be determined in the presence of 1000-fold excesses of chloride in bromide.

Iodide has been determined in groundwaters [2.128] by utilizing the reaction between cerium IV and arsenic III which is catalysed by iodide ion. This reaction is stopped by the addition of N,N'-bis-(2-hydroxypropyl)-o-phenylene diamine, which combines with cerium IV to give a red colour, the intensity of which is proportional to the iodide concentration. The standard deviation of the method ranges from 0.3 ng (10 ng iodide) to 3.8 ng (100 ng iodide).

In a further method [2.129] for determining iodide in borehole water, iodide is precipitated with nitron in sulphuric acid medium. The extinction of a dichlorethane extract of the reaction mixture is then measured. There is no interference from chloride or sulphate or from bromide if sufficient aqueous chlorine is added, and nitrate interferes only in so far as it consumes nitrogen.

Weil et al. [2.130] have described a rapid quantitative ultramicro method for the determination of iodide in concentrations of $1-100$ $\mu g l^{-1}$ based on the catalytic accelaration of the reaction between chloramine-T and N,N'-tetramethyldiamino-diphenylmethane, and the photometric measurement of the extinction of the reaction product at 600 nm. Iodate can also be determined after reduction with zinc. A polarographic method is also described for the determination of iodide after preliminary oxidation to iodate using bromine in alkaline solution. Typical results are given for samples of borehole water from sediments of the Mediterranean Sea.

2.3.1.3 Inductively Coupled Plasma Atomic Emission Spectrometry

Miyazaki and Brancho [2.131] preconcentrated iodide by conversion to iodine and extraction into xylene. The extract was determined at 172.28 nm using inductively coupled plasma atomic emission spectrometry. Iodate was reduced to iodide which was then treated in the same way to give a total iodide plus iodate content. Detection limits were, respectively, 8.3 and 21 μg^{-1} for iodate and iodide. Large concentrations of bromide interfere.

Michlewicz and Carnahan [2.132] give details of a method for the determination of iodide based on pneumatic nebulization and ultrasonic netrulization of the sample into a helium microwave induced plasma. The detection limit was 3 $mg l^{-1}$ bromide.

Potassium, magnesium, sodium and lithium interfered moderately with pneumatic nebulization but significantly with ultrasonic nebulization.

2.3.1.4 Ion-Selective Electrodes

Various workers [2.133–2.136] have used an iodide-ion electrode for the determination of iodide in natural waters. Butler and Gersshey [2.133] used an iodide-selective electrode to measure down to 1n $mol l^{-1}$ iodide separated from other anions by ion-exchange high performance liquid chromatography. Nakayama and Kimoto [2.135] used a flow-through electrode system. Weiss [2.136] determined both iodides and cyanides in micro μmol quantities using an iodide-selective electrode.

Zernalova and Senyavin [2.137] examined conditions for the potentiometric determination of iodide in natural water.

Yuan et al. [2.246] used schiff base complexes of cobalt II as neutral carriers for highly selective iodide electrodes. Thus electrode demonstrates excellent selectivity to iodine. The selectivity sequence is $I > CNS > NO_2 > ClO_4 > Br > NO_3 > Cl > SO_4$.

2.3.1.5 Gas Chromatography

Wu et al. [2.138] have described a method for the derivatization of iodide into pentafluorobenzyl iodide using pentafluorobenzyl bromide. The derivative was analysed at µg levels by gas chromatography with electron capture detection. The effects of solvent, water content, base or acid concentration, amount of pentafluorobenylbromide, reaction time and reaction temperture were examined. Interferences by common anions were minimal. The method was applicable to iodide determination in spring water.

2.3.1.6 High Performance Liquid Chromatography

Moss and Stephen [2.139] determined iodide, bromide and chloride in natural waters by conversion to alkyl mercury II halide and measurement by high performance liquid chromatography.

Stetzenbach and Thompson [2.140] employed high performance liquid chromatography on anion exchange columns to determine iodide in admixture with chloride, bromide, nitrate and thiocyanate in natural water. This method is discussed further under Multihalide Analysis (Sect. 2.5).

Iodide in natural waters has been determined [2.141] by ion exchange high performance liquid chromatography using an iodide ion selective electrode as a detector. On-line preconcentration of iodide on an anion guard cartridge allowed determinations down to $1n$ $mol\,l^{-1}$.

Verma et al. [2.239] determined iodide in natural waters by high performance liquid chromatography after precolumn derivatization of iodide to 4-iodo-2,6-dimethylphenol. The derivative was detected by an ultraviolet detector. Down to 2 µg l^{-1} iodide can be determined by this procedure.

2.3.1.7 Micelle Chromatography

Okada [2.244] has used micelle exclusion chromatography to determine iodide, bromide, nitrite, nitrate and iodate in natural waters.

Okada [2.142] has used micelle exclusion chromatography to determine iodide in the presence of other anions(bromide, iodate, nitrite and nitrate) in water. The method is based on partition of the anions to a cationic micelle phase and shows different selectively from ion exchange chromatography.

2.3.1.8 Ion Chromatography

Dils and Smeenk [2.106] have discussed the determination of iodide and phosphate in natural waters.

2.3.1.9 Polarography

In the polarographic determination [2.143] of 0.2 µmol quantities of iodide in natural waters, the iodide was oxidised to iodate with sodium hypochlorite and the excess oxidant destroyed with sodium sulphite.

Iodate was determined by differential pulse polarography and the concentration of iodide calculated as the difference between the concentrations of iodate in a sample before and after oxidation.

Luther et al. [2.242] carried out a direct determination of down to 12 ppt (u.s) of iodide in seawater using cathode stripping square wave voltammetry.

2.3.1.10 Potentiometric Method

Bardin and Tolstousov [2.145] used a flow-type concentration cell with two silver-silver iodide electrodes for the potentiometric determination of down to 3 µg l^{-1} of iodide in natural waters. At iodide concentrations of up to 0.6 µmol l^{-1} the variation in EMF of the cell is nearly proportional to the concentration of iodide.

2.3.1.11 Neutron Activation Analysis

Malvano et al. [2.146] determined iodide in water by this technique and compared results obtained with those obtained by the autoanalysed spectrometric method.

2.3.1.12 Radiochemical Method

Palagni [2.147] used the pulsed column bed technique to separate selectively and preconcentrate low levels of ^{127}I$^-$ in natural water. The pulsed bed consisted of a syringe containing a scintillation detector and a gamma ray spectrometer. The adsorbent consisted of open cell polyurethene foam impregnated with long chain tri-n-alkylamine which formed a complex with iodide.

2.3.1.13 Isotope Dilution Analysis

This technique combined with mass spectrometry has been used to determine down to 0.1 µg l^{-1} iodide in natural waters [2.148].

2.3.2 Rainwater

2.3.2.1 Spectrometric Method

Sub microgram amounts of iodide in rainwater have been determined spectrophotometrically [2.149] by its catalytic effect on the substitution reaction of the mercury II-4-(2-pyridyrazo) resorcinol complex with 1,2-diaminocyclohexane-NN-$N'N'$-tetraacetic acid. The sample solution is first freed from cations by passage a column of Amberlite IR 120-B. The eluate is buffered to pH 9, then treated with mercuric ions and 4-(2-pyridylazo) resorcinol, and absorbance at 500 nm recorded automatically as a function of time. The extinction after 30 is measured and compared with iodide standards which have been treated in the same way.

2.3.3 Potable Water

2.3.3.1 Spectrometric Methods

Manual [2.150] and automated [2.151] spectrometric methods have been described for the determination of iodide in potable water. Moxon [2.151] describes an automated procedure capable of determining iodide ion in potable water in amounts down to 0.4 $\mu g\, l^{-1}$ and total inorganic iodine (including iodate) in amounts down to 0.2 $\mu g\, l^{-1}$. The methods are based on the catalytic effect of iodide on the destruction of the thiocyanate ion by the nitrite ion:

$$2CNS^- + 3NO_2 + 3NO_3^- + 2H^+ \rightarrow 2CN^- + 2SO_4^{2-} + 6NO + H_2O.$$

The methods are applicable to normal drinking waters, with possible interfering ions having negligible effects. The system is easy to use and can handle 20 samples per hour.

When series of drinking waters were run by this method, the peaks were found to be irregular and erratic. It was found that there was a relationship between the hardness of the waters and the peak irregularity, and this was removed when the alkaline earth metals were precipitated out with potassium carbonate. Subsequently, samples and standards were made up in 0.3% potassium carbonate solution. This addition of alkali prolonged the stability of the standard solutions from 6 h to 8 d, but led to a decrease in the sensitivity of the method. Chloride at a level of 400 mg l^{-1} gave a 20% increase in the response to a 4$\mu g\ l^{-1}$ of iodide standard solution. To overcome this interference, a large excess of chloride in the form of sodium chloride solution was introduced into the sample stream and it was found that, as well as removing the interference effect of chloride, the sensitivity of the method was greatly increased.

A considerable proportion of the total inorganic iodide present in water can consist of iodate. System A, illustrated in Fig. 2.4, recovered iodate quantitatively

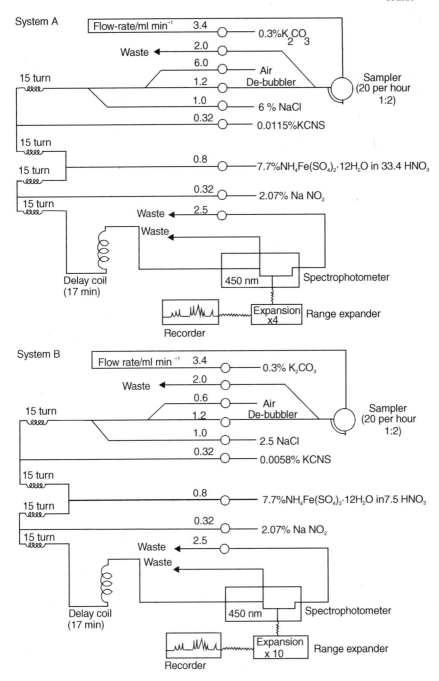

Fig. 2.4. Flow diagrams for the determination of total inorganic iodine in waters (*system A*) and free iodide in waters (*system B*) (From [2.151])

and gave measure of the total inorganic iodine in solution. In order to determine free iodide only, the oxidation-reduction potential of the reaction mixture was adjusted so that iodate was not reduced to iodine or iodide. This was achieved by (a) reducing the concentration of nitric acid in the ammonium iron (III) sulphate reagent, (b) reducing the concentration of the sodium chloride reagent and (c) reducing the concentration of the potassium thiocyanate reagent. These changes caused a corresponding decrease in sensitivity. The manifold system shown in Fig. 2.4 has two different sets of reagents. Set A was used for the determination of total inorganic iodine in water over the range 0.2 – 5.0 $\mu g\, l^{-1}$ of iodide and set B was used for the determination of free iodide in water over the range 0.4 – 5.0 $\mu g l^{-1}$ of iodide. A comparison of the concentration of free iodide and total inorganic iodine in a range of potable waters determined by the methods described below is shown in Table 2.14.

The results in Table 2.15 show that it is necessary to adopt a standardised sampling procedure. Samples were made up in 0.3% potassium carbonate solution in polypropylene calibrated flasks, centrifuged and then stored in polystyrene bottles (30-ml universal containers, Sterilin Ltd., Teddington, Middlesex). Fresh potable water was treated in the same way, but allowed to stand for 3 days before analysis.

Reagents for the Determination of Total Inorganic Iodine (System A). All chemicals used should be of analytical-reagent grade and glass distilled water should be used in preference to deionised water.

Standard iodide solution – 4 g l^{-1}. Dissolve 0.5232 g of potassium iodide, previously dried in an oven at 105 °C for 2 h, in distilled water and dilute to 100 ml in a calibrated flask.

Standard iodide solution – 40 mg l^{-1}. Dilute 10 ml of the standard iodide solution (4 g l^{-1} of I) to 1000 ml with distilled water in a calibrated flask (stable for 1 month).

Table 2.14. Amounts of free iodide and total inorganic iodine in a range of United Kingdom potable waters. (From [2.151]; Copyright 1984, Royal Society of Chemistry, UK)

Sample source	Total iodine/$\mu g\, l^{-1}$	Free iodide/$\mu g\, l^{-1}$
London (borehole)	28.0	28.0
Harrogate	2.2	0.9
File	7.8	7.8
Bristol	4.1	2.1
Nottingham	14	3.5
Amesbury	4.3	0.9
Oxford	4.3	1.3
Gloucester	3.2	3.2
Dunoon	1.1	0.9
Nuneaton	2.8	1.6

Table 2.15. Changes in the iodine concentration of standards, blanks and drinking water solutions made up in 0.3% potassium carbonate in different containers over a period of days. (From [2.151]; Copyright 1984, Royal Society of Chemistry, UK)

Test Solution	Container	Iodine concentration µg l^{-1}			
		After 1h	After 1day	After 3days	After 8days
4 µg l^{-1} solution	Glass calibrated flask	3.9	4.0	4.0	4.2
	Polyethylene bottle	3.9	3.9	4.0	4.0
	Polystyrene bottle	3.9	3.9	3.9	3.9
Blank solution	Glass calibrated flask	0.0	0.2	0.4	0.5
	Polyethylene bottle	0.0	0.1	0.2	0.3
	Polystyrene bottle	0.0	0.0	0.0	0.0
Laboratory drinking water straight from tap	Glass calibrated flask	4.7	5.5	5.7	5.8
	Polythylene bottle	4.5	5.5	5.8	5.6
	Polystyrene bottle	4.5	5.4	5.7	5.7
Cambridge drinking water	Glass calibrated flask	5.4	–	5.8	6.1
	Polystyrene bottle	5.4	–	5.5	5.4
Hertford drinking water	Glass calibrated flask	4.4	–	5.0	5.0
	Polystyrene bottle	4.5	–	4.6	4.5
Oxford drinking water	Glass calibrated flask	5.6	–	6.0	6.0
	Polystyrene bottle	5.7	–	5.9	5.8

Standard iodide solution – 200 µg l^{-1} of I. Dilute 5 ml of the standard iodide solution (40 mg l^{-1} of I) to 1000 ml with distilled water in a calibrated flask. Store in a polythene or polystyrene bottle (stable for 1 month).

Working solutions – into 200-calibrated flasks pipette 5, 4, 3, 2, 1 and 0 ml of standard iodide solution (200 µg l^{-1} of I). Add 2 ml of 30% m/V potassium carbonate solution and dilute to 200 ml with distilled water. These are the working standards. Store in polystyrene bottles and prepare freshly every 2 weeks.

Potassium carbonate solution – 30% w/v. Dissolve 300 g of potassium carbonate in distilled water and make up to 1 l.

Potassium thiocyanate solution – 0.0115% w/v. Dissolve 0.115 g of potassium thiocyanate in distilled water and make up to 1 l.

Sodium nitrite solution. Dissolve 4.14 g of sodium nitrite in distilled water and dilute to 200 ml (stable for 1 day only).

Sodium chloride solution – 6% w/v. Dissolve 60 g of sodium chloride in distilled water and dilute to 1 l.

Ammonium iron III sulphate reagent. Dissolve 77 g of ammonium iron III sulphate (NH$_4$Fe(SO$_4$)·12H$_2$O) in approximately 300 ml of distilled water. Add 334 ml of concentrated nitric acid. (sp. gr. 1.42) and make up to 1 l. Heat on a hot plate until all traces of solid dissolve.

Reagents for the Determination of Free Iodide (System B). As follows.

Standard iodide solutions. These are exactly the same as those described for system A.

Potassium carbonate solution – 30% w/v. Dissolve 300 g of potassium carbonate in water and dilute to 1 l.

Sodium nitrite solution – 2.07% w/v. Dissolve 4.14 g of sodium nitrite in distilled water and dilute to 200 ml (stable for 1 day only).

Sodium chloride solution – 2.5% w/v. Dissolve 25 g of sodium chloride in distilled water and dilute to 1 l.

Potassium thiocyanate solution – 0.0058% w/v. Dissolve 0.058 g of potassium thiocyanate in distilled water and dilute to 1 l.

Ammonium iron III sulphate reagent. Dissolve 77 g of ammonium iron III sulphate ($NH_4Fe(SO_4) \cdot 12H_2O$) in approximately 400 ml of distilled water. Add 75 ml of concentrated nitric acid (sp. gr. 1.42) and make upto 1 l. Warm untill all traces of solid dissolve.

Sodium oxalate reagent. Dissolve 5 g of sodium oxalate in 100 ml of 5 vol.% sulphuric acid (this reagent is toxic).

Apparatus. Centrifuge speed of 50 Hz, glass or polypropylene centrifuge tubes of 150 ml capacity, Polystyrene bottles of 30–50 ml capacity. AutoAnalyzer system for colorimetric analysis, e.g. the Technicon. Auto-Analyzer I system with a range expansion facility that was operated and maintained in accordance with the instructions given in the Operator Instruction Manual.

Procedure. Wash all glassware and polystyrene containers with concentrated nitric acid and rinse copiously with distilled water before use. Dispense 1.0 ml of 30% w/v potassium carbonate solution into a 100-ml calibrated flask. Make up to 100 ml with a water sample and shake well. Centrifuge the resulting solution for 5 min at 50 Hz. Decant about half of the solution into polystyrene bottles. Allow fresh tap water to stabilise for 3 days before analysis.

Set up the manifold system shown in the flow diagram (Fig. 2.4) and use the appropriate set of reagents for either total inorganic iodine determination (A) or free iodide determination (B). Load the sample tray with a set of working standards followed by 20 samples interspersed with a working standard every fifth sample. Complete the series with another set of working standards and run at a rate of 20 per hour.

If deposits of iron III thiocyanate occur, these may be removed by running two sample cups of sodium oxalate reagent through the system at the end of a run.

Calculation. Plot a calibration graph of the mean standard peak heights against their respective iodine concentrations. The iodine concentration of a sample is obtained by comparing its peak height with the calibration graph. Multiply the result by 1.01 to compensate for the addition of potassium carbonate.

The effects of ions commonly occurring in drinking waters that could cause possible interference were tested and the results, representing a mean of three determinations, are shown in Table 2.16. The results show that interferences do not present any major problems in the analysis of drinking waters for either total inorganic iodide or free iodide.

Table 2.16. Effect of added ions on the determination of (A) total inorganic iodine concentration and (B) free iodide concentration in a 4 µgl^{-1} standard iodide solution. (From [2.151]; Copyright 1984, Royal Society of Chemistry, UK)

Element added	Concentration mg l^{-1}	Maximum concentration in drinking waters mg l^{-1}	(A) Total I found µg l^{-1}	(B) Free I found µg l^{-1}
Zn^{2+}	1	0.69	4.0	3.9
Cu^{2+}	1	0.36	4.0	3.8
Li^{2+}	1	0.01	4.1	3.8
Pb^{2+}	0.1	0.046	4.0	3.9
Fe^{3+}	2	1.5	4.0	3.8
Mn^{3+}	1	0.06	3.9	3.9
Ni^{2+}	0.2	0.013	4.0	4.2
Hg^{2+}	0.005	0.001	3.9	4.2
Co^{2+}	0.2	0.011	4.0	4.2
Mg^{2+}	30	23	4.1	4.0
Cl^-	400	245	4.0	4.0
Br^-	1	–	4.0	4.1
F^-	1	–	4.1	4.3
IO_3^-	0.004	–	8.0	4.0
SO_4^{2-}	200	–	4.1	4.5
Humic acid	20	–	3.9	0.0
Iodoform	1	–	5.6	5.6
Methyliodide	0.1	–	4.2	4.2

Table 2.17. Comparison of the proposed method for total inorganic iodine with automated spectrophotometric cerium IV sulphate-arsenious and method [2.128] for a range of waters. (From [2.151]; Copyright 1984, Royal Society of Chemistry, UK)

Sample source	Iodine concentration µg l^{-1}	
	Proposed method	Keller's et al. method
Slough	11.5	12.7
Oakington	5.5	6.0
Catterick	0.7	0.7
Hexham	2.3	2.9
Royston	4.5	5.2
Harrogate	2.2	1.7
Braintree	13.7	13.9
Benson	5.8	6.5
London (borehole)	28.0	30.0
File	7.8	5.8
Nottingham	14.0	15.4
Dunoon	1.1	1.1
London (tap)	5.8	5.9

Table 2.17 shows a comparison of total iodine determinations obtained by the above automated method of Moxon [2.151] and an automated method which is based on the method of Dubravcic (cerium IV sulphate–arsenious acid) modified for an AutoAnalyzer system. The mean recovery of iodide by the total inorganic iodine method and the free iodide method are $100 \pm 5.4\%$ and

$102 \pm 7.5\%$ respectively. The results show reasonable agreement with those of the automated method.

2.3.3.2 Gas Chromatography

Backmann and Matussa [2.152] have described a gas chromatographic method involving the formation of halogenated derivatives of cyclohexanol for the determination of iodide, chloride and bromide in potable water. This method is discussed further under Multihalide Analysis in Sect. 2.5.

Grandet et al. [2.153] have described a gas chromatographic method for the determination of µg levels of iodide in potable water. This method is discussed further under Multihalide Analysis in Sect. 2.5.

2.3.3.3 High Performance Liquid Chromatography

Salov et al [2.243] determined iodide, chloride, bromide, chlorate, bromate and iodate in potable water by high performance liquid chromatography with an inductively coupled argon plasma mass spectrometric detector.

2.3.4 Industrial Effluents

2.3.4.1 Ion Chromatography

Rocklin and Johnson [2.118] have discussed the determination of iodide together with other anions (cyanide, sulphide and bromide) in industrial effluents by ion chromatography. These methods are discussed further in Sect. 11.4 (Ion Chromatography).

2.3.5 Cooling Waters

2.3.5.1 Titration Method

Erdey and Kozmutza [2.154] used xenon difluoride as a reagent for the determination of mg l^{-1} quantities of iodide and iodine in nuclear reactor cooling waters. The test solution is made 1 N in sulphuric acid and excess solid xenon difluoride added to convert iodide to periodate which is then determined by standard iodometric procedures.

2.3.6 Foodstuffs

2.3.6.1 Spectrophotometric Method

Moxon and Dixon [2.155] have described a semi-automatic method for the determination of iodide in foodstuffs such as butter, soups, cheese, salad cream, fruit juice, bacon, sausages, beefburgers, solid fat, coffee, meats, cooking oil, fruit, flour and fish. This method is based on the catalytic effect of iodide ions on the decomposition of thiocyanate by nitrite, with an accompanying decrease in colour of the iron III thiocyanate produced by the addition of iron III ions. The sample was first ashed with potassium carbonate at 400–600 °C for up to 3 h. The resulting solution was then analysed by an automated autoanalyser procedure.

When applied to foods, the method has a precision of about 10%, a detection limit of 10 µg kg^{-1} sample and a mean recovery of about 90%.

Stopped flow methods have been used to determine iodide in pharmaceutical and food samples [2.156].

2.3.7 Seaweed

2.3.7.1 Atomic Absorption Spectrometry

Chakabarty and Das [2.157] used cold vapour atomic absorption spectrometry to carry out indirect determinations of iodide in seaweed.

2.3.8 Animal Feeds

2.3.8.1 Ion-Selective Electrodes

Ion-selective electrodes have been used to determine iodide in water, animal feeds, plant tissues and fertilizers.

2.3.9 Table Salt

2.3.9.1 Polarography

Differential pulse polarography has been used to determine iodide in table salt [2.158].

2.3.9.2 High Performance Liquid Chromatography

The precolumn derivativization procedure involving conversion of iodide to iodo-2, 6-dimethyl phenol described in Sect. 2.3.1 [2.239] has been applied to the determination of iodide in table salt.

2.3.10 Milk

2.3.10.1 High Performance Liquid Chromatography

The precolumn derivativization procedure involving conversion of iodide to iodo-2, 6-dimethyl phenol described in Sect. 2.3.1 [2.239] has been applied to the determination of iodide in milk.

2.3.11 Pharmaceuticals

2.3.11.1 High Performance Liquid Chromatography

The precolumn derivativization procedure involving conversion of iodide to iodo-2, 6-dimethyl phenol described in Sect. 2.3.1 [2.239] has been applied to the determination of iodide in pharmaceuticals.

2.4 Fluoride

2.4.1 Natural Waters

2.4.1.1 Spectrophotometric Methods

Moraru and Suten [2.159] have reviewed spectrophotometric methods for the determination of fluorides. The methods have been described based on the use of zirconium-xylenol orange [2.160], lanthanum-alizarin [2.161], and zirconium solochrome cyanine R [2.162]. These methods are all capable of determining fluoride in µg quantities in water samples. In an extractive spectrophotometric determination of micro and submicro amounts (down to 50 µg l^{-1}) of fluoride [2.163], buffered alizarin complex is added to the sample which is then extracted with triethylamine-pentanol (1:19) prior to spectrophotometric evaluation of the organic phase at 570 nm.

A method for the determination of fluoride is based on the bleaching effect of fluoride on a column of the zirconium eriochrome organic R complex [2.164].

This method is applicable over the range 2–25 µg fluoride. Sulphate does not interfere in the determination. Lanthanum-alizarin complexone has been used to determine fluoride [2.165]. The main interfering ions are removed from the sample with a strongly acidic cation exchanger. The reagent (a buffered pH 4.3 mixture of 3-amino-methylalizarin-NN-diacetic acid and lanthanum nitrate with added acetone) is added, and the red chelate is allowed to develop in the dark for about 20 min and the extinction determined against a reagent blank at 620 nm. If very high sulphate and chloride concentrations are present in the sample, the results must be corrected by reference to a calibration graph. Water samples containing aluminium (greater than 0.1 mg l^{-1}) must be adjusted to pH 9 with potassium hydroxide solution before analysis, to ensure quantitative ion exchange.

Ion exchange spectrophotometry has been used to determine microgram levels of fluoride in water [2.166].

2.4.1.2 Segmented Flow Analysis

This technique discussed in further detail in Sect. 1.32, can be applied to the determination of fluoride in the concentration range 0–2 mg l^{-1} in water.

2.4.1.3 Inductively Coupled Plasma Atomic Emission Spectrometry

To preconcentrate fluoride, Miyazaki and Brancho [2.167] converted fluoride into the ternary lanthanum-alizarin complexone fluoride and extracted it into hexanol containing N,N-diethylaniline. The extract was analysed directly by inductively coupled plasma atomic emission spectrometry for the determination of fluoride. Measurement of the lanthanum (II) 333.75 nm, emission line and comparison with a calibration graph enabled fluoride concentrations as low as 0.59 µg l^{-1} to be determined in Japanese river water (polluted and unpolluted), coastal seawater, and potable water samples.

Manzoori and Miyazaki [2.240] applied indirect inductively coupled plasma atomic emission spectrometry to the determination of fluoride in natural waters by flow injection solvent extraction. A detection limit of 0.03 µg l^{-1} fluoride was achieved.

2.4.1.4 Ion-Selective Electrodes

Shiraishi et al. [2.168] showed that a fluoride ion-selective electrode could be used to determine fluoride in well water in the presence of aluminium. These workers showed that the de-masking of fluorine from its aluminium complexes is enhanced by dilution of the sample. The permissible limits of aluminium concentration increase markedly with decreasing fluoride concentration, hence fluoride could be determined by a successive dilution method.

Zernalova and Senyavin [2.169] examined conditions for the potentiometric determination of fluoride in natural water using an ion-selective electrode. Some micro constituents of water did interfere when determining fluorides.

Ion-selective electrodes have been extensively applied to the determination of fluoride at the $mg\, l^{-1}$ and $\mu g\, l^{-1}$ levels in waters [2.170–2.174]. Warner [2.170], Sekerka and Lechner [2.171] and Erdmann [2.172] have applied the technique to natural waters. Sekerka and Lechner [2.171] achieved a detection limit of $0.2\, \mu g\, l^{-1}$ fluoride. Their methods employ a total ionic strength adjustment buffer usually comprising sodium hydroxide, 1,2-diaminocyclohexane-N,N,N,N-tetraacetic acid and acetic acid in fixed proportions. Aluminium, iron, calcium and magnesium do not interfere in these procedures within prescribed concentration limits. Pakalns and Farrar [2.173] have discussed the effect of surfactants on the determination of fluoride. Erdmann [2.172] has described an automated ion-selective electrode method for determining fluoride in natural waters. This method is capable of analysing 30 samples per hour and utilizes an autoanalyser system.

Trojanowicz and Lewandowski [2.175] have described a multiple flow-through potentiometric system for the continuous determination of fluoride in admixture with chloride and nitrate in natural waters. This method is discussed further in Sect. 2.5.

2.4.1.5 Gas Chromatography Atmospheric Pressure Helium Microwave Induced Plasma Emission Spectrometry

This technique has been applied to the determination of picogram quantities of fluoride in water [2.176]. The emission line of fluorine at 685.6 nm is used for detection.

2.4.1.6 Ion Exchange Chromatography

Okabayashi et al. [2.177] collected fluoride ion selectively on an anion exchange resin loaded with alizarin fluorine blue sulphonate.

2.4.1.7 Column Coupling Capillary Isotachophoresis

This technique offers very similar advantages to ion chromatography (Chap. 11) in the determination of anions, namely multiple ion analysis i.e. little or no sample pretreatment, speed, sensitivity and automation. Very similar advantages are offered by capillary isotachophoresis. Zelinski et al. [2.178] applied the technique to the determination of 0.02–$0.1\, mg\, l^{-1}$ quantities of fluoride, chloride, nitrate, sulphate, nitrite and phosphate in river water. The technique is discussed in further detail in Sect. 11.1.

2.4.1.8 Ion Chromatography

This technique has been applied to the determination of fluoride with other anions in natural waters. Methods available are summarised in Table 2.18 and discussed more fully in Sect. 11.1.

2.4.2 Rainwater

2.4.2.1 Ion-Selective Electrodes

Slanina [2.179] has discussed the application of ion selective electrodes to the determination of fluoride in rainwater. Detection limits of 0.005 mg l^{-1} were achieved with a relative accuracy of 3–5% using a sample volume of 0.5 ml.

Warner and Bressan [2.180] carried out direct measurements of less than 1 µg l^{-1} fluoride in rain, fog and aerosols with an Orion lanthanum fluoride ion-selective electrode.

2.4.2.2 Ion Chromatography

Applications of this technique for the determination of fluoride in rainwater are summarised in Table 2.19 and discussed more fully in Sect. 11.2.

2.4.3 Potable Water

2.4.3.1 Spectrophotometric Methods

Wierzbicki et al. observed that free chlorine [2.182] and ozone [2.183] interfere in the determination of fluoride by the zirconium-Alizarin Red-S-spec-

Table 2.18. Determination of fluoride in natural waters by ion chromatography

Type of water	Codetermined anions	References
Natural	Cl, NO_3, SO_4	2.31
River	PO_4, Br	2.106

Table 2.19. Determination of fluoride in rain water by ion chromatography

Type of water	Codetermined anions	Reference
Rainwater	NO_3, PO_4, SO_4	2.47
Rainwater	Cl, Br, NO_2, NO_3, SO_4, SO_3, PO_4	2.50
Rainwater	Br, Cl, NO_2, NO_3, PO_4, BrO_3, SO_4	2.181

trophotometric method. Chlorine partially decolourizes the lake formed; this is prevented by the addition of sodium thiosulphate in an amount equivalent to the concentration of free chlorine present. Ozone interferes at concentrations as low as 0.1 mg l^{-1} and its effect can also be overcome by the addition of a slight excess of sodium thiosulphate.

Mehra and Lambert [2.184] used a solid ion-association reagent comprising Buffalo Black and Brilliant Green dyes supported on activated alumina. When this solid is added to the sample, Brilliant Green reacts with fluoride releasing an equivalent amount of Buffalo Black for spectrophotometric determination at 618 nm. The test solution is passed through a column of the reagent and any Brilliant green released is selectively decomposed by adding solid sodium bisulphite to the percolate before measurement of the extinction. The calibration graph is rectilinear for up to 18 μg l^{-1} fluoride and the detection limit is 1 μg. Interfering ions, especially sulphate and bicarbonate, must first be removed by adsorption on Amberlite IRA 400 (NO_3^- form) and preferential elution of fluoride with 0.1 mol l^{-1} sodium nitrate; this treatment also serves to concentrate fluoride.

Anosova [2.185] determined fluoride in highly demineralised water by separating fluoride ion as hydrofluoric acid by steam distillation of the sample (concentrated by evaporation and treated with silver nitrate solution to precipitate chloride ion) from sulphuric acid medium. The distillate is treated with Alizarin Red S (C.I. Mordant Red 3) solution, buffered at pH 2.5–3.5, and then allowed to react with a known volume of standard (thorium nitrate solution). After 30 min, the colour is compared with those of a series of standards covering the range 0.0–0.1 mg l^{-1} fluoride.

2.4.3.2 Ion-Selective Electrodes

Various workers have discussed the applications of ion-selective electrodes to the determination of fluoride in potable water [2.186–2.188]. Reported detection limits are in the range 0.1 to 1 mg l^{-1}. Nicolson and Duff [2.188] have carried out a detailed study of an optimum buffer system for use with a fluoride selective electrode. They examined 11 buffer systems to determine maximum complexing ability, with optimal reproducibility and sensitivity, in the determination of fluoride in water by the direct potentiometric electrode method. A buffer system containing triammonium citrate was found to have the best masking ability as well as satisfying the other criteria. Only aluminium and magnesium posed interference problems. It is recommended that a minimum decomplexing time of 20 min (and preferably 24 h) be allowed and fresh buffer solutions be prepared after a maximum storage period of two weeks.

The composition of TISAB III M buffer is as follows:

58 g sodium chloride,

57 ml glacial acetic acid,

4 g CDTA,

2.43 g tri-ammonium citrate.

The method employed in the determination of fluoride was that of direct potentiometry using a calibration curve constructed from standard solutions. Subsequent aliquots of the same standard or sample solution were analysed until two EMF readings within 0.5 mV were obtained. A fresh calibration curve was constructed for each batch of buffer and each temperature of analysis.

After insertion of the electrode into the solution, the system was allowed to achieve equilibrium for 5 min when the EMF readings was taken to the nearest 0.1 mV. Further EMF reading were taken at 1 min intervals until consecutive EMF readings did not differ by more than 0.5 mV. This is considered to be the equlibrium EMF.

In Table 2.20 interference limits are shown for various ions in the determination of fluoride using the tri-ammonium citrate (TISAB III) buffer. Only aluminium, magnesium and borate give cause for concern.

Of all the buffer masking agent combinations examined, tri-ammonium citrate buffer, employing CDTA and tri-ammonium citrate as complexing agents, was most efficient in terms of masking ability. At 1 mg l^{-1} fluoride, aluminium and magnesium up to 100 mg l^{-1} and 1000 mg l^{-1} respectively were tolerated for 95% fluoride recovery. This is greatly superior to TISAB II which, at the same fluoride concentration and recovery level, tolerated only 2.7 mg l^{-1} aluminium, 400 mg l^{-1} magnesium, and, additionally, 1000 mg l^{-1} calcium which is not a problem with the tri-ammonium citrate buffer.

The best temperature of analysis ($20\pm0.1\,°$C or $50\pm0.1\,°$C) for maximum decomplexing was dependent upon the species involved and the buffer employed. This prevents the development of any simple method which would exploit the optimum decomplexing conditions for each individual species. However, while the maximum possible decomplexing may not be achieved in all cases, samples analysed at 20°C, 24 h following buffer addition, will, for each species and each buffer, at 1 mg l^{-1} fluoride be beneficial in terms of an increased fluoride recovery over that obtained at 20°C soon after buffer addition.

Table 2.20. Interference limits (mg l^{-1}) of several species. (From [2.188]; Copyright 1981, Marcel Dekker Inc., USA)

Buffer	Ion	0.1 mg l^{-1} F$^-$ Recovery limit		1.0 mg l^{-1} F$^-$ Recovery limit	
		100%	95%	100%	95%
TISAB III M tri-ammonium citrate	Al	>500	>500	35	100
	BO$_2$	>100	>100	>100	>100
	BO$_3$	>100	>100	>100	>100
	Ca	>1000	>1000	>1000	>1000
	H$_2$BO$_3$	>500	>500	>500	>500
	Hg	>500	>500	100	1000
	PO$_4$	>500	>500	>500	>500

Classic ion-selection electrode methods for fluoride suffer from the disadvantages of excessive electrode drift, long response time and inadequate precision.

Kissa [2.189] has described modifications to the determination of fluoride using the fluoride ion-selective electrode, to enable the determination of concentration in the range 1–10 mg l^{-1} fluoride with relative errors from 0.2 to 0.6%.

Kissa [2.189] overcame the aforementioned problems by (a) limiting the fluoride concentration to which the electrode is exposed to a 0.01–0.1 mg l^{-1} fluoride or 0.05–1mg l^{-1} fluoride concentration range; (b) measuring the electrode potential in the analyte by approaching equilibrium in the same direction from a higher potential to a lower potential and (c) keeping the temperature of the solutions constant within $\pm 0.2\,^\circ$C. The fluoride concentration in the analyte is adjusted to the concentration range of the ion-selective electrode by dilution or fluoride addition.

Determination of Fluoride by the Analyte Addition Method. Experimental details are as follows.

The fluoride ion-selective electrode, Orion Model 94-09A, was used in combination with a double junction electrode, Orion Model 90-01. The cell potentials were measured with the Orion Ionalyzer, Model 901. The Orion electrode holder (Catalog No. 13-641-814) was provided with a stop to keep the immersion depth of the electrodes constant. Two sets of electrodes were used, one set being exposed only to solutions containing 10–100 µg l^{-1} fluoride and the other to solutions containing 50–1000 µg l^{-1} fluoride. All volumetric flasks and beakers used were made of "nalgene". Agitation was provided by a thermally insulated magnetic stirrer operating at a constant speed.

The standardizing solutions were prepared by successive dilutions of a stock solution containing 11.10 g l^{-1} of reagent-grade sodium fluoride (dried to a constant weight at 125 °C). The standardizing solutions and analytes contained an acetic acid-sodium acetate buffer (2 vol.%) prepared by adding 2800 ml of water, 480 ml of acetic acid, reagent grade, and 500 ml of 30% sodium hydroxide solution, made of ACS certified, electrolytic sodium hydroxide pellets, diluting to 3800 ml with water and adjusting to pH 5.0–5.2 with sodium hydroxide.

TISAB II (without CDTA) or TISAB III [2.190] were used for electrolyte containing analytes.

The equilibration time needed to attain a stable electrode potential decreases with increasing fluoride concentration and stiring rate and increases with the increasing concentration change resulting from successive immersions. If the concentration difference is small, the equilibrium time can be reasonably short even at low fluoride concentrations. Kissa [2.189] restricted, therefore, the exposure of the fluoride ion-selective electrode to solutions differing less than 10 or 20 times in fluoride concentration. By limiting the fluoride concentration range and using a programmed immersion sequence, the equilibration time in the analyte was reduced to 3–7 min even at fluoride concentrations as low as 20 µg l^{-1}. This is the lower practical limit for determining fluoride conveniently, because the electrode response is no longer linear below this concentration.

If the millivolt reading indicates that the fluoride concentration in the analyte exceeds the concentration range of the fluoride ion selective electrode (20–100 or 100–1000 µg l^{-1} fluoride) the electrode is withdrawn immediately from the analyte and the fluoride concentration of the analyte adjusted by appropriate dilutions or additions. The concentration restriction applies also to rinsing of electrodes and their storage. The electrodes are stored in a buffer containing fluoride in a concentration corresponding to the lower limit of the fluoride concentrations measured.

Between immersions, the electrode stems are wiped and the electrode tips gently blotted with tissue paper to remove adhering liquid drops. The electrodes are not rinsed with water or a buffer solution. The electrodes are first immersed in 40 ml of the solution to be measured (analyte or reference solution) which is agitated for a minute and discarded. A fresh 40-ml portion of the solution is then used for electropotential measurement.

The precision of the electrode potential measurement by this procedure depends mainly on temperature constancy. The standard deviation of temperature and the electrode potentials in a buffered solution containing 20 µg l^{-1} fluoride were 0.16 °C and 0.18 mV and 0.05 °C and 0.08 mV respectively. If the temperature is kept within ± 0.2 °C, the electrode can be operated for weeks without calibration drift.

It has been observed by previous workers [2.191] that the electrode response drifted while the fluoride solution was stirred, but, when stirring was stopped, the electrode assumed a stable potential within 5–30 min, depending on the fluoride concentration. Kissa [2.189] examined this procedure and concluded that the apparently stable potential resulted from a reduced transport of fluoride ions to the electrode and did not always represent an equilibrium potential in the bulk of the solution. Under the conditions outlined in this modified method, a stable electrode potential can be obtained in a short time in stirred solutions as dilute as 20 µg l^{-1} fluoride.

The modified test conditions proposed by Kissa [2.189] are outlined in Table 2.21.

Ballczo [2.192] carried out direct potentiometric determinations of fluoride in potable water. This was carried out with the aid of tiron (pyrocatechol 3,5-disulphonate) as a decomplexing agent. This substance releases fluoride from a wide range of complex fluoride ions (e.g. aluminium, iron, titanyl, magnesium, silica and boric acid) and thus permits the determination of total fluoride content, even in the presence of appreciable amounts of aluminium. As the fluoride electrode is also highly sensitive to hydroxyl ions, the pH value must be kept at or slightly below neutral, but not so low as to allow the formation of hydrochloric acid. The buffering action of tiron is also sufficient to enable the use of any additional buffers, such as those for ionic strength adjustment, to be dispensed with.

Crosby et al. [2.193] examined five spectrophotometric methods, titration with standard thorium nitrate, and an ion-selective electrode method, for the determination of fluoride in potable water. The methods were examined for

Table 2.21. Sequential program of fluoride determination. (From [2.189]; Copyright 1983, American Chemical Society)

	External standard method		Analyte addition method	
	Electrode A	Electrode B	Electrode A	Electrode B
F^- concn (mg of F^- l^{-1}) of standardizing solution used for electrode storage	0.02	0.10	0.02	0.10
Sequential program for F^- concentration determination				
(I) Conditioning concn, mg of F^- l^{-1}	0.01	0.05	0.01	0.05
time, min	3	3	3	3
(II) Reference solution concn, mg of F^- l^{-1}	0.02	0.10	0.025	0.10
time, min	1 + 10	1 + 10	1 + 10	1 + 10
(III) Analyte time, min	1 + 10[a]	1 + 10[a]	record mV 10 min after adding analyte	record mV 10 min after adding analyte
Concentration range of F^- determined				
Lower limit, mg of F^- l^{-1}	0.02[b]	0.10[b]	0.025	0.10
Higher limit, mg of F^- l^{-1}	0.10[c]	1.0[c]	depends on analyte volume added	

[a] One minute in 40 ml of solution, discard solution, 10 min in another 40 ml of solution, record mV reading.
[b] If F^- concentration of the analyte is below this limit, add NaF solution to increase F^- concentration.
[c] If F^- concentration is above this limit, dilute analyte with diluted buffer (20 ml l^{-1}).

reproducibility, sensitivity, range, specificity, and, in the case of the spectrophotometric methods, for colour stability and temperature effects. Of the spectrophotometric methods, the alizarin complexan procedure is particularly suitable for samples containing only small amounts of fluoride. The electrode method surpasses all the colorimetric methods for speed, accuracy and convenience.

2.4.3.3 Ion Chromatography

This technique has been applied to the determination of fluoride and other anions (nitrate, phosphate, and sulphate) [2.47] methods are further discussed in Sect. 11.4.

2.4.4 Industrial Effluents

2.4.4.1 High Performance Liquid Chromatography

Hannah [2.194] has described a high performance liquid chromatographic anion exclusion method for the determination of fluorides in complex trade effluents. Pharmaceutical industrial effluents were examined for fluoride using an Ion-100 anion exchange column coupled with a GA-100 guard column. Organics were removed from the samples at pH 8.2 to 8.3 by passing through disposable G18 extraction columns. Conductivity detection allowed determination of fluoride at low mg l^{-1} levels in the presence of high concentrations of chloride and sulphate through an ion exclusion mechanism in the polymeric liquid chromatographic column. The limit of detection was 0.2 mgl^{-1} fluoride with a working linear dynamic range of 0.2–100 mg l^{-1}.

2.4.4.2 Ion Chromatography

Ion Chromatography has been applied to the determination of fluoride (and chloride, bromide, nitrate, nitrite, sulphate and phosphate) in industrial effluents [2.57]. This method is discussed further in Sect. 11.4.

2.4.5 Waste Waters

2.4.5.1 Ion Chromatography

This technique has been applied to the determination of fluoride (and chloride, sulphate, and phosphate) in waste waters [2.61]. The method is discussed further in Sect. 11.4.

2.4.5.2 Microdiffusion Method

Microdiffusion has been applied [2.195] to the determination of down to 50 µg l^{-1} fluoride in waste waters, effluents and river waters. The sample (1 ml containing <25 µg of fluoride ion) is placed in the outer compartment of a polypropylene Conway-type diffusion cell containing 1 ml of 0.5 N sodium hydroxide in a polystyrene cup in the inner compartment, then 1 ml of 70 vol.% sulphuric acid is rapidly added to the sample and a polystyrene lid, greased with silicone, is placed on the cell, which is then heated in an oven at 65 °C for 5 h. To the dried residue in the centre cup is added 1 drop of phenolphthalein solution and the contents are washed with water into a volumetric flask. Fluoride is then determined by the lanthanum complexan or spectrophotometric procedure [2.196, 2.197].

2.4.6 Sewage

2.4.6.1 Spectrophotometric Method

Devine and Partington [2.198] have shown that errors in the determination of fluoride in sewage by the SPADNS colorimetric method are due to sulphate carried over during the preliminary distillation step. It is suggested that the colorimetric method should be replaced by the fluoride-ion electrode method following distillation.

2.4.6.2 Ion-Selective Electrodes

Rea [2.199] used an ion-selective electrode to determine fluoride in sewage sludge – a trisodium nitrate buffer was used. Calibration was achieved by the standard addition procedure.

2.4.7 Biological Materials

2.4.7.1 Spectrophotometric Method

Venkateswarlin [2.200] has described a new approach to the determination of fluoride in biological materials. It involves extraction of fluoride from an acidified sample as fluorosilane into an immiscible organic solvent, followed by reverse extraction of fluoride into an alkaline solution.

2.4.8 Plant Extracts and Vegetative Matter

2.4.8.1 Ion-Selective Electrodes

Vickery and Vickery [2.201] have investigated the interference by aluminium and iron in the ion-selective electrode method for the determination of fluoride in plant extracts. They demonstrated that plant ashes may contain sufficient of these two elements to interfere seriously in the determination of fluoride when using the fluoride-selective electrode. In the presence of these metals, the known additions method gives erroneous results, as did that involving the attempted formation of complexes with ethylenediamine tetracetic acid, (disodium salt) or 1,2-cyclohexylene-dinitriloctetraacetic acid.

Good recoveries of fluoride ion were obtained in the presence of aluminium, iron, magnesium or silicate using sodium citrate as the complexing agent. Greater than 90% recovery of fluoride was obtained in the analysis of ashes of commercial tea, high in aluminium (2000 mg Kg^{-1}) and iron (2800 mg Kg^{-1}).

Villa [2.202] determined fluoride in vegetative matter using an ion selective electrode. Fluoride was extracted from dried vegetation by stirring with 0.1N perchloric acid for 20 min at 20°C. The fluoride content of the extract was determined at pH1 using the method of standard additions, thus eliminating the need to decomplex fluoride prior to analysis. The method was applicable over the range 4–2000 mg Kg^{-1} fluoride in vegetative matter such as grass, apples, pine needles, alfalfa and soorghum.

2.4.9 Milk

2.4.9.1 Ion-Selective Electrode

Fluoride ion has been determined in bovine milk using a fluoride ion-selective electrode [2.203]. In this method the milk sample is treated with nitrate ion and then centrifuged. Sodium hydroxide is added to the supernatant layer prior to adjustment to pH 5.2. The fluoride content is measured by immersing the ion-selective electrode directly into the sample extract. Reproducibility was ±0.5 mg l^{-1} at the 5.0 mg l^{-1} level. The method is applicable at fluoride concentrations below 1 mg l^{-1}.

2.4.10 Coal

2.4.10.1 Ion Chromatography

Conrad and Brown' less [2.204] have described a hydropyrolytic-ion chromatographic method for the determination of fluoride in coal and geological materials. The short term precision of this method is 5%. Fluoride determinations are linear from the detection limit of 1 µg up to 500 µg.

2.5 Mixed Halides

2.5.1 Natural Waters

2.5.1.1 Ion-Selective Electrodes

Trojanowicz and Lewarndowski [2.205] have described a multiple potentiometric system for the continuous determination of chloride, fluoride, nitrate and ammonia in natural waters. They describe a flow-through system for the simultaneous determination of chloride (30–150 mg l^{-1}), fluoride (0.08–0.4 mg l^{-1}), nitrate (5–20 mg l^{-1} litre) and ammonia (0.05–0.5 mg l^{-1}). Solid state chloride and fluoride electrodes, a PVC membrane nitrate electrode and a gas-sensing Orion 95-10 ammonia electrode were employed, the air-segmented sample stream being mixed with appropriate buffering solutions before entering the measuring cell. During continuous operation, calibration twice-daily was necessary to eliminate the effects of potential drift. Best results were obtained when the chloride:nitrate ratio in the calibrating solution was similar to that in the natural water being analysed(14:1).

Apparatus. Peristaltic multichannel pump type DP2-2 VEB MLW (GDR) with tubings from Ismatec (Switzerland). Ionalyzer Orion (USA) model 801A and digital printer model 751. Home-made electronic electrode switch for 5 channels, controlled by digital printer. Thermostat model U-1 from VEB MLW (GDR). The following electrodes were used as sensors – Chloride electrode from MERA-ELWRO (Poland), home made fluoride electrode [2.206], nitrate electrode with solid silver contact [2.207] and PVC membrane containing bathophenenthroline-nickel (II) complex and 2-nitrophenyloctyl ether [2.208], ammonia electrode Orion (USA) model 95-10.

In the measurement of chloride, fluoride and nitrate a silver-silver chloride electrode was applied as reference electrode [2.209].

Measuring cells were manufactured from Perspex. Internal diameter of all channels was 2 mm.

Reagents. The compositions of solutions used for adjusting appropriate conditions of determination and of standard solutions are listed in Table 2.22.

Procedure. A general scheme of the whole flow-through system is given in Fig. 2.5. By changing the position of the stop-cock either standard solution or sample solution is introduced into the system. In the case of many days of continuous analysis of river water it was indispensable to separate suspended

Table 2.22. Composition of solutions used in flow measurements for adjusting appropriate conditions and for standardizing. (From [2.205]; Copyright 1981, Springer-Verlag GmbH, Germany)

Solution	Components	Concentrations mol l^{-1}
Solution I	CH_3COOH	0.1
	CH_3COONa	0.1
	H_2O_2	0.03
	DCTA	0.01
Solution II	H_3PO_4	1.0
Solution III	NaOH	0.2
	EDTA	0.01
Standard solution I	NaCl	0.01 (pCl 2.00)
	NaF	5×10^{-5} (pF 4.30)
	$NaNO_3$	7×10^{-4} (pNO$_3$ 3.15)
	$(NH_4)_2SO_4$	5×10^{-4} (pNH$_3$ 3.00)
Standard solution II	as above	10-fold diluted standard solution I

matter on a glass frit. The standard solutions and sample streams pass through coils thermostated at $25 \pm 0.5\,°$C. After passing the peristaltic pump the sample or standard solution stream is divided into two parts. One of them flows directly to the mixing chamber preceding the ammonia sensor, while the second one, after mixing with solution K (acetate buffer containing hydrogen peroxide and DCTA) and debubbling approaches the measuring cell with chloride and fluoride electrodes and a common reference electrode. Leaving this cell, the stream is mixed with solution II (phosphoric acid) and, after debubbling, enters the nitrate electrode cell. The automatic electrode switch allows the potential values of all electrodes to be printed with the frequency controlled by the printer using the Orion 751 printer. The frequency of changes is in the range 6 s – 1 h.

Chloride and fluoride determinations obtained in natural water by this procedure are compared with those obtained by conventional methods in Table 2.23. Generally, results agreed within $\pm 5\%$ with those obtained by the reference method.

2.5.1.2 High Performance Liquid Chromatography

Salov et al. [2.243] determined chloride, bromide, iodide, chlorate, bromate and iodate in natural waters by high performance liquid chromatography with inductively coupled oxygen plasma mass spectrometry as a detector.

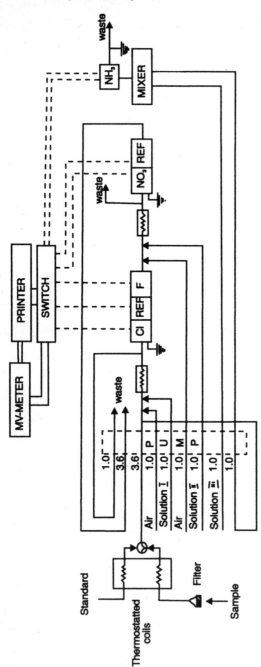

Fig. 2.5. Flow diagram of the four-parameter potentiometric system for the determination of chloride, fluoride, nitrate and ammonia in natural waters (From [2.205])

Table 2.23. Results of determinations in water with the flow-through system in comparison with reference methods (mg l^{-1}). (From [2.205]; Copyright 1981, Springer-Verlag, Germany)

Sample	Fluoride			Chloride		
	Flow through system	Potentiom. with TISAB	%	Flow through system	Titration with Ag$^-$	%
1	0.178	0.174	+2.2	115.7	112.8	+2.6
2	0.166	0.166	0	108.6	109.0	−0.4
3	0.198	0.199	−0.5	119.2	118.6	+0.5
4	0.197	0.209	−5.9	119.8	119.5	+8.7
5	0.169	0.174	−2.9	66.1	72.4	−8.7

2.5.1.3 Micelle Exclusion Chromatography

Okada [2.244] carried out micelle exclusion chromatography on mixtures of iodide, bromide, nitrite, nitrate and iodate.

2.5.2 Rainwater

2.5.2.1 Ion-Selective Electrodes

Slanina et al. [2.210] have discussed various methods for the determination of anions in rainwater. Methods discussed include ion-selective electrodes (chloride, fluoride), neutron activation analysis (chloride, bromide), ultraviolet spectroscopy (nitrate [2.212]) and nephelometry (sulphate [2.211]).

2.5.2.2 High Performance Liquid Chromatography

High performance liquid chromatography with a variable wavelength ultraviolet detector is a sensitive and precise method for the simultaneous measurement of chlorides, bromides, iodides, nitrates and thiocyanates in natural water samples. Stetzenbach and Thompson [2.213] applied the method employing anion exchange columns to ground water samples achieving detection limits of 50 µg l^{-1} for these ions.

These workers used a Spectra Physics (Sant Clara, CA) model SP–8780 liquid chromatograph using a Rheodyne (Cotati, CA) model 7125 injection valve with a 200-µl sample loop and with an Altex (Berkeley, CA) model 332 liquid chromatograph with a Valco (Houston TX) 7000 psi injection valve using a 200-µl sample loop. Detection was accomplished with a Hitachi model 10–40 variable wavelength detector and a Schoeffel model 770 detector from Kratos, Inc. (Westwoods, NJ). The chromatographic columns were: (1) a 25 cm long × 4.6 mm id packed with Whatman SAX 10 µ particle size and (2) 4-cm Brownlee anion exchange columns from Brownlees Labs (Santa Clara, CA). The mobile phase buffers were prepared with water that had been distilled over potassium permanganate and certified ACS potassium phosphate monobasic from Fisher Scientific Co. (Fair Lawn, NJ).

Thiocyanate, bromide, iodide, nitrate and nitrite have large enough extinction coefficients at 195–215 nm to make UV absorption detection useful for trace analysis at these wavelengths. An eluant of phosphate buffer prepared with HPLC grade water has no significant background absorption to 190 nm and could be used over the entire functional pH range of silica-based stationary phases. The choice of a suitable wavelength is generally limited only by the purity of the eluant and the limits of the detector. Figure 2.6 shows the area responses of a 10 mg l^{-1} bromide sample at several wavelengths. Wavelengths longer than 220 nm are not suitable for trace bromide analysis while those below 195 nm show only marginal increases in sensitivity. Thiocyanate demonstrates a similar wavelength absorptivity relationship but its usable wavelength limit extends to 235 nm. The absorption spectrum for the remaining anions are also shown in Fig. 2.6. Figure 2.7 shows the results of the analysis of water sample containing chloride at 4.4 mg l^{-1} and sub-mg l^{-1} concentrations of nitrate, bromide, thiocyanate and iodide. These results indicate the sensitivity of the measurement method and the time required to analyze all five anions. The practical limit of detection is about 50 mg l^{-1} for all of the anions except chloride which is about 1 mg l^{-1}.

Nitrite and nitrate are also easily separated and measured by this method.

System noise increases with decreasing wavelength and thereby reduces the precision and accuracy of the analysis. The standard deviation of three consecutive injections is equal to 1% at 1 mg l^{-1} and approximately 10% at 0.05% for both thiocyanate and bromide at 191 nm. At 220 nm the standard deviation for the method over a period of nine months at the 1 mg l^{-1} level is ± 0.02 mg l^{-1}.

Moss and Stephen [2.214] determined chloride, bromide and iodide by converting them to alkylmercury II halides and measurement by high performance liquid chromatography.

2.5.2.3 Ion Exchange Chromatography

Akaiwa et al. [2.215] have used ion exchange chromatography on hydrous zirconium oxide combined with a detection based on direct potentiometry with an ion selective electrode for the simulataneous determination of chloride and bromide in natural waters.

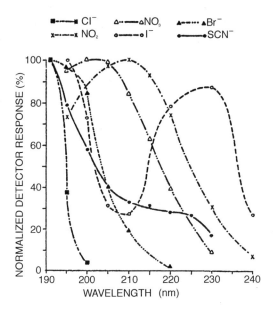

Fig. 2.6. Normalized detector response vs wavelength for Br^-, Cl^-, I^-, NO_2, NO_3, and SCN-showing the optimum absorption wavelengths (From [2.213])

Reagents. As follows.
redistilled water.
Hydrous zirconium oxide conditioned by immersion for one day in 0.5 mol l^{-1} sodium nitrate at about pH 2 before use [2.216, 2.217].

Apparatus. Orion Model 94.17 silver chloride electrode, Model 94–35 silver bromide electrode. Model 94–53 silver iodide electrode and Model 90–02 double junction reference electrode. Orion Model 701 digital PH meter combined with a Shimadzu R-111 M type recorder. Chromatographic column 6 mm in bore and 11 cm long. The apparatus for ion exchange chromatography was that described by Akaiwa et al. [2.218].

Preparation of Sample Solution. Take a suitable volume of sample solution in a 50-ml standard flask and add 5.0 ml of 5.0 mol l^{-1} sodium nitrate and 1.0 ml of 0.7 mol l^{-1} nitric acid to adjust the ionic strength and pH. Dilute the sample solution to volume. Prepare a standard solution in the same way, with a known amount of bromide and chloride solution.

Procedure. Inject about 5 ml of the sample solution into the column by syringe. Elute bromide and chloride in that order with 0.5 mol l^{-1} sodium nitrate, adjust to pH 2 at a flow rate of 0.55 ml min^{-1}. Determine bromide and chloride by comparing the peak heights of the chromatograms of the sample and the standard.

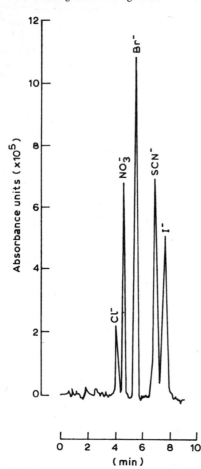

Fig. 2.7. Chromatogram of Cl^- at 4.4 mg l^{-1} NO_3^- at 0.85 mg l^{-1}, Br^- at 0.73 mg l^{-1} iodide at 3.08 mg l^{-1} and SCN^- at 0.88 mg l^{-1} in water; column Partisil 10 SAX 25 cm, solvent 0.15 KH_2PO_4, detector Schoeffel UV 195 nm, sample 50 μl and flow 3.0 ml min^{-1} (From [2.213])

Since hydrous zirconium oxide acts as an anion exchanger in acidic solution, the pH of both the eluent and the sample solution is adjusted to around 2. Under the conditions described, bromide and chloride can be quantitatively separated and the separation is not affected by variation of the pH in the range 1.5–2.5.

The silver chloride electrode gave poor response to iodide and bromide, and so did the silver bromide electrode to iodide. Although the silver iodide electrode responded to all three halides, the peaks were not sufficiently resolved and they were asymmetric. Further, there was a drift of the base line after detection of a halide ion which was not a component of the electrode and this drift caused disturbance in the following peak. This difficulty is eliminated by using hydrous zirconium oxide instead of the anion exchange resin for the chromatography, since it reverses the elution order for halide ions. The silver bromide electrode is then the most suitable as the detector for both bromide and iodide.

The standard additions method was used for accuracy tests. Table 2.24 shows very good agreement of the valued obtained by the calibration method with those from the addition method. The results show that the method is satisfactory for the sequential determination of bromide and chloride in natural water samples.

Akaiwa et al. [2.219] have also discussed the application of ion exchange chromatography to the analysis of mixture of bromide and chloride in water.

2.5.2.4 Column Coupling Capillary Isotachophoresis

This technique has been applied to the analysis of mixtures of chloride and fluoride. [2.220].

2.5.2.5 Micelle Chromatography

Okada [2.221] has used micelle exclusion chromatography to analyse mixtures of iodide, bromide, iodate and nitrite in water. The method is based on partition of the anions to a cationic micelle phase and shows different selectivity from ion exchange chromatography.

2.5.2.6 Ion Chromatography

This technique has been extensively applied to the determination of mixtures in natural water. Methods are reviewed in Table 2.25 and discussed further in Sects. 2.5 and 11.1. It has also been applied to the determination of mixed halides in rainwater. Methods are reviewed in Table 2.26 and discussed further in Sects. 2.5 and 11.2.

Table 2.24. Accuracy tests. (From [2.215]; Copyright 1982, Pergamon Publishing Co., UK)

Sample	Calibration curve method		Standard addition method	
	Br^-, µg l^{-1}	Cl^-, µg l^{-1}	Br^-, µg l^{-1}	Cl^-, µg l^{-1}
Pond water (Gunma Univ.)	27	8.6	27	9.0
Pond water (Kezouji park)	31	19.3	35	18.5
Tap water (Nitta Town)	–	38.5	–	39.0
Tap water (Gunma Univ.)	–	9.5	–	9.0

Table 2.25. Determination of halide mixtures in natural waters by ion chromatography

Type of water	Codetermined anions	References
Natural	Cl, Br, F	2.30, 2.31
Natural	F, Br	2.106

Table 2.26. Determination of halide mixtures in rain water by ion chromatography

Type of water	Codetermined anions	References
Rain water	Cl, F, Br	2.50, 2.51
Rain water	Br, Cl	2.222
Rain water	Cl, F	2.222

2.5.2.7 Gas Chromatography

Mack and Grimsrud [2.245] are described a photo-chemical modulated pulsed electron capture detector suitable for the gas chromatographic determination of chloride, bromide and iodide.

2.5.3 Potable Waters

2.5.3.1 Gas Chromatography

Grand et al. [2.223] have described a method for the determination of traces of bromide (and iodide) in potable water which permits determination of 50 µg bromide l^{-1} and 5 µg iodide l^{-1} without preconcentration of the sample, or 0.5 µg bromide l^{-1} and 0.2 µg iodide l^{-1} after preconcentration. The method is based on the transformation of the halides into 2-bromo- or 2-iodo-ethanol. These derivatives are extracted with ethyl acetate and determined with an electron capture detector. The difference in the retention periods enables both halides to be determined simultaneously in one sample. The method is substantially free of interferences and is suitable for use on many types of water.

Bachmann and Matusca [2.224] have described a gas chromatographic method for the determination of µg l^{-1} quantities of chloride, bromide and iodide in potable water. This method involves reaction of the halide with an acetone solution of 7-oxabicyclo-(4.1.0) heptane in the presence of nitric acid to form halogenated derivatives of cyclohexanol and cyclohexanol nitrate. The composition of the reagent is 1% 0.1 N nitric acid 7-oxabicyclo-(4.1.0) heptane, 25% water, 75% acetone. The column effluent passes through a pyrolysis chamber at 800 °C and then through a conductivity detector. The solution is injected onto a gas chromatographic column (OV-10-Chromosorb W-HP1 80–100-mesh), 150 cm long, 0.2 cm diameter, operated at 50 °C. Hydrogen is used as carrier gas. Chloride contents determined by this method were 5.0 ± 1.8 µg l^{-1} in potable water.

2.5.3.2 Ion Chromatography

This technique has been applied to the determination of chloride and fluoride in potable waters [2.47]. The method is discussed further in Sect. 11.3.

2.5.3.3 High Performance Liquid Chromatography

Salov et al. [2.243] determined chloride, bromide, chlorate, bromate and iodate in potable waters by high performance liquid chromatography with an inductively coupled argon plasma-mass spectrometric detector.

2.5.4 Industrial Effluents

2.5.4.1 Ion Chromatography

Methods for the determination of mixed halides in industrial effluents are reviewed in Table 2.27 and discussed move fully in Sects. 2.5 and 11.4.

2.5.5 Waste Waters

2.5.5.1 Ion Chromatography

This technique has been applied to the determination of chloride and fluoride in wastewaters [2.61]. The technique is discussed in Sect. 11.4.

Table 2.27. Determination of halide mixtures in industrial effluents by ion chromatography

Type of sample	Codetermined anions	References
Industrial Effluents	Cl, F	2.105
Industrial Effluents	Cl, F, Br	2.57
Industrial Effluents	Br, I	2.118

2.6 Chlorate, Chlorite, Perchlorate and Hypochlorite

2.6.1 Natural Waters

2.6.1.1 Titration Methods

Chlorate ion concentration at the sub mg l^{-1} level has been measured with high precision and accuracy by a modified iodometric method [2.225]. The chlorate is reduced by iodide ion in 6 mol l^{-1} hydrochloric acid

$$ClO_3^- + 6I^- + 6H^+ \rightarrow 3I_2 + Cl^- + 3H_2O.$$

Oxidation of iodide ion by air is prevented by removing the air with pure hexane and scrubbed nitrogen. Saturated disodium phosphate was used to neutralise the acid. The iodometric end-point was detected either with thyodene as the indicator or by the colour of iodine in the hexane layer. The precision of the method is better than 1%. For concentrations of chlorate below 3.5 mg l^{-1}, potentiometric titration is recommended.

The measurement of chlorate ion at various concentration levels with thyodene as the indicator was carried out as follows. The sample was delivered into a 60-ml glass stoppered flask. To this solution, 0.25 ml of potassium iodide whose concentration varied from 0.3 to 0.06 mol l^{-1} depending on the chlorate ion concentration (0.1 N to 1.0×10^{-4} N) was added. This solution was bubbled with nitrogen gas for 10 min. After deaeration, approximately 1 ml of hexane was added to the sample solution. Deaerated concentrated hydrochloric acid was added to the sample solution to adjust it to 6 mol l^{-1} in acid and immediately the flask was stoppered. After the reaction was complete (normally 20 min), deaerated saturated sodium phosphate solution was added to the sample solution to reduce the acidity to 3 mol l^{-1}. Finally, the liberated iodine was titrated with standard sodium thiosulphate solution and an appropriate amount of thyodene solution was added when the yellow colour of iodine in the aqueous phase became pale. For concentrated solutions, the thyodene colour change was used to determine the end-point. In the more dilute solutions, the titration end-point was determined by comparing the colour of the hexane layer of the sample solution with that of a blank solution consisting of only water and hexane. The reddish violet colour of iodine in the hexane layer makes it easier to detect a titration end-point because the limit of visibility of the colour of the hexane layer exceeds that of the thyodene-iodine colour. In each experiment, blank corrections were made for solutions of the same compositions as that of the sample solution except for using deionized distilled water in place of the sample.

In the most sensitive potentiometric titration procedure, an appropriate volume, 0.5 ml of 0.015 mol l^{-1} potassium iodide was added to the sample. Then 12 ml of deaerated hydrochloric acid and 23 ml of the deaerated saturated sodium phosphate solution were added for acidification and neutralization of sample, respectively. Under these conditions of lower chlorate ion concentration, the reaction between chlorate ion and iodide ion requires 40 min for completion. The end-point was detected from the first derivative titration curve obtained upon titration of liberated iodine with standard sodium thiosulphate using a combination redox electrode. At these low concentration levels, the blank volumes were nominally 20 and 30% of the sample titration volumes. The potentiometric titration method using hexane as a shielding agent is recommended for the determination of chlorate ion at low concentration levels (especially below 1 mg l^{-1}) because the blank is relatively small and no subjectivity by the investigator is involved.

Gordon [2.226] had described a procedure for the determination of down to 0.1 mg l^{-1} of chlorite and hypochlorite by a modified iodometric technique.

2.6.1.2 Flow Injection Analysis

Miller et al. [2.227] have described two approaches to the determination of trace levels of chlorate ion based on the reaction of chlorate ion with iodide to produce iodine. High hydrochloric acid concentrations accelerated the reaction. Flow injection analysis, which excluded air from the flow stream, was used. The first method could detect chlorate ions at levels of 10 $\mu mol\ l^{-1}$ and the second method, using a stopped flow condition, could detect levels of 1 $\mu mol\ l^{-1}$.

2.6.1.3 Ion Chromatography

Peschet and Tinet [2.229] have described an ion chromatographic method for the determination of hypochlorite (and cyanide and sulphide) in groundwaters. This method is discussed further in Chap. 11 (Ion Chromatography).

2.6.1.4 High Performance Liquid Chromatography

Salov et al. [2.243] determined chlorate, also chloride, bromide, iodate, bromate and iodide in natural waters by high performance liquid chromatography with an inductively coupled argon plasma mass spectrometric detector.

2.6.1.5 Ion-Selective Electrodes

Jain et al. [2.228] have described a liquid membrane perchlorate selective electrode.

2.6.2 Waste Waters

2.6.2.1 Spectrometric Method

Canelli [2.230] has described an autoanalyser-based procedure for the simultaneous determination of chlorite (and nitrite and nitrate) in waste water.

2.6.3 Potable Waters

2.6.3.1 Ion Chromatography

Dietrich et al. [2.241] used flow injection analysis and ion chromatography to determine chlorite and chlorate in potable waters.

2.6.3.2 High Performance Liquid Chromatography

This technique has been applied [2.243] to the determination of chlorite, iodide, bromide, chloride, bromate and iodate in potable waters.

2.6.4 Soil Extracts

2.6.4.1 Spectrophotometric Method

Bandereis [2.231] has described a method for the determination of chlorate in water extracts of soil based on its conversion to free chlorine upon reaction with hydrochloric acid, followed by spectrophotometric evaluation of chlorine by the spectrophotometric o-toluidine method. A correction is made for interference by iron III, nitrite, free chlorine derived from hypochlorites and strong oxidising agents by subtracting the absorbance of a modified blank, containing a lower concentration of hydrochloric acid, from that obtained in the test.

2.7 Bromate

2.7.1 Natural Waters

2.7.1.1 High Performance Liquid Chromatography

Salov et al. [2.243] determined bromate, chloride, iodide, bromide, chlorate, and iodate in natural waters by high performance liquid chromatography with an inductively couple argon plasma mass spectrometric detector.

2.7.2 Rainwater

2.7.2.1 Ion Chromatography

Pyen and Erdmann [2.51] have described an ion chromatographic method for the determination of bromate (and bromide, fluoride, chloride, nitrite, phosphate, nitrate and sulphate)in rainwater. This method is discussed further in Sect. 11.2.

2.7.3 Potable Waters

2.7.3.1 High Performance Liquid Chromatography

The method described in Section 2.7.1.1 [2.243] has been applied to potable water.

2.7.4 Foodstuffs

2.7.4.1 Flow Injection Analysis

The bread making qualities of flour tend to improve with storage of up to two months but this process can be accelerated by the addition of potassium bromate improver. The UK Bread and Flour Regulations (1984) permit the use of up to 50 mg kg^{-1} of potassium bromate in flour. Osborne [2.232] has described a flow injection method for the determination of bromate in flour in which the flour is extracted with zinc sulphate solution and a portion of the resulting clean phase treated with sodium hydroxide.

In the flow injection method [2.232], bromate is reacted with potassium iodide in acetic acid solution in the presence of ammonium molybdate to accelerate the reaction:

$$KBrO_3 + 6KI + 6CH_3COOH \rightarrow KBr + 3I_2 + 6CH_3COOK + 3H_2O.$$

The intensity of the blue colour produced with starch is evaluated spectrometrically. The analytical recovery was 90% and the range of the method 0.5–5 mg l^{-1}. The standard deviation was 0.11 mg l^{-1}.

2.8 Iodate

2.8.1 Natural Waters

2.8.1.1 Spectrometric Methods

Truesdale and Smith [2.233] described an automated method for the determination of iodate and iodide in river water. The method is based on catalytic spectrophotometry. The Standing Committee of Analysts [2.234] have described an automated method for the determination of iodate, iodide, iodine and bromide in river waters based on automated catalytic spectrophotometry. The method involves both the determination of total iodine in concentrations up to 5 µg l^{-1} and the separate determination of iodide and iodate. The second method was a two-step procedure in which total inorganic iodide was determined using the first method, then iodide extracted with tetraphenyl ammonium hydroxide, after which the first determination was repeated, the iodide being obtained by difference.

Jones et al. [2.235] give details of a procedure for the determination of traces of iodate, iodide and total iodine in fresh waters, based on catalytic determination of iodide using the reaction between ammonium cerium sulphate and arsenious acid; iodate is determined by the same method after extraction of iodide-iodine into chloroform as an ion-pair with the tetraphenylarsonium cation. Iodate can be determined in amounts down to 1 µg l^{-1} with a resolution of ±0.1 µg l^{-1} (95% confidence limits).

Apparatus. Automatic analysis was performed with a Technicon Auto Analyzer I Instrument equipped with a range expander, and a water-bath 2. Extractions with chloroform were conducted in 100-ml separating funnels fitted with stoppers and taps made of PTFE. An MSE Hi-Spin 21 centrifuge with glass centrifuge tubes was used to aid separation of the chloroform and aqueous phases. After extraction of iodide an aliquot of each aqueous phase was placed in a 2-ml glass sample cup used with the Auto-Analyzer; the chloroform present in the aqueous phase dissolves the standard polystyrene cups.

The AutoAnalyser manifold system used in these analyses is shown in Fig. 2.8.

Reagents. As follows. Ammonium cerium (IV) sulphate. Add 3.75 g of ceric ammonium sulphate $(2(NH_4)_2SO_4 - Ce(SO_4)_2 \cdot 3(1/2) H_2O)$ to 200 ml of distilled water. Add 52 ml of concentrated sulphuric acid (18 mol l^{-1}) stirring to dissolve the solid and make up to 1 l with distilled water. Filter the solution through a Whatman GF/C filter before use. To prepare the cerium reagent used for blank determinations, weigh out 1.5 g of the salt and then continue as before.

Arsenious acid. Dissolve 19.6 g of arsenic (II) oxide plus 14.0 g of sodium hydroxide in a small amount of water. Dilute the solution with a further 600 ml

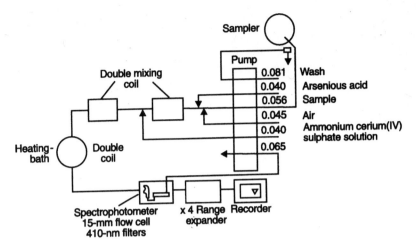

Fig. 2.8. Auto analyzer manifold. Numbers represent tube diameters in inches (From [2.235])

of distilled water. Add concentrated sulphuric acid (18 mol l^{-1}) dropwise until the solution is neutral to phenolphthalein, and add a further 56 ml of concentrated acid and 50 g of sodium chloride. Make the volume up to 1 l with distilled water and filter the solution through a Whatman GF/C filter before use.

Iodate and iodide standards. Dry sodium iodide or sodium iodate for about 1 h at 110 °C and dissolve 0.1191 g of the iodide salt or 0.1559 g of the iodate salt in 1 l of de-ionised, distilled water. Dilute these stock solutions, which contain 100 mg l^{-1} of iodine, with de-ionised, distilled water to give suitable working standards.

Tetraphenylarsonium hydroxide solution, 0.2 mol l^{-1}. Dissolve 4.55 g of solid AnalaR-grade tetraphenylarsonium chloride hydrochloride (Koch-Light Laboratories) in 50 ml of distilled water. Add 2.0 g of silver oxide and stir overnight, using an automatic stirrer. Filter through a Whatman GF/C glass-fibre filter before use and store in a dark glass bottle.

Chloroform, AnalaR grade.

Procedure I – Total Iodine. Assemble the AutoAnalyzer manifold according to the design given in Fig. 2.8. While aspirating only the arsenic reagent and deionised distilled water, set 0 and 100% transmission on the recorder. Introduce the cerium (IV) reagent and establish the base line between 31 and 35% transmission (absorbance 0.45 – 0.5) using a temperature of approximately 34 °C. When the base-line has settled within this transmission range, expand the 30–50% transmission interval to full-scale by using the X 4 range-expansion facility. Allow the base-line to settle at 5% apparent transmission and then introduce samples and standards at the rate of 50 per hour. Read off the peak height, relative to the base line, of the standards and construct a calibration graph. Remove arsenic and cerium reagents and pump out the manifold with the de-ionised distilled water and wash for a few minutes. When reducing agents are suspected to be present in the samples, introduce the cerium reagent used for blank determinations and re-establish a base line at 5% apparent transmission as described above. Aspirate samples at the rate of 50 per hour. Read off the peak heights of the sample blanks and substract these from the peak heights returned for the samples. By using the calibration graph and the blank-corrected peak heights, determine the total iodine concentration of each sample.

Procedure II–Iodate-Iodine. Pipette a 20-ml aliquot of sample into a clean glass separating funnel. Add 0.1 ml of the tetraphenylarsonium hydroxide reagent from an Eppendorf pipette. Shake, add 4.0 ml of chloroform, and shake for 2 min. Allow the two phases to separate for 10 min and then remove the organic phase. Add a further 4.0 ml of chloroform, shake for 1 min and repeat as above. Centrifuge the final aqueous phase at 4500 g for 15 min. Decant this aqueous phase into a 2.0-ml glass sample cup and aspirate, in duplicate, at the rate of 50 per hour into the AutoAnalyzer used for the total iodine method.

Prepare a calibration graph by feeding appropriate iodate-iodine standards directly to the AutoAnalyzer. Determine the reagent blank by treating a sample of de-ionised, distilled water according to the procedure above. Substract this reagent blank from the peak height returned for each sample. Where reducing agents are present, or a new water type is under investigation, determine the "catalytic method" blank as discussed above. Substract this blank from the peak height of the sample, and finally read off the iodate-iodine concentration from the calibration graph.

The addition of 100 mg l^{-1} of S(as SO_4^{2-}), HCO_3^-, P(as PO_4^{3-}), Ca^{2+}, Mg^{2+}, K^+ and Na^+, 30 mg l^{-1} of Cl^- and NO_3^-, 15 mg l^{-1} of Si(as SiO_3) and Br^- and 100 µg l^{-1} of Fe^{2+}, Fe^{3+}, Mn^{2+} and Ni^{2+} did not produce any significant inteference in the procedure for the determination of iodate. In these experiments a mixture containing 3.0 µg l^{-1} of iodide and 2.0 µg l^{-1} of iodate-iodine together with the other chemicals was analysed and in each instance analysis gave 2.0 ± 0.1 µg l^{-1} of iodate-iodine. The addition of up to 10^{-2} mol l^{-1} of either hydrogen or hydroxyl ion to the sample prior to its analysis also produced an insignificant interference.

2.8.1.2 Ion-Selective Electrodes

The determination of iodate and iodide has been carried out using an iodide-selective electrode in a solution at pH 9[2.233, 2.236]. No interference was caused by fluoride, chloride, bromide or thiosulphate. Only low concentrations of sulphide and cyanide can be tolerated.

Nakayama et al. [2.237] carried out automated determination of iodate in natural waters using a flow through electrode system. Following oxidation to iodine, quantitative concentration was obtained on a carbon wool electrode in a preconcentration cell. Residual chloride and bromide ions were then removed by rinsing the preconcentration cell with acetic acid. Iodine was eluted with reducing agent in a detection cell housing a polished silver electrode and iodate determined after being reduced to iodide. Current voltage curves obtained with the polished electrode are illustrated. The effect was studied of various salts on the peak current obtained in the analysis of filtered seawater. A detection limit of 5 ng of iodine is claimed for this method.

2.8.1.3 Micelle Exclusion Ion Chromatography

Okada [2.238] has used micelle exclusion chromatography to determine iodate in the presence of other anions (iodide, bromide, nitrite and nitrate) in water. The method is based on partition of the anions to a cationic micelle phase and shows different selectivity from ion exchange chromatography.

This technique has also been applied to the determination of iodate, nitrite, nitrate, bromide and iodide in natural waters [2.241].

2.8.1.4 High Performance Liquid Chromatography

Salov et al [2.243] determined iodate, chloride, iodide, chlorate and bromate in natural waters by high performance liquid chromatography with an inductively coupled argon plasma mass spectrometric detector.

2.8.2 Potable Waters

2.8.2.1 High Performance Liquid Chromatography

The procedure described in Sect. 2.8.1 [2.243] has also been applied to potable waters.

2.9 References

2.1 Society for Analytical Chemistry (UK), Analytical Quality Control (Harmonized Monitory) Committee (1979) Accuracy of determination of chloride in river waters: analytical quality control in the harmonized monitoring scheme. Analyst (London) 104: 190
2.2 Water Research Centre Technical Report TR 27 (1976) Accuracy of determination of chloride in river waters
2.3 Fishman MJ, Feist OJ (1970) Prof Paper U.S. Geological Survey No. 700-C, C226 – C228
2.4 Kuttel D, Szabadka O, Czakvori B, Mezaros K, Havas J, Pungor (1969) Magy Kem Foly 75: 181
2.5 Dojlido J, Bierwagen H (1969) Chemica Analit 14: 91
2.6 Schenbeck E, Ernst OZ (1970) Analyt Chem 249: 470
2.7 Rossner B, Schwedt G (1983) Fresenius Zeitschrift fur Analytische Chemie, 200: 315

2.8 Van Staden JF (1985) Fresenius Zeitschaft fur Analytische Chemie 322: 336
2.9 Basson WD, Van Staden JF (1981) Water Research 15: 333
2.10 Belcher R, Nadjafi A, Rodriguez-Vazques JA, Stephen WI (1972) Analyst (London) 97: 993
2.11 Tomkins DF, Frank CW (1972) Analytical Chemistry 44: 1451
2.12 Michlewicz KG, Carnahan W (1986) Analytica Chimica Acta 183: 275
2.13 Michlewicz KG, Carnahan W (1985) Analytical Chemistry 57: 1092
2.14 Michlewicz KG, Carnahan W (1986) Analytical Chemistry 58: 3122
2.15 Hara H, Okazaki S (1984) Analyst (London) 109: 1317
2.16 Ostrovidov EA, Bardin VV (1972) Zavod Lab 38: 1327
2.17 Andelman JB (1968) J Water Pollution Control Federation 40: 1844
2.18 Selwen-Olsen AR, Oeien A (1973) Analyst (London) 98: 412
2.19 Baszelle WE (1971) Analytica Chimica Acta 54: 29
2.20 Zeinalova E, Senyavin M (1975) Zhur Anal Khim 30: 2207
2.21 Trojanowicz M, Matuszewski W (1983) Analytica Chimica Acta 151: 77
2.22 Hara H, Wakizawa Y, Okazaki S (1985) Analyst (London) 110: 1087
2.23 Trojanowicz M, Lewardowski R (1981) Fresenius Zeuschrift für Analytische Chemie 308: 7
2.24 Belcher R, Major JR, Rodriguez-Vasquez JA, Stephen WI, Uden PC (1971) Analytica Chimica 57: 73
2.25 Stetzenbach K, Thompson GM (1983) Groundwater 21: 36
2.26 Moss PE, Stephen MI (1985) Analytical Proceedings (London) 22: 5
2.27 Akaiwa H, Kawamoto H, Osumi M (1982) Talanta 29: 689
2.28 Bocek P, Miedziak I, Demi M, Janak J (1987) J Chromatography 137: 83
2.29 Zelinski I, Kanmiansky D, Havassi P, Lednarova U (1984) Journal of Chromatography 294: 317
2.30 Van OS MJ, Slanina J, Deligny CL, Hammers WE, Agterdenbos J (1982) J Analytica Chimica Acta 144: 73
2.31 Smee BW, Hall GEM, Koop DJ (1978) Journal of Geochemical Exploration 10: 245
2.32 Wilken BD, Kock HH (1985) Fresenius Zeitschrift fur Analytische Chemie 320: 477
2.33 Montiel A, Dupont JJ (1974) Trib Cebedeau 27: 27
2.34 Jacobsen E, Tandberg G (1973) Analytica Chimica Acta 4: 280
2.35 Martheim FT, Peck EE, Lane M (1985) Society of Petroleum Journal, 25: 704
2.36 Henmann KG (1985) Fresenius Zeitschrift fur Analytische Chemie 320: 493
2.37 Johannesson JK (1970) Journal of Radioanalytical Chemistry 6: 27
2.38 Reijnders HFR, Melis PHAA, Griepuik B (1983) Fresenius Zeitschrift fur Analytische Chemie 314: 627
2.39 Slanina I, Bakker F, Bruyn-Hes A, Mols JJ (1980) Analytica Chimica Acta 113: 331
2.40 Slanina J, Mols JJ, Baard JH, Vandel Sloot HA, Van Raaphorst JG (1979) International Journal of Environmental Analytical Chemistry 7: 161
2.41 Oikawa K, Saitoh H (1982) Chemosphere 11: 933
2.42 Buchholz AE, Verplough CJ, Smith JL (1982) Journal of Oceanographic Science 20: 499
2.43 Matzuchita S, Baba S, Ikeshiga T (1984) Analytical Chemistry 56: 822
2.44 Matzuchita S, Tada Y, Baba D, Hoskano J (1983) J Chromatography 259: 459
2.45 Wetzel RA, Anderson CL, Schleichen H, Crook DG (1979) Analytical Chemistry 51: 1532
2.46 Jones UK, Tartar AG (1985) International laboratory 36 September
2.47 Schwabe R, Darimont T, Mohlman T, Pabel E, Sonneborn M (1983) International Journal of Environmental Analytical Chemistry 14: 169
2.48 Wagner F, Valenta P, Nurnberg W (1985) Fresenius Zeitschrift fur Analytische Chemie 320: 470
2.49 Rowland AP (1986) Analytical Proceedings (London) 23: 308
2.50 Oikawa K, Saitoh H (1982) Chemosphere 11: 933
2.51 Pyen GS, Erdmann DE (1983) Analytica Chimica Acta 149: 355
2.52 Henmann KG, Kifmann R, Schindmeier W, Unger M (1981) International Journal of Environmental Analytical Chemistry, 10: 39
2.53 Levy DP (1984) Analyst (London) 109: 103

2.54 Backmann K, Matusca P (1983) Fresenius Zeitschraft fur Analytische Chemie 315: 243
2.55 Department of the Environment/National Water Council Standing Committee of Analysts, H.M. Stationery Office, Methods for the cxaination of waters and associated materials. Chloride in waters, sewage and effluents (1982)
2.56 Ilyukhina PF, Polyakova KI, Kuz'mina VM, Il'icheva IA (1970) Zavod Lab 26: 1049
2.57 Moska J (1984) Analytical Chemistry 56: 629
2.58 Barrera AB, Martinez FB (1979) Tecnica Investigacion Tratamiento 1: 18
2.59 Tanaka K, Ishizuka T (1982) Water Research 16: 719
2.60 Holcombe DJ, Maserole FB (1981) Water Quality Bulletin 6: 37
2.61 Green LW, Woods JR (1981) Analytica Chemistry 53: 2187
2.62 Hindin E (1975) Water and Sewage Works: 122: 60
2.63 Goruyova NN, Chernova MA, Frumina NS (1973) Zavod Lab 39: 1041
2.64 Florence TM, Farren YJ (1971) Analytica Chimica Acta 54: 373
2.65 Rodabaugh RD, Upperman GT (1972) Analytica Chimica 60: 434
2.66 Florence TM, Farrer YJ (1971) Analytica Chimica Acta 54: 373
2.67 Rodabaugh RD, Upperman GT (1972) Analytica Chimica Acta 60: 434
2.68 Karyakin AV, Babicheva GG (1968) Zhur Analit Khim 23: 789
2.69 Tomlinson K, Torrance K (1976) Analyst (London) 101: 1
2.70 Pilipenko AT, Ol'khovich PF, Gakel RK (1973) Ukr Khim Zh 39: 376
2.71 Torrance K (1974) Analyst (London) 99: 203
2.72 Florence TM (1971) Journal of Electroanalytical Chemistry 31: 77
2.73 Marshall GB, Midgley D (1978) Analyst (London) 103: 784
2.74 Marshall GB, Midgley D (1978) Analyst (London) 103: 438
2.75 Roberts KM, Gjerde DT, Fritz JG (1981) Analytical Chemistry 53: 1691
2.76 Tretter H, Paul G, Blum I, Schreke H (1985) Fresenius Zeitschrift fur Analytische Chemie 321: 650
2.77 Davey BG, Bembrick MA (1969) Proceedings of the Soil Science Society of America 33: 385
2.78 Mc Leod S, Stace HTC, Tucker BM, Bakker P (1974) Analyst (London) 99: 193
2.79 Bradfield EG, Cooke DT (1985) Analyst (London) 110: 1409
2.80 Clarke FE (1950) Analytical Chemistry 22: 553
2.81 Chapman BR, Goldsmith IR (1982) Analyst (London) 107: 1014
2.82 Department of the Environment/National Water Council Standing Committee of Analysts, HM Stationary Office, London, Bromide in Waters, high level titrimetric method (1981)
2.83 Kuttel D, Szabadka O, Czanvari B, Mezaros B, Hezarosi K, Havas J, Pungor E (1969) Magy Kem Foly 75: 181
2.84 Babkin MP (1971) Zavod Lab 37: 525
2.85 Moldan V, Mand Zyka J (1968) Microchemical Journal 13: 357
2.86 Archimbaud M, Bertrand MR (1970) Chim Analyt 52: 531
2.87 Dobalyi HF (1984) Anal Chem 56: 2961
2.88 Schenbeck E, Ernst OZ (1970) Analyt Chem 249: 370
2.89 Palmarchenko LM, Toropova UF (1968) Zhur Analit Khim 23: 1028
2.90 Kochetkova TM, Novikova TM (1981) Soviete Journal of Water Chemistry and Technology 3: 70
2.91 Babkin MP (1971) Zavod Lab 37: 524
2.92 Fishman NJ, Skougstad MW (1963) Analytical Chemistry 35: 146
2.93 Skougstad MW, Fishman MJ, Erdmann DE, Duncan SS (1979) in 'Techniques of Water Resources Investigations of the US Geological Survey, Methods for the Determination of Inorganic Substances in Water and Fluvial Sediments', Book 5, Chapter A1, p. 329
2.94 Pyen GS, Fishman MJ, Hedley AG (1980) Analyst (London) 105: 657
2.95 Anfalt T, Twengstrom S (1986) Analytica Chimica Acta 179: 453
2.96 Michlewicz KG, Carnahan W (1986) Analytical Chemistry 58: 3122
2.97 Weiss D (1971) Chemiske Listy 65: 305
2.98 Nota G, Vernassi G, Acampora A, Sonnolo N (1979) Journal of Chromatography 173: 228
2.99 Anudo M, Sayato Y (1983) Water Researach 17: 1823

2.100 Stetzenbach K, Thompson GM (1983) Groundwater 21: 36
2.101 Moss PE, Stephen MI (1985) Analytical Proceedings (London) 22: 5
2.102 Akaiwa H, Kawamoto H, Osumi M (1982) Talanta 29: 689
2.103 Okada T (1988) Analytical Chemistry 60: 1511
2.104 Van Os MJ, Slanina J, De Ligny CL, Hammers WE, Agterdenbos J (1982) J. Analytica Chimica Acta 144: 73
2.105 Smee BW, Hall GEM, Koop DJ (1978) Journal of Geochemical Exploration 10: 245
2.106 Dils JS, Sweenk GMM (1985) H_2O 18: 7
2.107 Masschelein WJ, Denis M (1931) Water Research 15: 857
2.108 Slanina J, Mols JJ, Baard JH, Van der Sloot HA, Van Raaphorst JG (1979) International Journal of Environmental Analytical Chemistry 7: 161
2.109 Radcliffe D (1970) Analytical Letters 3: 573
2.110 Deutsch Y (1974) Anal Chem 46: 437
2.111 Fishman MJ, Schroeder LJ, Schroeder LC, Friedman CE, Arozarena CE, Hedley AG (1985) Water Research 19: 497
2.112 Henmann KG, Kifmann R, Schindmeier W, Unger M (1981) International Journal of Environmental Analytical Chemistry 10: 39
2.113 Henmann KG (1985) Fresenius Zeitschrift fur Analytische Chemie 320: 493
2.114 Moxon RED, Dixon EJ (1980) Journal of Automatic Chemistry 3: 139
2.115 Backmann K, Matusca P (1983) Fresenius Zeitschrift fur Analytische Chemie 315: 243
2.116 Grandet M, Weil L, Quentin KE (1983) Zeitschrift fur wasser and Abwasser 16: 66
2.117 Morrow CM, Minear RA (1984) Water Research 18: 1165
2.118 Rocklin RD, Johnson EL (1983) Analytical Chemistry 55: 4
2.119 Van Staden JF (1987) Analyst (London) 112: 595
2.120 Roughan JA, Roughan PA, Wilkins JPG (1983) Analyst (London) 108: 742
2.121 Panel on Fumigant Residues in Grain Committee for Analytical Methods for Residues of Pesticides and Vetinary Products in Foodstuffs of the Ministry of Agriculture, Fisheries and Food UK (1976) Analyst (London) 101: 386
2.122 Kuttel D, Szubadka O, Czakvori B, Mezaroz B, Hezaroz K, Havas J, Pungor E (1969) Magy Kem Foly 75: 181
2.123 Knapp G, Spitzu H (1969) Talanta 16: 1353
2.124 Bognor J (1970) Mikrochim Acta 1: 112
2.125 Tamarchenko LM (1970) Zhur Analit Khim 25: 567
2.126 Neverdanskiene Z, Romanuskas E, Bunikieve I (1972) Nauch Trudy Uyssh Ucheb Zavod Lit SSR Khim Khim Tekhnol 14: 25 Ref: Zhur. Khim., 19GD (9) Abstract No. 9G87 (1973)
2.127 Novak J, Slhama I (1972) Collin Czeck Chem Commun 37: 2907
2.128 Keller HE, Doenecke K, Weidler K, Leppla W (1973) Am New York Academy of Science 210: 1
2.129 Ganchev N, Kireva A (1972) Mikrochim Acta 6: 889
2.130 Weil L, Torkzadeh N, Quentin KE (1975) Zeitschrift fur Wasser and Abwasser, Forschung 8: 3
2.131 Miyazaki A, Brancho K (1987) Spectrochimica Acta 423: 227
2.132 Michlewicz KG, Carnahan W (1986) Analytical Chemistry 58: 3122
2.133 Butler ECV, Gershey RM (1984) Analytica Chimica Acta 164: 1523
2.134 Woodson JH, Lieb haltsky HA (1969) Analytical Chemistry 41: 1894
2.135 Nakayama E, Kimoto T (1985) Analytical Chemistry 57: 1157
2.136 Weiss D (1972) Chemicke Listy 66: 858
2.137 Zernalova E, Senyavin M (1975) Zhur Anal Chem 30: 2207
2.138 Wu HL, Liu SJ, Funazo K, Tanaka M, Shono T (1985) Fresenius Zeitschrift fur Analytische Chemie 322: 409
2.139 Moss PE, Stephen MI (1985) Analytical Procedings (London) 22: 5
2.140 Stetzenback K, Thompson GM (1983) Groundwater 21: 36
2.141 Butler ECV, Bershey R (1984) Analytica Chemica Acta 164: 153
2.142 Okada T (1988) Analytical Chemistry 60: 1511
2.143 Zentner H (1973) Chemy Ind 10: 480

2.145 Bardin VV, Tolstounsov UN (1970) I3v Yvsshucheb, Zavod Khim Technol 13, 165 (1970). Ref: Zhur Khim 19GD (15) Abstract No. 15G158
2.146 Malvano R, Buzziogoli G, Scarlattini M, Conderellim G, Gandalfi C, Grosso P (1972) Analytica Chimica Acta 61: 201
2.147 Palagni S (1983) International Journal of Applied Radiation and Isotopes 34: 755
2.148 Henmann KG (1985) Fresenius Zeitschrift fur Analytische Chemie 320: 493
2.149 Funghashi S, Tabata M, Tanaka M (1971) Analytica Chimica Acta 57: 311
2.150 Ghimicescu C, Stan M, Dragomir B (1973) Talanta 20: 246
2.151 Moxon RE (1984) Analyst (London) 109: 425
2.152 Backmann K, Matussa P (1983) Fresenius Zeitschrift fur Analytische Chemie 315: 243
2.153 Grandet M, Weil L, Quentin KE (1983) Fresenius Zeitschrift fur wasser and Abwasser Forschung 16: 66
2.154 Erdey A, Kozmutza K (1971) Acta Chim Hung 69: 9
2.155 Moxon RED, Dixon EJ (1980) Analyst (London) 105: 344
2.156 Gutierrez ML, Gomez-Hens A, Perez-Bendito D (1989) Analyst (London) 114: 89
2.157 Chakabarty D, Das AK (1988) Atomic Spectroscopy 9: 189
2.158 Daneshwar RG, Kulkarni AV, Zarapkar LR (1987) Analyst (London) 112: 1073
2.159 Moraru L, Suten A (1971) Studii Ceorc Chim 19: 757
2.160 Maejunas AG (1969) Journal of the American Water Works Association 61: 311
2.161 Quentin KE, Rosopulo AZ (1968) Analyt Chem 241: 241
2.162 Dixon EJ (1970) Analyst (London) 95: 272
2.163 Haarsama JPS, Agterdenbos J (1971) Talanta 18: 747
2.164 Gitsova S (1970) Khig Zdraevopazvane 13: 198 Ref: Zhur Khim 19GD (21) Abstract No. 21G155
2.165 Kempf T (1969) Z Analit Chem 244: 133
2.166 Capilan-Vallvey L F, Valencia M C, Bosque-Sendra M C (1988) Analyst (London) 113: 419
2.167 Miyazaki A, Brancho K (1987) Analytica Chimica Acta 198: 297
2.168 Shiraishi N, Murata Y, Nagagawa G, Kodama K (1973) Analytical Letters 6: 893
2.169 Zernalova E, Senyavin M (1975) Zhur Anal Chem 30: 2207
2.170 Warner TB (1971) Water Research 5: 459
2.171 Sekerka I, Lechner JF (1973) Talanta 20: 1167
2.172 Erdmann DE (1975) Environmental Science and Technology 9: 252
2.173 Pakalns P, Farrar YJ (1976) Analytica Chimica Acta 10: 1087
2.174 Villa AE (1988) Analyst (London) 113: 1299
2.175 Trojanowicz M, Lewandowski R (1981) Fresenius Zeitschrift fur Analytische Chemie 308: 7
2.176 Campanella L, Sbrilli R (1981) Metodi Analitici per le Acque, 1: No. 1
2.177 Okabayashi Y, Oh R, Nakagawa T, Tanaka H, Chicuma M (1988) Analyst (London) 113: 829
2.178 Zelinski I, Zelinska V, Kanmieisky D, Havassi P, Ledrarova U (1984) Journal of Chromatography 294: 317
2.179 Slanina J, Mols JJ, Baard JH, Van der Sloot HA, Van Raphorst JG (1979) International Journal of Environmental Analytical Chemistry 7: 161
2.180 Warner TB, Bressan DJ (1973) Analytica Chimica Acta 63: 165
2.181 Pyen G S, Erdmann D E (1983) Analytica Chimica Acta 149: 355
2.182 Wierzbicki T, Pawlita W (1969) Chemia Analit 14: 1193
2.183 Wierzbicki T, Pawlita W, Pietrzyk H (1971) Chemica Analit 16: 1079
2.184 Mehra MC, Lambert TL (1973) Microchemical Journal 18: 226
2.185 Anosova M (1970) Trudy Nizhni-Volzk. Nauchno-issled Inst Geol Geofiz 8: 297 Ref: Zhur Khim 199D (21) Abstract No. 21G161 (1970)
2.186 Light TS, Mannion RF, Fletcher KS (1969) Talanta 16: 1441
2.187 Callombel C, Bureau J, Cotte J (1971) Anal Pharm France 29: 541
2.188 Nicolson K, Duff EJ (1981) Analytical Letters 14: 493, 887
2.189 Kissa E (1983) Anal Chem 55: 1445
2.190 Instruction Manual – Fluoride Electrodes Model 94 – 09. Model 96 – 09 Instruction Manual, Model 901, Microprocessor Ionalyser Orion Research Inc, Cambridge MA (1978)

2.191 Phillips KA, Rix C (1981) Analytical Chemistry 53: 2141
2.192 Ballczo H (1979) Fresenius Zeitschrift fur Analytische Chemie 298: 382
2.193 Crosby NT, Dennis AL, Stevens JG (1968) Analyst (London) 93: 643
2.194 Hannah BE (1986) Journal of Chromatographic Science 24: 336
2.195 Hey AE, Jenkins SH (1969) Water Research 3: 901
2.196 Belcher R, West TS (1962) Analytical Abstracts 9: 2750
2.197 Greenhalgh F, Riley JP (1962) Analytical Abstracts 9: 1247
2.198 Devine RF, Partington GL (1975) Environmental Science and Technology 9: 678
2.199 Rea RE (1979) Water Pollution Control 78: 139
2.200 Venkateswarlin D (1974) Analytical Chemistry 46: 878
2.201 Vickery B, Vickery ML (1976) Analyst (London) 101: 445
2.202 Villa AE (1976) Analyst (London) 101: 445
2.203 Beddows CG, Kirk D (1981) Analyst (London) 106: 1341
2.204 Conrad VB, Brownless WD (1988) Analytical Chemistry 60: 365
2.205 Trojanowicz M, Lewardowski R (1981) Fresenius Zeitschrift fur Analytische Chemie 308: 7
2.206 Hulanicki A, Trojanowicz M, Sztander J (1979) Chemi Anal (Warwaw) 24: 617
2.207 Hulanikci A, Trojanowicz M Polish Patent 113239
2.208 Hulanicki A, Maj-Zurawska M, Lewandowski R (1978) Analytica Chimica Acta 98: 151
2.209 Sekerea I, Lechner JF (1975) Analytical Letters 8: 769
2.210 Slanina J, Mols JJ, Baard JH Van der Sloot HA, Van Raaphorst JG (1979) International journal of Environmental Analytical Chemistry 7: 161
2.211 Van Raaphorst JG, Slanina J, Borger D, Lingerak WA (1977) "National Bureau of Standards special publication 464" Proceedings of the 8th IMR Symposium
2.212 Slanina J, Bakker F, Bruyn-Hes AGM, Mols JJ (1978) Fresenius Zeitschrift fur Analytische Chemie 289: 38
2.213 Stetzenbach KK, Thompson GM (1983) Ground Water 21: 36
2.214 Moss PE, Stephen WI (1985) Analytical Proceedings 22: 5
2.215 Akaiwa H, Kawamoto H, Osumi M (1982) Talanta 29: 689
2.216 Amphlett CB, McDonald LA, Redman MJ (1985) J Inorg Nucl Chem 6: 236
2.217 Tustanowski J (1967) Chromatg 31: 268
2.218 Akaiwa H, Kawamoto H, Hasegawa K (1979) Talanta 26: 1027
2.219 Akaiwa H, Kawamoto H, Osumi M (1982) Talanta 29: 689
2.220 Zelensky I, Zelenska V, Kanmiensky D, Havassi P, Lednarova V (1984) Journal of Chromatography 294: 317
2.221 Okada T (1988) Analytical Chemistry 60: 1511
2.222 Jones UK, Tartar JG (1985) International Laboratory 36 September
2.223 Grandet M, Weil L, Quentin KE (1983) Zeitschrift fur Wasser und Abwasser Forschung 16: 66
2.224 Bachmann K, Matusca P (1983) Fresenius Zeitschrift fur Analytische Chemie 315: 243
2.225 Ikeda Y, Tang T, Gordon G (1984) Analytical Chemistry 86: 71
2.226 Gordon G (1984) Journal of the American Water Works Association 76: 98
2.227 Miller KG, Pacey GE, Gordon G (1985) Analytical Chemistry 57: 734
2.228 Jain AY, Jahan M, Tyagi V (1987) Analyst (London) 112: 1355
2.229 Peschet JL, Tinet C (1986) Techniques Sciences Methods 81: 351
2.230 Canelli E (1976) Water Air and Soil Pollution 5: 339
2.231 Banderis J (1965) J Science of Food and Agriculture 16: 558
2.232 Osborne BG (1987) Analyst (London) 112: 137
2.233 Truesdale VW, Smith PJ (1975) Analyst (London) 100: 111
2.234 Standing Committee of Analysts, Society for Analytical Chemistry London, HM Stationary Office, London. Methods for examination of water and associated materials 15 pp (p 22B ENV). The determination of iodine, iodate, iodide and traces of bromide in waters (1984, 1985)
2.235 Jones SD, Spencer CP, Truesdaee VW (1982) Analyst (London) 107: 1417
2.236 Paletta B (1969) Mikrochim Acta 6: 1210
2.237 Nakayama E, Kimuto T, Okazaki S (1985) Analytical Chemistry 57: 1157

2.238　Okada T (1988) Analytical Chemistry 60: 1511
2.239　Verma K, Jain A, Vernia A (1992) Analytical Chemistry 64: 1484
2.240　Manzoori JL, Miyazaki (1990) Analytical Chemistry 62: 2457
2.241　Dietrich AM, Ledder JD, Gallagher DL, Grabeel MN, Hoehn RC (1992) Analytical chemistry 64: 496
2.242　Luther GW, Swartz CB, Ullman CJ (1988) Analytical Chemistry 60: 1721
2.243　Salov VV, Yoshinaga J, Shibata Y, Moritana M (1992) Analytical Chemistry 64: 2425
2.244　Okada T (1988) Analytical Chemistry 60: 1511
2.245　Mack RS, Grimsrud EP (1988) Analytical Chemistry 60: 1684
2.246　Yuan R, Chai Ya Quin, Liu D, Gao D, Lij Z, Ya RQ (1993) Analytical Chemistry 65: 2572

3 Nitrogen-Containing Anions

3.1 Nitrate

3.1.1 Natural Waters

3.1.1.1 Spectrophotometric Methods

3.1.1.1.1 Direct Spectrophotometric Methods

Apart from cadmium reduction of nitrate to nitrite, other direct spectrophotometric methods for the direct determination of nitrate have been described and are reviewed in Table 3.1. None of these methods, however, has gained the popularity enjoyed by reduction procedures especially those employing cadmium.

3.1.1.1.2 Reduction Procedures

Reduction to Ammonia. Evans and Stevens [3.16] employed reduction of nitrate to ammonia with Devadas alloy followed by Nesslerization to estimate the ammonia produced.

Waughman [3.17] has described a procedure for determining nitrate in agricultural and environmental survey waters. Nitrate in the sample solution is reduced to ammonia by titanous sulphate and the ammonia is then released from the solution and diffused and absorbed onto a nylon square impregnated with dilute sulphuric acid. The nylon is then put into a solution which colours quantitatively when ammonia is present and a spectrophotometer is used to measure the colour.

Reduction to Nitrite. Reduction of nitrate to nitrite by cadmium or copperized cadmium followed by estimation of the nitrite produced by production of a diazo compound and spectrophotometric estimation is the most commonly used method for determining nitrate. It has the advantage of sensitivity, freedom from interference and automated analysis.

Table 3.1. Non cadmium reduction (i.e. direct) spectroscopic methods for the determination of nitrate in natural waters

Chromogenic reagent	D_{max} nm	Solvent extraction pre-concentration	Detection limit	Reference
Crystal violet	–	Chlorobenzene	< 60 μg l^{-1}	3.1
p-Fluorophenol	–	Toluene	–	3.2
Tetraphenyl-phosphonium chloride	269	Chloroform	< 20 mg l^{-1}	3.3
Salicylic acid	–	None	–	3.4
2, 4-Xylenol	–	None	–	3.5
9, 9'bianthracene-10, 10'dione	500	None	1 μg l^{-1}	3.6
Phenoldisulphonic acid	–	None	10 μg l^{-1}	3.7, 3.8
O-Toluidine	–	None	10 μg l^{-1}	3.9
Indigo carmine	–	None	6 mg l^{-1}	3.10, 3.11
Brucine	410	–	–	3.12
Chromotropic acid	420	–	–	3.13
3, 3'Dimethyl naphthidine	–	–	50 μg l^{-1}	3.14
Dihydroxy coumarin	–	Ethyl acetate	10 μg l^{-1}	3.15

Davison and Woof [3.18] compared different forms of cadmium as a reducing agent for the batch determination of nitrate. These workers [3.18, 3.19] have also carried out interference studies on the batch determination of nitrate in natural waters by the cadmium reduction procedure. They compared spongy cadmium with cadmium filings and found that reduction with spongy cadmium is less prone to interference than with cadmium filings. The optimum reduction time or the amount of cadmium filings for mixing reduction efficiency for natural samples or solutions containing interfering substances was obtained for standard solutions. Possible interfering ions in freshwaters were shown to be phosphate, silicate, hydrogen carbonate, sulphide and some organic compounds.

Earlier work by Davison and Woof [3.18], had shown that good recoveries on some natural freshwater samples could only be achieved with longer reduction times than had been found to be optimum for standard solutions. Many workers [3.19–3.23] have investigated the time dependence of the reduction step and observed that, for the more vigorously reducing forms of cadmium, the efficiency of the reduction attains a maximum and then declines. With less active cadmium preparations, the reduction efficiency slowly approaches a plateau. However, most of these experiments were performed on standard solutions. Conditions have been optimised for studying the time dependence of the reduction efficiency, as reflected by different column lengths at a given flow-rate [3.24]. As neither standards nor different water samples were studied in the same way, any change in reduction efficiency attributable to the sample water could not be assessed.

The mechanism of reduction must involve the oxidation of cadmium metal as given in the redox reactions 3.1 and 3.2:

$$NO_3^- + 2H^+ + 2e^- = NO_2^- + H_2O \qquad (3.1)$$

$$Cd = Cd^{2+} + 2e^-. \qquad (3.2)$$

Anything that can change the ultimate rate of electron transfer or the redox potential of reaction (3.2) can possibly interfere with the analytical method. Other metal ions and ligands can change the redox potential of reaction (3.2) and hence possibly decrease or increase the electron availability, which could show up as a positive or negative interference in the method. Alternatively, inorganic or organic complexing agents can interfere by associating with cadmium ions formed at the metal surface, so providing a block to the most active reducing sites.

Two types of cadmium were used in this study, filings and a sponge prepared by electrolytic deposition onto zinc rods [3.18].

Figure 3.1a shows a graph of absorbance vs time for the standards of 0.5 mg l^{-1} of nitrate-nitrogen that were measured, at different reduction times, in the same batches as some lake water samples. It was possible to calculate the confidence interval of each point because the measurements were performed in triplicate. However, owing to the confidence interval being fairly constant, and for purposes of clarity, only average values are shown in the figure. The cadmium filings showed a consistent reduction efficiency after 10 min for both masses of cadmium used. The spongy cadmium, however, attained a maximum reduction efficiency between 10 and 30 min which then declined. The same basic shape for the spongy cadmium results is shown in Fig. 3.1b, the graph of absorbance vs time for a lake water sample, and reference to Fig. 3.1c shows that the recoveries for spongy cadmium are near theoretical for reduction times in excess of 20 min. Figure 3.1b also shows that the reduction efficiency attained by

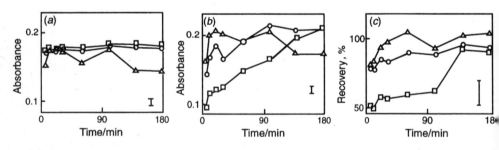

Fig. 3.1 a–c. Dependence on reduction time: **a** of absorbance of standards; **b** of absorbance of sample; **c** of recovery of spiked sample.
△ 0.5–0.6 g of spongy cadmium; ● 0.8–0.9 g of cadmium filings; ■ 2 g of cadmium filings. The error bar show representative confidence limits. The lake water sample had the following analysis (the units being mg l^{-1}). Calcium 7.8; magnesium 1.1; sodium 4.9; potassium 0.4; lead, copper, zinc and cadmium less than 0.005; hydrogen carbonate 13.4; chloride 7.3; sulphate 9.4; nitrate-nitrogen 0.53; ammonia nitrogen 0.008; phosphate phosphorus 0.014; total phosphorus 0.020; silicon 0.93. This analysis suggests that it is a soft water with no obvious signs of pollution (From [3.18])

cadmium filings in a lake water sample is greatly reduced for short reduction times, although this effect is must less pronounced when a large amount of cadmium is used for each reduction. Because of the consistent efficiencies that were obtained for the standards (Fig. 3.1a), the recoveries (Fig. 3.1c) have the same shape as the sample absorbances (Fig. 3.1b).

These results suggest the following. First, results obtained when using spongy cadmium are probably less prone to interference although they are less precise. Second, interferences in lake water slow down the reduction by cadmium filings. Third, the interferences can be reduced by either increasing the time of reduction or increasing the amount of cadmium filings used.

An increase in the reduction time or in the amount of cadmium will produce an increase in the effective reduction capacity to which the solution is exposed. That is to say, interferences appear to decrease the surface activity, an effect that can be overcome by either prolonging the exposure time or increasing the overall surface area of metal. Davison and Woof [3.19] studied only the effect of time.

For interference studies it was appropriate to use a reduction time, and an amount of cadmium, that had been shown to be prone to interference from lake water. The results of the interference studies is shown in Table 3.2. The common cations and anions do not interfere, except hydrogen carbonate, and its effect is negligible below 50 mg l^{-1}. Of the metal ions, only lead and copper interfered and then only at such high levels as would not be found in unpolluted waters. When 2 mg l^{-1} of copper were added the results tended to be erratic. This is probably because of an inconsistent induced change in the redox potential of reaction (3.2). Both silicate and phosphate ions, in their various forms, interfered. In order to test whether the interference was due to the complexation of cadmium ions, ethylenediamine-tetraacetic acid (EDTA) was added to complex the cadmium ions preferentially. This addition had no effect and the EDTA itself did not interfere. As expected, the strong oxidising agent chromium (VI) oxide did interfere in the redox process. Sulphite ions have reducing properties, but the interference observed was probably a result of the formation of cadmium sulphide which then blocked the active sites. Of the organic compounds tested, only cysteine interfered.

In order to check whether the interfering compounds were affecting the reduction step or the colour-development stage, phosphate, silicate, hydrogen carbonate, sulphide and cysteine were added to nitrite determinations with the reduction step omitted; only sulphide interfered. However, it is unlikely that it interferes at the same stage in the nitrate determination because all sulphide will have probably been previously removed as cadmium sulphide.

Figure 3.2a shows how the absorbance, in the presence of known interferences, varies with the time of reduction. The effect of all interfering substances is reduced on increasing the shaking time and near-theoretical yields are obtained in the presence of high concentrations of hydrogen carbonate and cysteine after 50 and 140 min, respectively. Longer times would be necessary to overcome the problems created by phosphate, silicate and sulphide.

Table 3.2. Effect of other substances on the determination of nitrate-nitrogen. (From [3.18]; Copyright 1978, Royal Society of Chemistry, UK)

Other substances	Concentration of other substance[a] mg l^{-1}	Added as	Error[b] in determination of nitrate-nitrogen (mg l^{-1})		
			0.000 mg l^{-1} present	0.140 mg l^{-1} present	2.100 mg l^{-1} present
Ca^{2+}......	100	$CaCl_2$			
Mg^{2+}......	50	$MgCl_2$			
Na^+......	25	NaCl	0.000	0.000	−0.016
K^+......	25	KCl			
HCO_3^-	250	$NaHCO_3$	0.000	−0.014	0.016−0.216
SO_4^{2-}......	50	Na_2SO_4			
HCO_3^-......	125	$NaHCO_3$	0.000	−	−0.233
HCO_3^-......	50	$NaHCO_3$	0.000	−	−0.077
Zn^{2+}......	2	$ZnSO_4$	0.000	0.000	−0.011
Cu^{2+}......	2	$CuSO_4$	0.000	0.000	−0.186
Cu^{2+}......	0.1	$CuSO_4$	−0.005	−0.003	+0.008
Fe^{3+}......	10	$NH_4Fe(SO_4)_2 \cdot 12H_2O$	+0.016	+0.025	+0.030
Mn^{2+}......	10	$MnSO_4$	0.000	+0.003	+0.019
Ni^{2+}......	2	$NiSO_4$	0.000	+0.003	+0.066
Co^{2+}......	2	$CoSO_4$			
Al^{3+}......	5	$AlCl_3$	0.000	0.000	−0.088
Al^{3+}......	2.5	$AlCl_3$	−	−	−0.033
Pb^{2+}......	2	Pbacetate	−0.003	−0.058	−0.786
Pb^{2+}......	0.1	Pbacetate	−0.008	−0.005	−0.041
Si......	19 m	Na_2SiF_6	−0.003	−0.055	−0.773
Si......	9	Na_2SiF_6	0.000	−0.030	−0.419
Si......	23	$Na_4Si(OH)_4$	0.000	−0.090	−1.236
Si......	2.3	$Na_4Si(OH)_4$	−0.003	−0.011	−0.132
Orthophosphate (as P)	40	KH_2PO_4	0.000	−0.093	−1.748
Orthophosphate (as P)	2	KH_2PO_4	+0.003	−0.093	−1.354
Pyrophosphate (as P)	2	$Na_4P_2O_7$	−0.000	−0.110	−1.696
Hexametaphosphate (as P)	2	$(NaPO_3)_6$	−0.003	−0.110	−1.649
EDTA......	5	Disodium salt	0.000	−0.005	+0.011
EDTA......	5	Disodium salt			
Orthophosphate (as P)	2	KH_2PO_4	0.000	−0.099	−1.422
Cr(VI)......	2	CrO_3	+0.003	−0.112	−1.850
S^{2-}......	1	Na^+	−0.00/5	−0.088	−0.482
S^{2-}......	0.5	Na^+	−0.005	−0.041	−0.167
Alanine......	1	Amino acid	+0.003	+0.0008	+0.019
Cysteine......	1	Cysteine hydrocholoride	−0.003	−0.055	−0.830
Acetate......	2	Na^+	0.000	−	+0.011

[a] The braces denote the simultaneous presence of the indicated substances in the test solutions.
[b] If the other substances had no effect, the results would be expected (95% confidence) to lie within the ranges 0.000 ± 0.005, 0.140 ± 0.007 and 2.100 ± 0.068 mg l^{-1} respectively.

The same time-dependent interference studies were conducted using spongy cadmium and the results are shown in Fig. 3.2b. The absorbance increased with shaking time up to 20 min, and after that it decreased slightly with time. Phosphate, hydrogen carbonate and cysteine did not show any appreciable interferences after this time. Although sulphide and silicate did interfere, their concentrations were much higher than those usually encountered in natural freshwater samples.

The analytical implications of this work in the determination of nitrite are as follows. Although cadmium filings are more easily and reproducibly prepared than spongy cadmium, and have been shown to give more precise results [3.18]

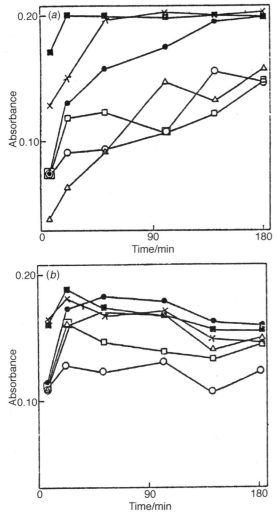

Fig. 3.2a,b. Plot of absorbance vs reduction time: a Using 0.5–0.6 g of cadmium filings; b Using 0.5–0.7 g of spongy cadmium for a standard solution of 0.56 mg l^{-1} of nitrate-nitrogen with the following additions: nil; △ 2 ml l^{-1} of phosphate phosphorus; □ 25 ml l^{-1} of silicate silicon; × 250 mg l^{-1} of hydrogen carbonate; ○ 1 mg l^{-1} of sulphide; ● 1 mg l^{-1} of cysteine hydrochloride (From [3.18])

for the batch procedure, reduction by spongy cadmium should, perhaps, be preferred because it is much less prone to interferences. Table 3.3 gives basic performance characteristics determined for this method by using 0.5–0.6 g of spongy cadmium and a shaking time of 20 min. Values for S_t showed that the precision of the method ranged from a relative standard deviation $S_r = 7.8\%$ at 0.14 mg l^{-1} to $S_r = 2\%$ at 2.1 mg l^{-1} of nitrate-nitrogen. S_W can be used to calculate the criterion of detection [3.25] using $ts_W \sqrt{2}$ where t is Student's t. A value of 0.006 g l^{-1} is obtained for the 95% confidence level.

Reduction conditions must always be optimised for natural samples rather than standards, and, ideally, for each individual water body that is measured. The results suggest a realistic quality control procedure, which can be used to assess the accuracy of measurements on a variety of water samples; two measurements could be made, each at different effective reduction capacities. This measurement technique could be very easily arranged in an automated system by having two channels with two different reduction column lengths. Disparity between the duplicate results would indicate the possible presence of an interfering species.

Koupparis et al. [3.26] have described an automated method for the determination of nitrate in river waters with a microcomputer-based stopped-flow mixing system. Nitrate is reduced to nitrite with a copperized cadmium-silver alloy or cadmium tube column fitted to the stopped-flow system, Nitrite is determined using fast kinetic, multi-point or single-point procedures with N-(1-naphthyl)ethylenediamine dihydrochloride as the colour reagent. Water samples in the range 0.025–3 mg l^{-1} of NO_3^--N can be processed with a throughput of up to 100 samples per hour and a detection limit of 0.013 mg l^{-1}. Interference studies showed that cyanide, dichromate, iodide, sulfide, copper and tin ions cause negative results.

Reagents. Standard stock solution, 1000 mg l^{-1} nitrate (6.07 g $NaNO_3$ l^{-1}) and 1000 mg l^{-1} nitrite (4.93 g $NaNO_2$ l^{-1}). Treat these with a few drops of chloroform and keep in a refrigerator. Add a pellet of sodium hydroxide to the nitrite stock solution to prevent liberation of nitrous acid. Prepare working standard

Table 3.3. Precision results for nitrate determination using spongy cadmium. (From [3.18]; Copyright 1978, Royal Society of Chemistry, UK)

Solution	Mean absorbance	Standard deviation[a] mg l^{-1}		
		S_W^b	S_b^c	S_t^d
0.000 mg l^{-1} nitrate-nitrogen	0.005	0.0024 (6)	–	–
0.140 mg l^{-1} nitrate-nitrogen	0.047	0.0042 (6)	0.0072 (5)	0.0110
2.100 mg l^{-1} nitrate-nitrogen	0.726	0.0279 (6)	NS[e] (5)	0.0416

[a] The degrees of freedom for the standard deviations are given in parentheses.
[b] Within-batch.
[c] Between-batch.
[d] Total standard deviation.
[e] Not statistically significant.

solutions in the range 0.025 – 3 mg l^{-1} nitrate and 0.025 – 2 mg l^{-1} nitrite daily by appropriate dilution.

Sulphanilamide. Dissolve 10 g of sulphanilamide 0.50 g of n-(1-naphthyl) ethylenediamine dihydrochloride and 30 ml of 85% phosphoric acid in water and dilute to 1 l. Store this solution in an amber bottle in a refrigerator.

Reduction buffer. 15 g of sodium tetraborate decahydrate, 80.2 g of ammonium chloride and 0.8 g of Na_2 EDTA per litre of deionized water. This buffer has a pH of around 8. Acetate, phosphate and tris buffers (0.025 mol l^{-1}) containing 1.50 mol l^{-1} ammonium chloride and 0.0024 mol l^{-1} EDTA in the pH range 4–9.

Apparatus. The automated microcomputer-based stopped-flow system (S.F.S.) has been described by Malmstadt et al. [3.27]. This system, interfaced with an AIM 65 (Rockwell International) microcomputer, provides for automatic aliquoting and mixing of sample and reagent, and delivery of the mixed solution into the measurement cuvette (1 cm pathlength). Each syringe delivers 150 μl of sample or reagent. The reduction column is fitted to the inlet port of the sample three-way valve using a 0.25 in – 28 thread tube end fitting. An interference filter with a bandpass of about 10 nm at 540 nm is used in the photometer unit.

Reduction Columns. The cadmium alloy reductor was made from teflon tubing (15 cm long, 2.0 mm i.d., 12 GA: Alpha Wire Corp.). A 1.6 mm o.d. wire of the same length made of 95% cadmium – 5% silver (Kapp Alloy and Wire Inc. Franklin, PA), was inserted into the tubing. The cadmium tube column was made from a cadmium metal rod (13 cm long, 1 cm o.d.) drilled to provide a 0.8 mm i.d. cylinder. Tube end fittings of 0.25 in – 28 thread were fitted to both ends for easy connection with the three-way sample valve and a short teflon tube for aliquoting. The reductors were copperized by using a syringe to force through the column 10 ml each of the following solutions: (1) 1 mol l^{-1} hydrochloric acid, (2) deionized water, (3) 2% (w/v) with copper sulfate solution, and (4) deionized water in that sequence at about 20 ml min^{-1}. Slower and prolonged flushing with copper solution causes the formation of red copper particles that may block the flow system of the analyzer.

Procedure. Determination of nitrate – premix 2 ml of nitrate standards of water samples with 0.5 ml of the reduction buffer solution in 5-ml disposable polystyrene micro beakers by means of a dispenser-dilutor (Labindustries). Use one channel of the stopped-flow system to aliquot the reagent solution and the other channel (the reductor) for the standards and samples on the turntable. Load the appropriate basic and machine language programs from the cassette recorder into computer memory. For the multipoint reaction rate procedure (concentration range 0.2 – 3 mg l^{-1} nitrate – N) use a 0.8 s delay time, 1.5 s measurement time and total reduction time of 15, 20 or 25 s, depending on the relevant nitrate concentration range. For the single point kinetic procedure (concentration range 0.025 – 3.0 mg l^{-1} nitrate – N) use a delay time of 10, 15 or 20 and a measurement time of 0.7 s (one integration). In the execution of the program, a 150 µl aliquot of each standard is held in the reductor column during the reduction time selected and is then mixed with an equal volume of colour reagent and transferred to the observation cell. The reaction rates or the absorbance values are printed, and, by utilizing the concentrations of standards provided by the operator, the calibration curve is constructed. Each sample is then measured and its concentration is printed out. This value should be automatically corrected for any predilution of the sample.

The nitrite determination in the range 0.025 – 2 mg l^{-1} is done first without the use of the reduction column. Then the reductor is fitted to the system and the samples, including three nitrite standards, are processed as described above. In the execution of the program, two calibration curves are constructed and the operator provides the nitrite concentration found for each sample. The computer calculates the nitrate concentration using the equation

$$\text{Rate of absorbance} = a + S_{NO_2} \times \text{mg l}^- \text{NO}_2 + S_{NO_3} \times \text{mg l}^{-1} \text{NO}_3$$

(rate for multipoint kinetic and absorbance for single point procedure) where a is the average of the intercepts of the two calibration curves and S the slope of the respective calibration curve.

Using this method, Koupparis et al. [3.26] carried out a recovery study on river and lake water samples by measuring the nitrate concentration before and after standard addition. Samples were also assayed for nitrite and the equation for the determination of nitrite and nitrate in the mixture was used. The nitrite content was found to be equivalent to less then 0.2 mg l^{-1}. Samples higher than 3 mg l^{-1} were diluted with distilled water. The results obtained (Table 3.4) show an average recovery of 100.2%.

Table 3.4. Recovery data for the determination of nitrate in lake and river water samples with the multipoint kinetic procedure and cadmium alloy column. (From [3.26]; Copyright 1982, Elsevier Science Publisher, BV, Netherlands)

Sample	Nitrate content (mg l^{-1} NO$_3^-$-N)			Recovery (%)
	Before std. addition	After std. addition		
		Expected	Determined	
1	1.90	6.90	7.05	102.2
2	10.50	13.00	13.00	100.0
3	11.95	14.45	14.25	98.6
4	0.10	1.10	1.11	100.9
5	0.40	1.40	1.41	100.7
6	10.70	13.20	13.15	99.6
7	10.90	13.40	13.55	101.1
8	12.90	14.15	14.20	100.4
9	5.20	7.70	7.65	99.4
10	0.39	1.39	1.37	98.6
		Average		100.2

The effect of various potential interferences was also investigated. The results of these experiments are shown in Table 3.5. Most of the common ions present in water samples do not interfere, even in high concentrations. Those ions that showed serious interference are rarely found in water samples at the highest concentrations examined.

Lake waters frequently contain a relatively high concentration of organic matter. In determinations of nitrate by the copperized cadmium reduction method, this organic matter can interfere by chelating with copper in the reduction system. Hilton and Rigg [3.28] overcame this difficulty by using an excess of zinc ions to protect the copper catalyst from chelation. They devised a discrete analyses procedure based on these principles for the determination of nitrate.

Reagents. As follows.
   Copper II sulphate solution, 2.28 g l^{-1} $CuSO_4$ $5H_2O$.
   Zinc II sulphate solution, 51.0 g l^{-1} $ZnSO_4 7H_2O$.
   Sodium hydroxide solution, 0.2 mol l^{-1} (8 g l^{-1}) sodium hydroxide.
   Reductant solution. Dissolve 0.218 g of hydrazine sulphate in 400 ml of de-ionized water. Add 5 ml of copper II sulphate solution and 5 ml of zinc II sulphate solution. Make up to 500 ml. Prepare freshly as required.
   Diazotisation reagent – dissolve 2.5 g of sulphanilamide in 400 ml of de-ionised water. Add 25 ml of concentrated phosphoric acid (check concentrated phosphoric acid for low nitrite content prior to use). Dissolve 0.125 g of N-1-naphthylethylethylenediamine dihydrochloride in the resulting solution and dilute to 500 ml. Store in a refrigerator.

Table 3.5. Effect of diverse ions on the determination of nitrate by the multipoint reaction-rate procedure with the cadmium alloy wire column. (From [3.26]; Copyright 1982, Elsevier Science Publisher, BV, Netherlands)

Interferent[a]	Ion: nitrate[b]	Error[c] (%)	Interferent[a]	Ion: nitrate[b]	Error[c] (%)
Cyanide	100	−99.5	Copper	100	−50.5
	10	−7.0		10	None
	1	None	Iron(III)	10	None
Dichromate[d]	10	−21.5	Magnesium	100	+6.5
	1	−10.4		10	None
Iodate	100	−7.5	Mercury	100	−2.5
Iodide	100	−14.1	Tin(II)	10	−88.5
	10	None		1	−15.2
Sulfide[d]	10	−12.0	Nickel[d]	100	−11.0
	1	−2.6		10	None
Cobalt	100	−3.5	Phenol	100	−6.1

[a] The following ions in 100 concentration ratio showed negligible interference: acetate, bromide, carbonate, fluoride, oxalate, perchlorate, sulfate, sulfite, cadmium, calcium, manganese, zinc, silicate, urea, and hypochlorite.
[b] Nitrate concentration 2 mg l^{-1}.
[c] None means < 2%
[d] Permanent decrease in the reduction efficiency of the column; regeneration is needed.

Apparatus. Pye Unicam URAS system (ACI) automatic chemistry unit with an SP6-550 ultraviolet visible spectrophotometer and a Hewlett Packard 9815A calculator or equivalent. The instrument settings were as given in Table 3.6.

Direct adaptation to the discrete analyser met with little success, owing to a non-zero intercept of approximately 50 µg l^{-1} on the concentration axis, although nitrate and nitrite standards of the same concentration gave the same absorbance. A possible cause of the concentration axis intercept was considered to be the presence of an excess of hydrazine, reducing nitrate to nitrite prior to diazotisation. The concentration of hydrazine had been set at 190 mg l^{-1} in the reduction step as recommended by Downes [3.29], which agree with other published procedures for optimised hydrazine-copper reductions, e.g. Kamphake et al. [3.30]. However, it was decided to reoptimise this parameter. The temperature, time of reduction and copper and zinc concentrations were fixed as in Downes method [3.29] and the hydrazine concentration was varied. The absorbances from standard nitrite and nitrate solutions were measured and the peak of the nitrate hydrazine concentration plot was found. The resulting hydrazine concentration of 75 mg l^{-1} in the reduction stage was then used in the optimisation of the other parameters. These were found to peak at the original settings. The calibration, using the reoptimised concentration for the reduction step, was now found to pass through zero.

The resulting method is linear up to 600 µg l^{-1} of nitrate as N with a sensitivity of 0.051 absorbance unit per 100 µg l^{-1}. Interpretation of the performance statistics given in Table 3.7 shows a limit of detection of 14 µg l^{-1} (95% confidence levels). A plot of the concentration of total oxidised nitrogen, determined in a variety of natural waters, vs the concentration, measured for the same samples using the manual cadmium reduction method [3.31] gave a good straight line ($r = 0.967$, $n = 107$) passing through the origin.

Jones [3.32] has described a cadmium reduction method involving shaking the sample with spongy cadmium as an alternative to the use of cadmium columns. Shaking is carried out for a period of 90 min followed by colorimetric determination of the nitrite formed. This technique has advantages over using a cadmium column in that sample/cadmium contact time can be more consistently controlled for a large series of samples and that problems of progressive deterioration in reduction capacity met in column methods are avoided. Recove-

Table 3.6. Instrument settings for the Pye Unicam AURA discrete analyser cycle time, 15 s; temperature, 33 °C; path length, 10 mm; and wavelength, 540 nm. (From [3.28]; Copyright 1983, Royal Society of Chemistry, UK)

Reagent	Dispenser colour	Position	Reaction time/min	Syringe size/µl	Syringe volume/µl	Stroke/mm
Sample		47		450	400	32.0
Sodium hydroxide solution	Blue	47	1.5	2000	800	13.8
Reductant solution	Orange	41[a]	1	1000	250	14.0
Diazotisation reagent	Green	13[a]	3.25	1000	400	22.5

[a]Stirrer.

Table 3.7. Precision measurements of nitrate in standard solutions. (From [3.28]; Copyright 1983, Royal Society of Chemistry, UK)

Solution concentration /µg l^{-1}	Mean absorbance	Standard deviation[a]/µg l^{-1}		
		S_w	S_b	S_t
0	0.005	3 (10)	—	—
200	0.101	5 (10)	13 (9)	14
400	0.202	6 (10)	21 (9)	22
600	0.209	11 (10)	26 (9)	28
800	0.395	18 (10)	33 (9)	38
1000	0.474	13 (9)	27 (8)	30

[a] The degrees of freedom for the within- (S_w) and the between-batch (S_b) standard deviations are given in parentheses; S_t is the total standard deviation.

ries of nitrate-spiked natural water samples with up to 12 parts per thousand salinity were 94 – 106%.

Apparatus. Centrifuge that accommodates 50-ml tubes. Mechanical flask shaker. Spectrophotometer equipped with 1- and 10-cm cells for use at 540 nm. pH meter.

Reagents. All chemicals are reagent grade. Distilled, deionized water is used in all solutions.

Ammonium hydroxide, sp. gr. 0.88.

Aluminium hydroxide suspension. Dissolve 125 g aluminium potassium sulfate in 1 l water. Warm to 60 °C and stir while slowly adding 55 ml ammonia. Permit the mixture to settle. Decant the supernatant from the concentrated suspension. Wash the precipitate, settle and decant four to six times.

Ammonium chloride, 0.7 mol l^{-1}, pH 8.5. Dissolve 37.4 g ammonium chloride in about 800 ml water. Add ammonia to adjust pH to 8.5. Dilute to 1000 ml.

Zinc metal sticks (e.g. Fisher Z-13).

Cadmium sulfate solution, 20% w/v.

Hydrochloric acid, 6N.

Spongy cadmium. Stand zinc metal sticks in 20% cadmium sulphate solution overnight. Reaction of two new sticks with about 80 ml of 20% cadmium sulphate solution in a slender 100-ml beaker yields about 15 g wet cadmium. Remove the zinc sticks from the precipitated spongy cadmium. Acidify the solution with a few drops of 6 N hydrochloric acid and drain it from the cadmium. Cover the cadmium with 6 N hydrochloric acid. Stir to wash the metal and to break it into small particles. Drain. Rinse with water (about ten times) until pH is above 5, and store under water. Keep the cadmium wet even when weighing. Used cadmium can be prepared for reuse by repeating the hydrochloric acid wash and rinse.

Colour reagent B (U.S. EPA, 1979). Dissolve 10 g sulfanilamide and 1 g N-(1-phthalyl)ethylenediamine dihydrochloride in a mixture of 100 ml 85% phosphoric acid and 800 ml of water. Dilute to 1000 ml. This solution is stable for several months when stored in a brown bottle in the dark.

Procedure. If sample is coloured, pour 50 ml into a 50-ml graduated screw cap centrifuge tube (e.g. Corning 25335). Add 2 ml aluminium hydroxide suspension (shake it frequently). Cap and shake the tube. Centrifuge at about 1100 g for 5 min. Carefully transfer 25 ml of supernatant, or of original sample if not coloured, to a clean 50-ml centrifuge tube. Add 5 ml of ammonium chloride solution and about 1 g wet spongy cadmium. Cap tubes and secure them in a rack on a mechanical shaker so that tubes are horizontal for maximum mixing. Shake samples about 100 times min^{-1} for about 90 min. Pipet 10 ml of reduced sample into a 25 ml-Erlenmeyer flask. Add 0.5 ml of colour reagent B. Allow colour to develop for 10 min at $20 \pm 5\,°C$ in dim light. Read absorbance at 540 nm within 2 h.

When sample concentration is greater than 2 µm nitrate, use a 10-cm optical path. The volumes specified in the procedure may be scaled up or down to fit the available volume of sample and the capacity of the available spectrophotometer cell. For low volume, low concentration samples, a 10-cm cell that requires only about 9 ml can be used (e.g. Hellma OS 165, Hellma Cells, Inc., Box 544, Borough Hall Station, Jamaica, NY).

Prepare a reagent blank and potassium nitrate standards that bracket the concentration range of the samples. Treat the blank and standards in the same manner as the samples, including the colour removal step, when used. Absorbance values reflect the combined concentrations of nitrate and nitrite. For nitrate concentration, determine nitrite separately, without the reduction step, and subtract that value from the combined value.

One of the most appealing features of this method employing a shaker instead of a cadmium column is that the cadmium-sample contact time for a number of samples can be consistently controlled simultaneously. Samples can be shaken in covered flasks, but racks of screw-capped centrifuge tubes are handled more efficiently between work bench and shaker table. Shaking for 60–90 min gave complete reduction at concentrations between 1 and 100 $\mu mol\ l^{-1}$ nitrate. There was slight but inconsistent evidence of reduction of nitrite when 1 $\mu mol\ l^{-1}$ samples were shaken for longer than 60 min.

Mean absorbance and coefficient of variation for potassium nitrate standards analyzed in triplicate by this method are given in Table 3.8. Absorptivity (ratio of absorbance to the product of optical path length and concentration) was 0.029 and 0.032 for the 10- and 1-cm optical paths, respectively.

Suspended solids and coloured organic matter interfere with the analysis and should be eliminated by filtration and precipitation as described in the procedure. To assess the extent to which other components of natural waters affect the accuracy of the proposed method, known amounts of nitrate were added to a number of sample types with salinities of 0–12% (Table 3.9).

Table 3.8. Absorbance and coefficients of variation (CV) for potassium nitrate standards (n = 3). (From [3.32]; Copyright 1984, Pergamon Publishing Co., UK)

$\mu mol\ l^{-1}$ Nitrate	Optical path (cm)	Mean absorbance	% CV
0.0	10	0.050	14
0.5	10	0.197	2.1
1.0	10	0.343	1.0
2.0	10	0.619	0.61
0.0	1	0.006	36
2.0	1	0.070	0.82
5.0	1	0.167	1.5
10	1	0.330	0.30
20	1	0.649	0.15

Table 3.9. Recovery of known additions of NO_2^- Nitrite. (From [3.32]; Copyright 1984, Pergamon Publishing Co., UK)

Sample	Salincty PP 10^3	n	Concentration mean ($\mu mol\ l^{-1}$)	CV (%)	% Recovery
A. Pamlico River	8	5	1.58	6.71	
$+1\ \mu mol\ l^{-1}\ NO_3^-$		5	2.55	2.20	97.0
$+2\ \mu mol\ l^{-1}\ NO_3^-$		5	3.47	3.77	94.5
B. Jacks Creek	0	5	0.918	17.9	
$+1\ \mu mol\ l^{-1}\ NO_3^-$		5	1.87	4.01	95.2
$+2\ \mu mol\ l^{-1}\ NO_3^-$		5	2.96	3.50	102
C. Peat bog pore	0	2	0.776	1.18	
$+1\ \mu mol\ l^{-1}\ NO_2^-$		2	1.74	3.54	96.4
$+2\ \mu mol\ l^{-1}\ NO_2^-$		2	2.90	2.12	106
D. Neuse River	0m	4	8.07	2.14	
$+5\ \mu mol\ l^{-1}\ NO_3^-$		4	13.1	3.79	101
$+10\ \mu mol\ l^{-1}\ NO_2^-$		4	18.1	1.22	100
E. Neuse River	12	4	0.160	12.5	
$+1\ \mu mol\ l^{-1}\ NO_2^-$		4	1.11	8.64	95.3
$+2\ \mu mol\ l^{-1}\ NO_3^-$		4	2.12	2.45	98.0
F. Pamilco pore $H_{-2}O$	0	3	1.07	1.62	
$+0.5\ \mu mol\ l^{-1}\ NO_2$		3	1.59	2.18	104
$+1\ \mu mol\ l^{-1}\ NO_3^-$		3	2.01	0.860	94.0

Recoveries for these nitrate-spiked samples were 94–106%. According to Nydahl [3.33], chloride ions can have a strongly retarding effect on the rate of nitrate reduction. However, Jones [3.32] found that, when the reduction solution was buffered above pH 8 and optimum cadmium contact was used, reduction of $20\ \mu mol\ l^{-1}$ nitrate to nitrite was more than 99% complete, even in the presence of sodium chloride concentration corresponding to that of seawater.

Certain metals and phosphate may interfere with nitrate reduction. Precipitation of metal salts onto the cadmium surface may be prevented by addition of EDTA to the buffer solution. EDTA was not added to the sample examined in

Table 3.9 because precipitation of the salts occurs only when concentrations of iron, copper or other metals are above several milligrams per litre [3.35]. Olsen [3.34] demonstrated that 2.5 – 25 μ mol l^{-1} phosphate may decrease reduction of nitrate by 10 – 40% in cadmium-copper columns, and stated that the effect of phosphate is gradual and varies with the age of the column. These and other interferences are much less likely in the method described by Jones [3.32] than in a column reduction method, because shaking has a continual scouring effect on the cadmium during the reduction process and the cadmium is cleaned before each reduction.

Gauguch and Heath [3.36] modified the cadmium reduction method for the determination of nitrate in small (5-ml) samples of river water.

Reagents. As follows.

Amalgamated cadmium filings and reagents prepared as specified for the cadmium reduction method for nitrate determination in the APHA method [3.37]. Cadmium filings were amalgamated with a 1% (w/v) solution of mercuric chloride for 3 min after which the mercuric chloride solution was decanted. The filings were then washed several times with distilled water and stored in the dark under a 0.10 mol l^{-1} ammonium chloride solution. Prepare reaction tubes by adding one small scoop (0.76 $\pm$ 0.07 g) of the amalgamated cadmium filings into 18 × 150-mm borosilicate glass test tubes. Rinse the reaction tubes three times with 0.10 mol l^{-1} hydrochloric acid between successive uses. Prepare sulphanilamide reagent by dissolving 5 g of sulphanilamide in a mixture of 50 ml of concentrated hydrochloric acid and about 300 ml of distilled water. Bring this mixture to a final volume of 500 ml with distilled water. Prepare the N-(1-naphthyl)-ethylenediamine dihydrochloride reagent by dissolving 0.50 g of the dihydrochloride in 500 ml of distilled water.

Procedure. If sample pH was above 9, adjust the pH to between 8 and 9 using dilute hydrochloric acid. Buffer sample 5 ml aliquots by the addition of 0.1 ml 20% (w/v) ammonium chloride and then add to the reaction tubes. Shake the reaction tubes vigorously in an upright position on an oscillatory shaker (about 150 oscillations min^{-1}). After shaking for 25 min, remove a 2.5-ml aliquot by pipette to another tube into which was mixed 50 μl of sulphanilamide reagent. After at least 2 min, but not more than 5 min, add 50 μl N-(1-naphthyl)-ethylenediamine dihydrochloride reagent, mix and allow to react for 10 min. Determine the absorbance in 1-cm glass cuvettes using a spectrophotometer.

As an alternative to cadmium reduction, Zhang [3.38] showed that nitrate is reduced with 93% efficiency to nitrite in 8 min by ultraviolet light under alkaline conditions. He applied a method based on these principles to the analysis of river water samples. A cadmium reduction column was used to convert the remaining nitrate present in solution to nitrite. Nitrite determinations involved adjustment to pH 7 and sequential addition of sulphanilamide solution (0.6%), alpha-naphthylamine solution (0.6%) and sodium acetate solution (2 mol l^{-1}). Absorbances at 520 nm were used to determine nitrite concentrations against a calibration graph. For total nitrogen determinations, river water samples were digested with hydrochloric acid prior to analysis. Addition of potassium persulphate (4.0%) and sodium hydroxide (6 mol l^{-1}) solutions and subsequent irradiation for 17 min facilitated total nitrogen determination. With this method the coefficient of variation was 1.94% for river waters for a total nitrogen concentration of 1.8 mg l^{-1}.

Automated cadmium reduction methods for the determination of nitrate have been reviewed [3.39].

A study has been carried out of the accuracy of determination of total oxidized nitrogen (nitrite plus nitrate and nitrite) in river waters [3.40].

Airey et al. [3.41] have described a method for the removal of sulphide prior to the determination of nitrate in anoxic estuarine waters. Mercury II chloride was used to precipitate sulphide from samples of anoxic water. The sulphide-free supernatant liquid was used to estimate sulphide by measuring the concentration of mercury II. Nitrate was determined by a spectrophotometric method in amounts down to 1 μg l^{-1}.

3.1.1.2 Spectrofluorimetric Methods

Resorcinol [3.42] and 2, 2-dihydroxy-4, 4'-dimethoxybenzophenone [3.43] have both been proposed as fluorimetric reagents for the determination of nitrate. Afghan and Ryan [3.43] have described an automated method employing 2.2-dihydroxy-4, 4'-benzophenone which is capable of performing analyses at a rate of 20 samples per hour and has a detection limit of 5 μg l^{-1} nitrate. The automated procedure incorporates steps to overcome interference by high concentrations of chloride, sulphide and humic acid substances.

Apparatus. All Auto Analyzer equipment consists of standard Technicon modules. A Neslab RTE-3 circulator, manufactured by Neslab Instruments Ltd., was used to maintain a temperature of 20 °C during fluorescence measurement in the flow cell. Fluorescence was measured using a Ratio Fluorometer-2 manufactured by Farrand Instruments Ltd., in conjunction with a Hewlett Packard 680 strip chart recorder.

Preparation of Column to Remove Organic Matter. The XAD-2 column consists of 15 cm × 0.32 cm i.d. glass tubing packed with Amberlite XAD-2 polymeric adsorbent. The glass tubing was plugged at one end with glass wool and an N-6 nipple was butted against the plugged end using a $\frac{1}{2}$ in. length of $\frac{1}{2}$ in. i.d. Tygon tubing for sleeving. A short length of Technicon $\frac{1}{2}$ in. i.d. transmission tubing was attached to the nipple and fitted with a hosecock. The other end of the column was connected to a 65 mm filtering funnel using 3/16 in. i.d. Tygon tubing. The column assembly was secured in a vertical position and filled with distilled deionized water, making sure no air bubbles were trapped below the funnel and hosecock. A washed aqueous slurry of the resin was poured into the funnel and packed into the column by draining water through the bottom of the column. The funnel and Tygon tubing were removed from the top of the column. The N-6 nipple assembly complete with the transmission tubing and hosecock was filled with distilled deionized water and forced over the opening of the column, making sure no air bubbles were trapped inside. The column was then inserted while running the manifold as shown in Fig. 3.3.

Reagents. As follows.

2, 2'-Dihydroxy 4, 4'-dimethoxybenzophenone. 10^{-3} mol l^{-1} dissolved in concentrated sulphuric acid.

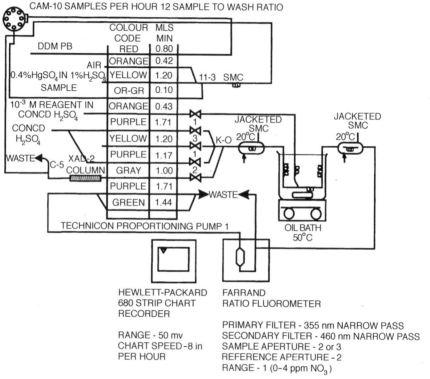

Fig. 3.3. Manifold for fluorometric method (From [3.43]) (DDM PB = 2, 2' dihydroxy 4, 4' dimethoxybenzophenone, 10^{-3} M)

Mercury sulfate solution, 0.4% solution. Dissolve 2 g of mercuric sulfate in distilled deionized water containing 5 ml of concentrated sulphuric acid. Dilute to 500 ml with distilled deionized water.

Nitrate stock solution. Dissolve 0.6850 g of sodium nitrate in distilled deionized water and dilute to 1 l.

Amberlite XAD-2-polymeric absorbent-20–50 mesh, B.D.H. Chemicals. Purify the resin by solvent extraction and store under methanol. Wash the resin several times prior to packing the column with distilled deionized water to remove the methanol.

Procedure. Connect the manifold and AutoAnalyzer equipment as shown in Fig. 3.3. This involves two steps: (i) the dilution of samples and the addition of mercuric sulfate to remove any interference due to chloride and sulfide, and (ii) the development of fluorescence for the determination of nitrate.

First dilute the sample and mix with mercuric sulfate to remove any possible interference due to a high concentration of chloride and sulfide. Because of the high sensitivity of the reaction, it was necessary to dilute the sample to cover the most common range, namely 0.025–4.0 mg l^{-1} nitrate, prior to introducing the sample for fluorescence development.

Analyse samples at the rate of 10 samples per hour using a 1:2 sample to wash ratio. It was possible to analyze samples containing as low as 0.5 µg l^{-1} of nitrate nitrogen by eliminating the dilution step prior to the determination. However, the above-mentioned range was chosen since the nitrate content of the majority of natural water samples lies in that range.

Figure 3.4 shows uncorrected excitation and emission spectra of 2,2'-dihydroxy-4;4'-dimethoxybenzophenone and its reaction product with nitrate under

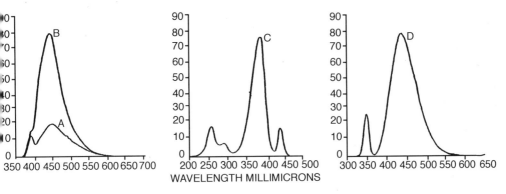

Fig. 3.4. Uncorrected fluorescence spectra of solutions run through the manifold in Fig. 3.3. *Curve A* emission spectrum of DHDMBP; *Curve B* emission spectrum of 2×10^{-6} mol l^{-1} nitrate + DMBMBP; *Curve C* excitation spectrum of 1×10^{-6} mol l^{-1} nitrate + DHDMBO; *Curve D* emission spectrum of 1×10^{-6} mol l^{-1} quinine sulfate in 0.1 N sulfuric acid. For all spectra, coarse sensitivity = 1, fine sensitivity = 80, excitation and emission = 2.0. For A and B excitation $\lambda = 385$, for C emission $\lambda = 440$ and D excitation $\lambda = 350$ (From [3.43])

similar experimental conditions as for Fig. 3.3. The sensitivity of the reaction was also compared with similar concentrations of quinine sulfate.

Excitation was carried out at 380 nm and emission at 445 nm. The fluorescence development did not take place in a medium containing less than 70% sulphuric acid. The sensitivity and the concentration range where the response was linear varied with the concentration of sulphuric acid. A maximal fluorescence was obtained between 83–86% sulphuric acid. Nonlinear calibration curves were obtained when the ratio of sulphuric acid to water exceeded 92%. Therefore, an 80% sulphuric acid medium was used. In the majority of automated methods, the most convenient way of eliminating interferences is to dilute the sample in the manifold prior to the determination. Because of the high sensitivity of the method described by Afghan and Ryan [3.43], it was possible to incorporate a dilution step into the manifold as shown in Fig. 3.3 to tolerate concentration as high as 100, 10 and 1 mg l^{-1} of chloride, fulvic acid and sulfide, respectively. The dilution step was sufficient for the removal of these interferences for the majority of natural waters. However, in some instances, concentrations of chloride, sulfide and fulvic acid substances may exceed the above levels. Therefore, it was necessary to incorporate an alternate method for the removal of these interferences prior to the determination of nitrate.

In the procedure the sample is diluted with mercuric sulphate to avoid interference by chloride and sulphide. An XAD-2 resin column was incorporated after the sample dilution step to eliminate interference from fulvic acid substances: ammonia, glycine, urea, amino acids and nitrite were without interference in the procedure.

3.1.1.3 Ultraviolet Spectroscopy

Munoz [3.44] adapted the ultraviolet procedure for the determination of nitrate to an automatic discrete-sample system, the AMA-40 system for the determination of nitrate in waters at low mg l^{-1} levels. Results indicate that the automated system is simple to set up and is easily applicable to routine determinations of nitrate in water samples, giving excellent repeatability of results.

Brown and Bellinger [3.45] have proposed an ultraviolet technique which is applicable to both polluted and unpolluted fresh waters and some estuarine waters. Humic acid and other organics are removed on an ion-exchange resin. Bromide interference in sea water samples can be minimised by suitable dilution of the sample but this raises the lower limit of detection such that only on relatively rich (0.5 mg l^{-1} nitrate) estuarine and inshore waters could the method be used. Chloride at concentrations in excess of 10 000 mg l^{-1} does not interfere.

The method is either not affected by, or can allow for, interference from phosphate, sulphate, carbonate, bicarbonate, nitrite, coloured metal complexes, ammonia, dyes, detergents, phenols and other ultraviolet absorbing substances. The method incorporates three features designed to reduce interferences.

(a) Humic acid interference is reduced by carrying out measurements at 225 nm, a higher wavelength than that used by previous workers (210–229 nm).
(b) Removal of inorganic interferences, particularly the removal of bromide interference in sea water, by diluting the sample fivefold, the removal of nitrite by the addition of sulphamic acid and the removal of metals by passage through Amberlite IR 120 cation exchange resin.
(c) Removal of ultraviolet absorbing organics by passage through a specific ion-exchange resin such as Amberlite XAD-2.

In Table 3.10, nitrate determination by this method in the presence of various interfering substances is compared with two alternate methods, namely phenoldisulphonic acid and ion selective electrode methods. It is seen that, in general, the method proposed by Brown and Bellinger [3.45] is less subject to interference.

3.1.1.4 Atomic Absorption Spectrometry

Cresser [3.46] showed that nitrate can be rapidly reduced to ammonia by the action of titanium(III)sulphate at room temperature. Subsequent displacement of the ammonia by a current of air enables the gas-phase molecular absorbance of the ammonia to be measured using an atomic absorption spectrophotometer and thus provides a sensitive and selective method for the determination of nitrate. Ions which affect the reduction of nitrate by titanium salts, such as cobalt, copper, iron, and zinc, interfere with this method and must be removed if present in large amounts.

3.1.1.5 Flow Injection Analysis

Various workers [3.47–3.52] have applied this technique to the determination of nitrate in natural waters.

Nakashima et al. [3.52] determined nitrate in natural waters by the flow-injection analysis technique, according to which the sample (650 µl) was injected into the carrier solution (a mixture of EDTA and ammonium chloride in aqueous solution at pH 8.1) and the nitrate first reduced to nitrite by passage through a column of copperized cadmium. The nitrite was then reacted with a mixture of p-aminoacetophenone and m-phenylene-diamine and the absorption measured at 456 nm. The influence of several variables on the accuracy and sensitivity of the method was investigated. The limit of detection was 1.5 µg l^{-1} and the sample rate about 40 samples per hour. Relative standard deviations for nitrate nitrogen in the range 0.1–0.3 mg l^{-1} were about 1%.

Al-Wehaird and Townsend [3.50] give details of a procedure for the determination of traces of nitrate (0.02–5 mg l^{-1} nitrogen) in water by flow-injection analysis. Trivalent titanium chloride was used to reduce nitrate to nitrite before diazotization with sulphanilamide and N-(1-naphthyl)ethylenediamine, and

Table 3.10. The effects of various interferences on the determination of nitrates by three methods. For all determinations (with the exception of natural river waters) the concentration of NO_3^- N was 1.125 mg l^{-1}. All concentrations given are in mg l^{-1}. (From [3.45]; Copyright 1978, Pergamon Publishing Co., UK)

Interference	Phenoldisulphonic acid		Method UV method		Selective ion electrode	
	Uncorrected	Corrected	Uncorrected	Corrected	Uncorrected	Corrected
Chloride						
100	0.62	1.07[a]	1.08		1.08	1.55[d]
1000			1.05			
Bicarbonate-carbonate						
100	1.13		1.06		7.87	1.24[e]
260	1.23					
1000			1.13			
Bicarbonate-chloride						
100 Cl:200 HCO$_3$						1.58[e]
100 Cl:100 HCO$_3$						1.46[e]
50 Cl:100 HCO$_3$						1.46[e]
Nitrate						
0.304	1.11		1.48	1.10[b]		
1.500						1.31[e]
3.0–4	1.13			1.06[b]		
Ammonium						
1.4	0.98					
14.0	0.66		1.08			1.13[e]
1400.0			1.08			
Phenol						
1.0	1.10		1.29	1.10[c]		1.13[e]
10.0			1.86	1.08[c]		
Daz[f]						
1.0			1.12			2.48[e]
10.0			1.56	1.05[c]		202.5[e]
Unpoluted river water humic acid-inorganic salts NO$_3^-$ N by Polarography 1.170	1.69		1.77	1.77[c]		3.15[e]

[a] Addition of exact amount of silver sulphate to precipate chloride with no excess silver.
[b] Addition of sulphamic acid (final conc. in sample approximately 20 mg l^{-1}).
[c] Treatment with Amberlite XAD-2 resin.
[d] Addition of saturated silver sulphate (1 ml to 10 ml sample). Not recalibrated
Addition of 2 mol l^{-1} acetic acid to sample (0.09 ml per 30 ml sample). Not recalibrated.
[e] Calibrate for presence of 3 ml saturated silver sulphate and 0.09 mol l^{-1} acetic acid in 30 ml sample.
[f] Daz-a commercially available household detergent.

spectrophotometric determination at 530 nm. Divalent copper was the only common ion that caused interference and this could be removed by an on-line cation-exchange mini-column.

Discrete injection segmented flow analysis has been used [3.51] to determine nitrate and phosphate in small volume samples.

3.1.1.6 Ion-Selective Electrodes

Ion-selective electrodes have been applied to the direct determination of nitrate in natural waters [3.53–3.66].

Nitrate has been determined by a nitrate-selective electrode [3.51] consisting of a porous polymer wick sealed with liquid ion-exchanger (tris (4,7-diphenyl -1, 10-phenanthrolinato) nickel II in 2-nitro-p-cymene). Chloride and bicarbonate interferences are overcome by the addition of silver sulphate and 0.5 mol l^{-1} phosphate PH$_2$ buffer, respectively.

Schechter and Gruener [3.58] developed a nitrate-selective electrode and applied it to highly mineralized borehole waters. Results obtained by this procedure were compared with those obtained by the standard phenoldisulphonic acid spectrophotometric method. These workers concluded that the electrode method is satisfactory under laboratory conditions but not as a field method because of the detailed calibration and complex controls required.

To evaluate the net influence of most common anions occuring in water, Schechter and Gruener [3.58] made measurements in a variety of synthetic solutions containing 1–50 mg l^{-1}-nitrate-nitrogen and varying amounts of nitrite, hydrogen carbonate, carbonate, chloride and sulphate. The results are presented in Table 3.11.

It is observed that the level of interfering ion, the total ionic strength and the level of nitrate do influence the measured potential. For most anions occuring in water, there are major additive influences from nitrite and chloride.

Table 3.11. Anions interference in nitrate-nitrogen determination by specific ion electrode. (From [3.58]; Copyright 1976, American Water Works Association)

Anion Added mg/l		Nitrate as nitrogen added mg l^{-1}					
		0	1	5	10	25	50
		Nitrate as nitrogen detected-mg l^{-1}					
Nitrite	5	1.20	1.30	5.00	10.50	26.10	53.20
	10	1.50	1.70	5.60	10.50	28.20	58.10
	50	4.40	4.50	8.40	13.50	28.70	60.40
Bicarbonate	50	0.25	1.10	5.80	11.50	25.00	50.00
	250	0.40	1.50	7.30	12.00	26.10	51.50
	500	0.50	1.60	7.50	13.80	27.00	52.50
Carbonate	50	1.00	1.50	6.00	11.40	27.00	54.00
	250	1.15	1.70	6.40	12.00	29.00	58.00
	500	1.20	1.70	6.70	12.50	29.00	59.50
Chloride	50	1.30	1.30	5.80	11.50	27.00	52.00
	250	2.50	2.10	7.00	13.20	28.00	56.00
	500	3.60	3.00	8.00	15.20	30.00	59.10
	1000	5.20	4.80	9.80	17.00	34.00	64.20
Sulfate	50	0.20	1.00	4.90	10.40	25.00	50.00
	200	0.30	1.10	5.00	10.10	25.10	50.00
	500	0.60	1.00	5.00	10.00	25.20	49.70
	1000	0.70	0.90	4.80	9.80	24.90	49.30

Nitrite is seldom present in high concentrations in borehole water. As to the strongly interfering chloride, it must be initially removed by precipitation with silver sulphate if more accurate determinations are to be made. The elimination of chlorides was also performed by treatment with cationic ion exchange resin, prepared in the silver form. In this study better results were obtained by silver sulphate precipitation

In Table 3.12 nitrate determinations on well water and spring water samples by ion selective electrodes and phenoldisulphonic acid spectrophotometry are compared. Good agreement was generally obtained between results obtained by the two methods.

Cox and Litwinski [3.59] described a voltammetric ion-selective electrode for the determination of nitrate down to 6.7×10^{-6} mol l^{-1}. The electrolysis cell used was equipped with an anion-exchange membrane sheath.

Potentiometric titration using an Orion 93–07 nitrate-selective electrode and an Orion 90-02 reference electrode has been used [3.60] to determine nitrate in lake waters. Of the varients of this procedure, the Gran method [3.61] was found to be most suitable. Six known additions (0.0, 0.10, 0.15, 0.20, 0.25 and 0.30 ml of 10^{-2} mol l^{-1} sodium nitrate) were added to 10 ml of the sample and after each addition the electrode potential was checked and recorded. Analysis was performed with constant stirring and in the presence of 0.1 ml of silver fluoride to each 10 ml of sample in order to maintain a constant ionic strength and to eliminate interferences from such as chlorides or sulphates.

For calculating the nitrate concentration by Gran's method, the known transformation of the Nernst equation was employed:

$$\text{antilog } E/S = \text{const} + a_{NO_3^-}$$

where E is the emf of the electrochemical cell, S the electrode function of the nitrate electrode (found to be 55 mV per decade). The graph obtained represents the dependence between antilog $E*S$ and the added quantity of standard

Table 3.12. Comparison of specific ion electrode and PDS[a] method for nitrate-nitrogen determination in well water. (From [3.58]; Copyright 1976, American Water Works Association)

Nitrate-nitrogen level range mg l^{-1}	Number of samples	Nitrate-nitrogen mg l^{-1}		Conductivity mmho		Chlorides mg l^{-1}	
		Electrode[b]	PDS[b]	Min	Max	Min	Max
<5.0	4	2.92	2.60	0.45	1.10	76	140
5.0–10.0	45	8.78	8.02	0.50	3.15	40	750
10.1–20.0	71	15.24	14.45	0.70	3.15	111	740
20.1–30.0	12	26.15	25.00	0.80	2.70	95	600
>30.0	16	35.94	34.86	1.80	2.90	500	580
Total	148						

[a]Phenoldisulphonic acid.
[b]mean values.

solution. The intersection of the graph with the horizontal axis represents the volume of standard solution containing a quantity of nitrate equal to that in the analysed solution. Table 3.13 shows the experimental results for eight different samples.

Ultratrace amounts ($\mu g\, l^{-1}$) of nitrate, nitrite (and ammonium 62) have been determined in pond waters by atmospheric pressure helium microwave induced emission spectrometry utilizing a gas generation technique. The instrument components are summarized in Table 3.14. The instrumental set-up consists primarily of the plasma system, the optical measurement system and the gas generation system.

Ammonium ion is oxidized to nitrogen gas by sodium hypochlorite in an alkaline medium and nitrite ion by sulphamic acid in an acidic medium. Nitrate ion is reduced to nitrite in a cadmium/copper column and then determined as above. The nitrogen produced by the chemical reduction is introduced into the

Table 3.13. Nitrate-nitrogen content ($mg\, l^{-1}$) in lake water samples analysed by Gran method. (From [3.60]; Copyright 1979, Springer-Verlag GmbH, Germany)

S.No	Gran
1 (standard) 1.06	1.06
2	1.87
3	1.93
4	2.80
5	2.94
6	2.89
7	2.05
8	1.82
9	2.35

Table 3.14. Instrumental components. (From [3.60]; Copyright 1979, Springer-Verlag GmbH, Germany)

Components	Model or size	Manufacturer
Microwave generator	MR-3S	Ito Chotanpa Co., Ltd.
Microwave cavity discharge tube	TM_{010} type 6 mm o.d., 3 mm i.d.	laboratory constructed laboratory constructed
Monochromator	JE-50E	Nippon Jarrell-Ash Co., Ltd.
Lens	f:60 mm, 25 mm ϕ	
Photomultiplier	R 787	Hamamatsu TV Co., Ltd.
Power supply picoammeter	part of ICAP-500	Nippon Jarrell-Ash Co., Ltd.
Chart recorder	B-181H	Rigakudenki Co., Ltd.

helium flow via a three-way valve system connecting the reaction vessel. Plasma is generated by a microwave generator and the emission signals are measured at 391.4 nm. Experimental procedures are summarized in Table 3.15.

Four wave-lengths listed in Table 3.16 were used for the determination of nitrogen. These peaks of nitrogen-containing species provided different detection limits and dynamic ranges. The detection limits and dynamic ranges obtained for ammonium and nitrite nitrogens are also summarised in Table 3.16. The detection limit and dynamic range for nitrate were very similar to those for nitrite. The detection limit was defined as the signal level corresponding to twice the standard deviation of the blank signals.

As can be seen in Table 3.16, the N_2^+ "main" system bandhead at 391.4 nm provided the lowest detection limit, while the N_2 "second positive" system bandhead at 337.1 nm and the nitrogen atomic line at 746.8 nm were observed to have wide dynamic ranges (more than three orders of magnitude). The emission signal of the NH band maximum ("A" system) at 336.0 nm was the largest of the four analytical wave-lengths but the detection limits of this band, especially for ammonium nitrogen, were not so good because of the significantly large blank

Table 3.15. Experimental procedures for nitrogen gas generation. (From [3.60]; Copyright 1979, Springer-Verlag GmbH, Germany)

Purging the air in the reaction vessel
↓
Injection of sample (2 ml)
↓
Injection of pH adjustment solution (0.5 ml)
↓ 0.25 N NaOH solution[a]
   1 N HCl solution[b]
Purging the air in the solution
↓
Injection of reaction agent solution (0.1 ml)
↓ 10% NaOBr solution[a]
   0.2 mol l^{-1} sulfamic acid solution[b]
Measurement

[a]In the case of ammonium nitrogen.
[b]In the case of nitrite-nitrogen.

Table 3.16. Detection limits and dynamic ranges for ammonium and nitrite-nitrogens at different emission lines and bands. (From [3.60]; Copyright 1979, Springer-Verlag GmbH, Germany)

Wavelength, nm	Ammonium N		Nitrite N	
	Detectn lim, mg l^{-1} N	Dynamic range	Detectn lim, mg l^{-1} N	Dynamic range
337.1 (N$_2$)	0.004	∼3000	0.008	∼1000
391.4 (N$_2^+$)	0.004	∼200	0.004	∼200
336.0 (NH)	0.02	∼50	0.006	∼100
746.8 (N)	0.009	∼5000	0.01	∼3000

signals at 336.0 nm. The relative standard deviation of eight replicate measurements was about 7% at the 10-fold concentration of the detection limit and was about 4% at the 100-fold concentration for all the analytical wavelengths.

The effect of various concomitants possibly coexisting in water samples was investigated. The samples containing 0.4 mg l^{-1} ammonium nitrogen or 0.3 mg l^{-1} nitrate nitrogen were analyzed in the presence of 100 – 20 000 mg l^{-1} of the inorganic ions. Of the organic compounds, amino acids and urea were examined where the analysis was done in the presence of 0.5 – 100-times concentration (in molar concentration) of the concomitants. The results, summarized in Table 3.17, show that only a few inorganic ions and organic compounds interfere with nitrogen analysis by helium microwave induced plasma emission spectrometry.

The results of nitrate determinations in pond water quoted in Table 3.18 show that results obtained by this procedure are consistent with those obtained by spectrophotometry. No nitrite was found in the sample/ by either method.

Table 3.17. Interferences of inorganic ions and organic compounds with determination of ammonium and nitrite-nitrogens (observed at 391.4 nm). (From [3.60]; Copyright 1979, Springer-Verlag GmbH, Germany)

Concomitant	Concn mg l^{-1}	Rel sensitivity	
		Ammonium N^a	Nitrite N^b
none		100^c	100^c
Na^+	13 000	100	115
K^+	20 000	98	105
Mg^{2+}	500	98	104
Ca^{2+}	1 000	97	102
Al^{3+}	100	103	102
Fe^{3+}	1 000	96	101
Fe^{2+}	100	114	71
Mn^{2+}	100	84	99
Co^{2+}	100		106
Cu^{2+}	100	103	107
Zn^{2+}	100	99	102
Cl^-	20 000	100	100
NO_3^-	1 500	105	107
SO_4^{2-}	4 000	107	99
PO_4^{3-}	500	102	70
Alanine	7 mmol l^{-1}		100
	1 mmol l^{-1}	97	
	7 mmol l^{-1}		95
Serine	1 mmol l^{-1}	90	
Glutamic acid	7 mmol l^{-1}		99
	1 mmol l^{-1}	94	
	7 mmol l^{-1}		99
Urea	0.015 mmol l^{-1}	114	

[a]The concentration of the test solution was 0.4 mg l^{-1} as N.
[b]The concentration of the test solution was 0.3 mg l^{-1} as N.
[c]Defined as 100.

Table 3.18. Analytical results in the determination of nitrate-nitrogen in pond water. (From [3.60]; Copyright 1979, Springer-Verlag GmbH, Germany)

Sample no.[a]	Nitrate nitrogen content, mg l^{-1} as N	
	Plasma emission[b]	Colorimetry
1	0.010 ± 0.002	0.013 ± 0.001
2	0.019 ± 0.002	0.021 ± 0.001

[a]Samples 1 and 2 refer to the surface water samples taken from the Sanshiro-ike pond at the University of Tokyo.
[b]Measured at 391.4 nm.

One instrument [3.63] for determining nitrate is based on ion selective electrode measurements, with an acidic buffer solution which ensures that alkaline earth carbonates do not deposit, and variations in sample ionic strength are decreased. A feature is the automatic checking and autocorrection for sensor drift, carried out at two concentrations at preset time intervals. Monitoring trials have shown that the instrument can produce accurate and consistent records almost indefinitely without manual intervention.

Synott et al. [3.64] compared ion-selective electrodes and gas-sensing electrode techniques for the measurement of nitrate in natural waters. The gas electrode method is based on the reduction of nitrate to ammonia by titanium trichloride and detection of the ammonia by an ammonia electrode. The results indicated satisfactory reproducibility and correlation between the two methods, except for sea water where the high concentration of dissolved salts made the membrane electrode method unreliable. The gas-sensing method, however, was unaffected by the presence of chloride in high concentrations and also enabled the operator to compensate for the presence of trace quantities of free ammonia in the sample.

Csiky et al. [3.65] used disposable clean-up columns of chemically bonded alkylaminosilica for the selective removal of humic substances prior to measurements with a nitrate ion-selective electrode. Using the clean-up procedure, recoveries averaged 95% and reproducibility was independent of humic acid concentration. Without clean-up, recovery was always greater than 100%, increased with humic acid concentration and changed with time. The nitrate concentrations found in natural waters after clean-up were in good agreement with determinations by ion chromatography.

O'Hara and Okazaki [3.66] studied the effect of surfactants on the determination of nitrate in river and stream waters using a nitrate ion-selective electrode. The interferences due to anionic surfactants and their elimination by the addition of cationic surfactants was examined. The interference of dodecylbenzenesulphonate below 1 mg l^{-1} was unimportant. The interference of 1-10 mg l^{-1} dodecylbenzenesulphonate could be mostly eliminated by the addition of a cationic surfactant such as cetyltrimethylammonium bromide. An anomalous response behaviour occurs in the presence of some cationic surfactants.

3.1.1.7 Ion Exchange Chromatography

Sherwood and Johnson [3.67] have described an ion exchange chromatographic determination of nitrate and amperometric detection at a copperized cadmium electrode. The chromatograms obtained in this procedure resolve nitrate from dissolved oxygen.

3.1.1.8 Ion Chromatography

Ion chromatography has been applied to the determination of nitrate in natural waters. These applications are reviewed in Table 3.19 and discussed more fully in Sect. 11.1.

3.1.1.9 Mass Spectrometry

Heumann and Unger [3.73] have described a mass spectrometric determination of nitrate in mineral waters. The method is based on the technique of isotopic dilution analysis. The formation of negatively charged nitrite thermal ions from the nitrate in the water permits the nitrate to be detected in the mass spectrometer. Using nitrogen-15 labelled nitrate as a radioactive tracer for the isotopic dilution method, nitrate could be quantitatively determined down to the $\mu g\ l^{-1}$ level. The method is also applicable to nitrite, after separation of nitrate ions, which can be performed by ion exchange in the alkaline pH range.

3.1.1.10 Enzymic Assay

Enzymic methods have been used for the determination of nitrate [3.74, 3.75]. These methods are based on the enzymic reduction of nitrate to nitrite or ammonia. In one method [3.74] the nitrite is then estimated by the sulphanilamide-N-(l-naphthyl)ethylenediamine dihydrochloride coupling procedure.

Kiang et al. [3.75] reduce nitrate with nitrate reductaze and nitrite with nitrite reductaze using an air gap electrode to monitor the ammonia produced

Table 3.19. Determination of nitrate in natural waters by ion chromatography

Type of water sample	Codetermined anions	References
Natural	NO_2	3.68
Natural	Cl, Br, SO_4, NO_2	3.69
Natural	Cl, Br, F, NO_2, SO_4, PO_4	3.70
Natural	SO_4	3.71
PORE	SO_4, PO_4, Cl, F, Br	3.72

by the reduction of nitrate. Between 5×10^{-5} and 1×10^{-2} mol l^{-1} nitrate can be determined by this procedure.

3.1.1.11 Miscellaneous

Other techniques for determining nitrite include paper chromatography of diazo dyes to reduce interference by metals [3.76], stopped flow analysis [3.77], direct injection enthalpimetry [3.78] and a nitrogen–15 isotopic tracking method [3.79].

Yoshizumi et al. [3.80] described the determination of down to 1 µg l^{-1} nitrate by a flow system with a chemiluminescent NO$_x$ analyser.

3.1.2 Rainwater and Snow

3.1.2.1 Spectrophotometric Methods

A requirement of methods for the determination of nitrate in rainwater is that it has sufficient sensitivity to determine the very low levels of this anion encountered in this type of sample. With this in mind, Osibanjo and Ajaya [3.81] have modified the usual distillation procedure for the determination of low levels of nitrate using 3, 4 xylenol as a chromogenic reagent. In this method the nitration product is extracted with toluene. The method is relatively free from interferences, is rapid and is highly sensitive, being capable of determining down to 5 µg l^{-1} nitrate in unpolluted surface and ground waters and rain with a recovery rate of 96 – 108%. The method was also applied to the analysis of soil, river sediments and mineral oils.

Nitration of 3, 4-xylenol by nitrate is carried out instantaneously at 0 °C in 80% sulphuric acid and the nitration layer extracted into toluene. The toluene layer is treated with sodium hydroxide to form a coloured product in the aqueous layer, the absorbance of which is measured at 432 nm. Between 0.5 and 2.2 mg l^{-1} of nitrate was found in rainwater samples by this procedure.

Spectrophotometric methods, based on the use of diphenylamine or diphenylbenzidine [3.82], and xylenol [3.83] have been used for the determination of nitrate in rainwater. The methods using amines are insufficiently sensitive for rainwater analysis (LD 0.5 mg l^{-1}) whereas the 3, 4-xylenol method is capable of analysing at the 5 µg l^{-1} nitrate level. In this method, nitration of 3, 4-xylenol by nitrate in the sample is carried out instantaneously at about 0 °C in 80% sulphuric acid and the nitration product is extracted into toluene, the excess of the reagent remaining in the aqueous layer. The toluene layer is then treated with sodium hydroxide solution to form a coloured product (the sodium salt of the nitrophenol) in the aqueous layer, the absorbance of which is measured at 432 nm.

Reagents. As follows.

Ethanol, re-distilled.

3, 4-Xylenol, 2% solution in ethanol. Dissolve 2 g of 3, 4-xylenol in ethanol in a 100-ml calibrated flask and dilute to the mark with ethanol.

Sodium hydroxide solution, 1%. Dissolve 1 g of sodium hydroxide pellets in distilled water in a 100-ml calibrated flask and dilute to the mark with water.

Sulphuric acid, 80%. Mix 100 ml of 98% sulphuric acid carefully with 22.5 ml of distilled water and allow the solution to cool to room temperature before use.

Toluene.

Nitrate standard solution. Prepare a 100 µg ml^{-1} nitrate solution by dissolving 0.1629 g of potassium nitrate in distilled water in a 1-l calibrated flask and diluting to the mark with distilled water. Solutions of lower concentration were prepared by serial dilution of this standard stock solution with distilled water.

Apparatus. Pye Unicam SP6 spectrophotometer (Model 200) with matched 1 cm silica cells.

Procedure. Prepare a calibration graph by transferring with a semi-micro burette 1.0 ml each of standard nitrate solutions, containing between 10 and 100 µg ml^{-1} of nitrate, into two-necked-100 ml round bottomed flasks. Evaporate each solution nearly to dryness, heating each flask gently on a hot plate. Cover the flask and cool in ice. Add 5 ml of 80% sulphuric acid followed by 1 ml of 2% 3, 4-xylenol solution. Take a reagent blank through the same procedure. Cover the flask and shake for 60 s. Carefully transfer the contents of the flask into a separating funnel with 80 ml of ice-cooled distilled water dispensed from a wash bottle.

Add 10 ml of toluene to each solution, shake the funnel for 60 s and allow the phases to separate. Discard the lower aqueous layer and wash the toluene extract with 10 ml of distilled water by shaking the flask for a further 1 min. Add 5 ml of 1% sodium hydroxide solution from a pipette to convert the phenol into a phenoxide. Shake the funnel for 1 min and allow the phases to separate. Rinse the stem of the funnel with few drops of the yellowish lower layer. Measure the absorbance of the remaining portion at 432 nm using matched 1-cm silica cells with distilled water as reference.

For the analysis of rainwater samples, proceed as follows. Pipette 10–50 ml of rainwater (filtered if necessary) into a two-necked 100-ml round bottomed flask and evaporate nearly to dryness on a hot plate. Cover the flask and cool in ice. Then proceed as described above for the preparation of the calibration graph.

The effects of various interferents results are shown in Table 3.20. There are no serious interferences from most of the anions studied, including nitrite and chloride, to which most methods for nitrate determination are normally intolerant. However, bromide interferes at the 1:100 level and above, and nitrite interferes slightly at the 1:10 level and above. These two interferents can be removed,

Table 3.20. Effects of Various ions on the relative absorbances of nitrate ion values given are percentage changes in absorbance. (From [3.83]; Copyright 1980, Royal Society of Chemistry, UK)

Ion	NO_3^- to ion ratio			
	1:1	1:10	1:100	1:1000
F^-	0.77	4.26	4.26	6.60
Cl^-	1.16	6.59	2.32	80.62
Br^-	4.65	9.30	29.45	75.96
I^-	4.65	2.71	3.88	39.92
CO_3^{2-}	4.65	4.65	9.30	2.30
PO_4^{3-}	0.77	1.16	3.87	3.10
NO_2^-	0.38	10.79	11.24	11.62
SO_3^{2-}	0.77	5.42	6.98	9.30
NH_4^+	0.77	2.71	9.30	9.30

however, by the addition of silver sulphate or mercury sulphate and sulphamic acid, respectively, to the sample solution prior to extraction with toluene.

3.1.2.2 Flow Injection Analysis

Flow injection analysis has been applied to the determination of nitrate in rainwater [3.84] (and chloride, nitrite, sulphate and phosphate).

Madsen [3.85] used flow injection analysis with hydrazine reduction and spectrophotometric detection at 540 nm to determine less than 1 mg l^{-1} nitrate in rain water. Forty samples per hour can be analysed with a precision of 30% in the range 1 – 10 mg l^{-1}.

Kauniansky et al. [3.86] determined nitrate and sulphate in rainwater by capillary isotachophoresis. Separation conditions were chosen to eliminate potential interference from formate, acetate and propionate. Sulphate and nitrate were separated at pH 6.0 using histidine or 1, 3-bis(tris(hydroxymethyl)methyl aminopropane as buffering agents. The complex-forming cations, magnesium and calcium, were also suitable for optimizing separating conditions. Sulphate was retarded, relative to nitrate, by using trimethylenetetramine as a co-counter ion in the leading electrolyte.

3.1.2.3 Continuous Flow Analysis

This technique has been applied to the determination of nitrate in rainwater [3.84] (and chloride, nitrite, phosphate and sulphate).

A computer controlled multichannel continuous flow analysis system has been applied to the measurement of nitrate (and chloride) in small samples of rainwater [3.87].

3.1.2.4 Gas Chromatography

Faigle and Klockow [3.88] applied gas chromatography to the determination of traces of nitrate and (sulphate and phosphate) in rainwater. The dissolved salts are freeze-dried and converted to the corresponding silver salts. These are then converted to n-butyl esters with the aid of n-butyl iodide, the n-butyl esters being determined by direct injection into a gas chromatographic column. Sulphate and phosphate may be determined simultaneously on a column containing 3% OV-17 on Chromosorb G, while nitrate is determined separately with 3% tri-p cresylphosphate as the stationary phase on a similar support.

3.1.2.5 Ion Chromatography

This technique has been applied to the determination of nitrate and other anions in rainwater (Table 3.21). Methods are discussed in further detail in Chap. 11.2.

3.1.2.6 Polarography

Hemmi et al. [3.97] used differential pulse polarography to determine down to 8×10^{-7} mol l^{-1} nitrate in fresh snow. This method utilizes the catalytic reaction between nitrate and uranyl ion in the presence of potassium sulphate. The differential pulse peak is proportional to nitrate concentration from 1 to 50 µmol l^{-1}. Between 350 and 700 µg g^{-1} nitrate was found in snow samples by this method. Nitrite and phosphate interfere seriously in this method and should be absent if accurate results are to be obtained.

3.1.2.7 Electrochemical Methods

Electrochemical methods that have been used to determine nitrate include polarography [3.98, 399]. Bodini and Sawyer [3.100] described a voltammetric determination of nitrate at the µg l^{-1} level. This electrocatalytic method utilizes

Table 3.21. Determination of nitrite in rain water by ion chromatography

Type of water sample	Codetermined anions	References
Rain	Cl, PO$_4$, SO$_4$	3.89, 3.90
Rain	Cl, SO$_4$	3.91, 3.92
Rain	Cl, NO$_2$, PO$_4$, SO$_4$	3.93
Rain	Cl, NO$_2$, SO$_4$	3.94
Rain	Cl, F, Br, NO$_2$, SO$_4$, SO$_3$, PO$_4$	3.95
Rain	Cl, Br, F, NO$_2$, PO$_4$, BrO$_3$, SO$_4$	3.96

the concomitant reduction of cupric chloride and cadmium chloride on pyrolytic graphite to catalyze the reduction of nitrate ion.

3.1.3 Potable Water

Groundwaters provide almost 40% of the domestic water supply in England and Wales and the concentration of nitrate present has an important bearing on potability. The current WHO standard for European drinking water [3.109] recommends that levels of nitrate should not exceed 11.3 mg l^{-1} of nitrate-nitrogen (50 mg l^{-1} of nitrate). High levels of nitrate can give rise to harmful physiological effects, such as methaemoglobinaemia, particularly in infants, and at present there is concern over evidence that suggests that nitrate concentrations in some major aquifers are increasing to an undesirably high level. Analytical methods for nitrate therefore need to be accurate so that threshold values can be determined with confidence, and precise so that small trends in concentration can be detected quickly.

3.1.3.1 Spectrophotometric Methods

Spectrophotometric methods based on the use of resorcinol [3.110] and 2,4-xylenol [3.111] have been described for the determination of nitrate in mineral waters and ground waters.

Miles and Espejoi [3.111] compared results for nitrate in ground waters of low salinity obtained by the 2,4-xylenol spectrophotometric method and by ultraviolet spectroscopy. The ultraviolet procedure is not subject to interference from the major cations present in typical ground waters, and interference by high levels of nitrite can be overcome by addition of sulphamic acid. There was good agreement between nitrate levels determined by both methods on a number of ground water samples.

Reagents. As follows.

Standard nitrate, 100 mg l^{-1} of nitrate-nitrogen. Prepare by dissolving 0.721 g of potassium nitrate (dried at 105 °C) in water and diluting the solution to 1 l.

Standard nitrite, 100 mg l^{-1} of nitrite-nitrogen. Prepare by dissolving 0.492.6 g of sodium nitrite in water and diluting the solution to 1 l.

Sodium hydroxide. Stock sodium hydroxide – dissolve 200 g of sodium hydroxide in water and dilute the solution to 1 l. For a working solution, dilute 40 ml of the stock solution to 200 ml with water.

Sulphuric acid, 85 vol.%. Prepare by the careful addition of 850 ml of Aristar sulphuric acid, sp. gr. 1.84 to 150 ml of water.

2,4-Xylenol solution in acetic acid, 1%. Prepare by dissolving 1 g of the reagent in 100 ml of glacial acetic acid.

Sulphamic acid, 1%. Prepare by dissolving 10 g of sulphamic acid in 1 l of water.

Mercury (II) sulphate solution, 2%. Prepare by dissolving 2 g of mercury II sulphate in a mixture of 80 ml of water and 20 ml of sulphuric acid, sp. gr. 1.84.

Toluene. AnalaR.

Iron solution. BDH iron III chloride standard solution for atomic absorption spectroscopy. 1 ml = 1.00 mg of Fe.

Apparatus. Unicam SP500 Series 2 or equivalent, ultraviolet visible light spectrophotometer and absorption spectra were plotted using a double beam Unicam SP1800 ultraviolet visible light scanning spectrophotometer coupled to a chart recorder.

Procedure. Filter the samples through 0.45 µm membrane filters on collection and acidify by addition of 2 ml of 50 vol.% Arister sulphuric acid per litre of sample.

Ultraviolet Spectrophotometric Procedure. Dilute the samples as necessary to bring their concentrations of nitrate into the range 0 – 1.00 mg l^{-1} of nitrate-nitrogen and add sufficient 1% sulphamic acid solution to give a final sulphamic acid concentration of 0.1% w/v if the concentration of nitrite in the final solution is likely to exceed 20 µg l^{-1} of nitrite-nitrogen. Measure absorbances at 200 nm in 1.0-cm cells.

2, 4-Xylenol Method. To 10 ml of sample, or an aliquot containing not more than 300 µg of nitrate-nitrogen, in a 100-ml conical flask, add 2 ml of 2% mercury (II) sulphate and 0.5 ml of 1% sulphamic acid solutions; allow the flask to stand for 5 min and then add 1 ml of 1% 2, 4-xylenol solution and mix. Slowly add 24.5 ml of 85 vol.% sulphuric acid, while cooling the flask in running water. Place the flask in a water bath at 30 °C for 30 min, then transfer the contents to a 125-ml separating funnel and rinse the flask with two successive 25-ml volumes of water. Add 10 ml of toluene, shake the funnel gently for 3 min, allow the phases to separate and run the aqueous layer to waste. Add 10 ml of working strength sodium hydroxide solution to the toluene extract, shake for 1 min, allow the phases to separate, run off the sodium hydroxide layer into a 1.0-cm cell and measure the absorbance at 445 nm. Run a blank and a set of standards in the range 1 – 20 mg l^{-1} of nitrate-nitrogen.

Nitrite, which absorbs in a similar part of the ultraviolet region, is a potential source of interference in this method. A solution containing 1 mg l^{-1} of nitrite-nitrogen was scanned between 190 and 260 nm and the spectrum is shown in Fig. 3.5 together with that of a 1 mg l^{-1} solution of nitrate-nitrogen for comparison. The nitrite absorption maximum is at about 211 nm and it absorbs less strongly than nitrate at a similar concentration. Solutions containing 0.5 mg l^{-1} of nitrate-nitrogen and up to 100 µg l^{-1} of nitrite-nitrogen were prepared and the interference from nitrite is significant only at concentrations above

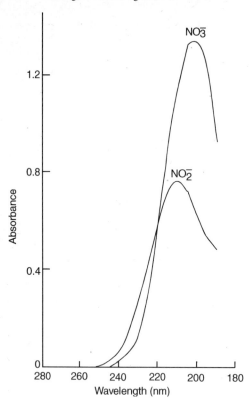

Fig. 3.5. Absorption spectra of nitrate and nitrite standard solutions containing 1 mg l^{-1} of nitrogen (From [3.111])

20 µg l^{-1}. Interference by higher concentrations of nitrite can be removed by the addition of 0.1% w/v of sulphamic acid.

Iron does not interfere at concentrations below 1000 µg l^{-1} and chloride is without effect at concentrations below 100 mg l^{-1}.

Results obtained for nitrate by ultraviolet spectroscopy and spectrophotometry are compared in Table 3.22. Good agreement was obtained for all ground water samples analysed.

3.1.3.2 Continuous Flow Analysis

Thompson and Blankley [3.101] have described an automatic ultraviolet continuous flow determination of nitrate in raw and potable water. The method has a linear calibration range of 0–30 µg l^{-1} nitrate N and a detection limit of 0.01 µg l^{-1}.

The incorporation of a dialysis membrane minimised the effect of humic acid type substances that did not freely diffuse through the membrane, and also avoided the necessity to filter water samples prior to measurement.

Table 3.22. Nitrate Concentrations determined by ultraviolet and 2, 4-xylenol methods. (From [3.111]; Copyright 1977, Royal Society of Chemistry, UK)

Type of water	Absorbances measured in 1-cm cells Nitrate-nitrogen/mg l^{-1}	
	Ultraviolet method	2, 4-Xylenol method
Bunter sandstone	0.04	0.04[a]
Bunter sandstone	8.3	8.2
Bunter sandstone	0.005[a]	0.04[a]
Bunter sandstone	0.54	0.55
Bunter sandstone	4.5	4.6
Lincolnshire limestone	7.9	7.9
Lincolnshire limestone	5.8	5.5
River Thames, Chiswick	11.0	10.9

[a]4-cm cell.

The method requires no sample pre-treatment or dilution. It utilises a conventional dialysis membrane to minimise interference effects from suspended, colloidal and organic matter.

Reagents. As follows.
  Distilled water.
  Potassium nitrate, 1000 μg ml^{-1} of N. Dissolve 7.221 g of potassium nitrate, dried for 1 h at 105 °C in 1 l of water.
  Humic acid (Aldrich Chemicals).
  Fulvic acid.

Reagent R_1. Dissolve 1.00 g of sulphamic acid in about 500 ml of water, carefully add 5.0 ml of sulphuric acid (98 wt%) followed by 4.00 g of hydroxylamine sulphate and 3 drops of a 30% w/v Brij 35 solution (BDH Chemicals). Dilute to 1 l with water. Prepare freshly every two weeks. The hydroxyl-amine sulphate was added to reduce any chromium VI, iron III, chromium III and iron II respectively. The lower oxidation states of these elements do not significantly absorb at 210 nm at the maximum concentrations likely to be encountered (see below).

Reagent R_2. Dissolve 10 g of anhydrous sodium sulphate in approximately 500 ml of water, add 3 drops of 30% w/v Brij 35 solution and dilute to 1 l.

Apparatus. Absorbance measurements were made at 210 nm with a Cecil Instruments 272 ultraviolet visible spectrophotometer, fitted with a 10-nm path length Spectrosil silica micro-flow cell (Type 75, Starna Ltd). The output trace was monitored on a Smiths Industries Servoscribe recorder. A Chemlab Instruments 15 channel peristaltic pump and 40 position autosampler were used. The continuous-flow manifold is depicted in Fig. 3.6.

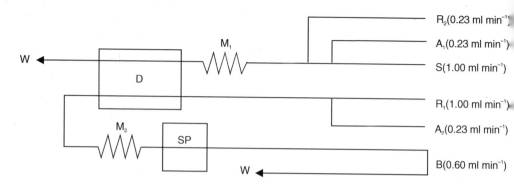

Fig. 3.6. Diagram of the continuous flow manifold. Air (A_1) is used to segment the flow of sample (S) that is mixed with reagent R_1 in a four-turn 20 mm diameter mixing coil M_1 and pumped through one half of the 150 mm diameter (D) to waste (W). Reagent R_2 is segmented with air (A_2) pumped through the other half of the dialyser and mixed in coil M_2 with dialysed sample. Mixing coils M_1 and M_2 are identical. Nitrate absorbance at 210 nm is measured by spectrosphotometer SP where part of the flow is pulled through a 10- mm silica flow cell by pump tube B (From [3.101])

Excellent precision is obtained by this procedure (Table 3.23).

The dialysis membrane minimised the ultraviolet absorption interference from organic matter present in water, as the efficiency of transfer of this matter through the membrane was significantly lower than that of nitrate ions. The nitrate concentrations of three highly coloured raw Pennine waters were determined by the conventional continuous-flow reduction method [3.103], the manual nitrate method [3.102] (using unfiltered samples), and the above dialysis method. The results (Table 3.24) show that the dialysis method eliminates the ultraviolet absorption contribution from organic matter observed with the manual method and that there is good agreement between the dialysis method and the continuous-flow reduction method. The manual method (no dialysis) is

Table 3.23. Precision of analytical results. (From [3.101]; Copyright 1984, Royal Society of Chemistry, UK)

Solution	Standard deviation[a]/µg ml^{-1} of N			Mean concentration found/µg ml^{-1} of N
	S_w	S_b	S_t	
0.00 µg ml^{-1} of N	0.032 (5)	–	–	0.031
2.00 µg ml^{-1} of N	0.10 (5)	N.s. (4)	0.11 (8)	2.03
15.00 µg ml^{-1} of N	0.17 (5)	N.s. (4)	0.17 (8)	14.95
18.00 µg ml^{-1} of N	0.16 (5)	0.25 (4)	0.30 (8)	18.04
River Dove water	0.061 (5)	N.s. (4)	0.071 (7)	4.58
River Dove water + 9.55 µg ml^{-1} of N	0.11 (5)	N.s. (4)	0.16 (6)	14.08[b]

[a] Numbers in parentheses correspond to the number of degrees of freedom. N.s. indicates that the result is not statistically significant at the 0.05 level of significance.
[b] Recovery of added nitrate was 99.48%.

Table 3.24. Comparison of methods for the determination of nitrate in some coloured raw Pennine water samples. (From [3.101]; Copyright 1984, Royal Society of Chemistry, UK)

Sample No.	Apparent colour A m^{-1} [a]	Nitrate concentration/µg ml^{-1} of N		
		Reduction method	Manual UV method[b]	Proposed method[c]
1	10.0	0.81	2.19	0.85
2	12.2	0.64	2.19	0.65
3	10.4	1.33	2.98	1.34

[a] Measured on unfiltered sample at 400 nm; A = absorbance units.
[b] No correction at 275 nm was applied to the manual method.
[c] Calibration range 0–2.5 µg ml^{-1} of N.

prone to very significant positive interference from the naturally occurring organic matter.

The interference effect of other substances is given in Table 3.25, and it can be seen that the method is relatively free of interference effects for its intended uses. There was a small positive bias from chromium VI, acetone and humic

Table 3.25. Effect of other substances on nitrate. If other substances did not interfere, the results would be expected (95‡ confidence) to be in the following ranges: 10.00 ± 0.33 µg ml^{-1} and 0.00 ± 0.05 µg ml^{-1}. Negligible interference was observed from the presence of Ca (200 µg ml^{-1}), Mg (200 µg ml^{-1}), K (200 µg ml^{-1}) and NH$_4$ (100 µg ml^{-1} N), all present as chlorides. (From [3.101]; Copyright 1984, Royal Society of Chemistry, UK)

Substance	Concentration µg ml^{-1}	Added as	Measured nitrate concentration[a] µg ml^{-1} of N	
Cl	1000	NaCl	10.19	0.02
SO$_4$	1000	Na$_2$SO$_4$	10.09	0.00
PO$_4$ (as P)	5	KH$_2$PO$_4$	9.74	0.01
F	5	NaF	9.70	0.00
Humic acid	250	–	10.15	0.33
Fulvic acid	250	–	11.58	1.36
Fe(II)	10	FeSO$_4$	10.27	0.06
Fe(III)	10	FeCl$_3$	9.89	0.00
NO$_2$ (as N)	50	NaNO$_2$	10.01	0.00
Cr(VI)	10	K$_2$Cr$_2$O$_7$	9.85	0.13
Persil[b]	10	–	9.76	0.00
Al	2	Al$_2$(SO$_4$)$_3$	9.92	0.00
Mn	2	MnSO$_4$	10.03	0.02
I	5	KI	10.17	0.00
Br	5	KBr	9.98	0.00
Glucose	100	–	10.25	0.00
Acetate	100	CH$_3$COONa	10.25	0.00
Benzoate	100	C$_6$H$_5$COONa	14.80	4.62
Acetone	100	–	10.75	0.45

[a] Results in the first column have an added nitrate concentration of 10 µg ml^{-1} of N.
[b] Proprietary detergent.

acid, and a somewhat larger bias from fulvic acid and the benzoate ion. However, the interference tests were carried out at concentrations of these determinands considerably in excess of levels likely to be encountered in potable waters.

3.1.3.3 Ultraviolet Spectroscopy

This technique has been investigated by several workers.

Rennie et al. [3.105] describe a method for determining down to 6 µg l^{-1} nitrate nitrogen in raw and potable waters which uses an activated carbon filter, at an elevated pH, to eliminate interference from organic matter. Interference from several cations that are precipitated out of solution is also removed by this system. The results obtained using the technique show that the method is precise.

Rennie et al. [3.105] utilised absorbances at 210 and 275 nm to indicate the presence of nitrate and organic matter respectively. The absorbance at 275 nm was used as an indicator of the presence or absence of organic matter that could also absorb at 210 nm.

Although some workers have corrected measurements of nitrate made at 210 nm for organic absorbance based on measurements at or near 275 nm, this method of correction becomes decreasingly reliable when low nitrate contents are being measured. Hence, in the method described below, organic matter is completely removed from the sample prior to ultraviolet spectroscopy. This method is based on the observation at that pH values above 12 one particular grade of carbon black ADC 33 (Sutcliffe Speakman Co.) will adsorb 98–100% of organic compounds from the water sample and, unlike many other grades of carbon black, will retain no nitrate. An analytical grade carbon filter paper (Schleicher and Schüll No. 58) was also satisfactory in this respect and permitted organics removal from the alkaline sample to be conducted in a filtration assembly.

Reagents. Reagents of analytical reagent grade are preferable. Distilled water should be used rather than deionised water, which could contain UV-absorbing materials derived from the ion exchangers and from organic matter present in the feedstock. It is advisable to check that the reference distilled water does not have an absorbance exceeding 0.01 at 210 nm. Using the 50-ml portion in the reference cell, the other portions and the residue should have an absorbance reading of less than 0.01 for the original distilled water to be considered acceptably free from UV-absorbing materials.

Sodium hydroxide solution, 3.5% w/v. Store in a polyethylene bottle (stable for about 3 months).

Mixed acid reagent. Dissolve 5 ± 0.1 g of sulphamic acid in 500 ml of vol.% sulphuric acid. This solution should be stored in borosilicate glass (stable for at least 2 months).

Standard nitrate solution, 2 mg l^{-1} N. Prepare freshly by 50-fold dilution of a 100 mg l^{-1} N stock standard solution (stable for 3 months) of potassium nitrate (0.7218 g l^{-1} of dried potassium nitrate). Dilute the standard solution as appropriate to produce 0.25, 0.5, 0.75, 1.0, 1.5 and 2.0 mg l^{-1} N standard solutions. Borosilicate glass containers should be used.

Apparatus. Ultraviolet speectrophotometer capable of measuring absorbances at 210 nm and a pair of matched ultraviolet-grade silica cells (path length 10 mm). The filtration apparatus shown in Fig. 3.7 consisting of 16 layers of 60 mm diameter Schleicher & Schüll No. 508 active carbon papers sandwiched between two 60 mm diameter Whatman GF/C papers in a Carlson-Ford filter holder (Gallenkamp & Co. Ltd. London), was used in conjunction with a suitable vacuum source and a Buchner flask.

To prepare a new carbon filter stack for use, wash with 500 ml of 3.5% w/v sodium hydroxide solution followed by 200 ml of distilled water. This treatment gives reagent blank values of the order of 0.025 absorbance unit. Between batches of analyses, contact with a laboratory atmosphere may result in contamination of the filter stack with UV-absorbing substances and it may be necessary to clean the filter stack by the above alkaline procedure in order to maintain satisfactory reagent blank values. The top GF/C paper may be changed between samples if the filter exhibits significant headloss. The operational life of the prepared filter stack is dependent on the organic content of the samples; a typical capacity is 100 samples with an absorbance of 0.045 at 275 nm before organic breakthrough occurs. Glass bottles should be used for samples; these bottles and other glassware used for the method should be cleaned by treatment with concentrated sulphuric acid followed by thorough rinsing with distilled water.

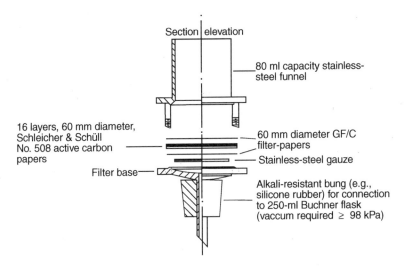

Fig. 3.7. Activated carbon filter assembly (From [3.105])

Procedure. Add 5 ± 0.1 ml of sodium hydroxide solution to 100 ml of sample containing not more than 2 mg l^{-1} of nitrate in a calibrated flask and mix. Pass 30 ml of this solution through the filter and discard it, then filter 50 ml of the solution and keep the filtrate. Pipette 40 ml of this filtrate into a 50-ml calibrated flask containing 5 ml of the mixed acid reagent, dilute to 50 ml with distilled water and measure the absorbance at 210 nm against water in the reference cell. Absorbances corrected for the reagent blank are then converted into milligrams of nitrogen per litre using a regressed equation from standard and a reagent blank treated in the same manner. A blank determination must be carried out with each batch of analyses by taking 100 ml of distilled water through the full procedure in place of the sample. The calibration graph is linear up to at least 2.0 mg l^{-1} N. Linear regressions of the calibration graph gave a mean correlation coefficient of 0.9996 for a $y = mx + c$ equation and the mean standard error estimate was 0.016 mg l^{-1} N. An absorbance reading of about 0.4 is given by a 1 mg l^{-1} N standard solution treated as described above.

The results given in Table 3.26 show that, for samples for which the absorbance at 275 nm is significant before carbon filtration, the absorbance falls to zero after carbon filtration. This indicates that the organic matter has been totally removed by the carbon. Now, if any species absorbing at 210 nm (other than nitrate) were present, the UV method would give a higher 'apparent' nitrate concentration than the automated method, which is specific for nitrate and is free from bias. There is no such bias between the two methods and so this validates the assumption that absorbances at 210 and 275 nm are inter-related such that absence of absorbance due to organic matter at 275 nm indicates absence of absorbance at 210 nm.

Nitrite interference is overcome by the presence of sulphamic acid in the mixed reagent.

The effects of certain concentrations of the adventitious contaminants are summarised in Table 3.27 and the concentration that gave an absorbance at 210 nm equivalent to 0.02 mg l^{-1} N is defined as the interference limit for the method (denoted by an asterisk) in Table 3.27. The lack of an interference effect, compared with that found by direct UV spectrophotometry, of the sodium salt of dodecylbenzene-sulphonic acid is due to the effectiveness of removal of organic matter by the filter. Whereas iron II and iron III interfere significantly in direct UV methods, such interference is eliminated in this method as a result of the formation of insoluble hydroxides at the elevated pH, the precipitates being removed by the GF/C paper on the filter. It is apparent that the interference effects of anions are not reduced by carbon treatment.

3.1.3.4 Ion Chromatography

Schwabe et al. [3.106] have described a method for the determination of nitrate (and chloride, fluoride, phosphate and sulphate) in potable waters. The method is discussed in Chap. 11.3.

Table 3.26. Comparison of results obtained by the proposed method and by an established automated method for raw and potable waters and final effluents. (From [3.105]; Copyright 1979, Royal Society of Chemistry, UK)

Sample	Nitrate content/mg l^{-1} N		Absorbance at 275 nm	
	Automated method	Proposed method	Without carbon filtration	With carbon filtration
Treated surface waters—				
Sutton Hall (1 + 1)	2.40	2.70	0.030	0.000
Lamaload ″	1.05	1.04	0.064	0.000
Alwen ″	0.40	0.52	0.044	0.000
Langthwaite ″	0.80	0.72	0.036	0.000
Burehole waters—				
Hooton (1 + 1)	3.20	3.20	0.006	−0.001
Haydock (1 + 3)	7.40	7.40	0.021	0.001
Slag Lane ″	0.05	0.06	0.023	0.004

Table 3.27. Interference effects. (From [3.105]; Copyright 1979, Royal Society of Chemistry, UK)

Interfering substance		Interference effect mg l^{-1} N	
Species	Concentration/ mg l^{-1}	With carbon filtration	Without carbon filtration
Chloride (Cl$^-$)	2500[a]	0.02	0.02
	5000	0.06	0.06
Bromide (Br$^-$)	1.0[a]	0.02	0.02
	2.0	0.04	0.04
Iodide (I$^-$)	0.65[a]	0.02	0.02
	1.30	0.05	0.05
Iron (Fe^{3+})	13.5	<0.02	<0.02
	27	<0.02	0.08
	1000[b]	<0.02	1.4
Iron (Fe^{3+})	0.2	<0.02	<0.02
	0.4	<0.02	0.08
	1000[b]	<0.02	1.75
Dichromate (Cr$_2$O$_7^{2-}$)	0.15[a]	0.02	0.02
	0.30	0.08	0.08
Manganese (Mn^{2+})	100	<0.02	<0.02
	1000[b]	<0.02	0.4
Dodecylbenzenesulphonic acid, sodium salt	3.5	<0.02	<0.02
	7.0	<0.02	0.06
	1000[b]	<0.02	1.8

[a]Indicates "Interference limit".
[b]Indicates Highest concentration investigated.

3.1.3.5 Polarography

Marecek et al. [3.107] investigated the feasibility of using a hanging drop electrode for the determination of nitrate, perchlorate, and iodide in potable water. A three-electrode system was used with a polarographic analyser. The potential range of the method was increased by using crystal violet dicarbollylcobaltate electrolyte in the nitrobenzene phase and magnesium sulphate in the aqueous phase, with a lead/lead sulphate reference electrode.

3.1.4 Industrial Effluents

3.1.4.1 Spectrophotometric Methods

Dodin et al. [3.108] observed that when a solution of nitrate at pH 2.0 is irradiated with ultraviolet light, the nitrate will oxidize methyl orange. They used this as the basis of a spectrophotometric method for determining nitrate in industrial effluents.

3.1.4.2 Ion Chromatography

Ion chromatography has been applied to the determination of nitrate in industrial effluents [3.112] (and chloride and bromide). This method is discussed in Chap. 11.4.

3.1.5 Waste Waters

3.1.5.1 Spectrophotometric Methods

Nitrates have been determined [3.113] in waste waters in the presence of nitrites by a spectrophotometric method using salicylane as the chromogenic agent. Nitrites interfere in the determination of nitrates using the salicylane method but can be removed by amido sulphonic acid. The spectrophotometric method with salicylane may be used for determining nitrate levels of up to 20 mg l^{-1} nitrate-nitrogen in the presence of equimolar concentrations of nitrite-nitrogen.

3.1.5.2 Raman Spectroscopy

Furaya et al. [3.114] have used resonance Raman spectroscopy to determine nitrite in waste waters. They also proposed a method for simultaneous determination of nitrite and nitrate. These workers [3.115] have also discussed a laser Raman spectroscopic method for the determination of down to 1 mg l^{-1} nitrate in waste waters.

Raman Spectra were recorded on a JASCO Model J-800 Laser Raman Spectrophotometer. A Spectra-Physics Model 164 argon ion laser was used as an exciting source. As a detector, an HTC-R-464 photomultiplier was used. The spectral data were processed by a data processor containing a micro-computer, supplied with the spectrophotometer. In order to improve the S/N ratios of the spectra, the data were accumulated and the number of repetitive scans was determined according to the noise levels for each of the samples. The sample solutions were illuminated with the 488 nm line at 200 mW output. A mirror served to double pass the laser beam through a cylindrical sample cell (0.3-ml) Pyrex glass. In order to obtain the largest and most reproducible intensities, it is important that the mirror system is adjusted carefully in such a way that the scattering lights focus into a clear image on the entrance slit. As a key band, the most intense Raman line at 1045 cm^{-1} was used.

In the case of waste and treated waters, the detection limit becomes about an order of magnitude higher than that in the case of pure water because of the strong luminescence of water samples. Furaya et al. [3.114] demonstrated that the addition of potassium iodide as a quencher makes the background markedly smaller. By this procedure, the sensitivity becomes comparable to that in the case of pure water samples.

A plot of the background intensities against the time of laser beam irradiation after the addition of various inorganic salts shows the effectiveness of potassium iodide.

Furaya et al. [3.114] recommended that Raman spectra were measured on sample solutions which were then exposed to the laser beam for about 30 min.

3.1.6 Sewage

3.1.6.1 Ion-Selective Electrodes

Nitrate levels sewage in amounts down to 1 mg l^{-1} have been determined by specific ion electrodes [3.116, 3.117]. Petts [3.117] used a non-porous plastic membrane nitrate selective electrode and compared results obtained with this electrode and those obtained by a standard spectrophotometric method in a sewage works effluent. The results indicate that the selective ion electrode is suitable both for laboratory and plant monitoring purposes. Only chloride and, to a lesser extent, nitrite and bicarbonate interfere in these measurements of nitrate.

Table 3.28 gives a survey of results obtained by both methods on sewage effluents with a high nitrate content.

Table 3.28. Nitrate analysis of effluents of sewage treatment plants[a]. (From [3.117]; Copyright 1975, Water Research Centre, Medmenham, UK)

Sample no.	Potentiometric NO_3^-N, mg l^{-1}	Spectrophotometric[b] NO_3^-N, mg l^{-1}
1	28	28
2	43	44
3	52	41
4	43	37
5	34	39

[a]The data represent a single analysis.
[b]As measured independently.

3.1.7 Plant and Soil Extracts

3.1.7.1 Ion Chromatography

Bradfield and Cooke [3.118] give details of a procedure for the determination of nitrate (and chloride, phosphate and sulphate) in aqueous extracts of soil by an ion chromatographic technique with indirect ultraviolet detection. Recoveries ranged from 84 to 108%. The technique is discussed further in Chap. 11.6.

3.1.8 Soils

3.1.8.1 Spectrophotometric Methods

Elton Both [3.119] determined nitrate in soils by a spectrophotometric method using a reaction based on nitration of 3.5-dinitrophenol.

Tecator [3.120] described a flow injection analysis for the determination of nitrate in 2 mol l^{-1} potassium chloride extracts of soil samples. Nitrate is reduced to nitrite with a copperized cadmium reductor and the nitrite is determined by a standard spectrophotometric procedure.

Hadjidemetriou [3.121] has carried out a comparative study of the determination of nitrates in calcifereous soils by the phenoldisulphonic acid and the chromotropic acid spectrophotometric methods, and by a nitrate ion-selective electrode method. He used 0.02 N cupric sulphate as soil extractant. Silver sulphate was added to remove chlorides. Nitrites, if present, were eliminated by acidifying the extract with N sulphuric acid. The most rapid methods were those based on chromotropic acid and the ion-selective electrode method. Correlation coefficients between results obtained by these methods were very consistent (0.9996–0.9998).

3.1.8.2 Ion-Selective Electrodes

Goodman [3.122] has described an automated procedure for the determination of nitrate in soils. The apparatus automatically extracts and analyses batches of up to 60 soil samples. Analysis is performed electrochemically by means of an ion-selective electrode and reference electrode. Corning ion-selective electrodes were found to be superior to those produced by Orion in this application. Recoveries of nitrate in this method were between 94 and 95%. The calibration curve was linear down to 2.5 mg l^{-1} nitrate.

Bremner et al. [3.123] used a nitrate-selective electrode for the measurement of nitrate in soil. Sulphate, phosphate, chloride and nitrite all interfere in this procedure.

3.1.8.3 Ion Chromatography

Bradfield and Cooke [3.124] give details of a procedure for the determination of nitrate and (chloride, phosphate and sulphate) in aqueous extracts of plant materials by an ion chromatographic technique with indirect ultraviolet detection. Recoveries ranged from 84 to 108%. This technique is discussed in further detail in Chap. 11.6.

3.1.9 Plants

3.1.9.1 Spectrophotometric Method

Tanaka et al. [3.125] have described a spectrophotometric method for the determination of nitrate in vegetable products. This procedure is based on the quantitative reaction of nitrate and 2-sec-butylphenol in sulphuric acid (5 + 7), and the subsequent extraction and measurement of the yellow complex formed in alkaline medium. The column reaction is sensitive and stable and absorbances measured at 418 nm obey Beers law for concentrations of nitrate-nitrogen between 0.13 and 2.5 µg ml^{-1}. In this procedure the vegetable matter is digested at 80 °C with a sodium hydroxide silver sulphate solution, concentrated sulphuric acid and 2-sec-butylphenol are added and after 15 min standing time the nitrated phenol is extracted with toluene. Finally the toluene layer is back extracted with aqueous sodium hydroxide and evaluated spectrophotometrically at 418 nm. The standard deviation of the whole procedure was 1.4% and analytical recoveries ranged between 91 and 98%.

3.1.9.2 Ion Chromatography

Bradfield and Looke [3.124] have applied ion chromatography with indirect ultraviolet detection to the determination of nitrate, together with chloride, phosphate and sulphate in plants and soils.

Using a revised phase system with a dynamically generated ion chromatographic column, retention times, respectively, of 7.9, 5.5, 12.6 and 18 min were obtained for the above anions. This technique is discussed further in Chap. 11.

3.1.9.3 Polarography

Hemmi et al. [3.126] have described a differential pulse polarographic procedure for the determination of nitrate in environmental samples such as silage, grass, plants, snow and water. This method utilizes the catalytic reaction between nitrate and uranyl ion in the presence of potassium sulphate.

The differential pulse polarographic peak is proportional to nitrate ion concentration from 1 to 50 $\mu mol\, l^{-1}$. The detection limit for nitrate in water is $8 \times 10^{-7}\, mol\, l^{-1}$.

Using this procedure, between 1 and 70 $mg\, g^{-1}$ nitrate was found in vegetation samples.

3.2 Nitrite

3.2.1 Natural Water

3.2.1.1 Diazotization Spectrophotometric Methods

Spectrophotometric methods for determining nitrite are generally based on conversion of nitrite to nitrous acid, then diazotization of an aromatic amine by the nitrous acid, followed by coupling of the diazotised product with an amine or phenol to produce a dye which is evaluated spectrophotometrically. Numerous diazotization–coupling reagent systems have been studied. There are reviewed in Table 3.29. Several of these methods have emerged as being particularly useful, some of which have been automated.

3.2.1.1.1 Diazotization Reagent: o-Nitroaniline

Coupling Reagent: N-(1-Naphthyl)*ethylene diamine dihydrochloride*. Chaube et al. [3.136] developed a method based on these reagents in which the coupled red-violet dye is extracted into isoamyl alcohol prior to spectroscopic evaluation at 545 nm.

Table 3.29. Diazotization-coupling systems for the spectrophotometric determination of nitrite in natural water

Diazotization agent	Coupling agent	max nm	Range	Limit of detection	Reference
4-Amino-1-naphthalene sulphonic acid	1-Naphthol	530	$3-10\ \mu g\ l^{-1}$	–	3.127
Naphthalene-2, 3-diamine		355			3.128
Sulphanilamide	n-(1-Naphthyl) ethylene-diamine	–	–	$0.05\ mol\ l^{-1}$	3.129
Composite Reagent-Sulphanilamide/ N-(-1 naphthyl) ethylene diamine hydrochloride OR		542	–	–	3.130, 3.131
Sulphanilic acid/ N-(-1 naphthyl) ethylene diamine hydrochloride OR		541	–	–	3.130, 3.131
4-nitroaniline/ N-(-1 naphthyl) ethylene diamine hydrochloride		542	–	–	3.130, 3.131
4o-Aminoaceto-phenone	N-Phenyl-1-naphthylamine	610	$0.1-4\ \mu g\ l^{-1}$	–	3.132
4-Nitroaniline	1-Naphthol	–	–	–	3.133
4-Nitroaniline	8-Quinolinol	550	–	–	3.134
4-Nitroaniline	2-Methyl-8-quinolinol	585	–	–	3.135
o-Nitroaniline	N-(-1 Naphthyl) ethylene diamine hydrochloride	525	–	–	3.137
Sulphanilic acid	8-Amino-naphthlene-2-sulphonic acid	525	–	–	3.137
Miscellaneous	Miscellaneous	–	–	–	3.138
4-Amino Salicylic acid	Naphthol-1-1-ol	520	$100-3000\ \mu g\ l^{-1}$	–	3.139
Orthanilic acid	Pesorcinol	426	$1-12\ \mu g\ l^{-1}$	–	3.140
P-Aminophenol	Pesorcinol	435	–	–	3.141
Aniline-4-Sulphonic acid	N-Ethyl naphthylamine	537	$1-3\ mg\ l^{-1}$	–	3.200
Aniline-4-Sulphonic acid	N-Phenyl-1-naphthylamine-8-Sulphonic acid	554	$1-3\ mg\ l^{-1}$	–	3.200

Reagents. All reagents used are of analytical-reagent grade, and are as follows. Standard 1000 μg ml^{-1} nitrite solution. Prepare in deaerated doubly distilled water from reagent dried for 4 h at 110 °C and standardized. Add a small amount of chloroform as stabilizer. Prepare a 1 μg ml^{-1} working standard by dilution with deaerated, doubly distilled water.

o—Nitroaniline solution, 0.001 mol l^{-1} in 20% aqueous ethanol.
N—(1-Naphthyl)ethylenediamine dihydrochloride solution, 0.1%.

Procedure. To a known volume of water sample (containing 2.5–1.5 µg of nitrite) in a 25-ml standard flask, add 1 ml of o-nitroaniline solution and adjust to 1 mol l^{-1} hydrochloric acid concentration. Shake the flask for the next 2 min to ensure complete diazotization. Add any masking agent needed and 2 ml of N-(1-naphthyl)ethylene diamine and make up to the mark with 5 mol l^{-1} hydrochloric acid. Measure the absorbance at 545 nm against distilled water after 5 min. Prepare a calibration graph for 2.5–15 µg of nitrite in a similar manner.

Solvent extraction. Place a known volume of sample (up to 100 ml, containing 2.5–15 µg of nitrite) in a 250-ml separatory funnel and form the red-violet dye as described above, except that the final acidity is adjusted to 2.5 mol l^{-1} hydrochloric acid. Extract the dye with two 10-ml portions of isoamyl alcohol, dry the combined extracts with anhydrous sodium sulphate and dilute to volume in a 25-ml standard flask with the solvent. Measure the absorbance at 545 nm against a reagent blank similarly treated. Prepare a calibration graph by treating standards in a similar manner.

The effects of foreign species are shown in Table 3.30, which gives concentrations of foreign species that cause 2% error in the determination of 0.1 mg l^{-1} of nitrite by the extraction method (100 ml samples). Copper II, iron II and sulphur dioxide interfere but the interference of up to 100 mg l^{-1} of sulphur dioxide can be eliminated by oxidizing it to sulphate with hydrogen peroxide. Bismuth and iron(III) require masking with tartrate.

3.2.1.1.2 Diazotization Reagent: p-Nitroaniline

Coupling Reagent:2-Methyl-8-quinolinol. Tsao and Underwood [3.135] give a procedure for the determination of low concentrations of nitrite in natural water

Table 3.30. Effect of diverse species on determination of 0.1 mg l^{-1} of nitrite (100 ml of solution extraction method). (From [3.136]; Copyright 1984, Pergamon Publishing Co., UK)

Tolerence limit ppm	Species
800	SO_4^{2-}, NO_3^-, PO_4^{3-}, HCO_3^-, SiO_3^{2-}
150	Mg^{2+}, Ca^{2+}, Sr^{2+}, Ba^{2+}, Zr^{4+}, Co^{2+} Zn^{2+}, Cd^{2+}, Hg^{2+}, Pb^{2+}
100	Li^+, Be^{2+}, Cr^{3+}, Mo(VI), W(VI), Ni^{2+} Sb^{3+}, Bi^{3-a}, Se(IV), Te(IV), F^-, Br^-
50	Fe^{3+}, Bi^{3+a}, Se(IV)
40	I^-, aniline, HCHO, phenol
10	$S_2O_3^{2-}$
2	Cu^{2+}

^{a}Masked with 1 ml of 10% sodium potassium tartrate solution.

which involves diazotization with p-nitroaniline and coupling of the diazonium salt with 2-methyl-8-quinolinol, followed by solubilization of the azo dye in aqueous hexadecyltrimethylammonium bromide:

$$O_2NC_6H_4NH_2 + NO_2^- + 2H^+ \rightarrow O_2N°C_6H_4N^+ = N + 2H_2O$$
$$O_2N°C_6H_4N^+ = N + C_{10}H_{10}NO \rightarrow O_2N°C_6H_4N = N°C_{10}H_9NO.$$

Apparatus. Cary Model 118 spectrophotometer and a Radiometer RMH 26 pH meter.

Reagents. As follows.

Hexadecyltrimethylammonium bromide (J. T. Baker Chemical Co Analyzed Reagent).

Reagent-grade sodium nitrite dried at 110 °C for 4 h.

p-Nitroaniline recrystallized from isopropanol and from petroleum ether.

2-Methyl-8-quinolinol (8-hydroxyquinaldine) recrystallized from isopropanol.

Procedure. Transfer an aliquot of less than 5.4 ml of the sample containing 0.02–4.0 µg of nitrite to a 10-ml volumetric flask. Add 0.5 ml of 1×10^{-2} mol l^{-1} p-nitroaniline in 2 mol l^{-1} hydrochloric acid and mix the solutions. Then add 0.4 ml of 1×10^{-2} mol l^{-1} 2-methyl-8-quinolinol in 1 mol l^{-1} hydrochloric acid and 0.5 ml of 1×10^{-1} mol l^{-1} disodium ethylenediaminetetraacetate solution and, after mixing 0.7 ml of 4 mol l^{-1} sodium hydroxide solution. Finally, add 2.5 ml of 0.05 mol l^{-1} hexadecyl trimethyl ammonium bromide in 0.05 mol l^{-1} sodium chloride, dilute the solution to the mark with water, and mix thoroughly. After 1 min measure the absorbance at 585 nm and determine the nitrite concentration from a standard graph obtained with known solutions.

Table 3.31 shows the concentrations of some species which caused errors of 2% in nitrite determinations at the 0.46 mg l^{-1} level.

Norwitz and Keliher [3.130] determined nitrite in natural water using composite reagents containing sulphanilamide, sulphanilic acid, or 4-nitroaniline as the diazotizable aromatic amine and N-(1-naphthyl)ethylenediamine as the coupling agent. As a result of this work they recommended the following composition for the composite coupling reagents: (i) 0.25% sulphanilamide, 0.0060% N-(1-naphthyl)ethylenediamine and 5.42% hydrochloric acid; (ii) 0.50% suphanilic acid, 0.0050% N-(1-naphthyl)ethylenediamine and 0.29% hydrochloric acid; and (iii) 0.25% 4-nitroaniline, 0.010% N-(1-naphthyl)ethylenediamine and 15.0%

Table 3.31. Species and concentrations causing errors of 2% in the determination of 0.46 mg l^{-1} of nitrite. (From [3.135]; Copyright 1982, Elsevier Science Publishers Co., BV, Netherlands)

PO_4^{3-} (2200), SCN^- (200), Ca^{2+} (300), Mg^{3+} (85), CrO_4^{2-} (3000), Cr^{3+} (50), MoO_4^{3-} (400), Mn^{2+} (110), Fe^{3+} (60), Co^{2+} (180), Ni^{2+} (290), Zn^{2+} (200), Cd^{2+} (560), Al^{3+} (540), Pb^{2+} (800), HCHO (100), C_6H_5OH (80)

sulphuric acid. The wavelengths of maximal absorption were 542, 541 and 542 nm respectively. The effects of varying the ratios of the reagents, the overall amount of the reagents, the amount of acid, and the reaction temperature were studied as was the effect of methanol on the colours. The importance of the ratios of reagents as well as the overall quantities was demonstrated.

These workers [3.131] also studied interferences by various organics on the spectrophotometric determination of nitrite using composite diazotization –coupling reagents. They used three composite reagents to determine nitrite according to published methods: sulphanilamide and N-(1-naphthyl) ethylenediamine (NED); sulphanilic acid and NED; and 4-nitroaniline and NED. The effect of 54 organic compounds on the determination of nitrite by each of the above methods was established; tolerance limits are listed. Compounds examined included aliphatic amines, aromatic amines, phenolic compounds and miscellaneous organic compounds such as acetamide, cholesterol, rennin, dodecyl sodium sulphate, acetophenone, citric acid, caffeine, morpholine, starch, casein and sorbic acid. The effect of detergents and soap was also examined. Water immiscible solvents did not affect the colour. Aliphatic amines caused low results for all three tests as did many aromatic amines. Primary aromatic amines interfered more than secondary or tertiary amines. Phenolic compounds induced low results.

Coupling Reagent: 8 Quinolinol. This method [3.134] is very similar to that of Tsao and Underwood [3.135] except that it does not include the use of hexadecyltrimethyl-ammonium bromide as a solubilization agent for the dye.

Apparatus. ECIL spectrophotometer Model GS 865 or a Carl Zeiss Spekol with matched 1-cm glass cells.

Reagents. As follows.

Standard sodium nitrite solution, 1 mg nitrite ml^{-1}. Place 150 mg of dried analytical-grade reagent in 100 ml of deaerated, doubly-distilled water. Add a little chloroform as stabilizer. Prepare working standards by appropriate dilution.

p-Nitroaniline solution. Commercially available reagent, crystallized twice before use, as a 1×10^{-3} mol l^{-1} solution in 2 mol l^{-1} hydrochloric acid.

Procedure. Transfer an aliquot (not more than 15 ml) of the water sample containing 2–28 µg nitrite to a 25-ml volumetric flask and add 2 ml of p-nitroaniline reagent. Shake for 1 min and add 1 ml of ethanolic 0.2% (w/v) 8-quinolinol solution and 1 ml of aqueous 10% (w/v) disodium–EDTA solution. Make alkaline (to about pH 12) with 2 mol l^{-1} sodium hydroxide solution (about 3 ml) and dilute to the mark with distilled water. Measure the absorbance at 550 nm against a reagent blank. Calculate the amount of nitrite from a calibration graph prepared from measurements done in the same way.

Table 3.32 shows the effect of foreign ions on the determination of nitrite by this procedure.

Table 3.32. Effects of some ions and compounds on the determination of 0.4 mg l^{-1} nitrite. (From [3.134]; Copyright 1979, Elsevier Science Publishers BV, Netherlands)

Ion (tolerence limit in mg l^{-1})
PO_4^{3-} (1000), NO_3^- (1000), SO_4^{2-} (1000), Br^- (400), I^- (1.6), SO_3^{2-} (10^a), K^+ (1000), Be^{2+} (80), Ca^{2+} (160^b), Mg^{2+} (40^b), Ba^{2+} (200^b), Cr^{6+} (1.6), Cr^{3+} (8), Mo^{6+} (40), W^{6+} (40), Mn^{2+} (20), Fe^{3+} (40^c), Co^{2+} (40^b), Ni^{2+} (60^b), Zn^{2+} (40^b), Cd^{2+} (80^b), Hg^{2+} (80^b), Al^{3+} (60^b), Pb^{2+} (80^b), NH_4^+ (1000), SCN^- (200), Bi^{3+} (60^c), As^{3+} (20), Sn^{2+} (40), aniline (40), Formaldehyde (40, phenol (40))

^{a}In the presence of 0.5 ml of 1% H_2O_2 (20 vol).
^{b}In the presence of 1 ml of 10% EDTA.
^{c}In the presence of 1 ml of 10% sodium potassium tartrate.
All masking agents were added after 8-quinolinol.

Metal ions forming hydroxides in alkaline medium were expected to interfere but a large number of these ions were masked with EDTA. Copper(II), iron(II) and sulphide ions caused serious interference. The tolerance limits of the foreign ions shown in Table 3.32 are the amounts that caused not more than 2% change in the absorbance during the determination of a fixed amount of nitrite.

3.2.1.1.3 Diazotization Reagent: 4-Aminosalicylic Acid

Coupling Reagent: Naphth-l-ol. Flamerz and Bashir [3.139] studied the effects on this method of foreign ions that often accompany nitrite in water samples (Table 3.33).

The method is more selective towards anions than cations; this is not unexpected because these azo dyes are commonly used as chromogens for metal ions. However, an ion-exchange procedure can be used to remove the seriously interfering ions.

Apparatus. Shimadzu UV-210A double-beam spectrophotometer using 10-mm glass cells.

Reagents. As follows.

Standard nitrite solution. Dissolve 1.4990 g of sodium nitrite in distilled water, add 1 ml of spectroscopic grade chloroform and a pellet of sodium hydroxide and dilute the solution to 1 l with distilled water. This solution contains 1000 mg l^{-1} of nitrite ion. Prepare less concentrated solutions by dilution with distilled water.

4-Aminosalicylic acid solution. Dissolve 0.25 g of the sodium salt of 4-aminosalicylic acid monohydrate in 0.14 mol l^{-1} hydrochloric acid and dilute to 100 ml with 0.14 mol l^{-1} hydrochloric acid.

Naphth-l-ol solution. Add 0.35 g of naphth-l-ol to distilled water containing 1.5 g sodium hydroxide, dissolve by stirring and dilute to 100 ml with distilled water.

Table 3.33. Effect of diverse ions on the determination of nitrite. (From [3.139]; Copyright 1981, Royal Society of Chemistry, UK)

Interferent	Permissible amount in the presence of 10 μg of nitrite μg^{-1}
Ammonia	150
Carbonate	50
Chloride	180
Cyanide	75
Fluoride	120
Iodide	140
Hydrogen carbonate	110
Nitrate	80
Phosphate	110
Sulphide	4
Sulphite	25
Calcium	60
Cadmium	9
Cobalt (II)	8
Copper (II)	10
Iron (III)	7
Magnesium	15
Mercury (II)	30
Lead (II)	30
Tin (II)	20
Phenol	170

[a] Amount of diverse ion causing an error of less than 2% in the determination.

Procedure. Transfer an aliquot of the sample solution containing 1–30 μg of nitrite into a series of 10-ml calibrated flasks, add 1 ml of 0.25% 4-aminosalicylic acid solution and 1 ml of 0.35% naphth-1-ol solution (in 5.5% sodium hydroxide solution), dilute to the mark with distilled water and mix thoroughly. Measure the absorbances against a reagent blank prepared in the same manner but containing no nitrite ion, at 520 nm using 10-mm glass cells.

The colour develops immediately and is stable for up to 36 h. The nitrite concentration can be determined from a calibration graph constructed by plotting known concentrations of nitrite against the corresponding absorbance values, which gives a straight line passing through the origin at nitrite concentrations in the range 0–3 μg ml^{-1}.

The azo dye formed in alkaline medium has a maximum absorption centred at 520 nm. The slight absorption of the reagent blank at 520 nm emphasised the need for measurements to be performed against the reagent blank.

3.2.1.1.4 Diazotization Reagent: *p*-Aminoacetophenone

Coupling Reagent: Resorcinol. Suggested reaction sequence is

$$CH_3-C(=O)-C_6H_4-NH_2 + NO_2^- + 2H^+ \longrightarrow CH_3-C(=O)-C_6H_4-N^+N + 2H_2O$$

$$CH_3-C(=O)-C_6H_4-N^+\equiv N + \underset{HO}{C_6H_4}-OH \xrightarrow{pH\ 9} CH_3-C(=O)-C_6H_4-N=N-+H$$

$$\downarrow KOH$$

$$CH_3-C(-O^-)=C_6H_4=N-N=C_6H_4=O + H_2O.$$

Wu and Liu [3.141] have described a spectrophotometric technique based on these reactions. The method is based on the diazotization of p-aminoacetophenone followed by coupling with resorcinol in a sodium carbonate/sodium acetate medium to form a bright gold water-soluble azo dye. Mercuric chloride is added to eliminate the interference of sulphide and a composite EDTA/sodium hexametaphosphate reagent is used to mask foreign ions.

Reagents. As follows. Colour developing solution. Dissolve 2g of p-acetophenone and 1 g of resorcinol in 100 ml of hydrochloric acid (1 + 4). Keep the solution in a brown bottle in the refrigerator.

Complexing agent solution. Dissolve 10 g of sodium hexametaphosphate and 10 g of EDTA in 90 ml of water, heating until completely dissolved. Cool and dilute to 100 ml with water.

Mixed alkali solution. Dissolve 10 g of anhydrous sodium carbonate and 8.6 g of anhydrous sodium acetate in 50 ml of water, add 17 ml of dimethylformamide and dilute to 100 ml with water.

Standard nitrite solution. Dissolve 0.1500 g of sodium nitrite in water and dilute to 1000 ml. Add 1 ml of chloroform to inhibit bacterial growth. Dilute 100-fold to give a 1.0 mg l^{-1} working standard solution.

All reagents used were of analytical grade, and demineralized water was used throughout.

Procedure. If sulphide or iodide present place 100 ml of water sample in a beaker, and add 2 ml of 5% mercuric chloride solution. Let stand for 5 min then filter to remove mercuric sulphide and iodide. Analyse a known volume of the filtrate by the procedure below.

To a 100-ml Nessler tube (graduated at 50 ml) add an appropriate volume of test solution, containing less than 20 µg of nitrite, make up to the 50-ml mark with water, add 2.0 ml of complexing agent solution, 2.0 ml of colour developing solution, and mix. After 5 min add 6.0 ml of mixed alkali solution. Measure the absorbance at 435 nm in a 2-cm cell against a reagent blank prepared in the same manner but containing no nitrite.

Wu and Liu [3.141] examined the effect of various amounts of foreign ions on the determination of nitrite (Table 3.34). Most of the cations and common anions do not interfere; sulphide, thiosulphate, tetrathionate and iodide interfere but their effect can be mitigated by adding mercuric chloride solution as indicated in the method above.

3.2.1.1.5 Diazotization Reagent: Orthanilic Acid

Coupling Reagent: Resorcinol. Bashir et al. [3.140] have described a sensitive method based on these reagents which operates over the range 1–12 µg nitrite with a relative standard deviation, depending on concentration, in the range 0.4–3.5%. Cobalt seriously interferes in this method at the 4 mg l^{-1} level whilst sulphate is without interference even when present in the sample at concentrations between 1 and 1.5 g l^{-1}.

Apparatus. Shimadzu UV-210A double-beam recording spectrophotometer with 1-cm optical glass matched cells.

Table 3.34. Effects of various ions on determinations of 10 µg of nitrite. (From [3.141]; Copyright 1983, Pergamon Publishing Co., UK)

Foreign ion	Taken mg	Ion/NO$_2^-$	NO$_2^-$ found, µg
Zn (II)	2.0	200	9.8
Ni (II)	2.0	200	10.0
Mn (II)	2.0	200	10.0
Mo (VI)	2.0	200	10.0
Hg (II)	74.0	7400	10.0
Fe (III)	0.5	50	9.8
Bi (III)	0.5	50	10.0
Zr (IV)	0.5	50	10.0
Pb (II)	0.25	25	10.0
Cu (II)	0.10	10	10.2
Ag (I)	0.10	10	10.0
Se (IV)	0.10	10	10.0
V (V)	0.025	2.5	10.0
S^{2-}	7.0	700	10.1[a]
S$_2$O$_3^{2-}$	3.3	330	9.8[a]
S$_2$O$_4^{2-}$	2.5	250	9.8[a]
SO$_3^{2-}$	1.0	100	10.0[a]
I$^-$	5.0	500	10.2[a]
F$^-$	2.0	200	9.9

[a] After addition of 2 ml of 5% mercuric chloride solution, letting stand for 5 min, filtering off and washing with water.

Nitrogen-Containing Anions

$$\text{C}_6\text{H}_4(\text{SO}_3\text{H})-\text{NH}_2 + 2\text{H}^+ + \text{NO}_2^- \longrightarrow \text{C}_6\text{H}_4(\text{SO}_3\text{H})-\text{N}_2^+ + 2\text{H}_2\text{O}$$

$$\text{C}_6\text{H}_4(\text{SO}_3\text{H})-\text{N}_2^+ + \text{C}_6\text{H}_4(\text{OH})-\text{OH} \longrightarrow \text{C}_6\text{H}_4(\text{SO}_3\text{H})-\text{N}=\text{N}-\text{C}_6\text{H}_3(\text{OH})-\text{OH} + \text{H}^+$$

$$\text{C}_6\text{H}_4(\text{SO}_3\text{H})-\text{N}=\text{N}-\text{C}_6\text{H}_3(\text{OH})-\text{OH} + 2\text{OH}^- \longrightarrow \text{C}_6\text{H}_4(\text{SO}_2^{2-})-\text{N}-\text{N}=\text{C}_6\text{H}_3(\text{O})-\text{OH}^-$$

Reagents. As follows.

Standard nitrite solution. Dissolve 0.1499 g of sodium nitrite in distilled water, containing a pellet of sodium hydroxide to prevent liberation of nitrous acid, and 1 ml of spectroscopic-grade chloroform to inhibit bacterial growth. Dilute to 1000 ml with distilled water.

Orthanilic acid solution. Dissolve 0.2 g solid in distilled water containing 3 ml of concentrated hydrochloric acid. Dilute to 100 ml with distilled water.

Resorcinol solution, 0.2% (w/v) aqueous solution.
Potassium hydroxide solution, 2% (w/v).

Procedure. To a series of 10-ml volumetric flasks transfer 1 to 12 µg of nitrite, 1 ml of the orthanilic acid solution, 1 ml of resorcinol solution and 2 ml of potassium hydroxide solution. Dilute to the mark with distilled water. Measure absorbance against a reagent blank prepared in the same manner but containing no nitrite, at 426 nm using 1-cm cells. The colour develops immediately and is stable for 5 h. A straight-line calibration curve passing through the origin is obtained.

The absorption spectrum of the dye produced under these conditions is shown in Fig. 3.8.

3.2.1.2 Non-Diazotization Spectrophotometric Methods

Spectrophotometric methods for the determination of nitrite which are dependent on colour forming reactions other than diazotization and coupling include reaction with 2,3-dimethyl-1-phenyl-pyrazolin-5-one [3.142], 2-mercaptoethanol [3.143], mercaptoacetic acid [3.144], dihydroxy-4-imino-2-oxochroman [3.145], zirconyl ion [3.146] and 3,3' dimethyl benzidine [3.201].

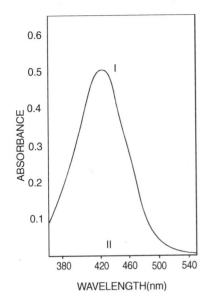

Fig. 3.8. Absorption spectra of (I) 6 μg of nitrite, treated as in the procedure, measured against reagent blank and (II) reagent blank measured against distilled water (From [3.140])

Nishimura et al. [3.147] carried out a study of the reaction between nitrite and the Greiss-Romijn reagent and, on the basis of this, devised a spectrophotometric method for the determination of nitrite in hot spring waters. Interference by iodide and sulphur compounds was avoided by the addition of mercuric chloride.

Lynch [3.148] has discussed the application of inverse spectrophotometry to the detection of nitrate in water using cerium IV as a chromophore.

Sanchez Pedreno et al. [3.149] have described a kinetic determination of traces of nitrite based on its inhibitory effect on the photochemical reaction between iodine and ethylenediamine tetraacetic acid.

Gabbay et al. [3.146] have described a rapid and particularly sensitive technique, for the determination of nitrite. Nitrites react with resorcinol in an acidic medium and the nitroso product forms a pale yellow chelate with the zirconyl ion. The absorbance of the chelate is measured at 347 nm. The method is suitable for the determination of nitrite ion in a 1-cm cell in the range from a few parts per 100 million to about 1 mg l^{-1}.

Proposed Reaction Mechanism. The reaction sequence involves two steps. In the presence of micro-concentrations of nitrites, resorcinol is presumed to react with nitrous acid as follows:

Nitrogen-Containing Anions

[Reaction: resorcinol (1,3-dihydroxybenzene with numbered positions 1,2,3,4) + HONO → nitroso-resorcinol (OH, OH, NO substituted benzene) + H$_2$O.]

The nitroso product exists in equilibrium with its tautomeric quinoidal oximes, which include a double-bond conjugated system, i.e.

[Reaction: quinoidal oxime (OH, =NOH, =O) + ½ZrO^{2+} → chelate with ZrO/2 + H$^+$]

The presence of zirconyl ions in the second step causes complete displacement of the equilibrium towards the 1-hydroxy quinoidal oxime, owing to the rapid formation of the coloured chelate, as follows:

[Tautomeric equilibria among three forms: (O, OH, NOH) ⇌ (OH, OH, NO) ⇌ (OH, O, NOH)]

Equipment. Spectrophotometer with 10- and 1-mm glass cells.

Reagents. Solutions A and B as follows.
   Solution A. Dissolve 1 g of resorcinol and 1.1 g of zirconyl chloride octahydrate in doubly distilled water containing 7.5 ml of concentrated hydrochloric acid. Dilute in doubly distilled water containing 7.5 ml of concentrated hydrochloric acid and dilute the solution to 500 ml.
   Solution B. Dissolve 1 g of sodium sulphate and 1.5 g of sodium acetate trihydrate in 500 ml of doubly distilled water.
   On the day of sampling, mix equal volumes of the two solutions. These solutions remain stable for several weeks if stored at 4 °C in a tightly stoppered amber-glass bottle.

Procedure. Pipette 2 ml of the sample containing nitrite in the range 0.3–5 mg l^{-1} into a 10-ml calibrated flask. Fill the flask to the mark with reagent solution and shake it thoroughly. After allowing 3 min for the reaction to take place, read the absorbance in a 1 cm cell at 347 nm against a similarly prepared blank. The concentration of nitrite can then be determined from a standard calibration graph, which is constructed by plotting known concentrations of nitrite against absorbance.

Figure 3.9 shows the absorption spectrum of the colour produced at pH 1.5, the maximum absorption occurring at 347 nm. Slight absorption of the reagent blank at 347 nm indicated the need for all measurements to be made vs the blank. The magnitude of the blank absorbance increased slightly with ageing of the reagent. The absorption of the reagent blank at 347 nm in a 1-cm cell, measured against water, was in the range 0.020–0.050, depending on the age of solution A.

Beer's law holds up to a concentration of 7 mg l^{-1} of nitrite, as is demonstrated by the calibration graph, which was a straight-line passing through the origin over this range.

The effects of various potential interferences was investigated (Table 3.35). Some of the problematic interfering ions in the diazotization methods, such as sulphite, copper(II), iron(II) or iodide, are tolerated in this procedure.

The interfering ions at pH 1.5 fall into two categories. The first group are ions that form, with the reagent, turbid products such as fluoride or phosphate. This interference is probably caused by the formation of a fluoride zirconate and diphosphatozirconic acid. The second group are ions that absorb light in the wavelength region of the reaction product, such as iron(III), dichromate and permanganate. These ions cause positive interference.

Among the ions that commonly occur in water samples, phosphate and iron(III) are the only ions that produce serious interference. However, as

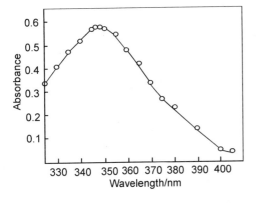

Fig. 3.9. Absorption spectrum of 1 mg l^{-1} of nitrite ion at pH 1.5 measured against reagent blank (From [3.146])

Table 3.35. Effect of diverse ions (nitrite concentration 1 mg l^{-1}). (From [3.146]; Copyright 1977, Royal Society of Chemistry, UK)

Ion investigated	Ratio of ion to nitrite (w/w)	Error %
Acetate	200	n
Bromide	200	n
Carbonate	200	n
Chlorate	200	−1.7
Dichromate	4	+5.1
Fluoride	50	n
Iodate	200	n
Iodide	150	−3.8
Manganate	12	+3.2
Oxalate	200	n
Perchlorate	200	+1.7
Persulphate	200	−1.8
Phosphate	8	+5.1
Sulphate	200	n
Sulphite	200	−1.7
Aluminium	200	n
Ammonium	200	n[a]
Cadmium	200	n
Calcium	200	n
Cerium (III)	200	n
Cobalt (II)	200	n
Copper (II)	200	−1.6
Iron (II)	200	−2.3
Iron (III)	8	+5.1
Lead (II)	200	n
Lithium	200	n
Magnesium	200	n
Manganese (II)	200	n
Potassium	200	n
Sodium	200	n
Strontium	200	n

n = negligible (less than the relative standard deviation).

interference from iron (III) constitutes an inherent disadvantage, its concentration must be reduced before analysis to within the tolerated concentration. This treatment can be carried out using a cation-exchange resin (Amberlite IR-120).

The interference by phosphate ion can be prevented by the addition of suitable amounts of sulphuric acid to the mixed solutions A and B. This procedure is accompanied by a gradual decrease in absorbance as the amount of acid added is increased.

3.2.1.3 Segmented Flow Analysis

This technique, discussed in Sec. 2.4.1.2, can be applied to the determination of nitrite in the 100–2000 $\mu g\, l^{-1}$ range in water.

3.2.1.4 Spectrofluorimetric Analysis

Rubio et al. [3.150] have described a spectrofluorimetric method for determining nitrite whereby the nitrite ion oxidizes pyridoxal-5-phosphate-2-pyridylhydrazone in acidic media to give a fluorescent product. Although less sensitive than other methods, it was selective, simple and rapid, and could be applied in the 0.1–1.0 $mg\, l^{-1}$ nitrite range.

3.2.1.5 Ultraviolet Spectroscopy

Espinola [3.151] has described a direct ultraviolet spectroscopic method for the determination of nitrite in natural waters. By using an addition technique and a reference nitrate/nitrite solution, he compensated for the interference caused by the overlapping of the nitrate and nitrite bands, which is normally a limiting factor in the analysis of mixtures of nitrite with large excesses of nitrate. The detection limit was 5×10^{-5} $mol\, l^{-1}$ nitrite, which corresponded to a minimal detectable amount of 2.3 $mg\, l^{-1}$ nitrite in the presence of up to 20,000 times greater an amount of nitrate.

The near-ultraviolet spectrum of sodium nitrite in aqueous solution is shown in Fig. 3.10. In aqueous solution the same general characteristic spectrum as in the nitrate medium is maintained, but a small blue shift is observed. At low concentrations, the very intense higher energy bands do not interfere with the lower energy bands of either nitrate or nitrite.

From spectra obtained independently it might be inferred that the broadening of the nitrate band would not strongly interfere with that of the nitrite at λ_{max} if the mutual concentrations did not exceed 1 $mol\, l^{-1}$ nitrate and 10^{-2} $mol\, l^{-1}$ nitrite. However, in reality the analysis of a mixture containing these proportions is unfeasible because the broadened nitrate band deforms the nitrite band causing a large error.

In the spectrophotometric analysis of nitrite in the presence of nitrate, it is fortunate that the nitrite ion in molten nitrate, as well as in aqueous solutions, has an absorption minimum in the region of the nitrate band, and an absorption maximum at 355 nm.

When the concentration of nitrite is greater than 10^{-2} $mol\, l^{-1}$, if the concentration of nitrate does not exceed 10^{-1} $mol\, l^{-1}$, the absorbances of nitrite in the presence of nitrate can be measured at the maximum wavelength vs a suitable reference solution. The reference solution must contain approximately the same concentration of nitrate as in the unknown sample. For concentrations

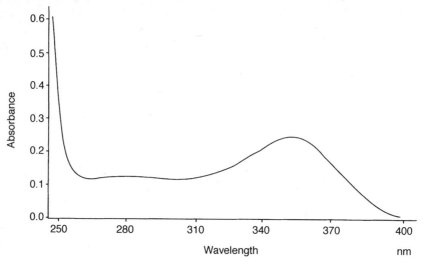

Fig. 3.10. Near ultraviolet spectrum of nitrite in aqueous solution. 10^{-3} mol l^{-1} NaNO$_2$; H$_2$O as reference. Mod. 15 Cary spectrophotometer; optical path 10 cm (From [3.151])

of nitrite 10^{-2} mol l^{-1} or less, and 1 mol l^{-1} nitrate, the resulting error caused in the molar absorptivity of nitrite is no longer negligible. This error can be avoided by using the addition technique outlined below. In this technique, all solutions are made with the same concentration of nitrate; then, a known and accurately measured amount of nitrite is added to all solutions except to the reference blank.

Addition Procedure. All solutions, including the blank, are prepared to contain the same concentration of nitrate. To all solutions, except the blank, a known and exactly measured amount of nitrite is introduced to bring its concentration to the 2×10^{-3} mol l^{-1} range; this concentration was calculated from the known molar absorptivity of nitrite ($E_{355} = 24.0$) in order to raise the absorbance of the sample into the region of minimum photometric error. Therefore, the solutions to be analysed for nitrite contain the same added background concentration of nitrate and nitrite.

The excellent results obtained under these conditions for a range of nitrite standards are shown in Fig. 3.11.

3.2.1.6 Flow Injection Analysis

This technique has been applied [3.152] to the fluorimetric determination of nitrite in river water using 3-aminonaphthalene-1,5-disulphonic acid as fluorescing agent. The reagent solution was prepared by dissolving sodium 3-

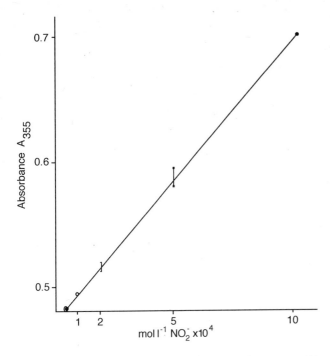

Fig. 3.11. Working curve for the nitrite analysis in the presence of large nitrate excess. Cary spectrophotometer mod. 15, with 50 Å/div., optical path 10 cm (From [3.151])

aminonaphthalene-1,5-disulphonate (0.035 g) in 0.01 hydrochloric acid (100 ml) and mixing 3 ml of this solution with 0.1 g EDTA, concentrated hydrochloric acid (50 ml) and distilled water (244 ml). The reagent stream was then mixed with a carrier stream of distilled water, passed through a mixing coil and mixed with an alkaline solution stream of sodium hydroxide (20%). The aromatic amine was derivatized, in acidic medium, with nitrite to form a diazonium salt which was converted, in alkaline medium, to a fluorescent azoic acid salt. Fluorescence intensity was recorded (470 mm) with a spectrofluorimeter. Nitrite levels as low as 0.01 µg l^{-1} were detected by this method.

Motomizu et al. [3.153] have performed a flow injection analysis – spectrophotometric determination of traces of nitrite in water based on the nitrosation reaction with N, N'-bis (2-hydroxypropyl) aniline.

Nakashima et al. [3.154] have described a spectrophotometric determination of nitrite at the µg l^{-1} level in natural waters by flow injection analysis. The limit of detection is 0.2 µg l^{-1} for sample injections of 650 µl. The sampling rate is about 30 per hour and the relative standard deviation is less than 1.3%.

Apparatus. A diagram of the flow system is shown in Fig. 3.12. The absorbance is measured with a Shimadzu double-beam spectrophotometer UV–140–02 with a 1$^-$ cm flow cell (8 µl) and recorded by a Rikadenki R022 recorder. A double-

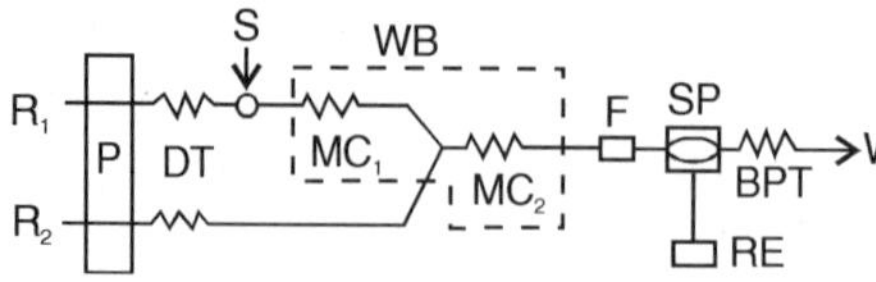

Fig. 3.12. Schematic flow diagram for the determination of nitrite R1, p-aminoacetophenone solution; R2, m-phenylenediamine solution; P, double plunger micro pump (1.7 ml min^{-1}); DT, damping coil (0.5 mm i.d. × 20 m); S, loop valve sample injector (sample volume 650 µl); WB, water bath 30°C); MC 1, mixing coil (1 mm i.d. × 150 cm); MC 2, mixing coil (1 mm i.d. × 100 cm); F, line filter; SP, spectrophotometer with flow cell (light path 10 mm, 8 µl); RE, recorder; BPT, back pressure coil (0.5 mm i.d. × 5 m); W, waste (From [3.154])

plunger micro pump (Kyowa Seimitzu KHU-W-52) is used; the same results are obtained by using a peristaltic pump (Tokyo Rikakikai MP-3). The sample solution (650 µl) is injected by a 6-way injection valve (Kyowa Seimitau KMH 6V) into the reagent R_1 stream. The temperature of the water bath, in which the mixing coils were heated to accelerate the reactions, is controlled by a Taiyo Thermo Minder Jr-10. The flow lines are made from polytetrafluoroethylene (PTFE) tubing 1 mm and 0.5 mm i.d.). The mixing coils (1 mm i.d.) MC_1 and MC_2 are optimally 1.5 and 1.0 m long, respectively, and are wound around plastic rods (6 mm o.d.). The damping coils (0.5 mm i.d.) are 20 m long, to cancel the pulse from the reciprocal pump. The back-pressure tubing (0.5 mm i.d.) are 5 m long, to prevent formation of bubbles.

Reagents. As follows.

Standard nitrite solution, 1 mg ml^{-1} nitrite-nitrogen. Sodium nitrite (4.93 g) dried at 105–110 °C diluted in 1 l of water. p-Aminoacetophenone stock solution. Dissolve 5 g of the reagent in 10 ml of concentrated hydrochloric acid diluted to 500 ml with water. Filter through a 0.45-µm Millipore filter and store in a brown glass bottle in a refrigerator.

m-Phenylenediamine stock solution. Dissolve the dihyrochloride (13.8 g) in 500 ml of water and filter through a 0.45-µm Millipore filter. Store in a brown glass bottle in a refrigerator.

Procedure. The reagent solutions R_1 and R_2 are prepared by diluting 40 ml of the appropriate stock solutions to 1 l; the pH of R_1 is adjusted to 1.3 and the pH of R_2 to 2.4. Both reagent solutions give the same response for three months (maximum period tested) if kept in a refrigerator when not in use.

The flow rates of R_1 and R_2 are 1.7 ml min^{-1} each. Sample solutions are filtered through a 0.45-µm Millipore filter. From the sample loop, 650 µl of sample solution is injected into the reagent R_1 stream. Peak heights are measured at 456 nm against water as reference. The concentration of nitrite is evaluated from the peak heights by using a calibration graph prepared from freshly prepared nitrite standards.

A calibration run was obtained using 0–30 µg l^{-1} nitrite-nitrogen standards. The method is one of the most sensitive flow injection procedures available for nitrite. The relative standard deviations of 10 injections each of solutions containing 10, 20 and 30 µg l^{-1} nitrite-nitrogen were 1.3, 1.1 and 0.39% respectively. Samples with higher concentrations of nitrite can be directly determined by decreasing the sample volume injected and the recorder sensitivity. Linear calibration graphs were then obtained up to 3 mg l^{-1} nitrite-nitrogen (the maximum concentration tested). The sampling rate was about 30 samples per hour.

The effect of various ions on the determination of nitrite was also investigated. No interference was caused by less than 50 mg l^{-1} Ca^{2+}, Mg^{2+}, K^+, NH_4^+, HCO_3^-, SO_4^{2-}, Cl^-, SiO_3^{2-}, NO_3^-, or $H_2PO_4^-$ in the determination of 30 µg l^{-1} nitrite-nitrogen.

In Table 3.36 some results obtained upon applying this method to various natural water samples are shown.

Other workers have applied flow injection analysis to the determination of nitrite [3.155, 3.156].

Table 3.36. Recovery of nitrite added to natural water samples[a]. (From [3.154]; Copyright 1983, Elsevier Science Publishers BV, Netherlands)

Sample	Amount of nitrite-N (µg l^{-1})			Recovery %
	Added	Found	Recovered	
River water	none	5.0		
	5.0	10.3	5.3	106
	10.0	15.2	10.2	102
River water	none	1.4		
	5.0	6.4	5.0	100
	10.0	10.8	9.4	94
Lake water	none	2.2		
	5.0	7.2	5.0	100
	0.0	12.5	10.3	103
Pond water	none	3.0		
	5.0	8.1	5.1	102
	10.0	12.8	9.8	98
Well water	none	2.4		
	5.0	7.6	5.2	104
	10.0	12.7	10.3	103
Rain water	none	2.6		
	5.0	7.8	5.2	104
	10.0	13.2	10.6	106
Irrigation water	none	4.9		
	5.0	10.0	5.1	102
	10.0	14.3	9.4	94
Irrigation water	none	5.3		
	5.0	10.3	5.0	100
	10.0	14.9	9.6	96

[a] Evaluation of recoveries of nitrite was based on the addition of nitrite to 16 ml of sample solution in a 20-ml calibrated flask.

3.2.1.7 Gas Chromatography

Funazo et al. [3.157] have given details of a procedure for the determination of nitrite in river water by gas chromatography after reaction with m-nitroaniline and hypophosphorous acid to convert nitrite to nitrobenzene.

It is well known that nitrite diazotises aromatic amines in acidic media and that the resulting diazonium ions are reduced by hypophosphorous acid to form benzene derivatives. The reactions are formulated as follows:

$$NO_2^- + Ar\text{-}NH_2 \xrightarrow{HX} Ar\text{-}N_2 + X^-$$

$$Ar\text{-}N_2^+ X^- + H_3PO_2 + H_2O \longrightarrow Ar\text{-}H + H_3PO_3 + HX + N_2.$$

An electron capture detector was used to monitor the gas chromatographic effluent. The detection limit and determination range of nitrite are from $0.5\ \mu g\ l^{-1}$ up to 1.00 and $1000\ \mu g\ l^{-1}$ respectively, which are much lower than those of the widely used colorimetric method for the determination of nitrite (detection limit $16\ \mu g\ l^{-1}$).

Reagents. All reagents were of analytical reagent grade and were used without further purification unless otherwise stated.

Hexane and de-ionised water, distilled before use.

Sodium nitrite, dried in an oven at $110\,°C$ for $1\,h$ before being accurately weighed.

Nitrite standard solution, 1.0%. Prepare by dissolving dried sodium nitrite in water. The titre was determined with standard potassium permanganate solution. Solutions containing nitrite at lower concentrations were prepared by dilution of this standard solution.

Nitroaniline, $4 \times 10^{-4}\ mol\ l^{-1}$.

Apparatus. Gas chromatographs: Shimadzu (Kyoto, Japan) Model GC-4BM or equivalent, equipped with a nickel-63 ECD and a Model GC-3BF equipped with dual flame-ionization detectors. Stainless-steel column ($3\ m \times 3\ mm$ i.d.) packed with 5% PEG–HT on 60–80 mesh Uniport HP Nitrogen was used as the carrier gas at a constant flow rate of $40\ ml\ min^{-1}$. The detector, injection port and column temperatures were maintained at 250, 250 and $180\,°C$ respectively.

Procedure. 100 to $1000\ \mu g\ l^{-1}$ nitrite, add a $0.5\ ml$ volume of $4.0 \times 10^{-4}\ mol\ l^{-1}$ m-nitroaniline solution in $1.0\ N$ hydrochloric acid to $1.0\ ml$ of aqueous sample in a reaction vessel (about 10-ml) fitted with a glass stopper. After a few minutes, add $0.3\ ml$ of $4.0\ mol\ l^{-1}$ hypophosphorous acid solution to the vessel. Allow the vessel to stand for $2\ h$ in a water bath controlled at $80\,°C$. At the end of the reaction period, add $1.0\ ml$ of hexane containing m-nitrotoluene (2.0×10^{-6} or $2.0 \times 10^{-5}\ mol\ l^{-1}$) as an internal standard. Extract the resulting nitrobenzene by shaking for 10 min at room temperature and separate the organic layer from the aqueous layer. Inject 2.5 or $0.6\ \mu l$ of the organic layer into the gas chromatograph equipped with an ECD. Measure the converted nitrobenzene by an internal standard method. When the nitrite concentration in the sample is lower than 100 µg l^{-1} use a 2.0×10^{-6} mol l^{-1} solution of m-nitrotoluene in hexane (injection volume 2.5 µl).

A calibration graph of nitrobenzene peak area vs nitrite concentration, for nitrite concentration ranges of 100–1000 and 10–100 µg l^{-1} is a straight line in the concentration range 200–1000 µg l^{-1} but not in the range 10–200 µg l^{-1}. Similarly, the relative peak area of nitrobenzene was plotted against the nitrobenzene concentration in a standard solution containing the internal standard. The graph was not linear at nitrobenzene concentrations below 4.35 µmol l^{-1}. From these results, it is concluded that the curvature of the nitrite calibration graphs is not due to the derivatisation reaction but to the response of the electron capture detector to nitrobenzene. The small unknown peak was given by the blank without nitrite.

The gas chromatographic method was tested in the presence of several ions normally found in environmental samples. In Table 3.37 the peak area of nitrobenzene produced from the standard solution containing 100 µg l^{-1} of nitrite was arbitrarily assigned a value of 100. The concentrations of the ions (100 mg l^{-1}) added to the standard nitrite solution were much higher than those in environmental samples. The results indicated that these anions, except for nitrate, do not interfere in this method. Moreover, chloride added as sodium chloride at a concentration of 2.0% to the standard solution has no effect, which suggests the possibility of applying this method to the analysis of seawater. However, nitrate interferes positively at a concentration of 100 mg l^{-1} although this interference is not found at 10 mg l^{-1}.

River water samples analysed both by the gas chromatographic method and by the spectrophotometric method gave the results shown in Table 3.38. These results indicates that the gas chromatographic methods tend to give slightly lower values than those measured by the colorimetric method (the mean bias is 0.03).

Table 3.37. Results of interference study the concentration of nitrite was 0.10 µg ml^{-1}. (From [3.157]; Copyright 1982, Royal Society of Chemistry, UK)

Ion	Concentration/ µg ml^{-1}	Relative peak area[a]	95% Confidence interval
Standard	–	100.0 ± 2.0	97.5–102.5
Cl$^-$	20000	101.2 ± 2.3	98.3 – 104.1
Cl$^-$	100	99.4 ± 2.1	96.8 – 102.2
Br$^-$	100	100.4 ± 2.2	97.7 – 103.1
F$^-$	100	98.8 ± 1.9	96.4 – 101.2
SO$_4^{2-}$	100	97.8 ± 1.3	96.2 – 99.4
HCO$_3^-$	100	99.1 ± 1.5	97.2 – 101.0
H$_2$PO$_4^-$	100	98.6 ± 2.5	95.5 – 101.7
NO$_3^-$	100	111.9 ± 1.3	110.3 – 113.5
NO$_3^-$	10	99.6 ± 1.8	97.4 – 101.8
NH$_4^+$	100	100.4 ± 1.1	99.0 – 101.8

[a] Mean ± standard deviation for five replicate analyses (between batch). The peak area of nitrobenzene converted from the standard solution was arbitrarily assigned a value of 100.

Table 3.38. Results of study of intercomparison of methods. (From [3.157]; Copyright 1982, Royal Society of Chemistry, UK)

Sample	Nitrite determined/$\mu g \, ml^{-1}$		
	Colorimetry	Gas chromatography	
		Mean ± s.d.[a]	Relative standard deviation %
A	0.54	0.53 ± 0.013	2.5
B	0.57	0.54 ± 0.006	1.1
C	0.46	0.42 ± 0.009	2.1
D	0.69	0.62 ± 0.007	1.1
E	0.58	0.54 ± 0.004	0.7
F	0.61	0.58 ± 0.007	1.2

[a]Mean ± standard deviation for five replicate analyses (between batch).

Derivatization-electron capture gas chromatography has been used to determine $\mu g \, l^{-1}$ quantities of nitrite in water without interference from halides, nitrate, phosphate, sulphate, bicarbonate, ammonium and alkali metals and alkaline earth metals [3.158].

3.2.1.8 Ion Chromatography

This technique has been used to determine nitrite in natural waters, particularly rainwater. Methods are reviewed in Table 3.39 and discussed more fully in Sects. 11.1 and 11.2.

3.2.1.9 Polarography

Nitrites have been determined by oxidation at a glassy carbon electrode [3.159, 3.160], electrocatalytically using a molybdenum catalyst [3.161] and polarographically using differential pulse polarography using the reaction between nitrite ion and diphenylamine to produce diphenyl-nitrosamine [3.162].

Differential pulse absorbtive volumetric analysis has been used to determine $\mu g \, l^{-1}$ levels of nitrate in water [3.163].

Table 3.39. Determination of nitrite in natural waters by ion chromatography

Type of water sample	Codetermined anions	References
Natural	Cl, Br, SO_4, NO_3	3.69
Natural	Br, F, NO_3, SO_4, PO_4	3.70

Barsotti et al. [3.164] determined nitrite in amounts down to 0.4 µg l^{-1} N and nitrite by differential pulse polarography. Nitrate is reduced on a cadmium column to diphenylnitrosamine and then determined polarographically. Cadmium interference is removed by adjustment of pH and complexing with EDTA. Nitrite is determined directly as the diphenylnitrosamine.

In order to separate the cadmium II peak from that of diphenyl nitrosamine, advantage was taken of the fact that organic polarograms are usually pH dependent, while those of reversible inorganic ions are not. Changing the pH from 1.0 to 4.0 shifts the diphenyl nitrosamine peak, without loss of peak height, hence sensitivity, but causes no change in the cadmium II peak. A pH of 1.0 is used for the formation of diphenyl nitrosamine since the rate is fast at that pH with no decomposition. At a pH of 4.0 a peak separation of 140 mV occurs which is sufficient to determine nitrite ion in the presence of cadmium II to a level of approximately 200 µg l^{-1}. At this pH the EDTA complex of cadmium II has a stability constant of approximately 10^{16}. Thus, the addition of EDTA at a pH of 4.0 results in the complete disappearance of the cadmium II peak and total removal of any interference. Therefore, by an adjustment of pH and the addition of EDTA to the nitrite ion procedure, nitrate could be determined.

3.2.2 Rain water

3.2.2.1 Continuous Flow Analysis

This technique has been used to determine nitrite in rain water [3.165] (and chloride nitrate, sulphate and phosphate).

3.2.2.2 Flow Injection Analysis

Flow injection analysis has been applied to the determination of nitrite in rain water [3.165]. (and chloride, nitrate, sulphate and phosphate).

3.2.2.3 Ion Chromatography

This technique has been applied to the determination of nitrite and other anions (Table 3.40). The methods are discussed in further detail in Sect. 11.2.

Table 3.40. Determination of nitrite in rain water by ion chromatography

Type of water sample	Codetermined anions	References
Rain	Cl, PO_4, NO_3, SO_4	3.93
Rain	Cl, F, Br, NO_3, SO_4, SO_3, PO_4	3.95
Rain	Cl, Br, F, NO_3, PO_4, BrO_3, SO_4	3.96
Rain	Cl, PO_4, SO_4	3.90

3.2.3 Industrial Effluents

3.2.3.1 Spectrophotometric Methods

Maly and Kosikova [3.166] have described two spectrophotometric methods for the determination of nitrites in industrial effluents. These workers compared their methods with the standard method using sulphanilic acid and N-(1-naphthyl)ethylene-diamine and a new modification of the process with sulphanilic acid and 1-naphthol. The advantage of the latter is the possibility of diluting coloured solutions where, owing to high nitrite concentrations, the absorbance exceeds the range of the calibration curve. Urea and glycine (up to 100 mg l^{-1}) do not interfere with these methods. Iron at levels about 5 mg l^{-1} can be eliminated by clarifying the sample with an aluminium salt.

3.2.3.2 Ion Chromatography

This technique has been used to determine nitrite (and chloride, sulphate, nitrate, phosphate, fluoride and bromide) in industrial effluents [3.112]. The method is discussed further in Sect. 11.4.

3.2.3.3 Paper Chromatography

Paper chromatography has been used to identify nitrites in industrial effluents [3.167].

3.2.3.4 Amperometry

Oxidation flow injection amperometry using an electrochemically pretreated glassy carbon electrode has been used to determine nitrite in water [3.168].

3.2.4 Waste Waters

3.2.4.1 Spectrophotometric Methods

Various spectrophotometric procedures have been described for the determination of nitrite in waste waters [3.169–3.171]. In the spectrophotometric method [3.169] based on the reaction between Griess Romijn reagent (i.e. the reaction of nitrite with a primary aromatic amine to form a diazonium salt, which is then coupled with another aromatic compound to form an azo dye whose absorbance is measured) and nitrite, interference by iodide, sulphide thiosulphate, sulphite or tetrathionate is avoided by the addition of mercuric chloride solution.

Nakamura and Mazuka [3.170] extracted nitrite in an ethyl acetate solution of 4, 5-dihydroxycoumarin. Beer's law was obeyed up to 0.75 mg l^{-1}. Only iodide, sulphide, iron III and chromium ions interfered seriously at the 10 μg l^{-1} level. Results agreed well with those obtained using the diazotizing coupling method.

Koupparis et al. [3.171] have described procedures for the determination of nitrite using an automatic microprocessor-based stopped-flow analyser. The reaction is based on diazotization of sulphanilamide, the product being coupled with N-(1-naphthyl)ethylenediamine dihydrochloride to form a coloured azo dye which is measured at 540 nm. The analysis should be carried out on fresh samples to avoid bacterial conversion of nitrite to nitrate or ammonia. However, samples can be preserved for 1–2 days, either by freezing at $-20\,°C$ or by addition of mercuric chloride and storage at 4 °C. The methods are fast, sensitive, accurate and precise, and without serious interference. The sample throughput for routine analysis can be up to 360 samples per hour in the range 0.025–2.00 mg l^{-1} of nitrite-nitrogen.

The automated microprocessor-based stopped flow analyser has been described by Koupparis et al. [3.172].

Reagents. All chemicals used were of analytical reagent grade and deionized water was used in all experiments.

Nitrite standard solution, 1000 mg l^{-1} of nitrite-nitrogen. Dissolve 4.926 g sodium nitrite (Baker Analysed, 99.7%) oven dried at 100–105 °C for 2 h in de-aerated, de-ionised water and dilute to 1 l. Add a pellet of sodium hydroxide and 1 ml of spectroscopic grade chloroform in order to prevent liberation of nitrous acid and bacterial growth. Keep in a refrigerator and replace every two weeks. Prepare working standard solutions in the range 0.025–2 mg l^{-1} of nitrite-nitrogen daily by appropriate dilution.

Reagent solution. Prepare by dissolving 10.0 g of sulphanilamide, 0.50 g of N-(l-naphthyl)ethylene diamine dihydrochloride and 30 ml of 85% orthophosphoric acid in water and dilute to 1 l. Store the solution in an amber glass bottle in a refrigerator.

Apparatus. The automated microprocessor-based stopped flow analyser [3.172]. The entire system is automated using a Rockwell AIM 65 microcomputer for control of all operations, data acquisition and reduction, display and printout of results. The sampling mixing module is used to sample 150-µl volumes of reagent and standards of samples on the turntable with a reproducibility better than 0.1%, to mix them efficiently and transfer them into the observation cell. The photometric system is an easily constructed module with a 1-cm flow cell, automatic shutter, light source, photomultiplier and interference filters. An interference filter with a 10-nm band pass at 540 nm was used in the photometer.

An investigative program was used to evaluate the optimum parameters for the analytical procedure and a routine reaction rate program for the calibration graph and the analysis of the samples.

All measurements were carried out in an air conditioned laboratory maintained at a nominal temperature of about 25 °C.

Procedure. Load the turntable with the blank (de-ionised water) standards in the range $0.025-2$ mg l^{-1} of nitrite-nitrogen and the samples in 5-ml disposable polystyrene micro-beakers. Use one channel of the stop flow analyser to take aliquots of the reagent solution and the other the standards and the samples. Load the appropriate BASIC and machine language programs from the cassette recorder into the computer's memory. Then, in execution of the program, inject the blank into the flow cell using an integration time of 0.7 s. The dark current and 100% transmittance are measured automatically. The program then prompts the operator to select the time parameters. A delay time of 10 s and a measurement time of 0.7 s (one integration) are used for the single point kinetic procedure, and a delay time of 0.8 s and a measurement time of 1.5 s for the multi point kinetic procedure. The number of standards and samples to be measured, the number of measurements to be averaged and the number of flushes are then requested by the computer. The program then sequences through each standard, flushing the system between each standard (four flushes are needed) and prompting the operator for its concentration. After the standards have been measured, the micro-computer calculates the linear least-squares regression line and prints its slope, intercept, correlation coefficient and the standard error of the estimate. Samples are automatically measured, after which the concentration of the nitrite in the sample is calculated and printed.

A dedicated program can also be used with all the information, time parameters and concentration of standards contained in the software. Once the operator has input the number of samples, the analysis will be completed automatically.

The accuracy of this method was examined by measuring the nitrite concentration of waste water samples before and after a standard addition. Recovery was calculated after addition of 100 µl of 10 mg l^{-1} nitrite-nitrogen standard to 10 ml of each assayed sample. The results obtained show an average recovery of 98.6%. Similar results were obtained for the analysis of samples.

Major interference is caused by reducing or oxidising ions. The low results caused by the copper II ion because of its catalysis of the decomposition of the diazonium salt in methods with long reaction times are avoided in this method. The serious sulphite interference is also almost eliminated in this method. The sulphide interference can be eliminated by adding excess of cadmium ions and filtering. Addition of 200 mg l^{-1} of Cd^{2+} to a waste water sample containing 1 mg l^{-1} of nitrite-nitrogen and 50 mg l^{-1} of sulphide eliminated the error.

3.2.4.2 Raman Spectroscopy

Furaya et al. [3.173] have developed a method for the determination of nitrite in amounts down to 0.5 µg l^{-1} in waste and treated waters by resonance Raman spectrometry after conversion of nitrite to coloured azo dyes. Nitrate can also be determined simultaneously by this method. Due to the resonance effect, the intensities of spectra often become five or six orders of magnitude larger than those of usual Raman spectra. Therefore, Furaya et al. [3.173] attempted to determine nitrite by use of resonance Raman spectrometry.

This method, however, can be applied only to coloured compounds and aqueous solutions of nitrite are completely colourless. Therefore, nitrite was converted to a coloured product by diazotization and after that the resonance Raman spectrum was measured.

$$HNO_2 + NH_2 C_6H_4 + HSO_3 \longrightarrow C_6H_4 \diagup \diagdown \begin{matrix} N=N \\ | \\ SO_2-O \end{matrix} + H_2O \text{ (diazo compounds)}$$

$$C_6H_5 \diagup \diagdown \begin{matrix} N=N \\ | \\ SO_2-O \end{matrix} + \text{(α-naphthylamine)} \longrightarrow \text{(azo compound)}$$

NH$_2$ α-naphthylamine   HSO$_3$ NH$_2$

Raman spectra were recorded on a JASCO Model J-800 Laser Raman Spectrophotometer. A Spectra-Physics Model 164 Argon Ion Laser was used as an exciting source. A HTC 464 photomultiplier was used as a detector. The spectral data were processed by a data processor containing a micro-computer in conjunction with the spectrophotometer. In order to improve the S/N ratios of the spectra, the data were accumulated and the number of repetitive scans was determined according to the noise levels for each of the spectra. The sample solutions were illuminated with the 488.0 nm line at 200 mW output. In order to minimize the effect of local heating of samples due to the laser beam irradiation and the possible decomposition of the coloured product, the sample cell was rotated using a JASCO Model R-10 Cell Rotator (2500 rpm^{-1}). Moreover, the sample solutions were irradiated very close to the cell wall to minimize the reabsorption of scattering lights.

Reagents. As follows.

Sulfanilic acid solution. Dissolve 0.6 g of sulfanilic acid in 70 ml of warm water. After cooling, add 20 ml of hydrochloric acid, and dilute to 100 ml with water.

α-Napthylamine hydrochloride solution. Dissolve 0.6 g of α-naphthylamine hydrochloride in approximately 80 ml of water containing 1 ml of hydrochloric acid and dilute to 100 ml with water.

Sodium acetate solution. Dissolve 28 g of sodium acetate trihydrate in approximately 80 ml of water and dilute to 100 ml.

Both the standard and the waste water samples were filtered through 0.22 µm Millipore filters before the measurements.

Procedure. Transfer the sample water containing nitrite up to 7 µg l^{-1} into a 100-ml conical beaker and dilute it to approximately 40 ml with water. Add 1.0 ml of sulfanilic acid solution and, after shaking, leave standing for about 10 min. Add 1.0 ml of α-naphthylamine hydrochloride solution and adjust pH to 2.0–2.5 with sodium acetate solution. After 10 min, transfer this solution into a 50-ml volumetric flask (1 g of sodium thiocyanate put in beforehand to reduce background intensities) and dilute to the mark with water.

The coloured product obtained has a broad adsorption band with the maximum absorption at 520 nm (λ max = 41,000 cm^{-1}, 1 mol^{-1}). The coloured product exhibits the resonance Raman effect against the 488.0 nm line of argon ion laser. In Fig. 3.13 the resonance Raman spectrum of this product solution is

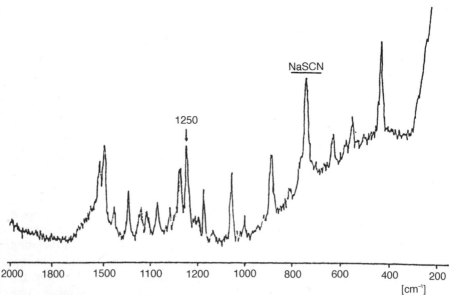

Fig. 3.13. Resonance Raman spectrum of the coloured product. Concentration of nitrite 0.2 mg l^{-1} (From [3.173])

shown. The concentration of the solution is 0.2 mg l^{-1} (4 × 10^{-6} mol l^{-1}). As is seen from this figure, the solution of the azo dye obtained gives a lot of bands in the 2000–200 cm^{-1} region in spite of remarkably low concentration. Of these bands, the one at 1250 cm^{-1} was used as a key band corresponding to nitrite since the band has the maximum peak intensity, and the background intensity in the vicinity of this band is comparatively low.

The sample solutions produced by the above mentioned reaction exhibit a comparatively strong background, which probably comes mainly from the fluorescence due to the laser beam irradiation, and, because of this, the required sensitivity cannot be attained unless this background intensity is made considerably lower by the addition of an inorganic salt as a quencher of fluorescence.

The addition of sodium thiocyanate leads to a marked decrease of the background. Moreover, it is seen that no distinct change of the reduced background is observed in this case, which means that the reduction rapidly reaches the equilibrium state.

In Fig. 3.14 the analytical curve obtained for aqueous solutions of nitrite in the concentration range 10–180 µg l^{-1} is shown. The curve can be fitted to a straight line in the concentration range below 140 µg l^{-1}. When the concentrations of the added nitrite are smaller, that is, from 0 to 10 µg l^{-1}, the relation obtained is also linear in this range. However, the line does not go through the origin and there is a blank value that comes from the contamination of the distilled water, probably in the atmosphere, etc. This blank value is calculated to

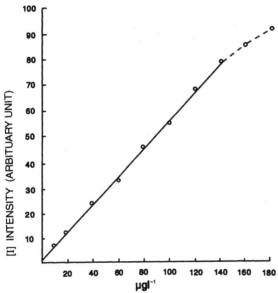

Fig. 3.14. Analytical curve obtained for the aqueous solutions of NO$_2$-in the concentration range from 10 to 180 mg l^{-1}. The empirical formula was determined by the least squares method: I = 1.45 + 0.55 C, S$_c$ = 2.70 µg l^{-1} (for the concentration) (From [3.173])

be about 1.3 $\mu g\, l^{-1}$. The existence of this blank causes no trouble in practice because, in the case of the analysis of nitrite in waste and untreated waters, the concentrations of nitrite are usually larger than 10 $\mu g\, l^{-1}$.

3.2.5 High Purity Water

3.2.5.1 Ion Chromotography

Ion chromatography has also been used to determine nitrate (also sulphate and chloride) in high purity water [3.174]. These techniques are discussed further in Sect. 11.5.

3.2.6 Soil

3.2.6.1 Spectrophotometric Methods

Bhuchar and Amar [3.175] determined nitrites in polluted water and soil by acidifying the sample to pH4 and adding mercapto acetic acid to produce a red coloured complex which is extracted into tributyl phosphate from a solution 2 N in acid. The red colour is evaluated spectrophotometrically at 322 nm. The method is applicable in the range 2–40 mg l^{-1} nitrite.

Wu and Liu [3.176] have described a spectrophotometric method for the determination of micro amounts of nitrite in water and soils. The chromogenic reagents were p-aminoacetophenone and resorcinol in sodium carbonate-sodium acetate medium at pH9 which form a golden coloured complex with nitrite at 435 nm:

$$CH_3CO-C_6H_4-NH_2^+ + NO_2^- + 2H^+ \rightarrow CH_3CO-C_6H_4-N^+ \equiv N + 2H_2O$$

$$CH_3-CO-C_6H_4-N^+ \equiv N + \underset{}{\overset{HO}{\bigcirc}}-OH \xrightarrow{pH\,9} CH_3CO-C_6H_4-N$$

$$= N-\bigcirc-OH + H^+.$$

Foreign ions are masked with a composite EDTA-sodium hexametaphosphate reagent and interference by sulphide is overcome by the addition of mercuric chloride. Soil samples are digested with cold water containing mercuric chloride and the precipitated mercuric sulphide filtered off prior to the addition of chromogenic reagents and spectrophotometric evaluation.

Beer's Law is obeyed up to 20 µg nitrite in 60 ml of test solution. The effect of 20 foreign ions was examined. Most of the cations and common anions do not interfere. Sulphide, thiosulphate, sulphite, tetrathionate and iodide interfere but their effect is mitigated by the addition of mercuric chloride.

Chaube et al. [3.177] investigated the determination of ultra trace concentrations of nitrite in polluted waters and soil. In their method the nitrite is used to diazotize o-nitroaniline and the o-nitrophenyldiazonium chloride produced is coupled with N-naphthylethylene diamine hydrochloride. The red-violet dye produced is extracted into isoamylalcohol and evaluated spectrophotometrically at 545 nm.

Beer's Law is obeyed in the range $0.1-0.6$ mg l^{-1} nitrite in the original sample and $0.02-0.15$ mg l^{-1} if the solvent extraction procedure is used. A wide range of foreign ions do not interfere in this procedure. Soil samples were acidified with sulphuric acid prior to filtration and analysis.

3.2.7 Milk

3.2.7.1 Flow Injection Analysis

Munksgaard and Thymark [3.178] used flow injection analysis to determine nitrate and nitrite in milk and milk products at levels below 10 mg kg^{-1}. The sample is dissolved in hot water, and fat and protein precipitated by dialysis. Nitrate in the filtrate is reduced to nitrite by means of copperised cadmium. Nitrite is then determined following conversion to the azo dye with N-napthylethylene diamine and spectrophotometric evaluation at 538 nm.

3.3 Nitrate and Nitrite

Methods are discussed below which cover the determination of both nitrate and nitrite in a single analysis.

3.3.1 Natural Waters

3.3.1.1 Spectrophotometric Methods

Two groups of workers [3.179, 3.180] have discussed methods for the determination of nitrite and nitrate in river waters.

Okada et al. [3.179] describe a technique for determining nitrite in river waters where the nitrite in the water diazotises p-aminoacetophenone, which is then coupled with *m*-phenylenediamine at pH 1.5 to 3. The 2,4-diamino-4'-acetylazobenzene formed is then measured using a spectrophotometer at 460 nm. Determination of nitrate is carried out using the same method, after the nitrate has been reduced to nitrite in a cadmium-copper column. By this method nitrite has been determined at levels of 1 μg l^{-1}.

Reagents. As follows. Standard nitrite solution. Dissolve about 1 g of dried sodium nitrite in 100 ml of re-distilled water. Determine its concentration by titration with standard potassium permanganate solution.

Standard nitrate solution. Potassium nitrate (1.1 g), dry at 105–110 °C weigh accurately and dissolve in 100 ml of distilled water.

p-Aminoacetophenone hydrochloride solution. Dissolve 0.5 g solid in 5 ml of concentrated hydrochloric acid and dilute with distilled water to 50 ml in a calibrated flask. The solution is stored in a brown glass bottle.

m-Phenylenediamine solution. Weigh 1.34 g solid *m*-phenylenediammonium chloride and dissolve in 0.08 ml of concentrated hydrochloric acid, dilute with distilled water to 50 ml in a calibrated flask and store in a brown glass bottle. This solution can be used for 10 days without deterioration.

Buffer solution (pH a.3). Mix ammonium chloride solution (0.1 mol l^{-1}) and aqueous ammonia (0.1 mol l^{-1}), adjust the pH with a pH meter.

Apparatus. Hitachi 139 spectrophotometer or equivalent with 1- or 10-cm glass cells, and cadmium-copper column with cadmium powder Merck, 20–60 mesh.

Determination of Nitrite. Pipette the water sample (25 ml) into a 50-ml calibrated flask and add 10–20 ml of distilled water, 0.5 ml of *p*-aminoacetophenone hydrochloride solution and mix well. The pH of the solution is 2.0–2.5. After 3–5 min add *m*-phenylenediamine solution (0.5 ml) and make the solution up to the mark with distilled water. After 30 min measure the absorbance at 400 nm in 10-cm cells. The calibration graph is a straight line passing through the origin and the molar absorptivity is 4.7×10^4 mol^{-1} cm^{-1}, whereas the molar absorptivity in toluene is 2.3×10^4 mol^{-1} cm^{-1}. The sensitivity is double that of the extraction method.

Determination of Nitrate. The following conditions were used to prepare the cadmium-copper column: grain size, 20–60 mesh; column cross-section, 0.8 cm^2; column length, 14 cm; flow-rate, 0.048 ml s^{-1}; pH, 9.0–10.5; reduction, 97%.

Pipette the sample solution (10 ml) into a 50-ml beaker and add 5 ml of a buffer solution (pH 9.3) and mix. Pass the solution through the column, which is then washed several times with 5 ml of 5-fold diluted buffer solution and 3–4 ml of distilled water. Collect the eluate in a 50-ml calibrated flask and make up to the mark with distilled water. Use this solution for the determination of nitrite mentioned above using 1-cm glass cells. Using this procedure the calibration graph was straight and the molar absorptivity was 4.6×10^4 mol^{-1} cm^{-1}, which was 97% of the absorptivity for pure nitrite (4.7×10^4 mol^{-1} cm^{-1}). At the same time nitrite was determined by this procedure and the recovery was 95%.

The ions normally present in river waters do not interfere in this procedure. No interference is caused by less than 10^{-2} mol l^{-1} calcium, magnesium, sodium, potassium, bicarbonate, sulphate, chloride or ammonium.

Workers at the Water Research Centre, UK [3.180, 3.181] have carried out detailed investigations of the accuracy of determination of total oxidized nitrogen and nitrite in river waters by the Griess–Ilosvay automated spectrophotometric procedure. Broadly, each of the 11 participating laboratories achieved total errors of not greater than ±20% of the concentration being determined or 0.1 mg l^{-1} of nitrogen (whichever was the larger for different sample concentrations).

The results obtained are summarised in Tables 3.41 and 3.42. The values of S_t for the two standard solutions, the river water and the spiked river water were compared with the appropriate target value using the F-test, and were accepted as satisfactory provided S_t was not significantly greater ($p = 0.05$) than the appropriate target.

For both total oxidized nitrogen and nitrite, all laboratories achieved a precision that was not significantly (0.05 probability level) greater than the target value. One laboratory (Laboratory 5) which initially failed to meet the precision target for total oxidised nitrogen, did so after modifying its analytical method and repeating the precision tests.

It is worth noting that the estimates of precision varied considerably between laboratories, even for the analysis of standard solutions with similar concentrations. Of course, some of these apparent differences reflect the random uncertainties of the estimates, but some differences are sufficiently large to indicate real differences in the precision of results obtained by individual laboratories, even when very similar analytical procedures are used.

Each laboratory also calculated the recovery, R, of the material determined from the spiked river water, where $R = 100\ (\bar{C}_S - \bar{C}_R)/A$ and $\bar{C}_S$ and $\bar{C}_R$ are the mean concentrations found for the spiked and unspiked river water respectively and A is the equivalent concentration of material determined added to the spiked sample, allowance being made for the slight dilution of the sample caused by the addition of the standard solution. It was agreed that recoveries would be considered acceptable if R was not significantly worse (t-test, $p = 0.05$) than the closer of the two values, 95 and 105%. These results are also summarised in Table 3.41, which shows recoveries of between 96 and 111% for oxidised nitrogen and 95 and 106% for nitrite with an overall mean of 101% for total oxidised nitrogen and 100% for nitrite. None of the individual recoveries were significantly worse than 95 to 105% and the results were, therefore, considered acceptable.

To ensure that differences in the determined ion concentration of the standard solutions used in different laboratories did not cause important between-laboratory bias, the following tests were made. The WRC prepared a standard solution for each determined material (i.e. total oxidised nitrogen, and nitrite) of

Table 3.41. Results from within-laboratory precision tests: Total oxidised nitrogen. Total standard deviation refers to the standard deviation of any one result in any one batch of analysis: the estimates shown have between 9 and 19 degrees of freedom. The target standard deviation is 5% of the concentration of the determinand, or 0.025 mg l^{-1} of nitrogen (whichever is the larger). (From [3.180]; Copyright 1982, Society for Analytical Chemistry)

Sample	Parameter	Laboratory										
		1	2	3	4	5	6	7	8	9	10	11
Standard solution 1	TON concentration/mg l^{-1} of nitrogen	0.4	1.0	4.0	1.5	0.5	1.0	1.0	1.0	0.5	1.0	2.0
	Relative total standard deviation %	2.3	5.1[a]	3.7	3.5	3.0	1.4	5.1[a]	4.5	3.4	5.9[a]	3.85
	Total standard deviation mg l^{-1} of nitrogen	0.013	0.051	0.147	0.052	0.015	0.014	0.051	0.015	0.017	0.059	0.077
Standard solution 2	TON concentration/mg l^{-1} of nitrogen	3.6	9.0	36.0	12.5	4.5	9.0	9.0	9.0	4.5	9.0	9.0
	Relative total standard deviation %	1.6	1.6	1.3	1.0	0.5	1.0	1.7	1.0	1.0	2.4	4.4
	Total standard deviation/mg l^{-1} of nitrogen	0.056	0.142	0.482	0.128	0.024	0.092	0.156	0.095	0.043	0.218	0.401
River water	TON concentration/mg l^{-1} of nitrogen	9.04	4.21	1.08	3.03	2.65	3.72	4.73	0.02	4.10	1.08	3.40
	Relative total standard deviation %	1.5	1.5	5.3[a]	1.7	1.2	1.4	1.8	5.3[a]	1.4	4.8	4.5
	Total standard deviation/mg l^{-1} of nitrogen	0.133	0.065	0.057	0.051	0.031	0.053	0.083	0.033	0.056	0.081	0.152
Spiked river water	TON concentration/mg l^{-1} of nitrogen	10.84	7.78	20.66	7.80	3.61	8.39	7.71	5.51	4.60	6.11	7.44
	Relative total standard deviation %	1.9	1.5	1.7	1.0	0.7	1.0	1.6	1.2	1.0	3.7	4.1
	Total standard deviation/mg l^{-1} of nitrogen	0.210	0.114	0.35	0.076	0.024	0.080	0.127	0.069	0.044	0.226	0.309
	Mean recovery of TON from spiked river water, %[b]	104.4 (±6.7)	97.1 (±1.6)	100.0 (±1.2)	103.3 (±0.9)	96.5 (±1.2)	96.0 (±0.7)	101.4 (±1.6)	98.1 (±0.5)	111.0 (±6.5)	97.4 (±2.6)	101.4 (±2.6)

[a] Not statistically significantly greater than the largest.
[b] 90% confidence intervals of the mean recoveries are shown in parentheses.

Table 3.42. Results from within-laboratory precision tests: nitrite. Total standard deviation refers to the standard deviation of any one result in any one batch of analysis; the estimates shown have between 9 and 19 degrees of freedom. The target standard deviation is 5% of the concentration of the determinand, or 0.025 mg l^{-1} of nitrogen (whichever is the larger). Laboratories 5 and 7 do not determine nitrite for routine Harmonised Monitoring purposes: Laboratory 7 did not participate in this work. (From [3.180]; Copyright 1982, Society for Analytical Chemistry)

Sample	Parameter	Laboratory										
		1	2	3	4	5	6	8	9	10	11
Standard solution 1	Nitrite concentration/mg l^{-1} of nitrogen	0.02	0.1	0.1	0.25	0.1	0.02	0.02	0.1	0.1	0.2
	Relative total standard deviation, %	5.8	9.3	5.2	0.9	10.0	4.0	10.0	6.7	7.3	4.7
	Total standard deviation/mg l^{-1} of nitrogen	0.0012	0.0093	0.0052	0.0022	0.010	0.0008	0.0020	0.0067	0.0073	0.009
Standard solution 2	Nitrite concentration/mg l^{-1} of nitrogen	0.18	0.9	0.9	2.0	0.9	0.18	0.18	0.9	0.9	1.8
	Relative total standard deviation, %	0.6	2.9	1.0	2.6	1.7	0.9	1.6	0.6	1.2	3.7
	Total standard deviation/mg l^{-1} of nitrogen	0.0011	0.026	0.0091	0.052	0.015	0.0016	0.0028	0.0055	0.011	0.066
River water	Nitrite concentration/mg l^{-1} of nitrogen	0.145	0.013	0.020	0.251	0.016	0.014	0.050	0.070	0.027	0.032
	Relative total standard deviation %	4.5	14.9	11.5	1.9	19.3	1.4	2.6	9.0	18.5	15.0
	Total standard deviation/mg l^{-1} of nitrogen	0.0065	0.0022	0.0023	0.0048	0.0089	0.0002	0.0013	0.0063	0.005	0.001
Spiked river water	Nitrite concentration/mg l^{-1} of nitrogen	0.006	0.569	0.534	1.258	0.519	0.171	0.146	0.526	0.504	1.61
	Relative total standard deviation, %	5.0	2.8	1.4	1.8	1.6	1.3	2.1	1.0	1.5	4.7
	Total standard deviation/mg l^{-1} of nitrogen	0.025	0.016	0.0073	0.0228	0.0087	0.0022	0.0031	0.0050	0.0078	0.076
	Mean recovery of nitrite from spiked river water, %[a]	98.0 ($\pm$ 0.4)	100.7 ($\pm$ 2.6)	105.0 ($\pm$ < 0.1)	100.1 ($\pm$ < 0.1)	106.1 ($\pm$ 1.3)	97.6 ($\pm$ < 0.1)	95.3 ($\pm$ 1.2)	98.0 ($\pm$ < 0.1)	100.5 ($\pm$ 1.2)	98.6 ($\pm$ 2.0)

[a] 90% confidence intervals of the mean recoveries are shown in parentheses.

1000 mg l^{-1} of nitrogen and portions of these solutions were sent to all laboratories. Each laboratory then accurately diluted portions of both the WRC standards and its own stock standard solutions to nominally the same concentration, which was chosen to be at or near the upper concentration limit of the laboratory's method so as to achieve the smallest relative standard deviation. Sufficient replicate determinations were made in a single batch of analyses on both sets of diluted standards so that, if there was a difference of 2% in their true concentrations, a statistically significant difference (at the 95% confidence level) would be detected. The required number of analyses on each standard was calculated statistically using the estimates of within-batch standard deviation obtained in the preceding stage of the work [3.182].

The maximum difference observed was 1.5% for both determined materials with an average difference over all the laboratories of 0.45% for total oxidised nitrogen and 0.44% for nitrite. This was considered satisfactory.

To complete this initial phase of AQC, direct checks of between-laboratory bias were made as follows. The WRC prepared and distributed portions of three standard solutions, (A), (B), and (C) containing both nitrite and nitrate, and two different river waters (D) and (E).

Both of the river water bulk samples were allowed to stand for 2 weeks to allow their total oxidized nitrogen and nitrite concentrations to stabilise before the distribution of samples. The residual determined material concentrations were determined at the WRC. On each of 5 days each laboratory made one determination of total oxidized nitrogen and nitrite in the following solutions: (i) standard solution (A); (ii) river water (D) which had been spiked with a specified amount of standard solution (B); and (iii) river water (E) spiked with a specified amount of standard solution (C). The spiked river waters were prepared freshly just before each batch of analyses. Each laboratory then calculated the mean of the five results and its 90% confidence limits (obtained from the five results only) for each solution and these results were returned to the WRC for examination for evidence of bias. The spiking procedure for the samples of river water was adopted in order to overcome problems caused by instability of the determined materials in the samples. All solutions were stored under refrigeration.

Those values for the maximum possible bias are given in Tables 3.43 and 3.44. It can be seen that the bias targets are met for all laboratories that reported results except for the following: total oxidised nitrogen, Laboratories 3 and 10 for solution (A) and Laboratory 2 for sample (C); and nitrite, Laboratory 2 for sample (C).

The results from Laboratory 10 (total oxidized nitrogen) and Laboratory 2 (nitrite) only marginally exceeded the targets. In the other instances the targets were not greatly exceeded; the laboratories were informed and sources of bias sought. Laboratory 3 found the bias to have arisen from base-line drift caused by changes in laboratory temperature. Calibration was then carried out more frequently and a subsequent check showed this modified method capable of meeting the bias target.

Table 3.43. Results of inter-laboratory bias tests: total oxidized nitrogen. Laboratory 5 results refer to original method. The target is a maximum possible bias of 10% of the concentration of the determinand, or 0.05 mg l^{-1} of nitrogen (whichever is larger). (From [3.180]; Copyright 1982, Society for Analytical Chemistry)

Laboratory	Standard solution (A)			Spiked river water (B)			Spiked river water (C)		
	Mean concentration found/mg l^{-1} of nitrogen	Deviation from expected value, %	Maximum possible bias,[a] %	Mean concentration found/mg l^{-1} of nitrogen	Deviation from average, %	Maximum possible bias,[a] %	Mean concentration found/mg l^{-1} of nitrogen	Deviation from average, %	Maximum possible bias,[a] %
1	10.43	+4.3	+6.9	4.99	−5.7	−8.7	14.00	−12.7	−16.1[b]
2	9.88	−1.2	−4.3	5.70	+7.8	+9.5	17.50	+9.2	+9.8
3	11.26	+12.6	+14.0[b]	5.30	+0.2	+2.4	16.34	+1.9	+3.6
4	10.20	+2.0	+3.5	5.30	+0.2	+3.8	16.60	+3.6	+5.7
5	9.96	−0.4	−2.46	5.36	+1.3	+2.3	13.70	−3.1	−2.9
6	10.25	+2.5	+3.3	5.37	+1.5	+2.3	16.68	+4.1	+5.1
7	10.41	+4.1	+5.0	5.14	−2.8	−4.0	16.14	+0.7	+3.4
8	10.39	+3.9	+7.8	5.19	−1.9	−2.4	15.19	−5.2	−5.9
9	10.15	+1.5	+1.7	5.26	−0.6	−1.6	16.12	+0.6	+3.0
10	10.60	+6.0	+11.2[b]						
Average (except Laboratory 3)	10.35			5.29			16.03		
Expected value[c]	10.00			5.22			16.12		

[a] At the 95% confidence level.
[b] Does not meet target.
[c] The expected value is: for solution (A), the true concentration; for samples (B) and (C), the sum of the concentration due to spiking plus the residual concentration present in the unspiked sample determined experimentally at the WRC.

Table 3.44. Results of inter-laboratory bias tests: nitrite. Laboratory 7 does not determine nitrite for Harmonised Monitoring purposes. The target is a maximum possible bias of 10% of the concentration of the determinand, or 0.05 mg l^{-1} of nitrogen (whichever is larger). All results are in milligrams per litre of nitrogen. (From [3.180]; Copyright 1982, Society for Analytical Chemistry)

Laboratory	Standard solution (A)			Spiked river water (B)			Spiked river water (C)		
	Mean concentration found	Deviation from expected value	Maximum possible bias[a]	Mean concentration found	Deviation from expected value	Maximum possible bias[a]	Mean concentration found	Deviation from average	Maximum possible bias[a]
1	0.207	+0.003	+0.007	0.096	−0.007	−0.012	0.547	−0.001	−0.061[b]
2	0.191	−0.013	−0.041	0.106	+0.003	+0.008	0.526	−0.022	−0.027
3	0.20	−0.004	−0.011	0.114	+0.011	+0.016	0.580	+0.032	+0.045
4	0.20	−0.004	−0.010	0.10	−0.003	−0.007	0.55	+0.002	+0.007
5	0.21	+0.006	+0.015	0.108	+0.005	+0.007	0.532	+0.034	+0.038
6	0.208	+0.004	+0.008	0.102	−0.001	−0.003	0.521	−0.027	−0.039
8	0.201	−0.003	−0.012	0.108	+0.005	+0.009	0.524	−0.024	−0.029
9	0.20	−0.004	−0.004	0.118	+0.015	+0.027	0.554	+0.006	+0.027
10	0.216	+0.012	+0.017	0.107			0.548		
Average	0.203			0.103			0.537		
Expected value[c]	0.204								

[a] At the 95% confidence level.
[b] Does not meet target.
[c] The expected value is: for solution (A) the true concentration: for sample (B) the concentration due to spiking.

Laboratory 2 showed a negative bias in analytical results for total oxidised nitrogen. This was investigated but no cause was identified. Subsequent work showed that this bias had been eliminated. To attempt to ensure that the required accuracy of results is maintained, analytical quality control is now an integral part of the routine analyses for the Harmonised Monitoring Scheme. Prime reliance for this purpose is placed on within-laboratory analytical quality control using statistical quality control charts. However, to obtain direct checks of between-laboratory bias, portions of a river water sample and a standard solution for spiking the sample with nitrite are distributed at intervals to all laboratories. The results of such tests are of value in indicating the efficiency of the analytical quality control work and are summarised in Tables 3.45 and 3.46. Each of these tests was carried out as described in the previous section except that the five replicate analyses for each material determined were made in one batch for the tests taking place after June 1977 and that natural levels of total oxidised nitrogen were determined, the samples being spiked only with nitrite.

Nagashima et al. [3.183] determined nitrate and nitrite by using second derivative spectrophotometry after conversion to nitrogen monoxide. Detection limits are 5×10^{-6} mol l^{-1} for nitrate and 1×10^{-6} mol l^{-1} for nitrite.

3.3.1.2 Segmented Flow Analysis

Marti and Hale [3.184] compared automated segmented flow and discrete analysers for the determination of nitrite, nitrate, ammonia and phosphate. The discrete analyser provided a 50% increase in the rate of analysis compared with segmented flow analysis. The accuracy and precision were comparable. Whilst both methods are in agreement, the rate of analysis (60 samples per hour) was faster by discrete analysis than segmented flow analysis (40 samples per hour).

This technique has been applied to the determination of nitrite and nitrate in concentration ranges 0–0.5, 0–5, 0–20, 0–40 and 0–100 mg l^{-1} (as N). See Sect. (2.1.1).

3.3.1.3 Ultraviolet Spectroscopy

Suzuki and Kuroda [3.185] used ultraviolet spectroscopy for the direct simultaneous determination of nitrite and nitrate.

3.3.1.4 Flow Injection Analysis

This technique has been used to determine nitrite and nitrate in natural waters [3.186, 3.187]. Van Staden [3.186] converted nitrate to nitrite using a pre-valve in-valve reduction technique before the sampling valve. Lerique and Lighetto [3.187] describe an analyser which depends on three principles – injection of

Table 3.45. Results from routine tests of between-laboratory bias: total oxidised nitrogen. A rule indicates that the particular laboratory did not participate. (From [3.180]; Copyright 1982, Society for Analytical Chemistry)

Laboratory	River water 1 (June, 1976)		River water 2 (June, 1977)		River water 3 (December, 1977)		River water 4 (October, 1978)		River water 5 (November, 1979)		River water 6 (July, 1980)		River water 7 (January, 1981)		River water 8 (July, 1981)	
	Mean TON content/ mg l^{-1} of nitrogen	Upper limit for bias, %	Mean TON content/ mg l^{-1} of nitrogen	Upper limit for bias, %	Mean TON content/ mg l^{-1} of nitrogen	Upper limit for bias, %	Mean TON content/ mg l^{-1} of nitrogen	Upper limit for bias, %	Mean TON content/ mg l^{-1} of nitrogen	Upper limit for bias, %	Mean TON content/ mg l^{-1} of nitrogen	Upper limit for bias, %	Mean TON content/ mg l^{-1} of nitrogen	Upper limit for bias, %	Mean TON content/ mg l^{-1} of nitrogen	Mean upper limit for bias, %
1	4.97	+7.9	—	—	5.63	+0.7	4.52	+1.8	3.47	−1.5	5.14	+1.8	4.59	+1.8	4.34	+1.6	+2.0
2	4.12	−13.9[a]	6.57	+4.5	5.70	+1.4	4.53	+1.1	3.63	+1.7	5.04	−1.5	4.71	+5.4	4.52	+6.3	+0.6
3	5.00	+7.9	6.2	−2.8	5.38	−6.0	4.45	−1.7	3.47	−1.9	5.11	+2.2	3.01[b]	−35.6[a]	4.30	+0.9	−4.6
4	—	—	6.41	+1.6	5.51	−1.6	4.51	+1.5	3.48	−1.9	5.70	+12.4[a]	4.52	−1.4	4.25	−2.4	+1.2
5	—	—	6.31	−1.6	5.62	−2.6	4.50	+0.8	3.65	+1.5	4.94	−3.6	4.50	−1.5	4.38	+3.3	−0.5
6	4.76	+4.1	6.17	−3.8	6.36[c]	+13.6[a]	4.75	+0.6	3.55	+3.4	5.89	+17.2[a]	4.68	+3.9	4.05	−10.2[a]	+4.3
7	4.78	+5.8	6.28	−3.5	5.50	−3.3	4.48	−1.3	3.30	−7.6	4.48	−13.0[a]	4.28	−6.7	4.26	−2.0	−3.0
8	4.60	−5.6	6.36	−0.8	5.91	+6.5	4.58	+1.6	3.29	−8.4	4.37	−16.9[a]	4.70	+3.9	4.25	−1.9	−2.7
9	4.19	−14.7[a]	6.65	+5.2	5.69	+1.4	4.06	−12.4[a]	3.60	0.0	5.06	−1.2	4.44	+5.1	4.48	+6.3	−1.3
10	4.76	+3.8	6.50	+2.0	5.98	+10.6[a]	4.57	+2.9	2.62	+4.9	5.26	+4.1	4.59	+1.6	4.40	+2.8	+4.2
11	—	—	—	—	5.31	−5.0	4.46	−2.4	3.65	+4.3	5.13	0.9	4.41	−3.3	4.00	−0.8	−2.0
Mean	4.64		6.38		5.62		4.49		3.50		5.10		4.54		4.20		

[a]Does not achieve the target for bias.
[b]Statistical outlier, not used in calculation of mean of laboratories.
[c]Not used in calculation of mean result because of doubts on accuracy of analytical system.

Table 3.46. Results from routine tests of between-laboratory bias: nitrite. A rule indicates that the particular laboratory did not participate. Laboratories 5 and 7 do not routinely determine nitrite for Harmonised Monitoring purposes. All results are in milligrams per litre of nitrogen. (From [3.180]; Copyright 1982, Society for Analytical Chemistry)

Laboratory	River water 1 (June, 1976)		River water 2 (December, 1977)		River water 3 (October, 1978)		River water 4 (November, 1979)		River water 5 (July, 1980)		River water 6 (February, 1981)		River water 7 (July, 1981)		Mean upper limit for bias
	Mean nitrite content	Upper limit for bias	Mean nitrite content	Upper limit for bias	Mean nitrite content	Upper limit for bias	Mean nitrite content	Upper limit for bias	Mean nitrite content	Upper limit for bias	Mean nitrite content	Upper limit for bias	Mean nitrite content	Upper limit for bias	
1	0.231	−0.033	0.409	−0.009	0.329	+0.011	0.300	−0.017	0.451	−0.010	0.293	−0.025	0.304	−0.021	−0.015
2	0.323[a]	+0.067[b]	0.456	−0.044	0.328	+0.010	0.352	+0.038	0.499	+0.043	0.349	+0.033	0.345	+0.022	+0.037
3	0.263	+0.008	0.40	−0.017	0.520	+0.001	0.280	−0.036	0.480	+0.020	0.338	+0.021	0.356	+0.035	+0.004
4	—	—	0.41	−0.007	0.340	+0.021	0.310	−0.013	0.432	−0.039	0.340	+0.022	0.316	−0.016	−0.005
6	0.263	+0.027	0.425	+0.010	0.305	−0.015	0.314	−0.011	0.477	+0.021	0.328	+0.020	0.309	−0.015	+0.005
8	0.242	−0.020	0.428	+0.015	0.305	−0.014	0.308	−0.012	0.444	−0.027	0.304	−0.020	0.340	+0.017	−0.009
9	0.248	−0.016	0.408	−0.013	0.308	−0.015	0.310	−0.006	0.468	+0.012	0.292	−0.031	0.298	−0.029	−0.014
10	0.246	−0.019	0.510	+0.105[b]	0.330	+0.011	0.350	+0.034	0.444	−0.022	0.302	−0.019	0.307	−0.017	+0.010
11	—	—	0.397	−0.021	0.308	−0.014	0.320	+0.004	0.449	−0.014	0.320	+0.002	0.333	+0.016	−0.004
Mean	0.260		0.417		0.319		0.316		0.460		0.318		0.323		

[a] Statistical outlier, not used in calculation of mean of laboratories.
[b] Does not achieve the target for bias.

sample into a continuous non-segmented flux of reagent, a reproducible residence time for the samples, and controlled dispersion of the sample. These objectives are achieved by independent peristaltic pumps for each of the four analysis channels, a very precise injection system, and a specially designed replaceable manifold. Nitrite determination is based on the sulphanilamide-naphthyl-ethylenediamine reaction with a colorimetric measurement at 540 nm. Nitrates are reduced to nitrites using a cadmium catalyst and the nitrites determined as above. The apparatus is connected to a microprocessor, can analyse 300 samples per hour, and results for each sample are printed out within about 20 s.

A computer controlled multichannel continuose flow analysis system has been applied to the measurement of nitrate and nitrite (and chloride, phosphate and sulphate) in small samples of rain water [3.165, 3.188].

3.3.1.5 High Performance Liquid Chromatography

Alawi [3.189] has discussed an indirect method for the determination of nitrite and nitrate in surface, ground and rain water by reaction with excess phenol (nitrite ions first being oxidized to nitrate) and extraction of the o-nitrophenol produced, followed by separation on a reversed phase HPLC column with amperometric detection in the reduction mode. Recoveries were 82% for nitrate and 77% for nitrite in the concentration range $10-1000$ $\mu g\, l^{-1}$. The method is claimed to be free of interference from other ions.

Using a radial compression C_{18} column and a mobile phase of aqueous tetramethylammonium phosphate, Kok et al. [3.190] have analysed mixtures of nitrate and nitrite at the 0.1 mg l^{-1} level and compared the results obtained with those found by ultraviolet spectroscopic screening methods.

They used a Waters Assoc. (Milford, MA. U.S.A.) Model 201 liquid chromatograph equipped with Model 441 absorbance detector set at 214 nm, a Model RMC Z radial compression module and a Radial-Pak C_{18} uBondapak column. Peaks were integrated by a Model 730 data module integrator. The mobile phase was 0.005 mol l^{-1} aqueous tetramethylammonium phosphate (Waters Assoc. Pic A UV grade reagent) which had been filtered through a Millex filter (Millipore HATF 01300) and degassed under vacuum. A flow rate of 3.0 ml min^{-1} was used. Water samples (20 µl) were injected directly after filtration as for the mobile phase.

Figure 3.15 shows typical chromatograms of standard nitrate-nitrite solution and of a water sample. It can be seen that, at a flow rate of 3.0 ml min^{-1}, each analysis was complete after 6 min.

Table 3.47 presents the nitrate and nitrite levels determined by HPLC and the nitrate levels found by the UV screening method in water samples from various sources. Low levels of nitrite were found in five of the samples analysed and were not detected in the others. Agreement between the nitrate concentrations found by HPLC and the UV screening method was poor. These

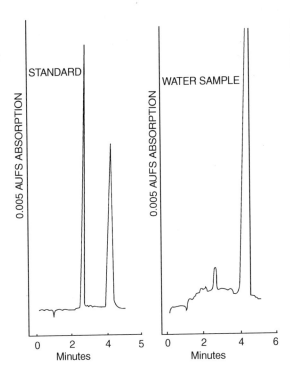

Fig. 3.15. Typical chromatograms of a standard nitrate-nitrite solution and a water sample (From [3.189])

discrepancies are probably a consequence of interferences by other materials in the UV method, which was intended as a rapid screening method.

To 50 ml of water, 1 mol l^{-1} hydrochloric acid (1 ml) was added, the sample was mixed thoroughly and the absorbance read at 220 and 275 nm in 10 mm cuvettes against distilled water as blank. The absorbance at 275 nm was doubled and subtracted from that at 220 nm to give the corrected nitrate absorbance. Nitrate was determined by reference to a calibration graph of nitrate concentration vs the corrected nitrate absorbance.

High performance liquid chromatography on a small bore column packed with microparticulate silica based ion exchange material has been used [3.191] to determine down to 0.01 mg l^{-1} nitrate in water without interference from other ions associated with potable, pond, river and stream water.

Detection of the separated peaks was achieved using a Pye LC-UV variable wavelength ultraviolet detector fitted with a 8-µl flow cell and set at 265 nm. The chromatographic system consisted of a Partisil 10-µm SAX column, and a mobile phase consisting of 10^{-3} mol l^{-1} potassium hydrogen phthalate (pH 3.95) in deionized water.

The method for the detection of non-ultraviolet absorbing ionic species relies on the technique of indirect photometric detection. In conventional HPLC where an ultraviolet detector is employed, the mobile phase is selected for many reasons and amongst them is the property of low absorbance at the monitoring wavelength. Compounds which elute and which have a significant absorbance at

Table 3.47. Levels of nitrate and nitrite found in water samples from various sources. (From [3.189]; Copyright 1984, Springer-Verlag GmbH, Germany) Values in mg l^{-1}

Sample No.	Source	HPLC method		UV screening method: nitrate[a]
		Nitrate	Nitrite	
1	Bore	3.1	—[a]	1.8
2	Bore	2.2	–	1.1
3	Bore	1.3	–	1.7
4	Bore	2.7	–	4.9
5	Bore	0.8	0.03	3.1
6	Bore	1.4	–	2.2
7	Reticulated	32.1	–	14.6
8	River	0.4	–	[b]
9	Dam	12.6	–	16.6
10	Dam	3.1	–	−0.8
11	Bore	2.9	–	2.9
12	Creek	0.8	–	1.0
13	River	4.0	–	–
14	River	0.3	–	–
15	Spring	6.3	0.8	10.2
16	Rain tank	3.3	–	3.7
17	Roof	3.2	–	3.5
18	Bore	0.5	–	1.6
19	Dam	4.3	0.5	6.0
20	Bore	3.7	–	2.6
21	Bore	2.9	–	1.1
22	River	9.1	0.10	8.9
23	Dam	0.7	–	0.8
24	Dam	0.9	–	–
25	Dam	0.5	–	1.4
26	Roof	3.6	–	4.4
27	Reticulated	4.2	0.04	0.5
28	Reticulated	0.5	–	0.4
29	River	2.4	–	–
30	Roof	0.6	–	1.2
31	Roof	3.3	–	4.6

[a] < 0.1 mg l^{-1} of nitrite detected by HPLC.
[b] < 0.2 mg l^{-1} of nitrate detected by UV method.

the set detector wavelength are thus observed as peaks of high absorbance against the low background absorbance of the solvent. This situation may be reversed however. Thus the mobile phase may contain an ultraviolet absorbing substance which will (after equilibration of the column) give a high, but constant, background absorbance. The presence of a non-absorbing species (such as an inorganic anion) in the mobile phase would thus be expected to cause a drop in absorbance. The elution of a non-absorbing species under these chromatographic conditions is therefore recognized as a peak corresponding to increased transmission, i.e. decreased absorption. Indirect detection of non-absorbing species may thus be achieved using a conventional variable wavelength ultraviolet HPLC detector.

In ion exchange chromatography a common component of the mobile phase is potassium hydrogen phthalate which possesses a strongly absorbing chromophore (the phthalate ion). Using a conventional anion exchange column, a mixture of inorganic anions such as chloride, nitrate, and sulfate may be separated and eluted using a potassium hydrogen phthalate buffer solution as the mobile phase. With an ultraviolet detector these non-absorbing anions may be detected indirectly. The principle of stoichiometric exchange applies and therefore when an anion such as nitrate desorbs from the column, phthalate ion absorbs to take its place. Hence the sensitivity of the detection system is directly related to the extinction coefficient, at the detector wavelength, of the absorbing species present in the mobile phase.

This mode of detection may be used to provide a simple and sensitive method for the determination of nitrate in natural waters. In this mode of chromatography the separation of the species of interest from other ions present depends upon many factors. Amongst the most important are the capacity of the ion exchange column and the concentration, i.e. the ionic strength, of the mobile phase. The simplest variables therefore are mobile phase strength and composition, and the pH of the mobile phase. Advantage may be taken of step-flow gradients to shorten analysis times.

The calibration curve obtained for a set of nitrate standard solutions was linear (correlation coefficient $= 0.9958$) and was defined by the equation $y = 4.07 \times + 12.5$. Using this calibration line a series of water samples was studied and the nitrate levels measured.

Lee and Field [3.192] have discussed a technique of post column fluorescence detection of nitrite, nitrate, thiosulphate and iodide anions in high performance liquid chromatography. These anions react with cerium IV to produce fluorescent species in a post column packed bed reactor.

Alawi [3.193] described a simple and specific high performance liquid chromatographic method for the determination of nitrite and nitrate in natural water based on the nitration of an excess of phenol by nitrate or oxidized nitrite ions; the 0-nitrophenol produced is extracted, separated on a reversed phase column and quantified using an amperometric detector in the reduction mode. Nitrite, if present, is first oxidized to nitrate by addition of 1% hydrogen peroxide. The method has been successfully applied to waters containing nitrate plus nitrite at the 5 µg l^{-1} level. By using a larger sample volume (500 ml) and injecting a larger aliquot (100 ml) onto the column, the sensitivity could be improved, giving a lower limit of about 1 µg l^{-1}. Recoveries were $82.6 \pm 2.37\%$ for nitrate and $76.9 \pm 1.94\%$ for nitrite in the $0.01 - 1$ µg ml^{-1} concentration range.

This method is free from interferences which, when coupled with the specificity and the high sensitivity of the electrochemical detection mode, renders it suitable for the determination of trace levels of nitrate and nitrite in surface, ground and rain water.

3.3.1.6 Column Coupling Capillary Isotachophoresis

This technique offers many similar advantages to ion chromatography in the determination of anions, namely multiple ion analysis, little or no sample preparation, speed, sensitivity and automation. Very similar advantages are offered by capillary isotachophoresis. Zelinski et al. [3.194] applied the technique to the determination of nitrite, nitrate, fluoride, chloride, sulphate and phosphate in river water. The technique is discussed in further detail in Sect. 11.1.

3.3.1.7 Micelle Exclusion Chromatography

Okada [3.195] has used micelle exclusion chromatography to determine nitrite and nitrate in the presence of other anions (bromide, iodide and iodate) in water. The method is based on partition of ions to a cationic micelle phase and shows different selectivity from ion exchange chromatography.

3.3.1.8 Reverse Phase Ion Interaction Liquid Chromatography

Ito et al. [3.283] used this technique employing octadecyl silane reverse phase columns coated with cetyltrimethyl ammonium chloride for the determination of nitrite and nitrate in seawater.

Marengo et al. [3.284] carried out an experimental design and partial least squares analysis for the optimisation of reversed phase ion interaction liquid chromatography and its application to the separation of nitrite, nitrate and phenylene diamine isomers. Down to 0.5 mg l^{-1} of these substances could be determined.

3.3.1.9 Ion Chromatography

Ion chromatography has been used to determine nitrite and nitrate in natural waters. Techniques are reviewed in Table 3.48 and discussed in more detail in Sect. 11.1.

3.3.1.10 Polarography

Polarography has been applied to the determination of nitrites and nitrates [3.196, 3.197]. In the polarographic technique [3.239], total nitrate and nitrite are determined. The nitrite is decomposed by the addition of methyl alcohol, and nitrate only is determined on a second polarogram. Boese et al. [3.197] described a method based on the enhancement of the differential pulse polarographic peak current of ytterbium by nitrate and nitrite.

Table 3.48. Determination of nitrite and nitrate in natural waters by ion chromatography

Type of water sample	Codetermined anions	References
Natural	None	3.68
Natural	Cl, Br, SO$_4$	3.69
Natural	Cl, Br, I, SO$_4$, PO$_4$	3.70

Nygaard [3.198] has described a pneumatoamperometric technique for the determination of nitrite and nitrate. These ions are reduced to nitric oxide with hydroquinoline and then nitric oxide purged from the sample and oxidized at a membrane covered anodically polarized platinum electrode.

Holak and Specchio [3.285] determined nitrite and nitrate in natural waters by differential pulse polarography with simultaneous nitrogen purging.

3.3.1.11 Resonance Raman Spectroscopy

Furaya et al. [3.173] investigated the simultaneous determination of nitrite and nitrate using resonance Raman spectroscopy. Figure 3.16 shows an example of the spectra of the sample solutions which contain much larger quantities of nitrate relative to those of nitrite. This spectrum demonstrates that the key band of nitrate can be measured almost independently of the adjacent band of the coloured product corresponding to nitrite. This result shows clearly the

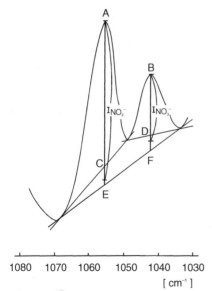

Fig. 3.16. The spectrum of the mixed solution of the coloured product and nitrate. The base lines for the analysis are determined as follows: $I_{NO_3} = BD + BF \times DF/AE + BF)$. $I_{NO_2} = AC + AE \times CEAE + BF)$. The concentrations of the $NO_2^- = 0.2$ mg l^{-1} and $NO_3^- = 100$ mg l^{-1} (From [3.173])

possibility of the simultaneous determination of nitrite and nitrate and also proves the excellent selectivity of Raman spectrometry.

Plots of $I_{NO_3^-}/I_{NO_2^-}$ against $C_{NO_3^-}/C_{NO_2^-}$ give the most satisfactory result ($I_{NO_3^-}, I_{NO_2^-}$ are peak heights of the nitrate band at 1045 cm^{-1} and that of the band at 1056 cm^{-1} corresponding to nitrite, respectively; $C_{NO_3^-}, C_{NO_2^-}$ are concentrations of nitrate and nitrite) $I_{NO_3^-}$ and $I_{NO_2^-}$ are estimated as shown in Fig. 3.16. The plots of peak heights or areas of the 1045 cm^{-1} band against $C_{NO_3^-}$ does not give such satisfactory results regardless of the selections of base lines. Thus the process of the simultaneous determination of nitrite and nitrate is as follows. (1) Sample waters are treated to form the coloured product from nitrite by the colour reaction. (2) From the resonance Raman band of the coloured product at 1250 cm^{-1}, the concentrations of nitrate are determined. (3) From the spectra in the 1030–1080 cm^{-1} region, $I_{NO_3^-}/I_{NO_2^-}$ is measured, and, using the analytical curve shown in Fig. 3.16 $C_{NO_2^-}$ is estimated by operation (2), and the concentrations of nitrate ($C_{NO_3^-}$) are determined.

3.3.2 Rainwater

3.3.2.1 High Performance Liquid Chromatography

The method described in Sect. 3.3.1 [3.189] for the determination of nitrite and nitrate in natural waters has been applied to rainwater.

3.3.2.2 Ion Chromatography

Methods for the determination of nitrite and nitrate in rain water are reviewed in Table 3.49 and discussed more fully in Sect. 11.2.

3.3.3 Potable Water

3.3.3.1 Spectrophotometric Methods

Zhou and Xie [3.199] give details of a rapid spectrophotometric method for the determination of traces of nitrite and nitrate in water. In this method, nitrate is quickly reduced to nitrite at pH 3.0 in 5 min with freshly prepared cadmium sponge, which is produced in-situ by the action of zinc powder in a dilute solution of cadmium chloride in the presence of ammonium chloride. At pH 2.0, nitrous acid diazotizes with p-amino-acetophenone, which is then coupled with N-(1-naphthyl)-ethylenediamine. The azo dye formed is extracted into n-butanol at pH 2.0 in the presence of β-naphthyl sulfonic acid and aluminium nitrate.

Table 3.49. Determination of nitrite and nitrate in rain water by ion chromatography

Type of water sample	Codetermined anions	References
Rain	Cl, PO_4, SO_4	3.93
Rain	Cl, PO_4	3.94
Rain	Cl, F, Br, SO_4, SO_3, PO_4	3.95
Rain	Cl, Br, F, PO_4, BrO_3, SO_4	3.96

The absorbance is measured at 550 nm. The molar absorptivity of the extract is about 4.8×10^4 mol^{-1} cm^{-1}. The ions normally present in water do not interfere when sodium metaphosphate is added as a masking agent.

Reagents. Unless stated otherwise, all reagents used are of analytical pure grade, and all solutions are diluted with nitrite-free water, which is obtained by distilling alkaline permanganate-distilled water using an all-glass distillation apparatus.

Standard nitrate solution. Dissolve 0.772 g anhydrous potassium nitrate and dilute with water to 100 ml (1.00 mg nitrogen or 4.43 mg nitrate ml^{-1}). Prepare a working solution by suitable dilution and store in plastic bottles.

Standard nitrite solution. Dissolve 0.492 g anhydrous sodium nitrite, having been dried in desiccator for 4 h, and dilute with water to 100 ml (1.00 mg nitrogen ml^{-1}). Prepare a working solution by suitable dilution and store in a brown bottle under refrigeration. Standardize iodometrically before using.

p-Aminoacetophenone hydrochloride solution. Prepare a 0.5% (w/v) solution in 1 + 9 hydrochloric acid, and store in a brown bottle.

N-(1-Naphthyl)-ethylenediamine dihydrochloride solution. Dissolve 0.5 g N(1-naphthyl)-ethylenediamine dihydrochloride in 100 ml of 1 + 99 hydrochloric acid.

Potassium chloride-hydrochloric acid buffer solution. Mix 50 ml of 0.2 mol l^{-1} potassium chloride with 17 ml of 0.2 mol l^{-1} hydrochloric acid, dilute to 200 ml with water and adjust pH to 1.8 with hydrochloric acid.

Ammonium chloride solution, 30% (w/v).

β-Naphthyl sulfonic acid, 0.5% (w/v). Dissolve 0.5 g β-naphthyl sulfonic acid in 100 ml water. Adjust to pH 6.0 with sodium hydroxide solution.

Sodium metaphosphate solution, 5.0% (w/v).

Sodium acetate solution, 30% (w/v).

Cadmium chloride solution, 0.1% (w/v).

Zinc powder, 100–200 mesh. Before use, wash with 0.2 mol l^{-1} hydrochloric acid, then rinse with water and dry [3.201].

Procedure. Transfer about 20 ml or less of the sample containing 0.5–10 μg nitrate-nitrogen to a 25-ml graduated flask. Add 1.0 ml of ammonium chloride,

1.0 ml of potassium chloride-hydrochloric acid buffer, 1.0 ml of cadmium chloride and 1.0 ml sodium metaphosphate, mix, adjust to pH 3.0, dilute to the mark with water and again mix thoroughly. After addition of 100 mg of zinc powder through a small glass funnel, mix the contents of the flask by inverting three times (each time swirling for 10 s) within 5 min and let it stand for 2 min for clarification. Pipette aliquots of the clear solution for determination of nitrite as follows.

Take about 30 ml or less of sample solution containing 0.1 – 2.0 µg nitritenitrogen in a 60-ml separatory funnel. Add 1.0 ml of p-aminoacetophenone hydrochloride solution and adjust to pH 2.0 with either sodium acetate solution or hydrochloric acid. After 15 min standing at room temperature, add 1.0 ml of N-(1-naphthyl)-ethylenediamine dihydrochloride solution and swirl. Add 1.0 ml of β-naphthyl sulfonic acid and 3 g of powdered aluminium nitrate, mix thoroughly again and extract the azo compound into 5.0 ml of n-butanol by shaking mechanically for 3 min. After clarification, measure the absorbance of the -n butanol layer at 550 nm in a 1-cm cell against a reagent blank.

3.3.3.2 Gas Chromatography

A method has been described [3.202, 3.203] for determining aqueous nitrates by conversion to nitrobenzene followed by electron capture gas chromatography. The procedure can also be used to determine aqueous nitrites and gaseous oxides of nitrogen if they are first converted to nitrates. The procedure was evaluated by analysing drinking water for nitrate over a period of one month. The method is sensitive and capable of measuring typical environmental levels of nitrogen compounds. The detection limit for nitrobenzene is about 10^{-12} g.

Reagents. As follows.
 Sulfuric acid, reagent grade.
 Benzene, thiophene-free reagent grade (J. T. Baker, Phillipsburg, N. J. USA).
 Hydrogen peroxide AR grade 30%.
 Nitrobenzene.
 Phenylmercuric acetate, Matheson, Coleman, and Bell (East Rutherford, N. J., USA).
 Potassium nitrate.
 Potassium nitrite.
 Silver sulfate.
 Sodium citrate.

Apparatus. Hewlett Packard Models 402 and 5750 gas chromatographs or equivalent, equipped with 63Ni and titanium tritide electron capture detectors. The columns used contained 1.5% SE-30 coated on 80–100 mesh Chromosorb G, acid-washed, packed in either 76 cm × 4 mm I.D. glass tubing or 51 cm × 3 mm I.D. heavy walled Teflon tubing. Isothermal column temperatures: between 100 and 150 °C, carrier gas flow rates of 40–60 ml min^{-1} of argon-methane (90:10 vlv).

Procedure. For the determination of nitrate, use a 1-dram vial with a polethylene stopper (Kimble No. 60975-L) as a reaction vessel. Introduce a 0.20 ml aliquot of aqueous sample into the vial, followed by 1.0 ml of benzene. Catalyse the reaction by addition of 1.0 ml of concentrated sulfuric acid. Shake the vial for 10 min. Remove the benzene layer immediately from the reaction vial with a Pasteur pipette, place it in a separate vial and analyse by gas chromatography with electron capture detection for the nitrobenzene concentration generated. Treat standard solutions of potassium nitrate in the same manner to generate a standard calibration plot relating nitrobenzene concentration to peak height. If higher precision is desired (approximately 4% relative standard deviation), add 2,5-dimethylnitrobenzene to the benzene prior to reaction as an internal standard (concentration 5×10^{-7} gm l^{-1} to correct for differences in injection size, detector sensitivity, fluctuations, etc. When this internal standard is used, normalize the peak heights of nitrobenzene to the peak heights of the standard before further data reduction.

Determine nitrite by difference, treating a second identical sample with an oxidant to convert nitrite to nitrate. Consquently, the concentration of nitrite plus nitrate is subsequently measured, so nitrite must be determined by difference. Specifically, the procedure used was to introduce 0.15 ml of the sample into the reaction vial, followed by addition of 0.05 ml of 0.1 N hydrogen peroxide. The rest of the procedure is the same as that for nitrate.

The results of applying the microtechnique to the analysis of nitrate in several samples of drinking waters are summarised in Table 3.50. In no cases were detectable amounts of nitrite found (the limit of detection for nitrite

Table 3.50. Comparison of analyses of drinking waters for NO_3^- concentration by GC-ECD and ISE. Numerical designations refer to selected municipal water supplies in Crowley County, Colo., U.S.A. Unless otherwise specified concentrations are expressed in ppm (w/w). (From [3.202]; Copyright 1975, Elsevier Science Publishers BV, Netherlands)

Site	Concentration NO_3^- (mg l^{-1})	
	GC-ECD[a]	ISE[b]
Boulder, Colo.	0.40	–
TW0011	44	48
TW0012	42	46
TW0013	51	52
TW0014	16	16
TW0015	35	36
TW0016	12	13

[a]Electron capture gas chromatography.
[b]Ion-selective electrode measurement.

depends on the nitrate concentration, since nitrite is determined by difference, generally a nitrite concentration which is 5% of the nitrate concentration can be detected). In Table 3.50 the results are shown of a comparative study of the electron capture gas chromatographic microtechnique vs ion-selective electrode measurements. The close agreement (usually within 5%) of the totally independent methods used for the analysis of identical samples is an excellent demonstration of the efficacy of the methods.

3.3.3.3 Ion Exchange Chromatography

Gerritse [3.204] separated nitrate and nitrite on a cellulose anion exchanger column and detected them in a spectrophotometric flow through cell at 210 nm. The detection limit is $1-5$ $\mu g\,l^{-1}$ as nitrogen with a sample volume of 100 to 200 µl and the linear range extends up to 20 $mg\,l^{-1}$ as nitrogen.

A stainless steel column (SS 316) of length 30 cm and i.d. 0.3 cm was filled with a mixture of Kieselguhr (BDH, Poole, Great Britain) and Ecteola ET 41 cellulose anion exchanger (Whatman: W & R Balston, Maidstone, Great Britain). The eluent used was 0.03 mol l^{-1} potassium sulphate solution and 0.01 mol l^{-1} Tris buffer at pH 7 in water. Simple filling of the column by tapping in small amounts of a mixture of kieselguhr (sieved fraction with particle size range $5-10$ µm) and Ecteola ET 41 was found to produce columns giving sufficient resolution of nitrate and nitrite. A ratio of kieselguhr to Ecteola of 1:2 (by weight) was used. The column was operated at a pressure of 40 bar. The eluent was monitored with a Cecil CF-212 spectrophotometric flow through detector at a wavelength of 210 nm. The temperature of the column was ambient. Samples were injected with the aid of a sample injection valve (Valvo C20) using a sample loop of 60 µl.

Examples of chromatograms of potable water and of sludge solution resulting from aeration of anaerobically digested sewage sludge are shown in Fig. 3.17. The sludge was centrifuged at 40000 g for 1 h, then the supernatant was diluted 1:100 with distilled water and injected into the column. Potable water was injected undiluted. Without dilution, samples with concentrations of up to 20 mg l^{-1} each (as nitrogen) of nitrite and nitrate could be injected.

Under the conditions used, the contribution of the column to dispersion of nitrite and nitrate is about 200 µl (expressed as the standard deviation of the output).

Davenport and Johnson [3.205] used ion exchange chromatography on Amberlite IRA-900 strongly basic resin to determine nitrate and nitrite in water. Perchloric acid (0.01 mol l^{-1}) was used as eluent and an electrochemical cadmium electrode detector was used.

Sherwood and Johnson [3.206] have described an ion exchange chromatographic determination of nitrate and amperometric detection at a copperized cadmium electrode. The chromatograms obtained in this procedure resolve nitrate from dissolved oxygen.

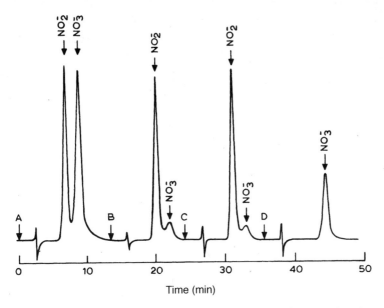

Fig. 3.17. Chromatograms resulting from the injection of samples containing nitrate and nitrite into a mixed bed column containing Kieselguhr and Ecteola ET 41 cellulose anion exchanger. Eluent: 0.03 mol l^{-1} K$_2$SO$_4$, 0.01 mol l^{-1} Tris buffer (pH 7); A, standard mixture containing 1.4 mg l^{-1} as nitrogen of both nitrite and nitrate (attenuation 0.1); B, sewage sludge sample, diluted 1:100 containing 2.7 mg l^{-1} as nitrogen of nitrite and 0.35 mg l^{-1} as nitrogen of nitrate (attenatuion 0.2); C, sewage sludge sample, diluted 1:100 containing 1.4 mg l^{-1} as nitrogen of nitrite and 0.1 ml l^{-1} as nitrogen of nitrite (attenuation 0.1); D, potable water containing 0.45 mg l^{-1} as nitrogen of nitrate (attenuation 0.1) (From [3.204])

3.3.4 Industrial Effluents

3.3.4.1 Ion Chromatography

This technique has been used to determine nitrate and nitrite (and chloride, sulphate, phosphate, fluoride and bromide) in industrial effluents [3.112]. The technique is discussed more fully in Sect. 11.4.

3.3.5 Sewage

Workers at the Water Research Centre, UK [3.207] have described detailed procedures based on the use of the Technicon Autoanalyser AA11 for the determination of nitrate and nitrite in sewage and sewage effluents. Measurements of nitrate in sewage at the 10–50 mg l^{-1} level were made with a within-batch standard deviation of 0.1 mg l^{-1} nitrate. Nitrate recoveries at the 12 mg l^{-1} level were in the range 100–100.7%. Standard deviations of nitrite determinations were in the range 0.003 (at 0.2 mg l^{-1} nitrite) to 0.01 (at 1 mg l^{-1} nitrite). Nitrite recoveries in the 0.4 mg l^{-1} region were between 98 and 102%.

3.3.6 Soil

3.3.6.1 Spectrophotometric Method

Henrickson and Selmer-Olson [3.208] applied an autoanalyser to the determination of nitrate and nitrite in water and soil extracts. In an autoanalyser the water sample, buffered to pH 8.6 with aqueous ammonia-ammonium chloride, is passed through a copperised cadmium reductor column. The nitrite formed is reacted with sulphanilic acid and N-1-naphthylethylene diamine and the extinction of the azo dye is measured at 520 nm. Nitrate can be determined in water in the range 0.01-1 mg l^{-1} with a standard deviation of 0.004 mg l^{-1}. For soil extracts the range and standard deviation are, respectively, 0.5–10 and 0.007 mg l^{-1}.

Garcia Gutierrez [3.209] has described an azo coupling spectrophotometric method for the determination of nitrite and nitrate in soils. Nitrite is determined spectrophotometrically at 550 nm after treatment with sulphanilic acid and N-1-naphthylethylene diamine to form an azo dye. In another portion of the sample, nitrate is reduced to nitrite by passing a pH 9.6 buffered solution through a cadmium reductor and proceeding as above. Soils were boiled with water and calcium carbonate, treated with freshly precipitated aluminium hydroxide and active carbon and filtered prior to analysis by the above procedure.

3.3.7 Biological Fluids

3.3.7.1 Spectrophotometric Method

Wegner [3.210] described a spectrophotometric procedure for the determination of nitrate and nitrite in biological fluids. Total nitrate and nitrite were determined by adding 1-pyrenamine in sulphuric acid and evaluation at 565 nm. Nitrite was determined by the standard sulphanilamide-N-1-naphthylenediamine diazotization method.

3.3.8 Meat

3.3.8.1 Ion-Selective Electrodes

Choi and Fung [3.211] have described a procedure for the determination of nitrate and nitrite in meat products by a direct potentiometric ion-selective electrode method using a nitrate-selective electrode. The soluble nitrate and nitrite are extracted from the meat with a Soxhlet extractor using a borax buffer solution of pH9. In this method the homogenised meat sample is extracted for 3 h and the extract made up to a standard volume. To determine nitrate, aluminium sulphate, sulphamic acid, silver sulphate and ammonium sulphate are added to the extract and the potential of the nitrate-selective electrode measured with respect to the double function reference electrode. The concentration of nitrate was measured by the standard additions method. Measurement of nitrite ion concentration was carried out on a separate sample portion using the same procedure for nitrate except that 0.01 mol l^{-1} sulphamic acid was replaced with 0.01 mol l^{-1} potassium permanganate solution to oxidize nitrite to nitrate. The increase in nitrate concentration was used to calculate the concentration of nitrite. Coefficients of variation for nitrate and nitrite were 0.96 and 3.8% respectively. Down to 3 µg kg^{-1} of nitrate and nitrite could be determined and results were in good agreement with those obtained spectrophotometrically.

3.3.9 Vegetables

3.3.9.1 Spectrophotometric Method

Garcia-Gutierrez [3.209] has described an azo coupling spectrophotometric method for the determination of nitrite and nitrate in vegetables.

Nitrite is determined spectrophotometrically at 550 nm after treatment with sulphanilic acid and N-1-naphthylethylene-diamine to form the azo dye. In another portion of the sample, nitrate is reduced to nitrite by passing the solution buffered to pH9.6 through a cadmium reductor and total nitrite is determined as above.

Vegetables were boiled with water and calcium carbonate, treated with freshly precipitated aluminium hydroxide and filtered prior to analysis as above.

3.4 Free Cyanide

3.4.1 Natural Waters

3.4.1.1 Spectrophotometric Methods

Broderius [3.212] determined free cyanide and hydrocyanide in natural waters and industrial waters directly by bubbling compressed air through the solution to displace cyanide, and collecting it in a glass bead concentration column. In a second method, hydrogen cyanide was allowed to diffuse from an enclosed solution into dilute sodium hydroxide in a dish suspended above the solution.

The separated and concentrated cyanide was determined spectrophotometrically by the chloramine-T method. The diffusion procedure was recommended for determining low concentrations of hydrogen cyanide.

Reagents.
  Sodium hydroxide, 1 N.
  Sodium hydroxide, 0.02 N.
  Acetic acid, 2.25 N.
  Phosphate buffer, chloramine-T [3.214].
  Chloramine-T, 1 g 1100 ml^{-1}.
  Standard cyanide solution. prepare concentrated sodium cyanide solutions, dilute to the required concentrations and adjust to pH 5.5 with hydrochloric acid to convert to free cyanide.

Apparatus. The apparatus used to isolate and concentrate hydrogen cyanide by the diffusion principle is shown in Fig. 3.18 [3.213].

Procedure – Air Purge Method. The method for determining hydrogen cyanide uses a gas regulator and flow meter, a temperature regulated sparging chamber, and a glass bubbler through which finely dispersed air is bubbled at 25 – 300 ml min^{-1} into 20 l of test solution. A glass bead concentration column coated with about 0.4 ml of 1.0 N sodium hydroxide is used to collect the displaced hydrogen cyanide and a water displacement bottle measures the volume of air dispersed in the sample and passed through the column. At the end of the collection period

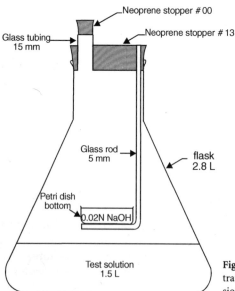

Fig. 3.18. Apparatus used to isolate and concentrate molecular hydrogen cyanide by the diffusion procedure (From [3.213])

wash off the cyanide absorbed in the concentration column with three 5-ml washings of deionized water into a 25-ml volumetric flask for colorimetric analysis.

The hydrogen cyanide concentration in a solution is calculated from the quantity of hydrogen cyanide collected in a column and by reference to a calibration curve relating the micrograms of hydrogen cyanide displaced per liter of air dispersed with the known hydrogen cyanide concentration in standard solutions of the same temperature.

Procedure – Diffusion Method. The apparatus used to determine hydrogen cyanide by the diffusion method consists of a 2.8-l Fernbach flask (Corning 4420) fitted with an airtight No. 13 neoprene rubber stopper and the bottom section of a petri dish (Corning 3160, 60 × 15 mm) attached by silicone cement to a solid glass support rod (Fig. 3.18). The isolation and concentration of hydrogen cyanide from an aqueous solution is accomplished by adding 1.5 l of sample to the flask and sealing with a stopper. After a short equilibration period pipette either 8.5 or 20.5 ml of 0.02 N sodium hydroxide through a small hole in the large stopper into the petri dish suspended in the air space above the sample. Seal the system and absorb the hydrogen cyanide diffusing from the surface of the test solution into the alkali solution. The air space between the sample and alkali serves to separate the nonvolatile and ionic species in solution from the collected hydrogen cyanide. A 2-h collection period is generally sufficient to allow for an analytically measurable amount of hydrogen cyanide to be transferred from the sample to the sodium hydroxide. Remove the small stopper in the lid and transfer an 8-or 20-ml aliquot of alkali absorbent containing the dissolved hydrogen cyanide to a 10- or 25-ml volumetric flask, respectively.

Procedure – Spectrophotometric Method. Acidify the 25-ml portions of solutions referred to above to about pH 5.6 with 3 or 6 drops of 2.25 N acetic acid, respectively. Add four drops of chloramine-T solution (1.0 g 100 ml^{-1}), mix, and allow to stand for about 2 min. A single solution combining a phosphate buffer and chloramine-T can be used to lower the pH of the 0.02 N sodium hydroxide and to halogenate the cyanide [3.214]. Then add 1.0 or 2.5 ml respectively of the pyridine-barbituric acid reagent [3.215] and mix and make the volume up to the mark with deionized water. After about 10 min, measure the intensity of colour produced at 582 nm against a blank. The cyanide content expressed as hydrogen cyanide is determined from a suitable calibration curve.

The concentration of hydrogen cyanide in a sample is determined from a calibration curve defined from tests performed with standardized sodium cyanide solutions prepared with deionized water and at about pH 6. The tests with the standard solutions are conducted at the same time and under the same conditions as those for the unknown samples. In this way standard solutions are treated as samples and slight changes in factors that may affect the diffusion of hydrogen cyanide from a test solution would be reflected in the calibration curve. A constant temperature water bath is used to ensure temperature control of the test solutions during the collection period.

The detection limit achieved in the diffusion method was approximately 1.3 µg cyanide l^{-1}. The amount of hydrogen cyanide collected by the diffusion procedure is temperature dependent. This effect is minimal, however, in the temperature rate of 15 °C to about 22 °C. At higher temperatures the rate of hydrogen cyanide displacement increases rapidly.

In the diffusion method, ammonia (50 mg l^{-1}) sulphide (1 mg l^{-1}) and phenol (50 mg l^{-1}) resulted, respectively, in only about an 8%, 13% and 0% reduction in the colour development by the pyridine-barbituric acid method [3.215] for solutions containing 40 µg cyanide l^{-1}. The displacement of hydrogen cyanide by diffusion from 1.5 l test solutions containing 25 µg cyanide l^{-1} was virtually unaffected by the individual presence of the following chemicals: 130 mg l^{-1} Cl^-, 7.4 mg l^{-1} SO_4^{2-}, 3.0 mg l^{-1} NO_3^-, 100 mg l^{-1} (as S) SCN^-, 50 mg l^{-1} NH_3^-, 0.1 mg l^{-1} Na_2S-S, 50 mg l^{-1} phenol, 50 mg l^{-1} o-cresol, 5 mg l^{-1} Zn^{2+}, 5 mg l^{-1} Cd^{2+}, 5 mg l^{-1} Pb^{2+}, 1.0 mg l^{-1} Fe $(CN)_6^{4-}$ or $Fe(CN)_6^{-3}$.

Sulfide at concentrations of 1.0 mg l^{-1} Na_2S-S extensively interfered with the determination of hydrogen cyanide. Since it was observed that such a concentration will only have minimal effect on the direct colorimetric procedure for cyanide determination, the interference process must occur in the test solution itself. Possibly sulfide oxidation products, especially polysulfides, react with cyanide to produce thiocyanate. Therefore, sulfide should be removed from a sample as soon as possible after collection to prevent its potential interaction with cyanide. Recommended procedures for sulfide removal suggest the addition of either lead or cadmium salts.

In Table 3.51 some results obtained in applying the diffusion procedure to several spiked natural water samples are shown.

It is seen that the type of sample has an effect on cyanide recovery. For natural water samples, with the exception of very hard well water (296 mg l^{-1} $CaCO_3$), the determined hydrogen cyanide concentrations were essentially equal to the expected values. In most cases the recoveries from spiked effluents were lower than from pure water samples, which resulted in lower determined hydrogen cyanide concentrations than expected. This response may be due to free cyanide-complexing components and other chemical and biological reactions

Table 3.51. Determination of HCN by the diffusion procedure for sodium cyanide spiked water samples at 20 °C. (From [3.212]; Copyright 1981, American Chemical Society)

Sample	pH value	Amt of total cyanide added (as HCN) µg l^{-1} Natural waters	HCN concn, µg l^{-1}	
			Expected[a]	Determined
Well	7.92	27.2	26.2	18.8
Spring	7.38	27.2	26.9	27.0
Creek	6.53	27.2	27.1	26.2
Lake	7.85	27.2	36.3	23.4
Pond	8.57	27.2	23.1	20.6

[a]Expected concentration is equal to that added as sodium cyanide plus that present in the effluent portion; all values adjusted for pH and temperature.

making a portion of the spiked cyanide absent or unavailable for measurement. Therefore measuring the recovery of cyanide added to certain samples is not necessarily a reliable means of studying the suitability of the method for natural water samples. In no instance, however, was a marked positive interference detected when hydrogen cyanide was expected to be absent or in low concentrations.

Nagashima [3.215] has discussed the use of reagents based on pyridine-pyrazalone and pyridine-barbituric acid for the spectrophotometric determination of cyanide in water.

Montgomery et al. [3.216] determined free hydrogen cyanide in river water by a method based on extraction of cyanide ion from water with 1:1:1 trichloroethane, then re-extraction of cyanide in aqueous sodium pyrophosphate solution. After the extraction of hydrogen cyanide into sodium pyrophosphate solution it is treated with saturated aqueous bromine, sodium arsenite and p-phenylene diamine and determined spectrophotometrically at 508 nm. Less than 0.01 mg l^{-1} of hydrogen cyanide can be detected.

Hangos Mahr et al. [3.217] separated and determined low concentrations of cyanide in water by an automated procedure. Separation was achieved by isothermal distillation. Two methods for the determination of cyanide in a flow type automatic analyser were compared (reaction with picric acid and the Aldridge method). A combination of isothermal distillation with automatic spectrophotometric determination based on the Aldridge method was recommended. The detection limit was 0.01 mg cyanide l^{-1}. Interference by sulphide and sulphite and methods for their removal are described.

A kinetic spectrophotometric method has been used [3.218] to determine low cyanide concentrations in the presence of large amounts of thiocyanate.

Fenong [3.219] has described an enzymatic method for the spectrophotometric determination of micro amounts of cyanide.

3.4.1.2 Segmented Flow Analysis

The technique described in Sect. 2.11 for the determination of chloride can also be applied to the determination of free cyanide in the concentration ranges 0 – 0.25 and 4 – 50 mg l^{-1} in water.

3.4.1.3 Flow Injection Analysis

Fogg and Alonso [3.220] have described an oxidative amperometric flow injection method for the determination of cyanide at an electrochemically pretreated glass carbon electrode.

3.4.1.4 Atomic Absorption Spectrometry

Cyanide has been determined indirectly by atomic absorption spectrometry [3.221].

3.4.1.5 Ion-Selective Electrodes

Ion-selective potentiometry in alkaline solution provides a suitable finish for the determination of cyanide [3.222 – 3.226]. Sulfide ion is an interferent, and a cadmium nitrate treatment has been found to remove sulfide without influencing the response of the silver iodide-silver sulfide electrode. If both the precipitate and solution from this treatment are colourless (sulfide absent) filtration of the mixture before the potentiometric measurement is unnecessary. Since the electrode response of the electrode is pH dependent, standards and samples must be brought to the same pH. It is noteworthy that the silver-silver sulfide electrode (e.g. Orion 94 – 16) is more sensitive to cyanide than the silver iodide - silver sulfide one [3.222, 3.224], but the response is affected by cadmium ion.

3.4.1.6 Gas Chromatography

Funazo et al. [3.227] have described a method for the determination of cyanide in water in which the cyanide ion is converted into benzonitrile by reaction with aniline, sodium nitrite and cupric sulphate. The benzonitrile is extracted into chloroform and determined by gas chromatography with a flame ionization detector. The detection limit for potassium cyanide is 3 mg l^{-1}. Lead, zinc and sulphide ions interfere at 100 mg l^{-1} but not at 10 mg l^{-1}.

$$C_6H_5NH_2 + NaNO_2 \longrightarrow C_6H_5N^+ = N + NaOH + OH^-$$

$$C_6H_5N^+ = N + CN^- \xrightarrow{Cu^{2+}} C_6H_5CN + N_2$$

To carry out the derivatization reaction, 0.25 ml of sodium nitrite (7.0×10^{-2} mol l^{-1} aqueous solution) and 1.0 ml of potassium cyanide (known concentration) were added to 0.25 ml of cupric sulphate (1.0×10^{-1} mol l^{-1}) in a 10 ml glass reaction tube with a glass stopper. Then 1.0 ml of chloroform solution was added and the tube was mechanically shaken for 30 min in an incubator kept at 55 °C. The chloroform solution contains aniline (3.0×10^{-1} mol l^{-1}) and p-dichlorobenzene (1.0×10^{-3} mol l^{-1}) as an internal standard. At the end of the reaction period, 0.5 ml of the organic layer was taken out of the tube and washed with 5.0 ml of 1 mol l^{-1} hydrochloric acid. The chloroform extract (1 µl) was injected into the gas chromatograph. The resulting benzonitrile was determined by an internal standard method.

A Shimadau GC-3BF gas chromatograph equipped with a dual flame ioniation detector was used. A glass column (1.7 m × 3 mm i.d.) was packed with

10% PEG-20 M on 60–80 mesh Shimalite W. The column and injection port temperatures were maintained at 140 and 250 °C respectively. The nitrogen flow rate was 25 ml min^{-1}. The peak areas were measured by a digital integrator.

Down to 3 mg l^{-1} free cyanide can be determined in waste waters by this procedure without interference from 100 mg l^{-1} of ammonium, cadmium II, chromium VI, fluoride, chloride, bromide, iodide, thiocyanate, nitrate, bicarbonate and phosphate. Lead II, zinc II and sulphide interfere at 100 mg l^{-1} but do not interfere at 10 mg l^{-1}.

3.4.1.7 Ion Chromatography

Bond et al. [3.228] have discussed the determination of free cyanide together with sulphide in natural waters by ion chromatography.

Peschet and Tinet [3.229] applied ion chromatography to the determination of free cyanide (and hypochlorite and sulphide) in ground waters. These methods are discussed further in Sect. 11.1.

3.4.1.8 Miscellaneous

Other techniques used for the determination of free cyanides in water samples include ultraviolet spectroscopy as the Ni(CN)$_4^{2-}$ complex [3.230]. Micro diffusion and potentiometry have also been used to determine cyanide [3.231].

3.4.2 Potable Waters

3.4.2.1 Ion Chromatography

The ion chromatographic procedure described in Sect. 11.3.2 is applicable to the determination of traces of cyanide in potable waters.

3.4.3 Industrial Effluents

3.4.3.1 Titration Methods

In their method for the determination of cyanide in industrial effluents, Hewitt and Austin [3.232] distil the acidified sample in the presence of zinc and lead acetates (to prevent decomposition of complex cyanides and sulphides) and collect the liberated hydrogen cyanide in sodium hydroxide. Cyanide is then estimated by titration with standard silver nitrate solution using 4-dimethylaminobenzyl-idenerhodamine as indicator.

Kogan et al. [3.233] titrated down to 1 mg l^{-1} cyanide in coke works effluents amperometrically in a basal electrolyte of 0.1 N sodium hydroxide using silver nitrate as titrant. Titration was carried out at - 0.2 V (vSCE) using a vibrating platinum electrode. Up to 500 mg l^{-1} thiocyanate, phenol and ammonia and 100 mg l^{-1} of chloride or sulphate do not interfere. Lead carbonate is added to avoid interference by sulphide.

Cadmium carbonate - magnesium chloride mixtures have also been used to prevent interference by sulphide [3.234] and theoretical yields are claimed for this procedure.

3.4.3.2 Spectrophotometric Methods

Dimethylaminobenzylidenerhodamine [3.235] and pyrazolone [3.236] have been used as chromogenic reagents in the determination of cyanide in industrial effluents. In the latter method [3.236], free cyanide is separated from complex cyanides by distillation under reduced pressure in the presence of zinc and lead acetates. The hydrogen cyanide evolved is absorbed in sodium hydroxide solution, and Chloramine T is added, after slight acidification with acetic acid to produce cyanogen chloride. A pyridine/pyrazolone reagent is added, and an optical absorbance measurement of the resultant blue dye is made at 620 nm wavelength. Volatile aldehydes and ketones interfere in the final colour development. The detection limit is about 10 µg l^{-1}.

The micro diffusion chloramine-T spectrophotometric method discussed in Sect. 3.4.1 [3.212] for the determination of free cyanide in natural waters has been applied to industrial effluents.

3.4.3.3 Continuous Flow Analyses

Pihlar et al. [3.237] developed the amperometric method of determining cyanides in industrial effluents to permit reliable and reproducible continuous flow measurements and flow injection analysis of discrete samples of galvanizing plant effluents at the µg l^{-1} level. The method is based on the measurement of the diffusion current arising from the oxidation of silver to dicyanoargenate (I). By use of the flow injection principle and a manifold arrangement, absolute amounts of less than one nanogram of cyanide can be determined precisely in volumes as small as 10 µl and at a rate exceeding 100 samples per hour.

Figure 3.19 outlines a straight manifold arrangement and cylindrical flow through electrode. For operation in the injection mode, the sample is introduced into the carrier stream by means of a microsyringe through a septum in the injector and passes through 20 cm of 0.5 mm bore tubing before entering the detector. The amperometric flow through electrode consists of a 0.9 mm diameter silver wire (99.9% Ag) mounted centrally by means of two Teflon stoppers

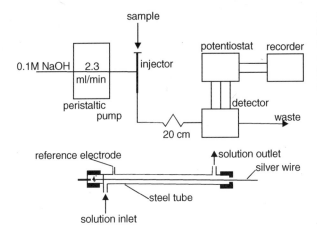

Fig. 3.19. Manifold arrangement and cylindrical flow through electrode used for the determination of cyanide. All tubes were 0.5 mm bore (From [3.237])

in a stainless steel (or nickel) tube of 1.9 mm bore which serves as a counter electrode. The cell hold-up volume is 130 µl. The mercurous sulphate reference electrode is bridged into the cell through capillary tubing as shown in Fig. 3.19. The potential of the silver electrode is controlled at -0.5 V vs the M.S.E. by means of a potentiostat. The current was continuously monitored by measuring the voltage drop across a 10 kΩ precision resistor connected in series with the counter electrode, with an electronic voltmeter (Radiometer GVM 30) coupled to a Varian A-25 strip chart recorder.

Current voltage curves were recorded with a PAR 174 polarographic analyser under hydrodynamic conditions and at a scan rate of 1 mV Sec^{-1}. The reelationship between limiting current and velocity of the fluid was measured with the aid of a waters M-5000 A pump at volume flow rates of between 0.1 and 10 ml min^{-1}.

Free cyanide was separated by micro diffusion in Conway cells. The volume of 0.1 mol l^{-1} sodium hydroxide used as absorbing solution was between 0.1 and 1.0 ml according to the amount of free cyanide present in the sample. The free cyanide released upon acidifying the samples (up to 5 ml) with 0.5 ml of 3 mol l^{-1} sulphuric acid was quantitatively recovered after 3 h of distillation at room temperature (20–25 °C). The absorption solution was subsequently injected directly into the analyser. A 0.1 mol l^{-1} solution of sodium hydroxide was used as the carrier stream.

Of the numerous substances tested, only sulphide, iodide and thiocyanate interfere seriously in this method and they can be eliminated by distillation or precipitation. Oxidants such as hypochlorite or hydrogen peroxide, which are common agents for detoxification of cyanide effluents, must be fixed by addition of arsenite or sulphite since they are reduced at the silver electrode and this results in a cathodic current. This is a disadvantage of the method compared with potentiometric techniques in which oxidants and reductants usually do not interfere. However, such compounds can be eliminated by simple chemical treatment ordinarily performed at the sampling stage.

3.4.3.4 Spectrofluorimetry

McKinney et al. [3.238] have described a rapid fluorimetric method for the determination of cyanide in mine waters. The method is based on the liberation of naphtho (2,3-c) (1,2,5) selenadiazole from the reagent $Pd_2I_2Cl_4$ by cyanide at 25 °C and pH 8. The naphthol (2,3-c) (1,2,5) selenadiazole is then extracted into hexane and its fluorescence is measured at 520 nm with excitation at 377 nm. The calibration graph is approximately rectilinear for up to 100 µg of cyanide per 3 ml of hexane. Mercuric ions and sulphide interfered.

3.4.3.5 Ion-Selective Electrodes

The application of cyanide selective electrodes has been discussed [3.239, 3.240]. Cuthbert [3.239] used the technique to analyse steel works effluents. For cyanide concentrations above 1 mg l^{-1} the electrode method and the standard argentimetric method gave similar results; for lower cyanide concentrations, the colorimetric pyridine-pyrazolone method gave higher results than either the potentiometric method or the colorimetric pyridine-barbituric acid method. Cuthbert [3.239] discusses the problems associated with the determination of cyanide in waste waters containing complex iron cyanides and sulphides.

Hafton [3.241] used a silver ion-selective electrode for the continuous determination of free cyanide in effluents in concentrations of 0.01 – 1.0 µg l^{-1}.

3.4.3.6 Ion Chromatography

Rocklin and Johnson [3.242] have discussed the determination of cyanide together with other anions (sulphide, bromide and iodide) in industrial effluents by ion chromatography. This method is discussed further in Sect. 11.4.

3.4.3.7 Polarography

Cyanide has been determined in industrial effluents by methods based on measurement of the potential at −0.45 V (vs SCE) in 0.01 mol l^{-1} sodium hydroxide medium [3.243]. There is no interference from phenol, acrylonitrile, urea or cyanide but sulphide and thiocyanate interfere.

Square wave polarography has also been used to determine 1–100 µg l^{-1} cyanide in industrial effluents [3.244].

3.4.3.8 Miscellaneous

Gregorowicz and Gorka [3.245] have reviewed methods for the determination of cyanide in industrial effluents.

3.4.4 Waste Waters

3.4.4.1 Spectrophotometric Methods

Fu-Sheng et al. [3.246] applied the copper-cadion 2B- Triton A-100 system to the spectrophotometric determination of micro amounts of free cyanide in waste water.

Nagashima and Ozawa [3.247] have given details for the determination of cyanide in aqueous samples by spectrophotometry, using isonicotinic acid and barbituric acid as reagents. Cyanide react with chloramine-T, and then the reagent is added to form a soluble violet-blue product, which is measured at 600 nm. Maximum absorbance is achieved in 15 min at 25 °C and remains constant for about 30 min. This method is applicable to water samples containing 0–5 mg l^{-1} cyanide. Results obtained on several plating waste waters are reported and show good agreement with those obtained by the conventional pyridine-pyrazolone method [3.248].

Details of the isonicotinic acid - barbituric acid procedure are given below.

In a dried reaction tube, place 10.0 ml of sample solution (pH 5–8) containing less than 5 µg of cyanide and free from sulfide and thiocyanate ions. Add 3.0 ml of potassium dihydrogen phosphate solution (1 mol l^{-1}) and 0.2 ml of 1% (w/v) chloramine-T solution, stopper the tube and shake gently. After 1–2 min, add 5.0 ml of isonicotinic acid - barbituric acid reagent, stopper the tube again, mix, and keep the mixture at 25 °C for 15 min. Measure the absorbance at 600 nm against a reagent blank.

The effect of diverse ions on the determination of 2 mg l^{-1} cyanide by the above method is shown in Table 3.52. A large amount of sulfide caused a negative error and 1 mg of sulfide did not develop colour. In the pyrazolone method, sulfite and ferricyanide also led to a negative error. Thiocyanate led to a significant positive error when sulfide or thiocyanate ions were present in a sample solution, and a suitable pretreatment is necessary for the removal.

The large positive error caused by thiocyanate suggested that the method could be used for determining this ion. This method can be used, as described above, for the determination of free cyanide or, with a suitable sample pretreatment, to the determination of total cyanide. The pretreatment involves distillation with phosphoric acid and an alkaline 10% EDTA solution, using 2% sodium hydroxide or 2% sodium acetate as absorbent. For the sodium hydroxide solution (2%) absorbent, the receiving solution was neutralized with acetic acid.

Table 3.52. Comparison of effect of diverse ions on the determination of 2.0 µg CN^-/10 ml by the isonicolinic acid-barbituric acid method. (From [3.247]; Copyright 1981, Gordon and Breach Science Publications, UK)

Ion	Added as	Amount added (µg/10 ml)	CN^- found (µg)	Error (µg)
Cl^-	NaCl	1000	2.0	0.0
NO_2^-	$NaNO_2$	1000	1.9	-0.1
NO_3^-	KNO_3	1000	2.0	0.0
SO_3^{2-}	Na_2SO_3	100		
		1000	2.0	0.0
SO_4^{2-}	Na_2SO_4	1000	2.0	0.0
CH_3COO-	$CH_3COONa \cdot 3H_2O$	100	2.0	0.0
OCN	KOCN	1000	1.9	-0.1
S^{2-}	$Na_2S \cdot 9H_2O$	1	2.0	0.0
		10	2.0	0.0
		100	0.9	-1.1
		1000		$-$
SCN^-	KSCN	2	2.9	$+0.9$
		5	4.3	$+2.3$
Fe^{3+}	$FeCl_3$	100	2.1	$+0.1$
$Fe(CN)_6^{4-}$	$K_4[Fe(CN)_6] \cdot 3H_2O$	100	2.0	0.0
$Fe(CN)_6^{3-}$	$K_3[Fe(CN)_6]$	10		
		100	2.0	0.0

Table 3.53. Determination of cyanide in plating waste waters by the isonicotinic acid-barbituric acid method and the pyridine-pyrazolone method in comparison. (From [3.247]; Copyright 1981, Gordon and Breach Science Publications, UK)

Sample	CN^- concentration (mg l^{-1})			
	Isonicotinic acid-barbituric acid method		Pyridine-pyrazolne method	
	2% NaOH	2% CH_3COONa	2% NaOH	2% CH_3COONa
I	0.04	0.04	0.03	0.04
II	0.11	0.12	0.10	0.12
III	0.19	0.18	0.19	0.18
IV	0.26		0.26	
V	0.35	0.34	0.34	0.35

Table 3.53 shows the good agreement in results for total cyanide obtained on wastewater samples obtained by this method and by the spectrophotometric pyridine - pyrazolone method [3.248].

Funazo et al. [3.261] have described a more sensitive gas chromatographic procedure capable of determining down to 3 µg l^{-1} total cyanide in waste waters. The method is based on the derivatization of cyanide to benzonitrile, which is extracted with benzene and determined by flame thermionic gas chromatography. In the derivatization reaction, aqueous cyanide reacts with aniline and sodium nitrite in the presence of copper II sulphate and forms benzonitrile.

Reagents. As follows.
Stock potassium cyanide solution (1 mg l^{-1}). Dissolve 2.51 g of potassium cyanide in 1.0 l distilled water; prepare this solution immediately before use and standardize by titration with silver nitrate. Add 10.0 ml of the solution to 2% (w/v) sodium hydroxide solution. Dilute the mixture to 1.0 l. Subsequently, dilute 10.0 ml of this alkaline solution to 100 ml with distilled water to give the stock cyanide solution (1.0 mg l^{-1}).

Sulphuric acid, IN, AR grade.
Aniline, aqueous 6.0×10^{-2} mol l^{-1}.
Sodium nitrite, aqueous 2.5×10^{-2} mol l^{-1}, AR grade.
Cupric sulphate, aqueous 3.0×10^{-1} mol l^{-1}, AR grade.
Benzene containing p-dichlorobenzene, 1.0×10^{-3} mol l^{-1}.
Hydrochloric acid, aqueous, 2.0 N, AR grade.

Apparatus. Shimadzu GC-6AM gas chromatograph or equivalent equipped with a flame thermionic detector. A stainless steel column (3 m × 3 mm i.d. packed with 5% PEG-HT on 60-80 mesh Uniport HP. Use nitrogen as carrier gas at a constant flow rate of 20 ml min^{-1}. Detector, injection port and column temperatures are 230, 230 and 170°C respectively. Measure peak areas with a digital integrator.

Procedure. Neutralize alkaline sample solutions with 1.0 N sulphuric acid before derivatization. To 2.0 ml of the neutralised solution in a 10-ml reaction vessel, add aqueous solutions of aniline (6.0×10^{-2} mol l^{-1}, 2.0 ml) sodium nitrite (2.5×10^{-2} mol l^{-1}, 0.5 ml) and copper II sulphate (3.0×10^{-1} mol l^{-1}, 0.5 ml). Allow the vessel to stand for 20 min in a water bath maintained at 80°C. At the end of the reaction period, add 1.0 ml of benzene containing p-dichlorobenzene (1.0×10^{-3} mol l^{-1} or 1.5×10^{-4} mol l^{-1}) as an internal standard and extract the derivatized benzonitrile by shaking the vessel for 10 min at room temperature (around 18°C).

For the determination of cyanide at concentrations lower than 0.2 mg l^{-1}, use internal standard concentrations of 1.5×10^{-4} mol l^{-1}. Centrifuge the benzene layer if it contains an insoluble solid (probably consisting of azo coupling compounds produced as by-products). Inject an aliquot (1.0 µl) of the clean benzene extract into the gas chromatograph after washing with hydrochloric acid (2.0 N, 3.0 ml), determine benzonitrile with a flame thermionic detector.

Funazo et al. [3.261] tested this method in the presence of several ions normally found in environmental samples (Table 3.54). The peak area of benzonitrile derivatized from the standard cyanide solution (0.5 mg l^{-1}) was arbitrarily assigned a value of 100.

Table 3.54. Results of interference study. (From [3.261]; Copyright 1982, Elsevier Science Publishers BV, Netherlands) CN^- concentration: 0.50 µg/ml

Ion	Concentration	Added as (mg l^{-1})	Peak area[a]
None			100.0 ± 2.0
F$^-$	500	NaF	99.2 ± 1.7
Cl$^-$	500	NaCl	99.4 ± 0.2
Br$^-$	500	NaBr	100.2 ± 1.8
I$^-$	500	KI	98.6 ± 1.3
SO$_4^{2-}$	500	Na$_2$SO$_4$	100.4 ± 3.2
NO$_3^-$	500	NaNO$_3$	100.8 ± 2.5
HCO$_3^-$	500	KHCO$_3$	99.9 ± 1.9
H$_2$PO$_4^-$	500	KH$_2$PO$_4$	99.4 ± 0.8
SCN$^-$	500	NaSCN	86.0 ± 2.1
SCN$^-$	100	NaSCN	99.0 ± 3.0
CNO$^-$	500	KCNO	104.1 ± 0.7
CNO$^-$	100	KCNO	102.8 ± 1.7
S^{2-}	0.1	Na$_2$S·9H$_2$O	90.9 ± 4.6
S^{2-}	0.05	Na$_2$S·9H$_2$O	100.2 ± 2.3
NH$_4^+$	500	(NH$_4$)$_2$SO$_4$	101.8 ± 1.7

[a]Mean ± S.D. of five replicate analyses.

None of the ions except thiocyanate, cyanate and sulphide interfered at a concentration of 500 mg l^{-1}. At this concentration, thiocyanate interferes negatively and cyanate positively. However, these interferences are not observed at the 100 mg l^{-1} level. Sulphide ion interferes at relatively low concentrations and even at 0.1 mg l^{-1} a negative interference is observed. Similar interferences of sulphide are well known in the pyridine pyrazolone method and in the method using the cyanide ion-selective electrode. However, sulphide can be removed from the sample solution by treating the alkaline sample at pH 11.0 with small amounts of powdered lead carbonate [3.262].

Funazo et al. [3.261] analysed several wastewater samples from various factories using cyanide, such as metal refining and plating factories by the gas chromatographic and by the spectrophotometric pyridine pyrazolone methods [3.248]. Immediately after collection, sodium hydroxide was added to the samples and the solution was adjusted to pH 12–13 in order to prevent the evolution of hydrogen cyanide.

The results are shown in Table 3.55 which reveals that the values measured by the two methods agree well.

3.4.4.2 Ion-Selective Electrodes

Fraut et al. [3.249] have described an electrode indicator technique for the determination of low levels of free cyanide in waste waters. Concentrations of

Table 3.55. Comparison of spectrophotometric and GC methods for the analysis of waste water samples. (From [3.261]; Copyright 1982, Elsevier Science Publishers BV, Netherlands)

Sample	Cyanide concentration (mg l^{-1})	
	Spetrophotometric method[a]	GC method[b]
A	0.48	0.51 ± 0.015
B	0.20	0.20 ± 0.006
C	0.13	0.16 ± 0.002
D	0.08	0.03 ± <0.0005
E	0.08	0.09 ± 0.002

[a]Pyridine-pyrazolone method.
[b]Mean = S.D. of five replicate analyses.

cyanide down to 30 µg l^{-1} are determined with use of the Ag$^+$ selective electrode and 10 µmol l^{-1} Ag(CN)$_2^-$ as indicator solution. The pH of the sample is adjusted to between 11 and 12 and an addition procedure is used, the concentration of cyanide in the sample being obtained by use of a Gran plot; samples containing more than 260 mg l^{-1} of cyanide must be diluted initially. Copper and nickel are masked with EDTA and any sulphide removed by addition of lead ions in slight excess. Ammonia does not interfere even in 1000-fold excess.

3.4.4.3 Gas Chromatography

Nota and Improta [3.250] determined cyanide in coke oven waste water by gas chromatography. The method is based on treatment of the sample with bromine and direct selective determination of the cyanogen bromide by gas-solid chromatography using a BrCN$^-$ selective electron capture detector. No preliminary treatment of the sample to remove interferences is necessary in this method, and in this sense it has distinct advantages over many of the earlier procedures. Bromine also oxidizes thiocyanate to cyanogen bromide. Previous treatment of the sample with aqueous formaldehyde destroys thiocyanate and prevents its interference.

Reagents. As Follows.
  Bromine Water.
  Phenol, 5% aqueous.
  Nitromethane, 5% aqueous.
  Buffer, borax, 0.5 mol l^{-1} pH 8.2.
  Formaldehyde, 40 wt% aqueous.
  Phosphoric acid, 20% w/v.
  Solution A. Transfer an amount of the sample containing 10–100 µg of cyanide to a 200-ml flask and add 5 ml of 20 wt% phosphoric acid solution. Add

bromine water until a persistent yellow colour is reached. Shake the flask for some time and, after a few minutes, destroy the excess of bromine by adding 2 ml of a 5% aqueous phenol solution and 10 ml of 5% aqueous nitromethane. Make the solution up to 200 ml with distilled water.

Solution B. To the same amount of the sample used for solution A, add 1 ml of Borax buffer 0.5 mol l^{-1} (pH 8.2) and 3 ml of a 40 wt% formaldehyde solution in a 200-ml flask. Then apply the treatment described above for solution A.

Standard solution. Obtain standard solutions of cyanogen bromide by treating an amount of stock potassium thiocyanate solution exactly as described for solution A. Obtain a standard curve covering the 0.05-0.5 mg l^{-1} cyanide range by diluting a 1.0 mol l^{-1} potassium thiocyanate solution corresponding to 26.0 mg l^{-1} cyanide. Plot the peak areas ratios of cyanogen bromide and nitromethane vs mg l^{-1} cyanide to obtain a calibration graph.

Apparatus. A "Fractovap" mod. 2400 gas chromatograph (C. Erba, Milan) equipped with an electron capture detector (63 Ni source, C. Erba, Milan) and with an electronic integrator (mod. 3380 A Hewlett Packard) or equivalent. The column was made from borosilicate glass (1m × 0.3 cm i.d.) and packed with Porapak Q 80-100 mesh, supplied by Waters Associates (Framingham, Mass.). Nitrogen was used as the carrier gas at a flow rate of 50 ml min^{-1}. The injector and detector temperatures were kept at 120 and 150 °C respectively. Retention times for cyanogen bromide and for nitromethane (internal standard) were about 4 and 7 min respectively with an oven temperature of 100 °C. Into the gas chromatograph 2 µl of the aqueous solutions described above were injected.

Calculations. The amount of cyanide in the original sample is calculated according to the relation

mg l^{-1} CN$^-$ = (a − b) v × 200

Where a and b are the amount of cyanide expressed in mg l^{-1}, obtained by relating the cyanogen bromide and nitromethane peak areas ratios of solutions A and B respectively to the ratios reported in the standard curve and v is the volume (ml) of the original sample.

Some typical results of cyanide analysis of coke oven water and coke oven water effluents are reported in Table 3.56, and are compared with the results obtained following the procedure suggested by the Standard Methods [3.251].

Generally, results obtained by the APHA standard method [3.251] are some 15% lower than those obtained by gas chromatography.

Nota et al. [3.252] have made the cyanogen bromide technique for determining cyanides and thiocyanate more sensitive and reproducible by adopting the headspace technique. This technique lends itself to automation.

Nota et al. [3.253] have also described a gas chromatographic headspace method based on different principles for the determination of 0.01–100 mg l^{-1} of cyanide and thiocyanate in coke oven waters and waste effluents. This method involves first transforming the cyanides, or the thiocyanates, into hydrogen

Table 3.56. CN⁻ determination in some typical samples of coke-oven water (A–E) and coke-oven water effluent into the sea (F–J). (From [3.250]; Copyright 1979 Elsevier Science Ltd., UK)

Sample	mg l^{-1}				
	Standard methods	Gas chromatographic method	CN⁻ added	Standard methods	CN⁻ recovered Gas chromatographic method
A	39.0	44.0	20.0	18.6	20.2
B	50.0	55.0	20.0	17.5	20.3
C	62.0	70.0	25.0	23.0	25.1
D	47.0	53.0	5.0	4.6	4.91
E	46.0	52.0	5.0	4.5	4.93
F	0.27	0.295	0.150	0.134	0.147
G	0.27	0.307	0.150	0.137	0.154
H	0.31	0.353	0.175	0.150	0.178
I	0.25	0.263	0.050	n.d.	0.047
J	0.37	0.414	0.050	n.d.	0.048

To each sample a known amount of KCN was added to check the recovery. Concentrations are expressed in ppm.

cyanide by acidification, then removing hydrogen cyanide from the aqueous sample by the headspace technique, and finally separating hydrogen cyanide by gas solid chromatography and selective detection with a nitrogen phosphorus detector.

A similar procedure is adopted for the determination of thiocyanate, the only difference being the quantitative transformation of thiocyanate into hydrogen cyanide according to the reactions:

$$SCN^- + Br_2 + 4H_2O \rightarrow BrCN + SO_4^{2-} + Br^- + 8H^+$$

$$BrCN + Red \rightarrow HCN + Br^- + Ox$$

(where Red = SO_2 or I' and Ox = SO_4^{2-} or I_3^-). If cyanide is present prior to the oxidation step, it must be transformed into unreactive cyanohydrin by an excess of formaldehyde or removed by boiling the solution previously acidified to pH 2.

Reagents. As Follows.
Phosphoric acid, aqueous 85% w/v.
Bromine water.
Phenol, aqueous 5%.
Sodium sulphite, aqueous 0.5 mol l^{-1}.
Potassium iodide, solid.
Phosphate buffer, pH 7.
Formaldehyde, aqueous 4% w/v.

Apparatus. F42 gas chromatograph (Perkin Elmer) equipped with a headspace device or equivalent and a nitrogen phosphorus detector. The column (2 m × 0.3 cm i.d.) was made of borosilicate glass and packed with Poropak Q

(80 – 100 mesh) supplied by Waters Associates (Milford, MA. USA). Nitrogen was used as the carrier gas at a flow rate of 40 ml min^{-1}. The flow rates of hydrogen and air were 4 and 150 ml min^{-1}, respectively. The injector and detector temperatures were 120 and 150°C respectively and the oven temperature was 100°C. The detector wire was heated as recommended by the manufacturer (Perkin Elmer).

The headspace conditions were as follows: needle temperature, 100°C; flushing of sampling capillary, 15 s; pressurization of sample vial, 60 s; sample withdrawal from headspace, 5 s; 20-ml vials thermostated at 80°C for 10 min were used. Under these conditions the retention time of hydrogen cyanide was 5 min.

Analysis of Cyanide. Adopt the procedure described below regardless of the presence of thiocyanate in the sample. Introduce 5.0 ml of aqueous sample containing between 0.01 and 100 mg l^{-1} of cyanide and 2 drops of 85% phosphoric acid (to give a pH of 2) into a 20-ml vial. Analyse the sample under the conditions given above. Any oxidizing agents present in the sample must be reduced with sodium sulphite prior to heating of the vial.

Analysis of thiocyanate in the Absence of Cyanide. Introduce 5.0 ml of aqueous sample containing between 0.02 and 200 mg l^{-1} of thiocyanate and 2 – 3 drops of 85% phosphoric acid (to give a pH of 2) into a 10 - ml volumetric flask. Add bromine water dropwise until a persistent yellow colour is obtained. Remove the excess of bromine after 5 min by the addition of 0.5 ml of 5% phenol solution. The quantitative removal of the excess of bromine is indicated by the attainment of a colourless solution.

Reduce cyanogen bromide by means of 0.5 ml of 0.1 mol l^{-1} sodium sulphite solution or a few crystals of potassium iodide. Add distilled water to the 10 ml mark. Introduce 5.0 ml of this solution into the headspace vial and analyse under the conditions given above.

Analysis of Thiocyanate in the Presence of Cyanide. Two procedures can be used for eliminating the interference of cyanide.

The first involves removal of cyanide by heating. Boil 50.0 ml of the sample, to which 1 ml of 85% phosphoric acid has been added, for 20 min in a beaker. After cooling treat the sample as described in the previous section.

The second involves treatment of the sample with formaldehyde. To 4.00 ml of sample in a 10-ml flask, add 0.2 ml of phosphate buffer (pH = 7) and 0.1 ml of 4.0% (w/v) formaldehyde. After 5 min add 0.5 ml of 85% phosphoric acid and 3–4 drops of bromine water. Allow the solution to stand 15 min (it must still be yellow) then remove the excess bromine by the addition of 0.5 ml of 5% phenol solution. Reduce the cyanogen bromide formed by adding 0.5 ml of 0.1 mol l^{-1} sodium sulphite solution or some crystals of potassium iodide. Dilute the solution to 10 ml with distilled water and introduce 5 ml into a 20-ml headspace vial and analyse as described above.

Calibration graphs for both cyanide and thiocyanate can be obtained by diluting solutions containing known concentrations of cyanide according to the method described above and plotting peak height against concentration.

Table 3.57 shows the effect of the salt concentration on the response for hydrogen cyanide. The matrix effect can be avoided by making use of the method of standard additions, as illustrated by the results in Table 3.58 for a solution containing 2 mg l^{-1} of cyanide. The presence of iron II, iron III and copper II in the sample decreases the response for hydrogen cyanide. The greatest effect is caused by copper II. Reducing agents do not interfere in the analysis of cyanide but oxidizing substances have to be reduced prior to heating of the sample. Oxidants and reducing agents do not interfere in the determination of thiocyanate.

Typical results for the analysis of cyanide in some coke oven waste waters and coke oven waste effluents are reported in Table 3.59.

The results for samples A-D analysed by the headspace method [3.251] and the cyanogen bromide method [3.250] are in good agreement. Samples E-H show good agreement when analysed by the headspace method and the standard additions method. The results obtained by using a calibration graph are appreciably higher owing to the different matrix of the standards and the samples.

The accuracy of the procedure was tested by spiking samples A-H with a known amount of cyanide; the recoveries, reported in the last two columns in Table 3.59 are excellent.

Table 3.57. Influence of sodium chloride concentration on the HCN response. (From [3.253]; Copyright 1981, Elsevier Science Publishers BV, Netherlands)

Amount of CN$^-$ (mg l^{-1})	NaCl added (%)	Peak height (cm)	Increase (%)
1.00	0	7.4	–
1.00	10	8.2	11
1.00	20	8.8	19
1.00	30	9.4	27

Table 3.58. Standard additions method applied to a 2 mg l^{-1} CN$^-$ sample. (From [3.253]; Copyright 1981, Elsevier Science Publishers BV, Netherlands)

Aliquot No.	Amount of CN$^-$ present (µg)	Sample volume (ml)	CN$^-$ added (µg)	Peak height (cm)
1	10	5.00	0.00	5.5
2	10	5.00	1.00	6.1
3	10	5.00	2.00	6.5
4	10	5.00	4.00	7.7
5	10	5.00	5.00	8.3

Table 3.59. Comparison with BrCN Method and accuracy of proposed procedure. (From [3.253]; Copyright 1981, Elsevier Science Publishers BV, Netherlands)

Sample	CN⁻ determined (mg l^{-1})			CN⁻ added mg l^{-1}	CN⁻ recovered by HS method mg l^{-1}	
	BrCN method	HS method				
		By calibration graph	By standard additions method		By calibration graph	By standard additions method
A	53	52	–	25.0	25.1	–
B	65	65	–	20.0	20.0	–
C	42	43	–	20.0	25.2	–
D	58	57	–	5.00	5.04	–
E	0.20	0.23	0.21	0.150	–	0.150
F	0.37	0.41	0.38	0.150	–	0.150
G	0.25	0.28	0.25	0.150	–	0.149
H	0.30	0.34	0.30	0.020	–	0.021

3.4.4.4 Chronopotentiometry

Procopio et al. [3.254] carried out an indirect determination of cyanide in waste waters by this technique. Cyanide was determined by measuring the displacement of the quarter wave potential of the chronopotentiometric graph of formaldehyde. After separation of the cyanide by distillation into alkaline media, formaldehyde solution was added and the resultant solution transferred to the cell of a constant current coulostat. A constant potential of -1.4 V was applied for 5 min while nitrogen flowed through continuously. The potential was removed and the solution allowed to become quiescent before applying a constant current to the working gold electrode. A transition time of 30 s was obtained. From the displacement of the formaldehyde curve with and without cyanide present, a detection limit of 10 μg l^{-1} cyanide was obtained with a relative error of less than 2% and a relative standard deviation of 1%.

3.4.5 Plant Materials

Cyanide in trace amounts is found in a large number of plants, mainly in the form of substituted glycosides. Plants so affected include grasses, fruit kernels, pulses, linseed and cassava.

Harris et al. [3.255] have described methods for the determination of cyanide in these materials based on either spectrophotometry using p-phenylene diamine-pyridine or gas chromatographically following conversion of cyanide to cyanogen bromide. Cyanide is extracted from the sample by digestion with phosphoric acid. Recoveries were in the range 96–99% (spectrophotometric method) and 90–96% (gas chromatographic method).

3.4.6 Feeds, Fertilizers and Plant Tissues

3.4.6.1 Ion-Selective Electrodes

Ion-selective electrodes have been used to determine free cyanide in water, animal feeds, plant tissues and fertilizers.

3.4.7 Soil

Tecator [3.256] produce apparatus based on distillation and titration or spectrophotometry for the determination of cyanides in waste waters, soil and sludges.

3.4.8 Biological Fluids

Spectrophotometry has been applied to the determination of hydrogen cyanide in biological fluids [3.257].

3.4.9 Air

Gelatin stabilized gold solution has been used to absorb cyanide from air samples prior to analysis [3.258].
Spectrophotometry has been applied to the determination of hydrogen cyanide in air [3.257].

3.4.10 Sewage Sludge

3.4.10.1 Spectrophotometric Method

Kodura and Lada [3.259] determined cyanide in sewage spectrophotometricaly using ferroin. Iron, copper sulphide, acrylonitrile, phenol, methanol, formaldehyde, urea, thiourea, caprolactam or hexamine did not interfere at concentrations up to 1 g l^{-1}.

3.4.10.2 Atomic Absorption Spectrometry

In an indirect atomic absorption procedure [3.260] for determining down to 20 $\mu \text{mol l}^{-1}$ cyanide, the sample is treated with sodium carbonate and a known excess of cupric sulphate to precipitate cupric cyanide. Excess copper in the filtrate is then determined by atomic absorption spectrometry and hence the cyanide content of the sample calculated. Iron III does not interfere except at low cyanide concentrations. Quantitative recovery of cyanide from sewage was obtained by this procedure.

3.5 Total Cyanide

Total cyanide includes free cyanide ions plus complexed metal cyanides, e.g. ferrocyanides. The methods discussed in the previous section determine only free uncomplexed cyanide. To determine uncomplexed plus complexed cyanide, i.e. total cyanide, it is necessary prior to analysis to decompose the complexed cyanides to free cyanide [3.262].

3.5.1 Natural Waters

3.5.1.1 Atomic Absorption Spectrometry

Rameyer and Janauer [3.263] have described a method using reactive ion exchange for the determination of complex iron cyanides in water in the mg l^{-1} range. The cyanide complexes are preconcentrated on shallow beds of sulphonated cation-exchange resin in copper II form by precipitation as copper hexacyanoferrate II or copper hexacyanoferrate III. Other cations, including contaminant iron species, are eluted with hydrochloric acid. Aqueous ammonia relatively releases and elutes the hexacyanoferrate by the formation of the copper ammine complex. Finally the complex cyanides are determined as iron by atomic absorption spectrometry.

3.5.1.2 Ion-Selective Electrodes

A procedure has been developed for determination of cyanide [3.264] in concentrations down to 2 mg l^{-1} by manual or automated potentiometry using a cyanide ion-selective electrode. Distinction is made between simple and complex cyanides by irradiation with ultraviolet light.

3.5.1.3 Gas Chromatography

Complex cyanides have been determined gas chromatographically in amounts down to 1 µg l^{-1} [3.265]. The complex cyanides are broken down by ultraviolet radiation, reacted with bromine water and the cyanogen bromide formed is separated and determined selectively by means of an electron capture detector.

3.5.1.4 Ion Chromatography

Rocklin and Johnson [3.242] have applied ion chromatography to the analysis of mixtures of the complex cyanides of cadmium, zinc, copper, nickel, gold, iron and cobalt. This is discussed further in Sect. 11.1.4.

3.5.2 Waste Waters

Various workers have discussed the determination of free and complexed cyanides in waste waters [3.266–3.270]. Bahensky [3.268] discussed inaccuracies in complex formation methods used to determine free and bound cyanide.

3.5.2.1 Spectrophotometric Methods

Leschber and Schlichting [3.269] discussed the decomposability of complex metal cyanides in the determination of cyanide in waste water. These workers found that complex cyanides of zinc, cadmium, copper, nickel, iron II and iron III corresponding to $1-100$ mg l^{-1} cyanide are easily decomposed by distilling the sample in the presence of dilute sulphuric acid and total cyanide can be determined in the distillate. However, $K_3(Co(CN)_6)$ (hexacyano cobalt III) is only partially decomposed by this treatment.

Hartzinger [3.270] critically examined the D 13/3 German standard method for the determination of free and complexed cyanides in waste water. He points out that interfering substances causing high results in the determination of cyanide in effluents by this pyridine-benzidine method (usually carried out at pH 7) appear to be unknown N-containing heterocyclic compounds, which, in the presence of cyanide, yield polymethine dyes. In cases of doubt the analysis is repeated after adjustment of the pH to 5.5-5.0 by use of an ion exchanger, thus avoiding the involved distillation procedure. The extinction at 480 nm is decreased by 70-75% at the lower pH, so that errors of greater than 0.1 mg l^{-1} in the cyanide concentration can be avoided.

Goulden et al. [3.271] determined simple and complex cyanides in waste water by modifying the APHA standard distillation colorimetric method [3.272] to ensure accurate results for less than 100 µg l^{-1} cyanide. In the manual procedure, the sample is distilled under reflux with sulphuric acid in the presence of aqueous magnesium chloride and mercuric chloride, the absorption system being modified by recirculating the absorbent (0.2 mol l^{-1} sodium hydroxide) by a peristaltic pump through a column packed with glass helices. The lower limit of detection is 5 µg l^{-1} cyanide but this is lowered to 1 µg l^{-1} by using an AutoAnalyser system in which is included a UV irradiation cell to decompose complex cyanides, e.g. those of iron and cobalt; this cell is by-passed when only simple cyanides are being determined. The hydrogen cyanide is evolved in a continuouse distillation apparatus, a mixture of phosphoric acid (1.2 N) and dilute H_3PO_2 being used, because these acids also inhibit oxidation of cyanides during irradiation or by any free chlorine present etc. The absorbent is 2% aqueous sodium acetate (pH 6.5), use of which eliminates one stage in the pyridine-pyrazolone colorimetric procedure. With a 7-ml sample the automated method has a detection limit of 7 ng, the coefficient of variation was 1.2% at the 40 µg l^{-1} cyanide level.

Royer et al. [3.273] also determined total cyanide in waste water by an Autoanalyser procedure. They too used magnesium chloride and mercuric chloride in the carrying solution to decompose complex metal cyanides. Down to 10 µg l^{-1} total cyanide could be determined by this procedure.

A spectrophotometric method has been described for the determination in waste waters of total cyanide, including free cyanide, complex cyanides and thiocyanates [3.274]. Thiocyanates, together with simple and complex cyanides, are reacted with acid cuprous chloride and the hydrogen cyanide formed is absorbed in a sodium hydroxide solution which is then made slightly acidic to prevent hydrolysis to cyanate. Chloramine-T is added to produce cyanogen chloride, a pyridine/pyrazolone reagent is added, and the optical absorbance of the resulting blue dye is measured at 620 nm. Volatile aldehydes and ketones interfere in the final colour development.

Kollau and Reidt [3.275] determined total cyanide and cyanogen chloride in amounts down to 0.1 mg l^{-1} in waste water. To determine total cyanide, hydrogen cyanide is liberated by distillation with sulphuric acid and collected in aqueous sodium hydroxide. Cyanide is determined spectrophotometrically at 578 nm after chloramine-T oxidation of cyanide to cyanogen chloride and formation of the coloured complex with pyridine and barbituric acid.

Drikas and Routley [3.276] have described a spectrophotometric method for the determination of total cyanide in waste water samples.

3.5.2.2 Ion-Selective Electrodes

Csikai and Barnard [3.277] used EDTA at pH 4 to displace cyanide from metal complexes and to avoid converting thiocyanate to free cyanide where oxidants are present. Sulfamic acid is added to prevent nitrite interference, sulphides are removed from the sample with calcium carbonate and from distillates with cadmium nitrate. Evolved cyanide is determined by pyridine–barbiturate photometry or potentiometrically with ion-selective electrodes.

Reagents. As follows.

Cyanide working standard solution. Prepare by dilution of 1000 mg l^{-1} $K_3Fe(CN)_6$ stock solution, standardize iodometrically and store in the dark.

Working solution, 0.2 mol l^{-1} in sodium hydroxide.

Acetate buffer. Prepare by adding 54 g of sodium acetate and 125 ml of acetic acid to water, dilute with water to 1000 ml and adjust to pH 4.0 with sodium hydroxide or acetic acid.

Phosphate buffer, pH 3.1. Dissolve 138 g of $NaH_2PO_4 \cdot H_2O$ in 900 ml of water, add 70 ml of acetic acid and dilute with water to 1000 ml.

Pyridine–barbituric acid reagent and the chloramine T solution [3.278].

Apparatus. Distillation apparatus with, preferably, ground glass joints. If rubber stoppers are used, wrap with polyfluoroethylene tape. Combined magnetic stirrer–heating mantle.

Sample Preservation and Sulfide Removal. Transfer at least 600 ml of the waste water to a well-rinsed screwcap polyolefin bottle and stabilize by adding 5 ml of 50% sodium hydroxide solution (this sample can be stored for several weeks without significant change in the total cyanide content). To remove sulfide, transfer 600 ml of the well-mixed sample to a 1-l beaker containing a magnetic stirring bar. Add 0.2 g of cadmium carbonate. Stir for only a few minutes. If a yellow precipitate forms, repeat the cadmium carbonate addition until no yellow precipitate is observed. Filter promptly through paper and collect 500 mL in a graduated cylinder.

Evolution of Cyanide. Transfer this filtered portion to a 1-l distillation flask. Add several glass beads unless magnetic stirring is used. To the absorption bottle add 50 ml of 1 mol l^{-1} sodium hydroxide and then water, if necessary, to a depth that ensures complete scrubbing of the air stream. Connect the absorber bottle to the condenser outlet tube. Start cooling water through the condenser. Connect the distillation flask to the condenser. Apply vaccum to the absorber outlet and adjust so that two to four bubbles sec^{-1} enter the liquid in the distillation flask.

Remove the air inlet tube, add 2 g of sulfamic acid to the distillation flask and swirl the assembly until the solid dissolves completely. Then add 5 g of solid EDTA, disodium salt, dihydrate, and swirl until most of it dissolves. Replace the air inlet tube, add four to five drops of methyl red solution (100 mg of indicator in 100 ml of alcohol), wash down the tube with water and then add, while swirling the assembly, portions of 2.5 mol l^{-1} sulfuric acid until the solution colour just changes from yellow to red. Then add 1 mol l^{-1} sodium hydroxide dropwise until the solution just turns orange. Now add 50 ml of pH 4.0 acetate buffer and swirl to mix.

Heat the mixture to boiling, but do not leave unattended during the initial heating to ensure that the air flow is maintained and that liquid neither backs up into the air inlet tube or the absorber inlet, nor emerges from the outlet of the absorber bottle. Reflux for 2 h with vapour rising three-quarters up the cooled length of the condenser. Stop the heating, but continue air flow for at least 15 min. Transfer the contents of the absorber to a 250-ml volumetric flask and dilute to the mark with water rinsings from the absorber and connecting tubes. If the cyanide content is not to be assessed within a few hours, transfer the solution to a well-cleaned screwcap polyolefin bottle.

Cadmium Salt Treatment of the Absorbing Liquid. Transfer about 10 ml of the diluted absorbing liquid to a small beaker and add a few drops of 3% cadmium nitrate tetrahydrate solution. If a yellow precipitate or solution forms, treat 75 ml of the diluted absorbing liquid with 0.5 ml of the cadmium nitrate solution. Heat, if necessary, to initiate sulfide precipitation and cool; filter through paper (Whatman No. 42 or equivalent). Test a portion of the filtrate with cadmium nitrate solution for the completeness of sulfide removal.

Photometric Determination of Cyanide. Pipette 10.0 ml of the clear, sulfide free, diluted absorbing liquid into a 100-ml volumetric flask. Add 40 ml of 0.2 mol l^{-1} sodium hydroxide. (For a 500 ml volume of the wastewater sample, if the total cyanide content is expected to be greater than 0.8 mg l^{-1}, use 5.0 ml of the diluted absorbing liquid and add 45 ml of 0.2 mol l^{-1} sodium hydroxide). In 100-ml volumetric flasks, prepare a blank (50 ml of 0.2 mol l^{-1} sodium hydroxide) and also a 10 µg cyanide standard diluted to 50 ml with 0.2 mol l^{-1} sodium hydroxide.

To each of the three flasks, add 15 ml of pH 3.1 phosphate buffer (pH is now 5.5–6.5) and 2.0 ml of chloramine-T solution and mix. After 1–2 min, add 5.0 ml of pyridine barbituric acid reagent, mix, dilute with water to mark and mix thoroughly. Allow at least 8 min for colour development and within 30 min read the absorbance of the sample and standard against the blank in a 10-mm cell at 580 nm. Calculate the total cyanide content of the sample, in mg l^{-1}, from either the absorbance of the 10 µg standard or the calibration curve.

For the calibration curve, pipette 0, 1.0, 2.0, 5.0, 10.0 and 15.0 ml of a working cyanide standard solution (1 mg l^{-1}) into 100-ml flasks and dilute with 0.2 mol l^{-1} sodium hydroxide to give a volume of 50 ml. Continue as for a sample. Plot the net absorbance vs micrograms of cyanide. Calculate the reciprocal sensitivity. Prepare a new curve whenever a new batch of pyridine–barbituric acid reagent is introduced.

Potentiometric Determination of Cyanide. For standardisation, place in a 100-ml beaker 40 ml of deionized water and 1.0 ml of 10 mol l^{-1} sodium hydroxide, bring to about 25 °C and stir magnetically at a rate that does not create a vortex. Insert a cyanide electrode (silver iodide silver sulfide, Orion 94–06 or equivalent) and a reference electrode (Orion 94–02 double junction or equivalent) connected to an expanded scale pH meter. After 5 min of mixing, record the millivolt value. In succession, add 0.05, 0.05, 0.04 and 0.50 ml of a cyanide standard (0.1 mg ml^{-1}) and then 0.40 and 0.50 ml of the stock cyanide solution (1 mg ml^{-1}) and record the equilibrium millivolt value. The seven readings correspond to 0, 0.005, 0.010, 0.050, 0.100, 0.50 and 1.00 mg of cyanide. On three-cycle semilogarithmic paper, plot the millivolt values (linear scale) vs milligrams of cyanide taken and draw a straight line through the points. Prepare a new curve each time a series of samples is to be run.

For the absorbing liquid from the treatment of wastewater samples, test for sulfide. If present, add 3% cadmium nitrate tetrahydrate solution, heat if necessary to initiate sulfide precipitation, cool, and filter through paper. Place a portion of the sulfide-free absorbing liquid in a beaker, bring to 25 °C, insert the electrodes, stir, and record the equilibrium millivolt value. Read the milligrams of cyanide from the calibration plot.

In many of the previously described procedures for total cyanide, high acidity is used as a means of decomposing complex cyanides prior to the determination of free cyanide ion. Csikai and Barnard [3.277] use a low acidity decomposition procedure (EDTA displacement at pH 4) and, in so doing, avoid interference by thiocyanate as illustrated in Table 3.60.

Table 3.61 presents typical results for wastewater samples containing thiocyanate. The values clearly indicate that the thiocyanate interference encountered in high-acidity procedure is avoided with the EDTA displacement procedure.

Table 3.60. Total cyanide values for waste water samples by high-acidity and EDTA displacement procedures. (From [3.277]; Copyright 1983, American Chemical Society)

Sample nature[a]	Thiocyanate present, mg l^{-1}	Nitrate present, mg l^{-1}	Total cyanide found, mg l^{-1}			
			High-acidity procedure		EDTA displacement procedure Sulfamic acid	
			Not added	Added	Not added	Added
WW H	<0.2	71	<0.02		<0.02	
WW I	<0.2	97	0.02		0.01	
WW J	<0.2	105	<0.02		<0.02	
WW K	<0.2	173	<0.02		<0.02	
WW L	1.2	90	0.37		0.07	
WW M	2.0	31	0.66		<0.02	
WW N	3.0	294	0.40		0.02	
WW O	4.0	9	0.60		0.07	<0.02
WW P	4.2	100	1.5		0.12	
WW Q	4.6	105	1.10		0.02	
WW R	5	102	2.2		<0.02	
WW S	5	33	1.20		0.02	
WW T	6	66	3.4		0.01	
WW C	6	127	2.0	0.26	0.12	0.02
WW F	9	140	2.8	0.16		<0.02
WW U	12	64	4.9		0.03	
WW V	15	91	4.7		0.18	<0.02
WW W	17	139	0.8		0.02	

[a] WW = wastewater.

Czikai and Barnard [3.277] evaluated their procedure by adding both free cyanide and complex metal cyanides to both deionized water solutions and wastewater samples containing or spiked with nitrate and thiocyanate. The recovery of cyanide was virtually complete. Typical findings for complex metal cyanides are presented in Table 3.61. For the cyanide complexes of chromium, copper, iron and nickel, the recovery of cyanide was usually greater than 95%. As mentioned previously, the failure to disrupt hexacyanocobaltate II due to its kinetic inertness is a limitation of most total cyanide procedures.

In Table 3.62 some results obtained by both the spectrophotometric and the cyanide electrode potentiometric finishes are presented. The similarity of cyanide determinations obtained by these two different procedures confirms that cyanide only is being determined.

3.5.2.3 Microdiffusion Method

Owerbach [3.279] discussed problems encountered in the determination of low concentrations of cyanide in photographic processing waste waters. Determination of both total and free cyanide in the same sample differentiates between the

Table 3.61. Recovery of cyanide from complex metal cyanides by EDTA displacement procedure without sulfamate added. (From [3.277]; Copyright 1983, American Chemical Society)

Sample nature[a]	Thiocyanate nitrate (present + added)		Cyanide (total)				Cyanide added as complex[c] of
	mg l^{-1}	mg l^{-1}	mg l^{-1} present	mg l^{-1} added	mg l^{-1} found	% Recovered	
DIW	0	0	0	1.00	1.00	100	Fe
DIW	0	0	0	1.00	1.00	100	Fe
DIW	0	0	0	1.00	0.99	99	Fe
DIW	0	0	0	1.00	0.96	96	Fe
DIW	0	0	0	5.00	4.95	99	Fe
DIW	0	0	0	5.00	4.85	97	Fe
DIW	0	0	0	1.00	1.00	100	Fe
DIW	0	0	0	1.04	1.07	103	Ni
DIW	0	0	0	1.04	0.81	78	Ni
DIW	10	100	0	5.00	4.95	99	Fe
DIW	10	100	0	1.00	1.01	101	Fe
DIW	10[b]	100	0	1.00	0.91	91	Fe
DIW	10	100	0	1.00	1.01	101	Fe
DIW	10	100	0	5.00	5.04	101	Fe
DIW	10	100	0	5.00	5.17	103	Fe
DIW	50	200	0	1.00	1.00	100	Fe
DIW	50	1000	0	1.00	1.00	100	Fe
WW AA	<0.2	131	<0.02	1.00	0.98	98	Cr
WW AB	<0.2	74	<0.02	1.00	<0.02	<2	Co
WW AB	<0.2	74	<0.02	1.00	1.00	100	Cu
WW AC	<0.3	120	<0.02	2.00	1.84	92	Fe + Ni
WW AD	1.0	41	0.02	2.00	2.02	100	Fe + Ni
WW AE	5.0	11	<0.02	1.00	1.01	101	Fe
WW AF	10	100	0.04	1.04	0.96	96	Ni
WW AF	10	100	0.04	1.04	0.96	96	Ni
WW AG	19	55	<0.02	0.40	0.37	93	Fe
WW AG	19	55	<0.02	0.40	0.36	90	Fe
WW AG	19	55	<0.02	0.40	0.37	93	Fe
WW AG	19	55	<0.02	0.40	0.38	95	Fe
WW AG	19	55	<0.02	0.40	0.36	90	Fe

[a] DIW = deionized water; WW = waste water.
[b] 10 mg l^{-1} sulfide also added.
[c] Added as Fe(CN)$_4^{3-}$, Ni(CN)$_4^{2-}$, Cu(CN)$_4^{3-}$, Cr(CN)$_4^{3-}$, and Co(CN)$_4^{3-}$.

toxic free cyanide and the complexed ferrocyanide. Because of possible instability of the cyanide, he recommended that samples should be analysed on-site, for instance by the microdiffusion technique.

3.5.2.4 Miscellaneous

Pohlandt et al. [3.280] have critically evaluated 92 methods applicable to the separate determination of free cyanides and complex cyanides in waste waters

Table 3.62. Total cyanide values by photometric and ISE-potentiometric finishes for samples treated by EDTA displacement procedure (values expressed in terms of original samples taken for distillation). (From [3.277]; Copyright 1983, American Chemical Society)

Nature of sample distilled[a]	Total cyanide found, mg l^{-1}	
	by photometry	by ISE-potentiometry
DIW	0.07	0.07
DIW	0.40	0.50
DIW	0.50	0.43
DIW	0.70	0.80
DIW	0.95	0.80
DIW	1.01	1.02
DIW	1.08	1.20
DIW	1.30	1.10
DIW	5.1	5.2
DIW	5.1	5.1
WW CA	0.06	0.06
WW CB	0.13	0.10
WW CC	0.31	0.29
WW BE	0.41	0.33
WW CD	0.66	0.64
WW CE	0.93	0.95
WW CF	0.96	0.97
WW CG	1.9	2.2

[a] DIW = deionized water; WW = wastewater.

and process streams. These include titrimetric, spectrophotometric, potentiometric, amperometric, polarographic, voltammetric, ion-selective electrodes, indirect atomic absorption spectrometric, gas chromatographic and automated techniques. From this work it appears that most titrimetric, colorimetric and electrochemical methods (including potentiometry and the use of cyanide-selective electrodes) for the determination of ionic cyanide are liable to interference from ionizable metal cyanide complexes, e.g. $Zn(CN)_4^{2-}$. These methods can therefore be employed only if such complexes are known to be absent. Potentially accurate methods for the determination of ionic cyanide include ion chromatography (because of its ability to effect separations) and indirect atomic absorption spectrophotometry based on the selective formation of $(Ag(CN)_2)^-$.

Similarly, none of the methods mentioned gives a reliable result for the total amount of cyanide present in a sample. The decomposition of stable complexes and the separation from interfering substances by distillation is therefore necessary. However, most distillation procedures cannot effect the decomposition of all the metal cyanide complexes and they suffer from interferences. The most useful distillation procedure appears to be the ligand displacement technique [3.280]. With this method, interference from sulphide and thiocyanate can be avoided and all the ionic and coordinated cyanide-except cyanide from $(Co(CN)_4)^{3-}$-present in a sample can be distilled out.

Because of the speed in decomposing cyanide complexes, irradiation with ultraviolet light warrants further attention. A combination of this method with the separation, by ion chromatography, of interfering species might result in a fast method for the determination of total cyanide.

3.6 Cyanate

3.6.1 Natural Waters

3.6.1.1 Paper Chromatography

Thieleman [3.281] attempted the separations of cyanate, thiocyanate and cyanide ions by thin-layer and paper chromatography.

The three ions were successfully separated by paper chromatography using methanol: pyridine: dioxan (7:2:1) as solvent in 8 h.

Cyanate was identified using a bromocresol purple spray reagent. Carbonate ions interfered.

Cyanate and cyanide in waste water have also been separated by paper chromatography [3.282]. Separation was achieved by the use of isopropyl alcohol: ethanol: water (9:4:3) as solvent with Filtrak FN8 paper. Bromocresol green solution was used as spray reaagent which produced clear green spots on a blue background with cyanate (Rf 0.29) and cyanide (Rf 0.04).

3.7 References

3.1   Baca P, Freiser H (1977) Anal Chem 49: 2249
3.2   Wagner R, Ruck W (1981) Zeitschrift fur Wasser and Abwasser, Forschung 14: 99
3.3   Burns DT, Fogg AG, Wilcox A (1971) Mikrochim Acta 1: 205
3.4   Ebel S, Herold G (1971) Deutsch Lebensmitt Rundsechau 67: 301
3.5   Downes MT (1978) Water Res 12: 673
3.6   Nawratil B, Marcantonatos M, Monner D (1974) Anal Chim Acta 68: 217
3.7   Seidowski S (1969) Roczn panst Zakl Hig 20: 527
3.8   Ceaucescu D, Sarbu A (1972) Chim Analit 2: 187
3.9   Maly J, Fadrus H (1975) Journal of the American Water Works Association 67: 395
3.10   Malhotra SK, Zanoni AE (1970) Journal of the American Water Works Association 62: 567
3.11   Holty JG, Patworowski HS (1972) Environmental Science and Technology 6: 835
3.12   Monselise JI (1973) Isreal Journal of Technology 11: 163
3.13   Bachhausen P, Buchholz N, Hartkamp H (1985) Fresenius Zeitschrift fur Analytische Chemie 320: 495
3.14   Ghimicesu C, Dorneanu V (1972) Talanta 19: 1474

3.15 Nakamura N (1983) Mikrochimica Acta 2: 69
3.16 Evans WH, Stevens JG (1972) Water Pollution Control
3.17 Waughman GJ (1981) Environmental Research 26: 529
3.18 Davison W, Woof C (1978) Analyst (London) 103: 403
3.19 Davison W, Woof C (1979) Analyst (London) 104: 385
3.20 Gauguch RF, Heath RT (1984) Water Research 18: 449
3.21 Elliott RJ, Porter AG (1971) Analyst 96: 522
3.22 Grasshoff K (1964) Kieler Meeresforsch 20: 5
3.23 Wood ED, Armstrong FAJ, Richards FA (1967) J Mar Biol Ass UK 47: 23
3.24 Morris AW, Riley JP (1963) Analytica Chimica Acta 929: 272
3.25 Wilson AL (1973) Talanta 20: 725
3.26 Koupparis MA, Walczak KM, Malmstadt HV (1982) Anal Chim Acta 142: 119
3.27 Malmstadt HV, Koupparus MA, Walczak KM (1980) Journal of Automated Chemistry 2: 66
3.28 Hilton J, Rigg E (1983) Analyst (London) 108: 1026
3.29 Downes MT (1978) Water Research 12: 673
3.30 Kamphake LJ, Hannah SA, Cohen JM (1967) Water Research 1: 205
3.31 Davison W, Woof C (1978) Analyst (London) 103: 404
3.32 Jones MH (1984) Water Research 18: 643
3.33 Nydahl (1976) Talanta 23: 349
3.34 Olsen RJ (1980) Limnology and Oceanography 25: 758
3.35 APHA Standard Methods for the Examination of Water and Wastewater, 13th edition, 1134 pp, Americal Public Health Association, New York (1980).
3.36 Gauguch RF, Heath RT (1984) Water Research 18: 449
3.37 American Public Health Authority Standard Methods for the examination of water and wastewater, 14th edition, American Public Health Association, Washington D.C. (1975)
3.38 Zhang Y (1986) Analyst (London) 111: 767
3.39 Henriksen A, Selmen-Olsen AR (1970) Analyst (London) 95: 514
3.40 The Accuracy of determination of total oxidized nitrogen and nitrite in river water: Analytical Quality control in the harmonized monitoring scheme. Analytical Quality Control Panel (Harmonized monitoring Committee), Analyst (London), (1982) 107: 1407
3.41 Airey D, Dal Pont G, Sanders G (1984) Analytica Chimica Acta 166: 79
3.42 Nakamura M (1980) Analytical Letters 13: 771
3.43 Afghan B, Ryan J (1975) Analytical Chemistry 47: 2347
3.44 Munoz JR (1974) Analytica Chimica Acta 72: 437
3.45 Brown L, Bellionger EL (1978) Water Research 223: 12
3.46 Cresser EMS (1977) Analyst (London) 102: 99
3.47 Anderson L (1979) Analytica Chimica Acta 110: 123
3.48 Gine MF, Bergamin H, Zagalto EAG, Reis BF (1980) Analytica Chimica Acta (1980) 114: 191
3.49 Fogg AG, Chamsi AY, Abdalia AA (1983) Analyst (London) 108: 464
3.50 Al-Wehaid A, Townshead A (1986) Analytica Chimica Acta 186: 289
3.51 Gardner WW, Malczyk JM (1983) Chemistry 55: 1645
3.52 Nakashima S, Yagi M, Zenik M, Takahashi A, Toei K (1984) Fresenius Zeitschrift fur Analytishe Chemie 319: 506
3.53 Hulanicki A, Lewandowski R, Maj M (1974) Analytica Chimica Acta 69: 409
3.54 Hansen EH, Ghose HA, Ruzieka J (1977) Analyst (London) 102: 705
3.55 Milham PJ, Bull JH (1970) Analyst (London) 95: 751
3.56 Keeney DR, Byrnes BH, Genson JJ (1970) Analyst (London) 95: 383
3.57 Hulanicki A, Lewandowski R, Maj M (1974) Analytica Chimica Acta 69: 409
3.58 Schechter H, Gruener N (1976) Journal of the American Water Works Association 68: 543
3.59 Cox JA, Litwinski GR (1979) Analytical Chemistry 51: 554
3.60 Simeonov V, Andrew G, Stoianov A (1979) Fresenius Zeitschrift fur Analytische Chem 297: 418
3.61 Gran G (1952) Analyst (London) 77: 661
3.62 Tanaka K, Matsumoto K, Haraguichi H, Fuwa K (1980) Anal Chem 52: 2361

3.63 New Monitor for Nitrate Effluent and Water Treatment Journal (1982) 22: 39
3.64 Synott JC, West SJ, Ross JW (1984) Proceedings of the International Conference - Chemistry for Protection of the Environment, Toulouse, France, 19 – 25th September 1984 (edited by L. Pawlowski, A.J. Verdier and W.I. Lacey), Elsevier 143 – 153
3.65 Csiky I, Marko-Varga G, Jonsson JA (1985) Analytica Chimica Acta (1985) 128: 307
3.66 O'Hara H, Okazaki S (1985) Analyst (London) 110: 11
3.67 Sherwood GA, Johnson DC (1981) Analytica Chimica Acta 129: 101
3.68 Iskandarani Z, Pietrzyk DJ (1982) Analytical Chemistry 54: 2601
3.69 Van OS, MJ Slanina J, De Ligny CL, Hammers WE, Agterdenbos J (1987) Analytica Chimica Acta 144: 73
3.70 Smee BW, Hall GEM, Koop DJ (1978) Journal of Geochemical Exploration 10: 245
3.71 Marko Varga G, Csiky I, Jonsson JA (1984) Analytical Chemistry 56: 2066
3.72 Wilkin RD, Koch HH (1985) Fresenius Zeitschrift fur Analytische Chemie 320: 477
3.73 Heumann KG, Unger M (1983) Fresenius Zeitschrift fur Analytische Chemie 315: 454
3.74 Senn DR, Carr PW, Klatt LN (1976) Analytical Chemistry 48: 954
3.75 Kiang CH, Kuan SS, Guilbault GP (1978) Analytical Chemistry 50: 1319
3.76 Balica G, Varzaru A (1973) Revta Chim 24: 1001
3.77 Koupparis MA, Walczak KM, Malmstadt HV (1982) Analyst (London) 107: 1309
3.78 Hansen LD, Richter BE, Eatough DJ (1977) Analytical Chemistry 49: 1779
3.79 Blaison N, LeTolle R, Dassonville G (1985) Aqua., No. 1: 33
3.80 Yoshizumi K, Aoka K, Matsuoka T, Asekura S (1985) Analytical Chemistry 57: 737
3.81 Osibanjo O, Ajayi SO (1980) Analyst (London) 105: 908
3.82 Parker BC, Thomson WJ, Zeller EJ (1981) Analyst (London) 106: 898
3.83 Osibanjo O, Ajaki SO (1980) Analyst (London) 105: 908
3.84 Reijnders HFR, Melis HPAA, Griepuik B (1983) Fresenius Zertschrift fur Analytische Chemie 314: 627
3.85 Madsen BC (1981) Analytica Chimica Acta 124: 437 (1981)
3.86 Kauniansky D, Zelensky I, Havasi I, Cerovsky M (1986) Journal of Chromatography 367: 274
3.87 Slanina I, Bakker F, Bruyn-Hes A, Mols JJ (1980) Analytica Chimica Acta 113: 331
3.88 Faigie W, Klockow D (1981) Fresenius Zeitschrift fur Analytishe Chemie 306: 190
3.89 Yamamoto S, Toda Y, Baba D, Hosako JJ (1983) Chromatography 259: 459
3.90 Wetzel RA, Anderson CL, Schleicher H, Crook DG (1979) Analytical Chemistry 51: 1532
3.91 Wagner F, Valenta P, Nurnberg W (1985) Zeitschrift fur Analytische Chemie 320: 470
3.92 Rowland AP (1986) Analytical Proceedings Chemical Society (London) 23: 308
3.93 Yamamoto M, Yamamoto H, Yamamoto Y, Matzuchita S, Baba N, Ikeshiya T (1984) Analytical Chemistry 56: 822
3.94 Buclholz AE, Verplough CJ, Smith JL (1982) Journal of Oceanographic Science 20: 499
3.95 Oikawa K, Saitoh H (1982) Chemosphere 11: 933
3.96 Pyen GS, Erdmann DE (1983) Analytica Chimica Acta 149: 355
3.97 Hemmi H, Hasche K, Ohzeki K, Kambar T (1984) Talanta 31: 119
3.98 Marko-Varga G, Jonsson JA (1984) Analytical Chemistry 56: 2066
3.99 Hwang CP, Forsberg CR (1973) Water Sewage Works 120: 71
3.100 Bodinc ME, Sawyer DT (1977) Analytical Chemistry 49: 485
3.101 Thompson KC, Blankley M (1984) Analyst (London) 109: 1053
3.102 Oxidized Nitrogen in Waters, 1981 HM Stationary Office London 1: 837
3.103 Methods of Analyses Yorkshire Water Authority, Leeds, UK (1981)
3.104 Moore JC (1975) Effluent and Water Treatment Journal, 155: 17
3.105 Rennie RJ, Sumner AW, Basketter FB (1979) Analyst (London), 104: 837
3.106 Schwabe R, Darimout T, Mohlman T, Pabel E, Sonneborn M (1983) International Journal of Environment Analytical Chemistry 14: 169
3.107 Marecek V, Janchenova H, Samec Z, Brezina M (1986) Analytica Chimica Acta 185: 359
3.108 Dodin EI, Makarenko LS, Tsvetkov VK, Kharlamov IP, Pavlova AM (1973) Zavod. Lab., 39: 1050
3.109 International Standards for Drinking Water, Third Edition, World Health Organisation, Geneva

3.110 Velghe N, Claeys A (1985) Analyst (London) 110: 313
3.111 Miles DL, Espejou C (1977) Analyst (London) 102: 104
3.112 Mosko J (1984) Analytical Chemistry 56: 629
3.113 Maly J, Fadrus H (1984) Vodni Hospodarstrz., 34: 275
3.114 Furaya N, Matsuyuhi N, Higuchi S, Tanaka S (1980) Water Research 14: 747
3.115 Furaya N, Matsuyuki A, Higuchi S, Tanaka S (1979) Water Research 13: 371
3.116 Langmuir D, Jacobson RL (1970) Environmental Science and Technology 4: 834
3.117 Petts KW (1975) Water Research Centre, Stevenage, UK. Technical Memoraddum No 108. Determination of nitrogen compounds by Technicon Autanalyser AA 11
3.118 Bradfield EG, Cooke DT (1985) Analyst (London) 110: 1409
3.119 Elton Both RR (1977) Progress in water Technology 8: 215
3.120 Tecator Ltd. Box 70, Hoganes, Sweden, Application Note, AN65/83. Determination of nitrate and ammonia in soil samples extractable with 2M potassium chloride (1983) and Application Note. ASN65-31, 183 Determination of nitrate in soil samples extractable with 2M potassium chloride using flow injection analysis (1983)
3.121 Hadjidemetriou DG (1982) Analyst (London) 107: 25
3.122 Goodman D (1976) Analyst (London) 101: 943
3.123 Bremner JM, Bundy LG, Agarwal AS (1968) Analytical Letters (London) 1: 837
3.124 Bradfield EG, Cooke DT (1985) Analyst (London) 110: 1409
3.125 Tanaka A, Nose N, Iwasaki H (1982) Analyst (London) 107: 190
3.126 Hemmi H, Hasebe K, Ohzeki K, Kambara T, P 319. Hakodate Technical College, Tokura-Cho, 226. Hakodata 042, Japan
3.127 Flamerz S, Bashir WA (1985) Analyst (London) 110: 1513
3.128 Wiersma JA (1970) Analytical Letters 3: 123
3.129 Anderson L (1979) Analytica Chimica Acta 110: 123
3.130 Norwitz G, Keliher PN (1984) Analyst (London) 109: 1281
3.131 Norwitz G, Keliher PN (1986) Analyst (London) 111: 1033
3.132 Celardin F, Morcantonatos M, Monnier D (1974) Analytica Chimica Acta 68: 61
3.133 Baveja AK, Nair J, Gupta VK (1981) Analyst (London) 106: 955
3.134 Nair J, Gupta VK (1979) Analytica Chimica Acta 111: 311
3.135 Tsao FP, Underwood AL (1982) Anal Chim Acta 136: 129
3.136 Chaube A, Baveja AK, Gupta KV (1984) Talanta 31: 391
3.137 Burton NG, Crosby NT, Patterson SJ (1969) Analyst (London) 94: 585
3.138 Toei K, Kiyose T (1977) Analytica Chimica Acta 88: 125
3.139 Flamerz S, Bashir WA (1981) Analyst (London) 106: 243
3.140 Bashir WA, Flamerz S, Ibrahim SK (1983) International Journal of Environmental and Analytical Chemistry 15: 65
3.141 Wu QF, Liu PF (1983) Talanta 30: 374
3.142 Bermenjo O, Mortinez F, Zunzungui Perez M (1972) Infeion Quim, Analit Pura apl Ind 26: 163
3.143 Singh DV, Mukherjee PP (1972) Indian Journal of Technology 10: 469
3.144 Bhuchar VM, Amar VK (1972) Indian Journal of Technology 10: 433
3.145 Nakamura M, Mazuka T, Yamashita M (1984) Anal Chem 56: 2242
3.146 Gabbay J, Almog Y, Davidson M, Donagi AE (1977) Analyst (London) 102: 371
3.147 Nishimura M, Matsumaga K, Kanazawa M (1969) Japan Analyst 18: 1372
3.148 Lynch TP (1988) Analyst (London) 113: 1597
3.149 Sanchez-Pedreno C, Sierra MT, Sierra MI, Sanz A (1987) Analyst (London) 112: 837
3.150 Rubio S, Gomez-Heans A, Valcarcel M (1984) Analytical Letters 17: 651 (1984)
3.151 Espinola A (1975) Analytical Letters 8: 627
3.152 Motomizu S, Mikasa H, Toei K (1986) Talanta 33: 729
3.153 Motomizu S, Rui SC, Oshima M, Toei K (1987) Analyst (London) 112: 1261
3.154 Nakashima S, Yagz M, Zenki M, Takahashi A, Toei K (1983) Analytica Chimica Acta 155: 263
3.155 Zagatto EAG, Jacintho AO, Mortatti J, Bengamin H (1980) Analytica Chimica Acta 120: 399
3.156 Trojanek A, Bruckenstein S (1986) Analytical Chemistry 58: 866
3.157 Funazo K, Kusano K, Tanaka M, Shano T (1982) Analyst (London) 107: 82
3.158 Funazo K, Tanaka M, Shono T (1980) Analytical Chemistry 52: 1222
3.159 Newberry JE, Lopez de Haddad MP (1985) Analyst (London) 110: 81
3.160 Trojanek A, Bruckenstein S (1986) Analytical Chemistry 58: 866
3.161 Cox JA, Brayter AF (1979) Analytical Chemistry 51: 2230
3.162 Chang SK, Kozenianskas R, Harrington GW (1977) Analytical Chemistry 49: 2272

3.163 Fogg AG, Alonso RM (1988) Analyst (London) 113: 1337
3.164 Barsotti DJ, Pylypiw HM, Harrington GW (1982) Analytical Letters 15: 1811
3.165 Reijnders HFR, Melis PHAA, Griepuik B (1983) Fresenius Zeitscheift fur Analytische Chemie 314: 627
3.166 Maly J, Kosikova M (1983) Vodni Hospodarstvi Series B 33: 137
3.167 Balica G, Varzaru A (1973) Revta Chem 24: 1001
3.168 Chamsi AY, Fogg AG (1988) Analyst (London) 113: 1723
3.169 Nishimura M, Matsunaga K, Kanazawa H (1969) Japan Analyst 18: 1372
3.170 Nakamura M, Mazuka T (1983) Analytical Letters 16: 811
3.171 Koupparis MA, Walczak KM, Malmstadt HV (1982) Analyst (London) 107: 1309
3.172 Koupparis MA, Walczak KM, Malmstadt HV (1980) Journal of Automatic Chemistry 2: 66
3.173 Furaya N, Matsuyuki N, Higuchi S, Tanaka S (1980) Water Research 14: 747
3.174 Tretter H, Paul G, Blum I, Schreke H (1985) Fresenius Zeitschrift fur Analytische Chemie 321: 650
3.175 Bhuchar VM, Amar UK (1972) Indian Journal of Technology 10: 433
3.176 Wu QF, Hiu PF (1983) Talanta 30: 374
3.177 Chaube A, Baveja AK, Gupta UK (1984) Talanta 11: 391
3.178 Munksgaard L, Thymark L (1987) FIA Star News Letter No 6
3.179 Okada M, Miyjata H, Toei K (1979) Analyst (London) 104: 1195
3.180 Water Research Centre (1982) Analyst (London) 107: 1407
3.181 Water Research Centre (1977) Technical Report TR 63, Accuracy of determination of total oxidized nitrogen and nitrite in river waters. Water Research Centre, Medmenham, Bucks UK November
3.182 Wilson AL (1979) Analyst (London) 104: 273
3.183 Nagashima K, Matsumoto M, Suzuki S (1985) Analytical Chemistry 57: 2065
3.184 Marti VC, Hale DR (1981) Environmental Science and Technology 15: 711
3.185 Suzuki N, Kuroda R (1987) Analyst (London) 112: 1077
3.186 Van Staden JF (1982) Analytica Chimica Acta 138: 403
3.187 Lerique D, Gighetto L (1983) Eau Industrie Nuisances No. 75 64
3.188 Slanina I, Bakker F, Bruyn-Hes A, Mols JJ (1980) Analytica Chimica Acta 113: 331
3.189 Alawi MA (1984) Fresenius Zeitschrift fur Analytische Chemie 317: 372
3.190 Kok SH, Buckle KA, Wootton M (1983) Journal of Chromatography 260: 189
3.191 Cooke M (1983) Journal of High Resolution Chromatography and Chromatography Communications 6: 383
3.192 Lee SH, Field LR (1984) Analytical Chemistry 56: 2647
3.193 Alawi MA (1982) Fresenius Zeitschrift fur Analytishe Chemie 313: 239
3.194 Zelinski I, Zelinska V, Kanmiensky D, Havasii P, Lednarova V (1984) Journal of Chromatography 294: 317
3.195 Okada T (1988) Analytical Chemistry 60: 1511
3.196 Montenegro MI, Cruz MJ, Matias ML (1971) Revta Port Quim 13: 217
3.197 Boese SW, Archer VS, O'Laughlin JW (1977) Analytical Chemistry 49: 479
3.198 Nygaard DD (1981) Analytica Chimica Acta 130: 391
3.199 Zhou TZ, Xie YM (1983) International Journal of Environmental Analytical Chemistry 15: 213
3.200 Kratochvil V (1984) Vodni Hospodarstvi 34: 303
3.201 Ghimicescu C, Dorncauu V (1972) Talanta 19: 1474 (1972)
3.202 Ross WD, Buttler GW, Tuff YDG, Rehg WR, Winiger MT (1975) Journal of Chromatography 112: 719
3.203 Tesch JW, Rehg WR, Sievers RE (1976) Journal of Chromatography 126: 743
3.204 Gerritse RG (1979) Journal of Chromatography 171: 527
3.205 Davenport RJ, Johnson DC (1974) Analytical Chemistry 46: 1971
3.206 Sherwood GA, Johnson DC (1981) Analytical Chimica 129: 101
3.207 Petts KW (1975) Water Research Centre Stevenage Herts U.K. Technical Memorandum No 108. Determination of nitrogen compounds by Technicon Autoanalyser AA11
3.208 Henrickson A, Selmer-Olson AR (1970) Analyst (London) 95: 514
3.209 Garcia Gutierrez G (1973) Infeion Quim analct pura apl Ind 27: 171
3.210 Wegner TN (1972) J Dairy Science 55: 642
3.211 Choi KK, Fung KW (1980) Analyst (London) 105: 241
3.212 Broderius SJ (1981) Analytical Chemistry 53: 1472
3.213 Broderius JJ, Smith LL (1977) Analytical Chemistry 49: 424

3.214 American Public Health Association "StandardsMethods of Analysis of Water and Wastewater" 14th edition, American Public Health Association, Washington DC (1975)
3.215 Nagashima S (1983) Water Research 17: 833
3.216 Montgomery HAC, Gardiner DK, Gregory JGG (1969) Analyst (London) 94: 284
3.217 Hangos - Mahr M, Pungor E, Kuznecov V (1985) Analytica Chimica Acta 178: 289
3.218 Rueda FJMV, Diez LMP, Perez RP (1988) Analyst (London) 113: 573
3.219 Fenong T (1987) Analyst (London) 112: 1033
3.220 Fogg AG, Alonso RM (1987) Analyst (London) 112: 1071
3.221 Razmilic B (1989) Atomic Spectroscopy 10: 74
3.222 Frant MS, Ross JW, Riseman HJ (1972) Analytical Chemistry 44: 2227
3.223 Schleuter A (1976) EPA Report 600/4-76-020, June 1976, 20 pp (PB-255852)
3.224 Hofton M (1976) Environ Sci Technol 10: 277
3.225 Owerbach D (1980) J Water Pollution Control Fed 52: 2647
3.226 Gussbert PJ (1978) Analytica Chimica Acta 87: 429
3.227 Funazo K, Kusano K, Tanaka M, Shono T (1980) Analytical Letters 13: 751
3.228 Bond AM, Heritage ID, Wallace AC, McCormick MJ (1982) Analytical Chemistry 54: 582
3.229 Peschet JL, Tinet C (1986) Techniques Sciences, Methods 81: 351
3.230 Stoggins MW (1972) Analytical Chemistry 44: 1294
3.231 Rubio R, Sanz J, Rauret G (1987) Analyst (London) 112: 1705
3.232 Hewitt PK, Austin HB (1972) Analytica Chimica Acta 71: 381
3.233 Kogan LA, Kuzovatova VN, Gagarinova LM, Zavorokhim LI Kohs Khim 3 42 (1970). Ref: Zhur Khim 19GD (15). Abstract No. 15G 232 (1970)
3.234 Barton PJ, Hammer CA, Kennedy DC (1978) Journal of Water Pollution Control Federation 50: 234
3.235 Thielemann H (1972) Mikrochim Acta 1: 28
3.236 Anon. Effluent and Water Treatment Journal 16: 309
3.237 Pihlar B, Kosta L, Hristouski B (1979) Talanta 26: 805
3.238 McKinney GE, Lau HKY, Lott PF (1972) Microchemical Journal 17: 375 (1972)
3.239 Cuthbert TJ (1976) Analytica Chimica Acta 87: 429
3.240 Fleet B, Von Storp H (1971) Analytical Chemistry 43: 1575
3.241 Hafton M (1976) Environmental Science and Technology 10: 277
3.242 Rocklin RD, Johnson EL (1983) Analytical Chemistry 55: 4
3.243 Gregorowicz Z, Gorke P (1970) Chemica Analit 15: 219
3.244 Komatsu M, Kakiyama H (1973) Japan Analyst 21: 315
3.245 Gregorowicz Z, Gorka P (1971) Chemia Analit 16: 703
3.246 Fu-Sheng W, Bai H, Nai-kui S (1984) Analyst (London) 109: 167
3.247 Nagashima S, Ozawa T (1981) International Journal of Environmental Analytical Chemistry 10: 99
3.248 Epstein J (1947) Analytical Chemistry 19: 272
3.249 Fraut MS, Ross JW, Riseman JH (1972) Analytical Chemistry 44: 2227
3.250 Nota G, Improta C (1979) Water Research 13: 177
3.251 APHA- AWWA - WPCF Standard Methods for the Examination of Water and Wastewater, 13th edition, 397 – 406 APHA New York (1971)
3.252 Nota G, Improta C, Mzraglia VR (1982) Journal of Chromatography 242: 359
3.253 Nota G, Miraglia VR, Improta C, Acampora A (1981) Journal of Chromatography 207: 47
3.254 Procopio JR, Macias JMP, Hernandez LH (1986) Analyst (London) 611: 11
3.255 Harris JR, Merson GHJ, Hardy MJ, Curtis DJ (1980) Analyst (London) 105: 974
3.256 Tecator Ltd, Box 70 5-26321, Hoganes, Sweden, Application Notes AN 89/87 and AN 86/87. Cyanides in waste waters, soils and sludges using the 1026 distilling unit (1987)
3.257 Kaur MP, Upadhyay S, Gupta VK (1987) Analyst (London) 112: 1681
3.258 Pal T, Gauguly A (1987) Analyst (London) 111: 1327
3.259 Kodura I, Lada Z (1972) Chemia Analit 17: 871
3.260 Manahan SE, Kunkel R (1973) Analytical Letters 6: 547
3.261 Funazo K, Kusano K, Wu HL, Tanaka M, Shono T (1982) Journal of Chromatography 245: 93
3.262 Standard Methods for the examination of water and waste water, American Public Health Association, American Water Works Association and Wastewater Control Federation, New York, 13 th ed., p. 397 (1971)
3.263 Rameyer GO, Janauer GE (1975) Analytica Chimical Acta 77: 133

3.264 Sekerka I, Lechner JE (1976) Water Research 10: 479
3.265 Nota G, Palombari R, Imperota C (1976) Journal of Chromatography 123: 411 (1976)
3.266 Nanomura M (1987) Analytical Chemistry 59: 2073
3.267 Weiner R, Leiss C (1971) Galvotechnik 62: 366
3.268 Bakensky V (1971) Koroze Ochr Mater 15: 21
3.269 Leschber R, Schlichting H (1969) Analyt Chem 245: 300
3.270 Hartzinger L (1970) Hetalloberflaeche 24: 281
3.271 Goulden PD, Afghan BK, Brooksbank P (1972) Analytical Chemistry 44: 1845
3.272 American Public Health Association, New York. Standard Water and Waste Water. No. 207C (1971)
3.273 Royer JL, Twichell JL, Muir SM (1973) Analytical Letters 6: 619 (1973)
3.274 Imperial Chemical Industries Ltd., Effluent and Water Treatment Journal 17: 77 (1977)
3.275 Kollau KN, Reidt MJ (1969) TNO News 24: 465
3.276 Drikas M, Routley BI (1988) Analyst (London) 113: 1273
3.277 Csikai NJ, Barnard AJ (1983) Analytical Chemistry 55: 1677
3.278 Methods for Chemical Analysis of water and wastes. EPA-600/479-020 STORET NO. 00720: Environmental Protection Agency, Cincinnati, OH (1979)
3.279 Owerbach D (1980) Journal of Water Pollution Control Federation 52: 2647
3.280 Pohlandt C, Jones EA, Lee AC (1983) Journal South African Institute of Mineral and Metallurgy 83: 11
3.281 Thieleman H (1970) Mikrochim Acta 3: 645
3.282 Thicleman H (1970) Pharmazie 25: 271
3.283 Ito K, Ariyoshi Y, Tanabiki F, Sunabara H (1991) Analytical Chemistry 63: 273
3.284 Marengo E, Genaro MC, Abrigo C (1992) Analytical Chemistry 64: 1885
3.285 Holak W, Speccio JJ (1992) Analytical Chemistry 64: 1313

4 Sulphur Containing Anions

4.1 Sulphate

4.1.1 Natural Waters

4.1.1.1 Titration Methods

Various titration procedures have been described for the estimation of sulphate. These include reaction with excess barium ions and back titration of excess barium, direct titration of sulphate with barium ions, and direct titration of sulphate with lead ions.

4.1.1.1.1 Reaction with Excess Barium Ions and Back Titration of Unconsumed Barium

In one method [4.1] for the determination of sulphate in natural waters, the sample is passed through a strongly acidic cation-exchange resin (H+form), aerated to remove carbon dioxide, then titrated with 0.1 N sodium hydroxide to the methyl orange end-point to obtain the sum of sulphate, chloride and nitrate.

Acidify the titrated solution with 0.2 N nitric acid and add excess of 0.05 N barium nitrate and then ethanol to precipitate barium sulphate. Neutralise the solution with 0.1 N sodium hydroxide, make alkaline with aqueous ammonia, add a solution of metalphthalein and naphthol green (C.I. Acid Green 1), and titrate unconsumed barium ions with 0.05 N EDTA to a green colour to obtain the content of sulphate. Make the solution acid with 0.2 N nitric acid, add a small amount of aluminium sulphate and then add naphthol green solution and ethanolic diphenylcarbazone, and titrate with 0.05 N mercuric nitrate until the solution is violet. Calculate the amount of chloride and obtain the amount of nitrate by difference. Determine correction factors by analogously treating water (but omitting the addition of barium nitrate and aluminium sulphate).

Duvivier [4.2] has described a potentiometric method which can be used for the continuous determination of sulphate ions in natural waters. It is based on the precipitation of sulphate ions with barium and the complexation of excess barium with EDTA at pH 10.5. A calcium electrode was used to indicate the end-point to avoid ionic exchange. The electrode was regularly reconditioned in slightly alkaline calcium chloride. The principal interference came from phosphates, but they were seldom present at high enough concentrations to cause serious errors. The relative error was 3% for 0.0001 N sulphate and 0.66% for 0.01 N.

MacKellar et al. [4.3] determined sulphate in natural water indirectly by spectrophotometric titration of excess barium ions with ethylene-diaminetetraacetate. Parts per million of sulphate could be determined with a precision of ± 3%.

Pagenkopf et al. [4.4] have described a procedure for the determination of sulphate in mineral waters which uses back titration of excess barium with a standard solution of sulphate using nitrosulphonazo-III indicator. The method was developed for the analysis of water samples for the mining areas of the northern Great Plains of USA which are often high in sulphate. Calcium and magnesium must be removed but phosphate, chloride, nitrate and fluoride do not interfere in amounts up to 200 mg l^{-1}. Ten samples h^{-1} can be analysed.

Reagents. As follows.
Barium chloride, 10^{-3} mol l^{-1}.
Nitrosulphanazo III (2,7-bis(4'-nitro-2' sulphophenylazo), 8-dihydroxy naphthalene-3, 6-disulphonic acid, sodium salt), 10^{-4} mol l^{-1}.
Acetone.

Apparatus. pH meter, magnetic stirrer, micropipette, fluorescent lamp. Ion-exchange resin, Dowex 50-X5, 20–50 mesh, H' form.

Procedure. Mix a 10.00-ml aliquot of the sample with 10 g of the ion exchange resin. Transfer a 5.00-ml aliquot of the supernatant liquid (filter or decant, avoiding fines) to a beaker, add 5.00 ml of the barium chloride solution and adjust to pH 1.90. At this point, add 1.0 ml of the indicator solution and 10 ml of acetone, and titrate the resulting mixture with the standard sulphate solution. Titrate rapidly initially but slowly near the end-point to permit the development of a faint pink color in the blue solution. Determine a blank by mixing 5.00 ml of distilled water with 5.00 ml of barium chloride solution, adjusting the pH, adding the indicator and acetone, and titrating with standard sulphate.

The use of a discontinuous titration is convenient for end-point detection when more than one sample is being titrated. The use of a fluorescent lamp and comparison with a solution that has been titrated to the equivalence point are helpful in assigning the end-point. The sample should have a sulphate concentration range of 20–90 mg l^{-1} for best results, so dilution of a mineral water sample may be necessary.

Analysis of sodium sulfate solutions which had a sulfate concentration range of 10–140 mg l^{-1} by this method exhibited a positive bias of 2.42% with a standard deviation of 4.38%. The positive bias results from a slightly premature assignment of the end-point in the titrations, and probably reflects the magnitude of the colour differences that the eye can detect.

Standard addition studies were made to evaluate the method and gave values listed in Table 4.1. The least squares correlation between sulphate observed and added is 0.997, with an interception value of 164 mg l^{-1} which is 1% above the expected amount. The removal of calcium and magnesium by ion exchange is critical since these ions cause a premature indicator change. At least ten samples can be analyzed per hour with results comparable to, or better than, those obtained with the conventional turbidimetric procedure. The procedure is not affected by phosphate, chloride, nitrate and fluoride in amounts up to 200 mg l^{-1}, although these anions can result in slightly different indicator colors.

4.1.1.1.2 Direct Titration of Sulphate with Barium Ions

Sulphate in natural waters has been determined [4.5] by direct titration with 0.01 N barium cyanate in acetone-water medium to the carboxyarsenazo (if phosphates absent) or nitchromazo (if phosphates present) end-point Cations contained in natural water were found to interfere and must be removed by prior cation exchange of the sample. In another method [4.6], sulphate is separated from interfering ions by passing the sample through a column of KU-2 resin (H$^+$ form), then the column is washed with water, the eluate is neutralised to the yellow colour of 4-nitrophenol with hydrochloric acid and 2 drops of acid in excess are added. To this solution is added an equal volume of acetone and 10 drops of 0.05% aqueous orthanilic K (3-(2-carboxyphenylazo)-6-(2-sulphophenylazo) chromotropic acid), and sulphate is titrated with 0.02 N barium chloride to a colour change from blue-violet to pale blue.

To overcome interference due to calcium ions in natural water in the determination of sulphate, Akos-Szabo and Inczedy [4.7] first mixed the sample with a cation exchange resin (H$^+$ form). To an aliquot of the clear liquid was added an equal volume of 96% ethanol and 30 mg solid barium chloride. Sulphate was then titrated with 10 mmol l^{-1} barium acetate solution using a radio frequency titration technique.

Table 4.1. Sulfate analysis for standard addition study. (From [4.4]; Copyright 1976, Elsevier Science Publishers BV, Netherlands)

Sulfate added (mg l^{-1})	Sulfate found (mg l^{-1})	Dilution 1:x	Recovery (%)	Sulfate added (mg l^{-1})	Sulfate found (mg l^{-1})	Dilution 1:x	Recovery (%)
0[a]	162	5	100.6	217[c]	378	5	100.0
217	392	5	103.4	327[c]	487	8.33	99.8
327[b]	487	8.33	99.8	434[c]	596	10	100.2
434	579	10	97.3	545[c]	704	10	99.7
545	719	10	101.8	653[c]	813	12	99.8
635	819	10	102.9				

[a]Turbidimetric analysis gives 171 mg l^{-1}.
[b]Turbidimetric analysis gives 460 mg l^{-1}.
[c]These samples were treated with Dowex 50-X8 before addition of Na$_2$SO$_4$.

Savvin et al. [4.8] titrated sulphate in natural water with standard barium chloride using 3-(2-carboxyphenylazo)-6-(2-sulphophenylazo) chromotropic acid as metallochromic indicator. Cations are first removed using a cation exchange column (H^+ form) and the detection limit is 1 mg l^{-1}.

4.1.1.1.3 Direct Titration with Lead Salts

Duvivier [4.9] titrated sulphate with standard lead nitrate solution in methanol medium and monitored the reaction with a lead-selective electrode. There was a maximal error of 1% in measuring 19.2 mg l^{-1} sulphate. Interference from carbonate ions could be eliminated by lowering pH to 4-4.3 using nitric or perchloric acid. Interference from calcium ions could be removed by an ion exchange process followed by titration with sodium hydroxide in the presence of methanol to pH 5.0.

Archer [4.10] titrated sulphate with standard lead nitrate using dithizone as indicator. Pakalns and Farrer [4.11] investigated the effect of several surfactants (cationic, anionic and non-ionic) and detergent builders (phosphates and nitriloacetic acid) in this method.

The method can tolerate quite high concentrations (up to 2500 mg l^{-1}) of all surfactants, particularly anionics, tripolyphosphates and soaps. Interference by high concentrations of metals can be eliminated by passing the acidified sample (pH 1.5) through a strongly basic cation exchange column prior to titration of sulphate with lead nitrate. Chloride concentrations up to 400 mg l^{-1} did not interfere. Pakalns and Farrer also titrated sulphate with 0.01 mol l^{-1} lead acetate in 1:1 acetone: water medium to determine sulphate in the range 0 to 50 µg. The extinction is measured at 430 nm after 20 min standing.

4.1.1.1.4 Spectrophotometric Titration

Askne [4.12] has described a photometric titration procedure for determining low concentrations ($2 \times 10^{-5} - 20 \times 10^{-5}\text{ mol l}^{-1}$) of sulphate in natural waters by measuring the intensity of the orange-coloured complex formed between thorin and excess barium ions from a solution of barium perchlorate.

The thorin method is based on the fact that sulphate ions in the sample solution are precipitated as barium sulphate through the addition of barium perchlorate via a burette. After the whole amount of sulphate is precipitated, further additions of barium ions form an orange-coloured complex with added thorin, the absorbance of which is measured at 520 nm.

The titration is carried out in a medium comprising aqueous sample (20% by volume)-acetone (80% by volume). By addition of perchloric acid, the pH is adjusted to between 2.5 and 4. The sample should be cation exchanged (Dowex 50 W $\times$ 8 50-100 mesh) before titration to prevent cations from interfering by forming coloured complexes with thorin.

The titration can be carried out either manually using a microburette or automatically using an autotitrator unit linked to the spectrophotometer, (Fig. 4.1).

The automatic equipment consists of a control unit connected to the titrator. This contains a photocell, the resistance of which is 2 and 250 k, respectively, in the directions forward–backward. The cell is connected via a switch either to the mirror galvanometer (Ri = 420 MΩ, current for maximum scale deflection = 0.57 µA) (manual titration), or to a follower (Ri = 400 MΩ) with an accompanying linear amplifier in the control unit (automatic titration). The follower and the linear amplifier are built from operational amplifiers.

The current from the photocell gives a voltage drop over the input of the follower which is amplified in the subsequent step. The amplified signal may be somewhat unstable and an RC-step is therefore used as an attenuation circuit.

Ion-exchange with barium iodate achieved by passing the sample through a column of homologously precipitated barium iodate, leads to the following reaction [4.13].

$$Na_2SO_4 + Ba(IO_3)_2 = NaIO_3 + BaSO_4.$$

Reaction of sodium iodate in the percolate with sodium iodide in hydrochloric acid medium leads to the production of an amount of iodine equivalent to the

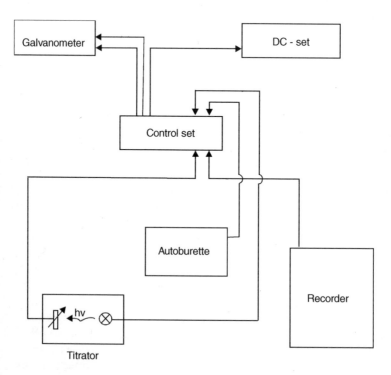

Fig. 4.1. Block diagram

sulphate content of the sample. Iodine is titrated with 0.05 N sodium thiosulphate to the starch end-point. In the determination of 250 mg l^{-1} sulphate in natural water by this method, 250 mg l^{-1} of sodium, potassium, calcium, chloride or nitrate did not interfere. Phosphate (100 mg l^{-1}) or bicarbonate (200 mg l^{-1}) caused an error of $+10\%$.

Some of the chromophoric reagents that have been used to estimate sulphate are reviewed in Table 4.2. Two preferred methods are discussed below.

Reijnders et al. [4.21] have described the following batchwise photometric method for the determination of sulphate in water. In this method metal ions are removed from the sample by ion exchange and the sulphate reacts with the barium complex of dimethylsulphonazo (III) in an ethanolic medium. The optical absorbance of the complex indicates the sulphate concentration.

Apparatus. Ion-exchange column (i.d.: 0.5 cm; length: 20 cm) filled with a cation exchanger (Dowex 50 W-X8, 20–50 mesh) in the H$^+$-form. Regenerate this exchanger with a hydrochloric acid solution (2 mol l^{-1}). The spectrophotometer used should allow a read-out reproducibility of 0.001 unit of optical absorbance and should have a bandwidth of maximal 5 nm at a wavelength of 654 nm. The cuvettes should have an optical pathway of 10.0 nm.

Table 4.2. Chromogenic reagent, for the spectrophotometric determination of sulphate

Chromophore	λ max nm	Interferences	Limit of detection	Reference
Excess 2-amino perimadine HCl added, unused reagent estimated	305	Slight with 500 mgl^{-1} acetate Cl, Br, I and 100 mg l^{-1} of PO$_4$, NO$_3$, F. Interference by IO$_3$, SO$_3$, SeO$_4$	4–120 mg l^{-1}	4.14
Molybdosulphate complex	–	–	–	4.33
6(-*p*-acetyl-phenylazol)-2-amino perimadine	–	–	micro	4.15
FeSO$_4^+$ Complex	325–360	–	0.1 nm	4.16
Thorin-thymolblue	–	Interference by humic acid	–	4.17
Thorin and throium (measure decrease in extinction)	410	–	0.2–0.5 mg l^{-1}	4.18
Thorin	520	All cations interfere, HBO$_3$, As, ClO$_4$, NO$_3$, Cl, F, CN, PO$_4$, S do not interfere	0–8 mg l^{-1}	4.19, 4.20
Barium dimethyl Sulphonzo III	654 –	No interference Cl $< 2 \times 10^{-2}$ mol l^{-1} NO$_3 < 3 \times 10^{-2}$ mol l^{-1} PO$_4 < 1.5 \times 10^{-2}$ mol l^{-1}	–	4.21
Reduction of SO$_4$ to sulphide by HCl-acetic anhydrie and spectrophotometric determination of sulphide as Fe III phenanthroline complex	510	–	–	4.22

Reagents. Prepare the colour reagent by mixing 150 ml water, 10.0 ml of concentrated acetic acid, 10.0 ml of a potassium nitrate solution (1.00 mol l^{-1}), 2.00 ml of a barium perchlorate solution (0.0100 mol l^{-1}) and 3.00 ml of a dimethylsulphonazo (III) solution (0.0100 mol l^{-1}). Make this aqueous solution up to 1 l with ethanol.

Standard sulphate solutions (e.g. 0.10 mol l^{-1} sulphate).

Procedure. Prior to measurement, percolate the sample over the ion exchanger. Pipette an aliquot of 1.00 ml from the treated sample, which may have a maximum sulphate content of 62 nmol. Transfer this aliquot into a stoppered test tube together with 5.00 ml of the colour reagent and 50 µl of a sulphate standard (0.10 mmol l^{-1} sulphate) Mix the contents of the tube.

Measure the absorbance at 654 nm in a 10.0-nm cuvette between 2 and 20 h after mixing. Use the mean value of four blank determinations using water as a sample to correct the measured values.

The results of determinations of pure sulphate solutions, containing 1–60 µmol l^{-1} sulphate are presented in Table 4.3. Table 4.4 shows the effect of different amounts of interfering substances on the results of sulphate standards. The lower limit of determination, defined as the concentration at which the coefficient of variance is 24%, was 2.6 µmol l^{-1} of sulphate. Chloride, nitrate and phosphate do not interfere below concentrations of 2.10^{-2}, 3.10^{-2} and $1.5 \cdot 10^{-3}$ mol l^{-1} respectively.

4.1.1.1.5 Automated Titration Procedures

Fishmann and Pascoe [4.23] studied the application of various automated titrimetric procedures to the determination of sulphate in natural waters.

Henrikksen and Paulsen [4.19] have described an automated procedure for the determination of sulphate in natural salt water which involves precipitation

Table 4.3. Results of analyses of pure sulphate solutions. (From [4.21]; Copyright 1979, Springer-Verlag GmbH, Germany)

Sulphate concentration given, µ mol l^{-1}	Sulphate concentration found	Stand. dev. µ mol l^{-1}	% rel
1.0	1.3	0.7	54
2.0	2.3	0.7	30
3.0	3.2	0.5	16
5.0	4.5	0.4	9
7.5	7.4	0.4	5
10.0	9.8	0.4	4
20.0	19.2	0.6	3
40.0	39.8	0.9	2.3
50.0	50.0	1.0	2.0
60.0	60.4	1.4	2.3

[a]Calculated from six determinations.

Table 4.4. Effect of different amounts of interfering substances on the results of sulphate standards. (From [4.21]; Copyright 1979, Springer-Verlag GmbH, Germany)

Sulphate given μ mol l^{-1}	Interference	Sulphate found μ mol l^{-1}
0	1	2.6
0	2	4
0	3	8
50	1	51
50	2	53
50	3	53
100	1	100
100	2	100
100	3	102

Key to interference:
1. No interferences.
2. Sulphate and interference as specified in 3 in 1:1 dilution.
3. Sulphate and 0.02 mol l^{-1} NaCl, 0.0015 mol KCl, 0.01 mol l^{-1} 1-1 NaHCO$_3$, 0.005 mol l^{-1} Fe(NO$_3$)$_3$, 0.01 mol l^{-1} Ca(NO$_3$)$_2$.

of sulphate with barium perchlorate dissolved in isopropanol, and determination of excess barium by its complex with thorin. Cations are removed by an acidic cation exchange resin. The precision of the method is ± 0.05 mg l^{-1} of sulphate over the range 0–8 mg l^{-1}.

Reagents. As follows.
Colour reagent. Dissolve 100 mg of thorin (Merck, Art.8294) in 500 ml of distilled water.
   Barium perchlorate (Ba(ClO$_4$)$_2$) solution. Dissolve 900 mg Ba(ClO$_4$)$_2$ in 1000 ml of distilled water and add 8.6 ml of concentrated perchloric acid (S.G. 1.67).
   Sodium acetate buffer. To 100 ml of 1 N sodium acetate add 1 N acetic acid to pH 5.6.
   Precipitating reagent. To 1000 ml of isopropanol add 10 ml of barium perchlorate reagent and 4 ml of sodium acetate buffer and mix well.

Apparatus. Use Dowex 50 W × 8, 50/100 mesh strong acidic cation exchange resin, or similar. The cation exchange resin contains significant amounts of free acid. Wash the resin several times (up to 12 times) with portions of distilled water until the pH of the wash water is the same as the distilled water used. Prepare the column as shown in Fig. 4.2. It is important to remove the air bubbles both before and after the sample has passed the resin. Any air bubble entering the manifold will disturb the peaks.

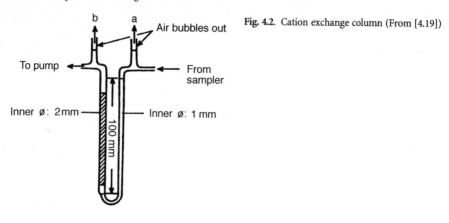

Fig. 4.2. Cation exchange column (From [4.19])

Procedure. Assemble an automatic analyser consisting of a sampler [4.20], an Ismatec MP-13 proportioning pump, a Vitatron UC 200 S photometer equipped with a 1-cm flow-through cell and a Vitatron UR 401 recorder with lin-log converter and expansion facilities similar to an AutoAnalyzer.

Incorporate an ion-exchanger column in the final analytical system. The flow scheme of the automatic sulfate method is shown in Fig. 4.3.

Aspirate the sample from the sampler and pump it through a cation exchange column. Mix the ion exchanged sample with the barium perchlorate

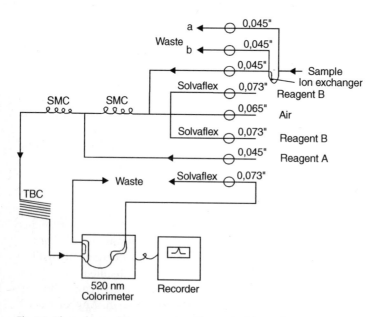

Fig. 4.3. Flow scheme of the automatic sulfate method (From [4.19])

solution and add the thorin reagent. After a time delay of about 4–5 min, measure the colour at 520 nm.

The flow scheme given in Fig. 4.3 is suitable for the range 0–8 mg l^{-1} sulphate. The range may be varied by proper choice of sample tube, light path of cell or recorder expansion. The lowest detectable concentration is 0.1 mg l^{-1} sulphate.

The sampling rate is 17 h^{-1} using 2 min for sampling and 1.5 min for washing. The isopropanol affects the Solvaflex tubing so that it gradually loses its flexibility, resulting in a drift in baseline and a change in peak height of the standards. The tubes may be used for about 1–2 weeks without serious problems. It is however, advisable to run the standard curve 2–3 times a day and to run a standard every 10 samples to control the stability of the analytical system.

All cations interfere. The following anions do not interfere below the concentrations (mg l^{-1}) given in parenthesis: H_3BO_3 (1000), Ac^- (1000), ClO_4^- (60), NO_3^- (60), Cl^- (25), F^- (10), HCN (1), PO_4^{3-} (1), HS^- (1). The interference from cations is eliminated by incorporating a strong acidic cation exchange column in the analytical system.

The precision of this method is ± 0.05 mg l^{-1} SO_4 in the range 0–8 mg l^{-1} SO_4.

Airey et al. [4.24] have described a method for removing sulphide prior to the determination of sulphate anions in anoxic estuarine waters. Mercury (II) chloride was used to precipitate free sulphide from samples of anoxic water. The sulphide-free supernatant was used to estimate sulphide by measuring the concentration of unreacted mercury (II), as well as to determine sulphate by a spectrophotometric method in which sulphide interferes. Sulphide concentrations in the range 0.5–180 000 µg l^{-1} sulphur could be measured, while the lower limits for sulphate was 0.024 µg l^{-1}.

4.1.1.2 Turbidimetric Methods

Coleman et al. [4.25] compared barium chloride and 4-amino-4'-chlorobiphenyl hydrochloride as precipitants in the turbidmetric determination of sulphate at 430 nm with barium chloride. Good results (coefficient of variation 1–4%) were obtained 30 s after mixing. The performance with 4-amino-4'-chlorobiphenyl hydrochloride for measurements 20 s after mixing was better (coefficient of variation 0.33 and 1.0% for 18 and 4 µg ml^{-1} of sulphate respectively); internal calibration for each run is recommended. In this method the reagent is stable for at least 1 month and there is no interference from up to 500 mg l^{-1} of phosphate.

Bonda et al. [4.26] described a column turbidimetric method for determining sulphate. The method uses solid barium chloride in crystals of 0.3–0.6 mm diameter. The method is precise and reproducibility is good. The advantage of the method is the stability of the reagents and the speed, economy and simplicity of performance. Photometric determination is carried out over the range

460–540 nm and the method may be used to determine sulphate in the range 30–150 mg l^{-1} with a correlation coefficient of 0.99.

Cationic, anionic and non-ionic surfactants, polyphosphates and nitrilo-acetic acid are without interference in barium-based turbidimetric methods for the determination of sulphate (Table 4.5).

Krug et al. [4.27] have used continuous flow injection turbidimetry for the rapid determination of sulphate in natural waters.

Slanina et al. [4.28] and Van Raaphorst et al. [4.29] claim a detection limit of 0.5 mg l^{-1} with a relative accuracy of 5% in the turbidimetric determination of sulphate in rain water.

4.1.1.3 Nephelometry

2-Aminoperimidine hydrochloride yields a microcrystalline precipitate of the corresponding sulphate with solutions containing down to 0.05 mg l^{-1} sulphate [4.30]. For the determination of sulphate, 0.5% aqueous reagent is added to the test solution, the mixture is diluted and a nephelometer reading is made after 5 to 10 min and compared with the readings obtained with use of standard potassium sulphate solution. Many other anions also yield precipitates but, with the exception of iodide, fluoride, silicofluoride and phosphate, at least ten-fold amounts of most common anions are tolerated.

4.1.1.4 Spectrofluorimetry

Nasu et al. [4.31] used the hydroxyflavone-thorium complex. The extinction at the absorption maximum at 390 nm of the 1:1 complex decreases in the presence

Table 4.5. The effect of surfactants on turbidimetric sulphate determination at various sulphate concentrations

Surfactant	mg l^{-1}	% Recovery		
		5 mg l^{-1} SO$_4$	10 mg l^{-1} SO$_4$	20 mg l^{-1} SO$_4$
Cationic detergent	1000	101	103	104
Linear alkyl	0.2	99	103	100
Sulphate	0.5	107	103	100
	2.0	144	120	110
Non-ionic detergent	30	112	107	104
Formulated detergent	5	109	105	102
Mixed detergent	5	106	102	100
	20	120	118	105
Sodium pyrophosphate	100	103	107	110
Sodium tripolyphosphate	2	100	105	101
	5	101	106	108
Soap	2	112	110	103
NTA	500	100	99	103

of sulphate owing to the formation of the sulphate complex of thorium. The blue fluorescence of the hydroxyflavone thorium complex at 470 nm also decreases on addition of sulphate. The calibration graphs are almost rectilinear, and reproducible for 0.2–4.0 and 0.02–0.8 mg l^{-1} of sulphate in 30% aqueous ethanol at pH 2.4–2.7 and pH 2.5–3.0 by the spectrophotometric and fluorimetric method, respectively. Interference is caused by iron III, aluminium, zirconium, barium, fluoride and phosphate, organic anions, and other oxy-anions of sulphur.

A further fluorescence method [4.32] is based on the quenching by sulphate of the fluorescence of the thorin-thorin complex. Interfering ions are first removed by ion-exchange. Down to 25 µg sulphate can be determined.

4.1.1.5 Flow Injection Analysis

Kondo et al. [4.34] determined sulphate ion in river water by flow injection analysis at a rate of 30 samples per hour. The method was applied to samples whose sulfate contents were typically less than 30 mg l^{-1}. The reagent solution contains dimethylsulfonazo-II, barium chloride, potassium nitrate and chloroacetate buffer in 70 vol.% ethanol, and is saturated with barium sulphate. The aqueous carrier stream is also saturated with barium sulphate. The sample is filtered and treated with Amberlite IR120 B cation exchanger before injection into the carrier stream, and the decoloration of the barium-dimethylsulfonazo III complex by sulphate is measured at 662 nm. The calibration graph is linear over the range 0–30 mg l^{-1} for sulphate in water.

Reagents. As follows.
Dimethylsulphonazo III solution. Transfer 5.6 ml of 10^{-2} mol l^{-1} dimethylsulfonazo III solution, 5 ml of 1 mol l^{-1} potassium nitrate, 3.9 ml of 10^{-2} mol l^{-1} barium chloride, 20 ml of 1 mol l^{-1} chloroacatic acid - sodium chloroacetate buffer (pH 2.8) and 700 ml of ethanol into a 1-l volumetric flask and dilute to the mark with distilled water. Saturate the solution with barium sulfate in an ultrasonic bath and filter it through a Millipore filter (pore size 0.45 µm).

Barium sulfate solution. Agitate distilled water with barium sulphate in an ultrasonic bath and filter through a Millipore filter (pore size 0.45 µm).

Other reagents used were of analytical reagent grade.

Apparatus. Shimadzu double beam spectrophotometer UV-140-02 or equivalent, with 1-cm micro flow cell (8 µl), recorded by a Toa Dempa FBR-251A recorder. A double plunger micro pump (Kyowa Seimitsu KHU-W-104) was used for the reagent and barium sulfate solutions. Sample solution (80 µl) was injected by a 6-way injection valve (Kyowa Seimitsu KMH-6V) into the carrier stream.

The flow lines were made from teflon tubing (1 mm and 0.5 mm i.d.). The reaction coil (1 mm i.d.) was optimally 5.5 m long and was wound around a glass rod (1.2 cm o.d.) to achieve complete mixing. The damping coils (0.5 mm i.d.) were 20 m long to cancel the pulse from the reciprocal pump. The back-pressure tubing (0.5 mm i.d.) was 10 m long to prevent formation of air bubbles.

A diagram of the flow system is shown in Fig. 4.4.

Procedure. The flow rates of the reagent solution and barium sulfate solutions are optimally 2 ml min^{-1} each. Filter sample solutions through a Millipore filter (pore size 0.45 µm) and pass through an Amberline IR120-B (H-form) column (1 cm i.d, 15 cm long) before injection. From the sample loop, inject 80 µl of treated sample solution into the carrier stream. Measure peak heights at 662 nm. Measure the concentration of sulphate ion from a calibration graph prepared with potassium sulphate standards.

The absorption spectra of DMSA-III and its barium chelate are shown in Fig. 4.5. The absorption maxima are 586 nm and 660 nm for DMSA-III and its chelate, respectively. Measurement at 662 nm provides the largest difference for the sulphate determination.

Maximum constant absorbance is achieved between pH 3.5 and 5.5, so the monochloroacetate buffer (pH 4.2) is again appropriate.

A calibration run obtained with 0–mg l^{-1} sulphate standards had good reproducibility. Linear calibration graphs can be obtained up to 30 mg l^{-1} sulphate, with a correlation coefficient of 0.999.

The effects of co-existing ions are listed in Table 4.6. Calcium ion in river water interferes with the determination. To prevent this, samples are first passed through a column of cation exchange resin. The first 20 ml of effluent is discarded and the next portion is used.

Van Staden [4.35] has described an automated turbidimetric determination of sulphate in natural waters by flow injection analysis. One sampling loop of a two-position sampling valve is used to sample an alkaline buffer/EDTA solution, while the other loop is used alternately to sample water, thus avoiding acidification of water samples and lessening baseline drift. Residual precipitate is redissolved from the walls of the flow cell. The coefficient of variation for standard sulphate solutions and for water samples, containing 50–1400 mg l^{-1} sulphate, is less than 0.95%. The procedure is suitable for carrying out 60 analyses per hour.

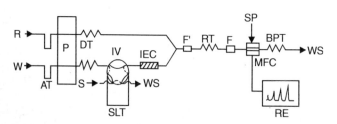

Fig. 4.4. Schematic flow diagram. R, reagent solution; W, water saturated with BaSO$_4$; P, double plunger micro pump (2.0 ml min^{-1}); AT, air trap; DT, damping coil (0.5 mm × 20 m); IV, injection valve; SLT, sample injector (1.0 mm tubing 80 µl); F, fine filter; RT, reaction coil 1mm × 5.5 m); MFC, micro flow cell (1 mm × 10 mm); BPT, back pressure coil (0.5 mm × 10 m); W, waste; SP, spectrophotometer; R, recorder (From [4.34])

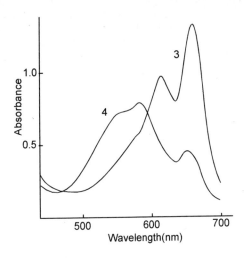

Fig. 4.5. Absorption spectra measured in 50 vol.% ethanolic solution against a water reference: (3) 2.5×10^{-5} mol l^{-1} barium-dimethyl-sulfonazo-III; (4) 2.5×10^{-5} mol l^{-1} dimethyl-sulphonazo-III (From [4.34])

Table 4.6. Effect of co-existing ions. (From [4.34]; Copyright 1982, Elsevier Science Publishers BV, Netherlands)

Ion	Added as	Conc. (mg l^{-1})	Found SO$_4^{2-}$ (mg l^{-1})	Ion	Added as	Conc. (mg l^{-1})	Found SO$_4^{2-}$ (mg l^{-1})
None	–	–	5.0	K$^+$	KCl	5×10^{-4}	5.1
Cl$^-$	KCl	5×10^{-4}	5.1	Na$^+$	NaCl	1×10^{-3}	5.0
HCO$_3^-$	KHCO$_3$	1×10^{-4}	5.0	NH$_4^+$	NH$_4$Cl	5×10^{-4}	5.0
NO$_3^-$	KNO$_3$	1×10^{-4}	5.0	Mg^{2+}	MgCl$_2\cdot$6H$_2$O	1×10^{-4}	5.0
H$_2$PO$_4^-$	KH$_2$PO$_4$	1×10^{-4}	5.1	Ca^{2+}	CaCl$_2\cdot$2H$_2$O	5×10^{-5}	4.9
SiO$_3^{2-}$	Na$_2$SiO$_3$	5×10^{-4}	5.0	Fe^{3+}	FeCl$_3\cdot$6H$_2$O	5×10^{-4}	5.0

Apparatus. Carle microvolume two-position sampling valve (Carle Cat. No. 2014) with two sampling loops. Water samples were constantly sampled by the one 60 µl loop, which were alternated by using the other 100 µl loop constantly to sample an alkaline buffer EDTA solution. The carrier stream was supplied by a peristaltic pump and the valve system was synchronised with the sampler unit. Cenco sampler. Cenco peristaltic pump operating at 10 rev min^{-1}. Manifold (see Fig. 4.6). Bausch and Lomb Spectronic 21 DV spectrophotometer (Rochester, New York) equipped with a 10-mm Helma type flow-through cell (volume 80 µl) and the sensitivity switch selected to high sensitivity. Mettler Recorder Model 1A 12 with a recorder range of 500 mV and recorder paper speed of 30 cm/h.

Reagents. As follows.
Barium chloride solution. Dissolve 0.20 g of thymol crystals in 500 ml of 0.005 mol l^{-1} hydrochloric acid solution at a temperature of about 80 °C. Cool to 40 °C and add 1.5 l of 0.005 mol l^{-1} hydrochloric acid solution. Add 4 g of gelatin very slowly and swirl until dissolved. When dissolved, add 20 g of barium chloride dihydrate and dissolve. Filter, if necessary.

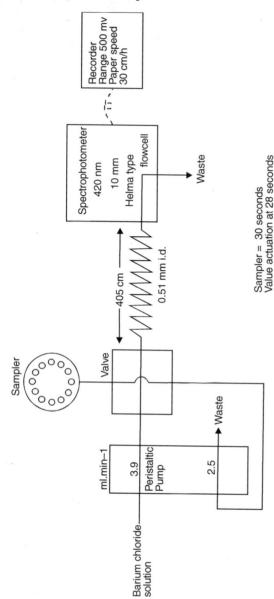

Fig. 4.6. Manifold and flow diagram. Tube i.d. = 0.51 mm. Tube length given in cm as indicated. Valve loops size 60 μl for water samples and 100 μl for buffer EDTA solution (From [4.35])

Buffer Solution. Dissolve 40 g of EDTA (disodium salt), 7 g of ammonium chloride and 57 ml of concentrated ammonia solution (sp. gr. 0.88) in 500 ml of distilled water. Dilute to 1 l with distilled water.

Standard sulphate solution. Prepare stock solution containing 1 000 mg l^{-1} of sulphate by dissolving 1.4787 g of anhydrous sodium sulphate, dried at 120 °C for 2 h, in distilled water and diluting to 1 l. Prepare working standard solutions in the range of 50–200 mg l^{-1} by suitable dilution of the stock solution.

Procedure. A schematic flow diagram for the determination of sulphate is shown in Fig. 4.6. The manifold consists of Tygon tubing with an i.d. of 0.51 nm cut into the required lengths and wound around suitable glass tubes with an o.d. of 15 mm. A barium chloride reagent carrier stream was provided at a constant flow rate of 3.9 ml min^{-1} by means of a peristaltic pump.

Inject samples automatically from a 60 µl sampling loop into the reagent stream by means of a Carle micro-volume two-position sampling valve. In between the samples inject automatically 100 µl of alkaline buffer-EDTA solution into the reagent stream from the other sampling loop. The above-mentioned process is achieved by placing the water samples in every second cup of the automated sampler, alternating the samples by placing alkaline buffer-EDTA solution in the cups between samples. Use a 30 s sampling cycle between sampling a water sample and sampling buffer-EDTA solution, giving the system a total sampling capacity of 120 per hour, i.e. a sampling capacity of 60 water samples per hour. Actuate the valve system on a time basis which is correlated with the sampler unit. The sampling valve actuates every 28 s after movement of the sampler to the next sample.

Krug et al. [4.36] have used flow injection analysis for the turbidimetric determination of sulphate in natural waters. They give details of equipment and procedures for a flow injection system with automated alternating streams of reagents for the turbidimetric determination of sulphate. The method is suitable for 120 samples per hour with a relative standard deviation less than 1% for sulphate concentrations in the range 1–30 mg l^{-1}.

Samples are injected into an inert carrier stream which is mixed with barium chloride to form a barium sulphate suspension. The range of the method can be extended to low concentrations by continuously adding sulphate to the sample carrier stream. System performance is improved by automatic alternate pumping of the reagent stream and an alkaline EDTA solution at high flow rate. All operations are controlled by an electronically-operated proportional injector-commutator.

Reagents. As follows.

R_1 reagent (Fig. 4.7). A 5% (w/v) barium chloride dihydrate solution in 0.05% (w/v) poly(vinyl alcohol) solution, prepared daily.

R_2 reagent. A 0.3% (w/v) EDTA (disodium salt) solution in 0.2 mol l^{-1} sodium hydroxide.

R_3 reagent. A 100 mg l^{-1} sulphate solution in 0.5 mol l^{-1} nitric acid.

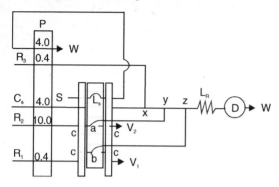

Fig. 4.7. Flow diagram of the systems for plant and water analysis. Flow rates given are in ml min^{-1}. Lines $xy = yz = 5$ cm and $bz = 20$ cm are made from 0.87 mm i.d. tubing; $ay = 20$ cm is made from 0.5 min i.d. tubing. L_R is the reaction coil (100 cm). Further details are given in the text (From [4.36])

The sample carrier stream C_S was 0.25 mol l^{-1} perchloric acid to minimize pH gradients in the sample zone.

Apparatus. Peristaltic pump and spectrophotometer with flow cell and recorder. Alternatively, an Ismatec mp 13 peristaltic pump could be used with a 634 Varian spectrophotometer and a REC 61 recorder with a REA 112 high- sensitivity unit (Radiometer). The wavelength was set at 140 nm. The manifold was made from 0.8 mm i.d. polyethylene tubing, except where stated otherwise. The connectors were made from perspex and the reaction coils and sample loops were built by winding the polyethylene tubes around PVC tubes (5 cm long, 2 cm diameter). The injector commutator had three 2:3:2 sections and was operated electronically.

The flow diagram of the system used for the sulphate determination is shown in Fig. 4.7, which indicates the injector commutator (I) in the sampling position, closed lines being denoted by c. The sample (S) is pumped (L) to fill the sample loop (LS), which is 100 cm long and defines exactly the volume to be injected, the excess of sample going to waste. Meanwhile, the barium chloride reagent (R_1) is directed towards its recovery vessel (V_1) while the alkaline EDTA reagent (R_2) pumped at high speed washes the analytical path. Commutation of the injector to the injection position introduces the selected sample volume into the sample carrier stream (C_S), and simultaneously switches the streams R_1 and R_2, so that R_2 is pumped to its recovery vessel (V_2) and reagent R_1 is directed to the analytical path. At confluence point x (Fig. 4.7), sulphate is added to the sample zone to permit determinations of 10 mg l^{-1} sulphate. At confluence point y, leakage of R_1 to the analytical path is insignificant. Reagent R_2 is added at point z, and the precipitation reaction occurs in the L_R coil. The sample zone then reaches the detection unit (D) where the turbidity is measured at 410 nm and recorded. The sample then goes to waste (W), and the injector commutator is switched back to the position indicated in Fig. 4.7, starting a new cycle. Thus the R_2 wash completely removes any residual precipitate which might act as a primary precipitation nucleus or clog the system.

After 1200 determinations, no baseline drift was noticed and changes in the slope of the calibration curves were very small (less than 5%), indicating good long-term stability. Also, relative standard deviations of measurements were in general about 0.5% when the sulphate content exceeded 10 mg l^{-1}.

When 500 mg l^{-1} of calcium (as calcium chloride), hydrogen carbonate (as sodium hydrogen carbonate) or magnesium (as magnesium chloride) was added to the sulphate standards, no interference was detected for the water analysis systems. Also, recoveries ranging from 97 to 102% were assessed for the water system; interference from organic matter thus seems unlikely. It can be seen from Table 4.7 that the results obtained by the proposed methods compare well with those obtained by the standard turbidimetric procedure for water samples [4.37]. Colour interference can be corrected by running blanks, if necessary; this is easily done in the flow injection system by withdrawing the barium chloride from the R_1 reagent.

Non-turbidmetric flow injection analysis has also been used to determine sulphate in water. One such system employs dimethylsulphonazo(III) as the reagent [4.38]. It was not necessary to saturate the reagent solution with barium sulphate. However, it was necessary to saturate the carrier solution with barium sulphate to obtain high sensitivity and good reproducibility. Moreover, when not in use, it was necessary to fill the reaction coil with ethanol: water (1:1) to obtain high sensitivity.

Calcium interference was a serious problem when this method was applied to waters. This was overcome by passing the samples through a cation exchange resin, located just after the sample injection valve. The suppressive effect of calcium ion was almost eliminated by inserting the cation exchange column. Columns either 8 or 15 cm in length showed almost similar effects. For the determination of 10 mg l^{-1} sulphate, other ions did not interfere within 5% negative errors up to the following concentrations : 30 mg l^{-1} of magnesium and ammonium, 50 mg l^{-1} of sodium, 80 mg l^{-1} of potassium and 100 mg l^{-1} (maximum concentrations tested) of chloride, nitrate, phosphate, hydrogen carbonate and silicate. These ions therefore did not interfere with sulphate determination at levels which are normally present in natural waters. The lifetime of the cation exchange resin column was found to be at least two months when regularly used.

Table 4.7. Comparison of procedures for the determination of sulphate in plant digests and natural waters. (From [4.36]; Copyright 1983, Elsevier Science Publishers BV, Netherlands)

Plant Sample	% S (dry matter)		Water Sample	mg l^{-1} SO$_4^{2-}$	
	Flow injection analysis	Manual gravimetric method		Flow injection analysis	Manual turbidimetric method
01	0.30	0.25	109	11	10
12	0.25	0.24	114	18	18
16	0.32	0.29	122	25	23
22	0.30	0.27	125	21	23
36	0.26	0.26	134	16	15
79	0.46	0.46	192	32	29

Calibration graphs were linear over the sulphate range 0–14 mg l^{-1} and slightly curved over the range 14–32 mg l^{-1}. The relative standard deviations of 20 analyses of solutions containing 6 and 10 mg l^{-1} of sulphate were 0.94 and 1.2% respectively. The detection limit was found to be about 0.2 mg l^{-1} of sulphate ion. The sampling rate was 20–30 samples per hour.

Sonne and Dasgupta [4.198] carried out a simultaneous photometric flow injection determination of sulphate, sulphide, polysulphide, sulphite and thiosulphate.

4.1.1.6 Atomic Absorption Spectrometry

In an indirect flow photometric method [4.39], sulphate is precipitated in hydrochloric acid medium by addition of a known amount of aqueous barium chloride, followed by flame photometric determination of excess barium in the filtrate at 493 nm. Atomic absorption spectrophotometry has also been used to determine sulphate. Little et al. [4.40] precipitated sulphate as lead sulphate from 40% ethanol medium and unconsumed lead was determined by atomic absorption spectrometry. Siemer et al. [4.41] precipitated sulphate as lead sulphate. The precipitate was filtered off on a porous graphite cup which was then placed in a constant temperature Woodriff furnace for analysis of lead by non-resonance line atomic absorption spectrometry. Calcium and chloride had no effect on the determination but high phosphate levels enhance the lead signal. Natural waters were analysed by this method and results compared well with those obtained by turbidimetric and gravimetric methods.

Montiel [4.42] reacted sulphate in a buffered medium with excess standard barium chloride. Unreacted barium was then determined by atomic absorption spectrometry at 553.55 µm and the concentration of sulphate in the sample calculated. Errors due to the presence of alkali and alkaline earth metals are corrected by the incorporation of calcium in the standard solution and by the presence of sodium in the buffer solution.

Kokkenonen et al. [4.43] have carried out the indirect determination of sulphate (and sulphite) in natural waters by flame atomic absorption spectrometry.

4.1.1.7 Inductively Coupled Plasma Atomic Absorption Spectrometry

Inductively coupled plasma emission spectrometry has also been used to determine sulphate directly in natural waters [4.44].

By monitoring the 180.73 nm sulphur line, it was possible to detect 0.08 mg l^{-1} sulphate with a precision of 0.8% RID at the 200 mg l^{-1} sulphate level. Instrument operating conditions are seen in Table 4.8.

Of the elements characteristically present as major components of natural waters, only calcium produced a slight interference. A weak calcium line at approximately 190.734 nm partially overlaps the sulphur line such that $1000\, mg\, l^{-1}$ Ca produces an apparent sulphate signal of $25\, mg\, l^{-1}$ sulphate. Correction for this interference is easily achieved by establishing the relationship between calcium concentration and apparent sulphate signal and inserting this information in the controlling software. The effect of calcium was then automatically substracted during the measurement of sample.

Natural water samples are normally acidified to stabilise them during storage. There is an effect due to hydrochloric acid concentration on the sulphur emission signal. This effect is conveninently overcome by making the acid content of samples and standards identical at, for example 1 vol.%.

The accuracy of the inductively-coupled plasma procedure was assessed by analysing waters of known sulphate composition, and by comparing measured sulphate values for a wide range of samples with those obtained for the same waters by an automated spectrophotometric procedure. Table 4.9 shows good agreement between the sulphate measurements obtained and the nominal values for International Standard Sea Water and an EPA Quality Control Standard.

4.1.1.8 Ion-Selective Electrodes

In an indirect method [4.45], excess ferric ions are added to the sample to complex the sulphate. The solution is then titrated with barium chloride and

Table 4.8. Plasma System Components. and operating Conditions. (From [4.44]; Copyright 1982, Elsevier Science Publishers BV, Netherlands)

Spectrometer	ARL 3400 C Quantovac; 1 m Paschen-Runge mounting; grating ruled 1080 lines mm^{-1}, 0.309 nm mm^{-1} reciprocal linear dispersion (3rd order). Primary slit width 20 µm, secondary slit 50 µm, Hamamatsu R-306 photomultiplier tube
Wavelength	180.73 nm. 3rd order
Readout	Digital readout of mean signal from three 10 s integrations
Frequency	27.12 MHz
Forward power	1080 W
Reflected power	5 W
Observation height	17 mm above load coil
Argon flow rates	Coolant 11 l min^{-1}, plasma 1.2 l min^{-1}, carrier 1 l min^{-1}, light path purge 1.5 l min^{-1}
Nebuliser	Concentric glass, J.E. Meinhard model TR-30-A3

Table 4.9. Comparison of results for sulphate by inductively-coupled plasma emission spectrometry with certified values for standard waters. (From [4.44]; Copyright 1982, Elsevier Science Publishers BV, Netherlands)

Sample	Sulphate content ($mg\, l^{-1}$)	
	I.C.P.E.S.	Certified value
EPA Quality Control Standard 2	7.3	7.2
IOS Standard Seawater	2800	2781

liberated ferric iron detected with an iron-selective electrode. In another method [4.46], sulphate is titrated directly with $0.01\,\text{mol}\,l^{-1}$ lead perchlorate in the presence of 50% dioxan, the end-point being detected with a lead-selective electrode. The indirect method is subject to several interference effects. The direct method is subject to interference effects by metal ions and phosphate. Hulanicki et al. [4.47] eliminated ionic interferences in the determination of sulphates using a lead sensitized ion-selective electrode. Jones et al. [4.48] studied interferences of a barium ion-selective electrode used in the potentiometric titration of sulphate.

Hara et al. [4.49] have applied a continuous flow method using a lead-selective electrode for the determination of sulphate.

A solid state lead selective electrode has been used to estimate sulphate in mineral waters [4.50]. The sample is passed through two columns of strong cation-exchange resins, the first in the Ag^+ form (to remove chloride) and the second in the H^+ form (to remove bicarbonate). After adjustment of the pH of the effluent to between 5 and 6, measure a volume, dilute with an equal volume of 1, 4-dioxen and titrate with, e.g., $5\,\text{mmol}\,l^{-1}$ lead nitrate, using the lead-selective electrode to indicate the end-point. Down to $10\,\text{mg}\,l^{-1}$ sulphate could be determined in the presence of up to a ten-fold excess of chloride. Phosphate interferes in the procedure.

Lukianets et al. [4.51] have described a rapid method for the determination of sulphate in river water by potentiometry with a lead selective electrode using a hard membrane of lead sulphide and silver sulphide. The titrant was lead II perchlorate. Acetone was the most effective solvent for accelerating the precipitation process. Adsorption of sulphate ions on to the precipitant was prevented by addition of crystal violet solution. The method was unaffected by pH in the range 2–8. Calcium, magnesium, sodium and chloride did not interfere with sulphate determination. Carbonate interference was eliminated by adding acetic, perchloric and nitric acid. Analysis of river water using the method compared well with analysis by high frequency titration with barium acetate.

4.1.1.9 Column Coupling Capillary Isotachophoresis

Bocek et al. [4.52] determined sulphate (and chloride) in mineral waters by this technique and Zelinski et al. [4.53] applied the technique to the determination of 0.02 to $0.1\,\text{mg}\,l^{-1}$ quantities of sulphate, chloride, fluoride, nitrate, nitrite and phosphate in river waters. The technique is discussed in further detail in Chap. 11.

4.1.1.10 Ion Exchange Chromatoqraphy

Stainton [4.54] has described an automated method for the determination of sulphate and chloride in natural waters. An ion exchange resin is used to convert the sulphates and chlorides to their free acids. Detection is achieved by electrical conductance. The use of silver-saturated cation exchange resin to precipitate chloride permits distinction between chloride and sulphate. High levels of nitrate, orthophosphate and fluoride give positive interference for sulphate; bromide and iodide similarly interfere with chloride estimates.

4.1.1.11 Ion Chromatography

This technique has been applied to the determination of sulphate in natural waters. Methods are reviewed in Table 4.10 and discussed more fully in Sect. 11.1.

Singh et al. [4.199] determined sulphate in deep sub surface waters by suppressed ion chromatography.

4.1.1.12 Miscellaneous

Sulphate has been determined in high purity water by molecular emission cavity analysis [4.60].

A recent contribution to the classical gravimetric barium precipitation method for the determination of sulphate is that of Ferus and Torrades [4.61]. These workers showed that the detection limit of the method could be extended below the usually accepted levels. They achieved a detection limit of $1 mg l^{-1}$ using 200 ml water samples. It was essential to use a large excess of barium chloride to achieve these detection limits.

Ligon and Dorn [4.62] determined sulphate and nitrate in rain water by mass spectrometry.

Meehan and Tariq [4.63] determined sulphate in the range 0.05 to 1.0 $mg l^{-1}$ by infrared spectroscopy by the KBr-pellet technique. Spectra were scanned from 1000 to 400 cm^{-1} and the extinction at 619 cm^{-1} was measured. Nitrate and nitrite do not absorb significantly between 800 and 600 cm^{-1}, and sulphate could be determined with a coefficient of variation of 5% in the presence of 400- and 60-fold molar excesses of nitrate and nitrite respectively.

Table 4.10. Determination of sulpahte in natural waters by ion chromatography

Type of water sample	Codetermined anions	References
Natural	NO_3	4.55
Natural	F, Br, Cl, NO_2, NO_3, PO_4	4.56
Natural	Br, Cl, NO_2, NO_3	4.57
Natural	None	4.58
Pure	Cl, NO_3	4.59

Five different methods have been compared for the determination of sulphate in river waters [4.64]. These are a gravimetric method involving barium sulphate precipitation, two titrimetric methods involving either indirect titrimetry with EDTA or direct barium titrimetry, a method employing indirect atomic absorption spectrometry, and a method using excess 2-aminoperimidine to precipitate sulphate, with two possible variants for estimation of the excess reagent. Details of the range of application, possible interferences and sensitivity of the method are given in each case.

4.1.2 Rainwater

Sulfate is present in rainwater at concentrations between 1 and $10\,mg\,l^{-1}$ and its determination may provide valuable information on global pollution levels, atmospheric circulation and geophysical phenomena. The recent adoption of the catalytic converter by the automobile industry has also increased the interest in sulfate determinations in rainwater. The sulphate ion is considered [4.65] to be the best indicator of change in aqueous precipitation composition due to anthropogenic activities. Levels of this ion were 2–16 times greater in eastern North American precipitation than in that from remote areas.

4.1.2.1   Spectrophotometric Methods

Burns et al. [4.66] determined down to $2\,mg\,l^{-1}$ sulphate indirectly in rainwater by a spectrophotometric method using an excess of 2-aminopyridine hydrochloride which forms an insoluble complex with sulphate. After titration the excess reagent is determined spectrophotometrically at 305.5 nm. Unfortunately, only up to $10\,mg\,l^{-1}$ iron can be tolerated by this method, otherwise interferences is negligible.

Suzuki et al. [4.67] have described a simple procedure for the spectrophotometric determination of sulphate in snow. It involves reaction of sulphate with amino perimidine to form perimidinyl ammonium sulphate, and collection of this precipitate on a teflon filter followed by dissolution in nitric acid. The resulting solution is made alkaline and the absorbance of the violet colour is measured at 550 nm. Both phosphate and nitrate interfere in the procedure.

Reagent. As follows.
2-Perimidinyl ammonium bromide, aqueous 0.3% prepared daily.

Procedure. Place in the funnel a 5-ml portion of sample solution containing up to 50 µg of sulphate followed by 3 ml of the 2-perimidinyl ammonium bromide solution. After 2 min, remove the solution by gentle suction. Wash the funnel and the precipitate with fine water droplets from a spray to remove excess of reagent. Add about 2 ml of concentrated nitric acid dropwise to the precipitate until the washings become colourless. Collect all the washings in a 50-ml beaker. Make the solution basic with the addition of 3 ml of 10 mol l^{-1} sodium hydroxide, transfer to a 25-ml volumetric flask and dilute to the mark with deionized water. Measure the absorbance at 550 nm against water. About 9 mg l^{-1} sulphate were found in snow samples by this method.

4.1.2.2 Continuous Flow Analysis

This technique has been applied to the determination of sulphate and chloride, nitrite, nitrate and phosphate in rainwater [4.68].

4.1.2.3 Flow Injection Analysis

Madsen and Murphy [4.69] adapted the automated methylthymol blue method for the determination of sulphate in rainwater to flow injection analysis processing. The calibration curve is linear up to 6 mg l^{-1}, sensitivity is approximately 0.1 mg l^{-1} and sampling rate is 20 h^{-1}. Average precision and accuracy are 4.1% and 97% respectively.

The flow injection system used is shown in Fig. 4.8. The sodium hydroxide reagent comprised 0.036 mol l^{-1} sodium hydroxide in 40 vol. % ethanol water. The methylthymol blue reagent which gave the highest sensitivity was prepared as follows:

0.116 g methylthymol blue was dissolved in a solution which contained 6.0 ml of 1.0 mol l^{-1} hydrochloric acid, 80 ml of deionized water and 14.0 ml of 1.526 g/l barium chloride dihydrate solution. Final dilution with 95% ethanol was to 1.0 l. Fresh reagent must be prepared daily.

The indirect determination of sulphate based on competitive reaction of sulphate and methylthymol blue with barium in solution can be accomplished by measurement of absorbance due to either uncomplexed methylthymol blue or the methylthymol blue barium complex. Absorbance at 460 nm due to uncomplexed methylthymol blue will increase while absorbance due to the methylthymol blue barium complex will decrease as sulphate concentration increases.

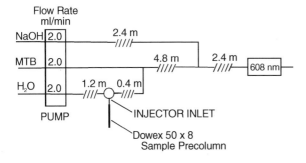

Fig. 4.8. Components of flow injection system (From [4.69])

The composition of reagent solution used will influence the applicability of the method. It was observed that the absorbance decrease at 608 µm due to methylthymol blue barium complex was approximately three times greater than the corresponding absorbance increase at 460 nm due to uncomplexed methylthymol blue. Absorbance base line stability measured in the absence of sulphate was essentially the same at both wavelengths. Measurement of the absorbance decrease (ΔA) at 608 nm was therefore adopted in the method.

In Table 4.11 are some sulphate determinations obtained by Madsen and Murphy [4.69] on rainwater samples using their flow injection analysis system and by ion chromatography are compared. Differences in results by the two procedures are insignificant at the 95% confidence level.

Many cations interfere with the determination of sulfates using methylthymol blue. Sample pretreatment with a cation exchange resin in the H^+ form will eliminate this interference. Rainwater samples typically contain 0.2 mg l^{-1} calcium and 0.2 mg l^{-1} magnesium. The addition of 1.0 mg l^{-1} calcium to two different rainwater samples did not alter results for sulfate by the flow injection method as can be seen from results presented in Table 4.11.

Reijnders et al. [4.70] compared results obtained by the continuous flow method with those obtained using a continuous flow apparatus and a flow through titrimeter. In each case dimethyl-sulphonazo III was used as a chromogenic reagent. Results obtained by these three procedures were compared with those based on the thorin-barium reaction in a continuous flow analyer. The chemical system employed in the first three methods mentioned above basically involves a complex of barium ions with the indicator dimethylsulphonazo III. Sulphate ions from the sample react with barium and set free the uncomplexed dimethyl-sulphonazo III. The absorption spectra of complexed and free dimethylsulphonazo III are different.

Table 4.11. Determination of sulfate in rainwater samples. (From [4.69]; Copyright 1981, American Chemical Society)

Sample no.	mg l^{-1} sulfate						
	Flow injection analysis	s	% s	Ion chromatography	Δ	% Δ	
1	0.78	0.08	10.3	0.91	−0.13	−14.3
2	3.75	0.08	2.1	3.90	−0.15	−3.8
3	3.50	0.02	0.6	3.52	−0.02	−0.6
4	2.60	0.04	1.5	2.51	+0.09	+3.6
5	1.75	0.05	2.9	1.67	+0.08	+4.8
6	0.82	0.01	1.2	0.83	−0.01	−1.2
6A[a]	0.80	0.03	3.8			
7	1.23	0.14	11.4	1.31	−0.08	−6.1
7A[a]	1.24	0.04	3.2			
8	0.67	0.03	4.5	0.69	−0.02	−2.9
9	2.23	0.07	3.1	2.37	−0.15	−6.3

[a]Samples 6A and 7A were spiked with an additional 1.0 mg l^{-1} calcium. Standard deviations are based on $N = 3$ except for sample 7A where $N = 6$

Results obtained by the four methods are summarized in Table 4.12, which shows that there is no significant differences between the results from the various methods.

Significantly higher sulphate values were returned by the thorin based spectrophotometric method than by the three methods involving dimethylsulphonazo III when the samples contained water matrix components (sodium chloride, potassium chloride, sodium bicarbonate, ammonium chloride, calcium nitrate and ferric nitrate) (Table 4.13).

Reijnders et al. [4.70] concluded that segmented continuous flow analysis and flow through titrimetry are suitable methods for the determination of sulphate in real samples. Results of (segmented) flow injection analysis and flow through titrimetry are less affected by interfering ions. The dimethylsulphonazo III chemical system is superior to the system with thorin.

Table 4.12. Comparison of the results of the examined methods in rainwater samples. (From [4.70]; Copyright 1980, Springer-Verlag GmbH, Germany)

Found μ mol l^{-1} sulphate

Cont. flow photometry; thorin		Cont. flow photometry; DMSA (III)		Flow-injection photometry; DMSA (III)		Cont. Flow titrimetry; DMSA (III)	
	s.d.		s.d.		s.d.		s.d.
63	2	65.0	0.6	61.9	0.7	64.0	0.5
85	3	81.0	1.0	79.2	0.8	83.2	0.7
60	2	61.0	0.7	61.5	0.9	61.2	0.6

DMSA III = Dimethyl Sulphonazo III.

Table 4.13. Comparison of interferences with respect to the examined methods. (From [4.70]; Copyright 1980, Springer-Verlag GmbH, Germany)

Sulphate given μ mol l^{-1}	Found μ mol l^{-1} sulphate				
	Interference	Cont. flow photometry; throin	Cont. flow photometry; DMSA (III)	Flow-injection photometry; DMSA (III)	Cont. flow titrimetry; DMSA (III)
0.00	1	3	<1.4	<2.6	<5
0.00	2	140	3	7	<5
0.00	3	270	8	14	8
50.0	1	51	51	50	50
50.0	2	190	45	50	51
50.0	3	300	34	48	53
100.0	1	100	101	100	100
100.0	2	240	95	96	104
100.0	3	350	77	90	106

Key to interference:
1. No infereference.
2. Sulphate and interferences as specified in 3 in a 1:1 dilution.
3. Sulphate and 0.02 mol l^{-1} NaCl, 0.0015 mol l^{-1} KCl, 0.01 mol l^{-1} NaHCO$_3$, 0.005 mol l^{-1} NH$_4$Cl, 0.0002 mol l^{-1} Fe(NO$_3$)$_3$, 0.01 mol l^{-1} Ca(NO$_3$)$_2$.

A computer controlled multichannel continuous flow analysis system has been applied to the measurement of sulphate (and nitrate, nitrate, chloride and phosphate) in small samples of rainwater [4.71]. Continuous flow methods (flow injection analysis and continuous flow analysis) have been used [4.72] to determine the principal constituents of rainwater. While the two methods are similar, they differ in certain important respects, such as the rate of flow, presence of air bubbles, diameter of the tubes and the reactions which are permissible within the retention period limits set by the physical constraints.

4.1.2.4 Flame Emission Spectroscopy

Wagner and Steel [4.73] describe a technique for analysing $0-4\,mg\,l^{-1}$ of sulphate in rainwater, which is based on the change in the solubility of powdered barium sulphate in rainwater, caused by the common ion (sulphate) at a constant temperature and ionic strength. The barium ion concentration, which measures barium sulphate solubility is determined by flame emission spectrometry.

4.1.2.5 Ion-Selective Electrodes

Scheide and Durst [4.202] have described an indirect method for determining $2-100\,mg\,l^{-1}$ sulphate in rainwater and natural waters by an ion-selective electrode. Isopropanol (80%) is added to the sample which is then titrated with lead nitrate solution, the end-point being determined with a lead ion-selective electrode. The reproducibility of results is better than 2%.

It is seen in Table 4.14 that sulphates determined by this method are in good agreement with gravimetric and turbidimetric results. Rainwater often contains copper and cadmium and these ions interfere with and deactivate the lead electrode membrane surface. However, they were present in such low

Table 4.14. Analysis of sulfate in rainwater. (From [4.202]; Copyright 1977, Marcel Dekker Inc., USA)

Sample mg l^{-1} added		Ion-selective electrode Found (mg l^{-1}) %	
B	3.57	3.57 ± 0.02	100.0 ± 0.6
C	5.02	4.93 ± 0.08	98.2 ± 1.6
D	10.00	10.12 ± 0.13	101.2 ± 1.3
400[a]	4.00	4.06 ± 0.04	101.5 ± 1.0
500[a]	8.33	8.24 ± 0.08	99.0 ± 1.0
		Turbidimetric Found % mg l^{-1}	
B	3.57	3.5 ± 0.35	98.0 ± 10.0
		Gravimetric Found % mg l^{-1}	
D	10.00	10.03[b]	100.3[b]

[a] Samples #400 and #500 were dilutions made from sample "D".
[b] Only one analysis was performed.

concentrations and reacted slowly enough that the rainwater samples could be analyzed directly with no sample pre-treatment. Because of this interference, however, the electrode has to be repolished after each run. This repolishing process takes about 30 s.

4.1.2.6 Gas Chromatography

Faiqle and Klochow [4.74] applied gas chromatography to the determination of traces of sulphate (and nitrate and phosphate) in rainwater. The dissolved salts are freeze dried and converted to the corresponding silver salts. These are then converted to the n-butyl esters with the aid of n-butyl iodide, the butyl esters being determined by gas chromatography. Sulphate (and phosphate) are determined simultaneously on a column containing 3% OV-17 on Chromasorb, whilst nitrate is determined separately with 3% of tri-p-cresol phosphate on Chromasorb.

4.1.2.7 Ion Chromatography

This technique has been applied to the determination of sulphate in rainwater. Methods are reviewed in Table 4.15 and discussed more fully in Sect. 11.2.

4.1.3 Potable Water

4.1.3.1 Spectrophotometric Method

Regnet and Quentin [4.85] describe a nephelometric method for determining up to 50 mg l^{-1} sulphate in potable waters in which a suspension of barium sulphate is produced by addition of a specially prepared barium chloride reagent, and the turbidity measured at 490 nm after a reaction time of 45 min. The calibration

Table 4.15. Determination of sulphate in rainwater by ion chromatography

Type of Water sample	Codetermined anions	References
Rain	Br, F, Cl, NO$_2$, PO$_4$, BrO$_3$, NO$_3$	4.75
Rain	Cl, PO$_4$, NO$_3$	4.76
Rain	Cl, Br	4.77
Rain	Cl, F, NO$_3$, PO$_4$	4.78
Rain	Cl, PO$_4$, NO$_2$, NO$_3$	4.79
Rain	Cl, NO$_3$	4.80, 4.82, 4.83
Rain	F, Cl, Br, NO$_2$, NO$_3$, SO$_3$, PO$_4$	4.81
Rain	None	4.84

curve can be considered as made up of two linear portions of different slope, which intersect at about 30 mg l^{-1} sulphate ion. The procedures for calibration and determination, taking about 2 h, are outlined. Other ionic constituents do not interfere at concentrations normally encountered, although nitrate can cause discrepancies if present in excess of 100 mg l^{-1}.

4.1.3.2 Flow Injection Analysis

Van Staden [4.86] reported the application of an active prevalve carbon filter for the turbidimetric determination of sulphate in potable waters. The filter is incorporated into an automatic flow injection system, between the sampler and the sampling valve system, to remove suspended and coloured materials which interfere with the spectrophotometric determination of turbidity. The method has been applied to the analysis of sulphate at concentrations up to 200 mg l^{-1} at a sampling rate of up to 60 samples per hour, with a coefficient of variation less than 1%. The accuracy is similar to that of a standard automated segmented method and standard flow injection methods.

4.1.3.3 Atomic Absorption Spectrometry

Siemer et al. [4.87] determined sulphate in potable waters by non-resonance line furnace atomic absorption spectroscopy. Direct sulfur determination is not routinely performed by flame or non-flame atomic absorption because of difficulties associated with the use of the far-UV resonance lines of that element. Therefore, efforts were directed at precipitation of the sulfate with a metal ion followed by determination of the metal retained on a porous graphite filter after washing off excess precipitant. Siemer et al. [4.87] discovered that non-resonance atomic absorption at 405.7 nm of the lead in a lead sulfate precipitate served the purpose.

At a furnace temperature of 1550 °C lead sulphate gives reasonable absorbance values for the amounts of sulphate in potable water. Between 50 and 75% of ethyl alcohol was incorporated during the lead sulphate precipitation and washing steps to decrease solubility errors. An excess of lead ions was added to the sample in this medium and excess lead ions determined.

In Table 4.16, sulphate determinations obtained by this technique are compared with those obtained turbidimetrically and gravimetrically.

4.1.3.4 Ion Chromatography

Schwabe et al. [4.88] determined sulphate (and chloride, fluoride, nitrate and phosphate) in potable waters by ion chromatography.

Table 4.16. Comparison of results of sulfate determinations with three different methods of analysis. (From [4.87]; Copyright 1977, Society for Applied Spectroscopy, USA)

Sample no.	Sulfate ion (mgl^{-1})		
	Gravimetric $n=1$	Atomic absorption spectrometry $n=5\,(\pm 1s)$	Turbidimetric $n=2$
10000 − 1	926	1077 ($\pm$ 31)	1180, 1168
6891 − 1	2800	2640 ($\pm$ 87)	2897, 2885
7036 − 1	52	57 ($\pm$ 4)	62.4, 60.5
7062 − 1	40	32 ($\pm$ 2.7)	29.0, 28.0

4.1.4 Industrial Effluents

4.1.4.1 Titration Methods

Methods have been described for determining sulphate in electroplating works effluents [4.89], chromium plating bath effluents [4.90] and industrial effluents [4.91]. For electroplating works effluents [4.89], the sample was first passed down a column of ion exchange resin in the H$^+$ form to remove interfering cations. Sulphate in the eluate is then determined by titration at pH 5.5 to 6.5 with 0.01 N barium chloride to the carboxyarsenazo end-point. 3-(2-Carboxylphenylazo), chromotropic acid and 3,6-bis (4-chloro-2-phosphorophenylazo) chromotropic acid, and, in the presence of phosphate or arsenate, nitroorthaonilic S (3,6-bis-(4-nitro-2-sulphophenylazo) chromotropic acid have also been used as indicators in the direct titration of sulphate with barium chloride. An alternate method is to precipitate sulphate as barium sulphate, dissolve the precipitate in excess 0.02 mol l^{-1} ammoniacal EDTA and titrate unconsumed EDTA with standard magnesium chloride solution to the Eriochrome Black T end-point.

4.1.4.2 Spectrophotometric Methods

Pilipenko et al. [4.92] compared various methods for the determination of sulphate in mine waters. These include a direct photometric method using octamine-u-amino-ol-dicobaltisol, which gave reproducible results in the concentration range 0.05–40 g l^{-1}. Chelatometric methods using barium rhodizonate or acidic chrome blue indicators were both rapid and convenient, although chloride concentrations of greater than 1 g l^{-1} interfered with barium rhodizonate determinations. For desalinated water, turbidimetric methods proved satisfactory.

Bhat et al. [4.93] reported methods for the determination of low levels of sulphides, sulphites and sulphates in waste waters. The method for sulphite and sulphide involves the reduction of the bis-2, 9-dimethyl-1-10-phenanthroline copper II ion by sulphide in the presence of formaldehyde and by sulphite in its absence. In the method for sulphate, stannous chloride in hydrochloric acid is used to reduce sulphate ions to hydrogen sulphide which is reacted with bis-2, 9-dimethyl-1-10 phenanthroline copper II ion.

All analyses are completed by measuring the absorbances due to the coloured copper I complex formed by chemical reduction. Whilst the methods are less than ideal from the point of view of sensitivity, simplicity and freedom from interferences, the colours formed are stable and the sulphate reduction system can be used repeatedly.

4.1.4.3 Ion Chromatography

This technique has been applied to the determination of sulphate in treated waters. Methods are reviewed in Table 4.17 and discussed more fully in Sect. 11.4.

4.1.5 Wastewaters

4.1.5.1 Ion-Selective Electrodes

Srivastava and Jain [4.95] investigated the performance of a sulphate ion-selective solid state membrane electrode prepared from hydrous thorium oxide gel with polystyrene as binder. The membrane showed good selectivity for sulphate ions in the range $0.1-100\,\mu\text{mol}\,l^{-1}$ and there was practically no interference from a large number of anions and cations. The electrode could be used at pH 6-10 and could be used in partially non-aqueous systems. This electrode was used to determine sulphate in waste waters from the pulp and paper and leather processing industries.

The electrode is non-Nernstian in nature. The useful pH range for this assembly is 6-10. The membrane electrode shows good selectivity to sulphate ions and there is practically no interference from a large number of anions and cations.

The electrode has also been used as an end-point indicator in potentiometric titrations involving sulphate ions. The titration of potassium sulphate ($20\,\text{ml}, 10^{-3}\,\text{mol}\,l^{-1}$) was performed with barium chloride solution ($5\times 10^{-3}\,\text{mol}\,l^{-1}$). The break in the curve is quite sharp and represents a perfectly stoichiometric end-point. Compared with direct potentiometric methods (based

Table 4.17. Determination of sulphate in treated waters by ion chromatography

Type of water sample	Co determined anions	Reference
Industrial effluents	Cl, NO_2, NO_3, PO_4, F, Br	4.94

on Nernst equation) potentiometric titrations offer substantial improvement in accuracy and precision. The standard deviation (ten replicates) of this particular titration was 0.01 ml.

4.1.5.2 Column Chromatography

This technique has been used [4.97] to determine sulphate in waste water. A post-column solid phase reaction detector was employed in conjunction with an anion exchange separation column to determine 5 to 40 mgl^{-1} sulphate.

4.1.5.3 Ion Exclusion Chromatography

This technique has been applied [4.98] to the determination of sulphate (and chloride, phosphate and carbonate) in waste waters. The technique is discussed in more detail in Sect. 5.1.6.

4.1.5.4 Ion Chromatography

This technique has been used to determine sulphate (and fluoride, chloride and phosphate) in waste waters [4.99].

4.1.6 Boiler Feed Water

4.1.6.1 Ion Chromatography

This technique has been used to determine very low concentrations of sulphate and chloride [4.100] and sulphate, chloride and nitrite [4.101] in boiler feed waters.

4.1.7 Soil

4.1.7.1 Titration Method

A volumetric method based on addition of excess barium ions and back titration with potassium sulphate to the sodium rhodizonate end-point has been used to determine water soluble sulphates in soil and irrigation water [4.102].

4.1.7.2 Molecular Emission Cavity Analysis

Molecular emission cavity analysis has been used to determine soluble sulphate in waters, soil and dusts [4.103].

4.1.7.3 Ion Chromatography

Ion chromatography has been used to determine sulphate in freshwater sediments [4.104].

4.1.8 Plant and Soil Extracts

4.1.8.1 Flow Injection Analysis

Krug et al. [4.105] used flow injection turbidimetry to determine sulphate in natural waters and plant digests. They described an improved flow injection system with alternative streams of reagents. Samples were injected into an inert carrier comprising 0.3% EDTA disodium salt and 0.2 mol l^{-1} sodium hydroxide. The inert carrier is mixed with 5% barium chloride containing 0.05% polyvinyl alcohol to form a barium sulphate suspension. The range of the method can be extended to low concentrations by continuously adding sulphate to the sample carrier stream. System performance is improved by automatic pumping of the reagent stream and an alkaline EDTA solution at high flow rate. All operations were controlled by an electronically operated proportional injector commutator. No baseline drift was observed even after analysis of 3000 samples. The method is capable of analysing 120 samples per hour with a relative standard deviation of less than 1% for sulphate concentrations in the ranges 1–30 mg l^{-1} or 5–200 mg l^{-1} (plant digests). Analytical recovery was 97–102%.

Plant samples were digested with a nitric acid perchloric acid mixture without further treatment [4.106].

4.1.8.2 Atomic Absorption Spectrometry

In an indirect method for determining sulphate in surface water and plant and soil extracts, Little et al. [4.107] precipitate sulphate as the lead salt in 40% ethanol medium. Unconsumed soluble lead is determined by atomic absorption spectrometry. The method is applicable to soil samples containing as little as 4 mg Kg^{-1} sulphate.

4.1.8.3 Ion Chromatography

Bradfield and Cooke [4.108] give details of a procedure for the determination of chloride, nitrate, phosphate and sulphate in aqueous extracts of plant materials and in soil solutions by an ion chromatographic technique with indirect ultraviolet detection. Recoveries ranged from 84 to 108%.

4.1.8.4 Autoanalyser Method

Ogner and Haugen [4.109] have described a technique for the automated determination of sulphate in water samples and soil extracts containing large amounts of humic compounds. This technique, can be applied to the determination of sulphate in concentration ranges of 0–60 and 0–3000 mg l^{-1} (as sulphate) in the aqueous extract.

4.1.9 Grain

4.1.9.1 Spectrophotometric Method

Basargin et al. [4.110] have described a spectrophotometric procedure for the determination of sulphate in grain. This method is based on the formation of a coloured complex with an absorption maximum at 640 nm between sulphate and 3,6-bis-(4-nitro-2-sulphophenylazo)-chromatropic acid. Down to 2 mg l^{-1} of sulphate in grain can be determined by this procedure with a relative error of ±1.3%. Borate, chloride, nitrate, perchlorate, arsenate and chromate do not interfere.

4.2 Sulphite

4.2.1 Natural Waters

4.2.1.1 Titration Methods

Bruno et al. [4.111] have described a procedure for the titration of sulphite in natural water using standard cerium (IV) as titrant.

Sulphites can be determined in natural waters by titration with mercury II solutions [4.112]. Sulphides, polysulphides, sulphates, thiosulphates, halide ions and thiols interfere in this procedure.

Sulphite has been determined in water [4.113] by a process involving acidifying the sample and purging with nitrogen to strip out sulphur dioxide, which is then absorbed in a solution containing ferric iron and 1,10-phenanthroline. The ferric iron is reduced to the ferrous state by the sulphur dioxide, and an orange tris-(1,10-phenanthroline)iron complex is formed, which can be quantified spectrophotometrically at 510 nm after removal of excess ferric iron with ammonium bifluoride. The effects of temperature and the removal of interferences are described. The detection limit is 0.01 mg l^{-1} sulphite.

Apparatus. The apparatus (shown in Fig. 4.9) consisted of a nitrogen gas source, stopcock or valve, 2 l min^{-1} flow meter, 250-ml gas washing bottle to purge the sample and a 100-ml Nessler tube with a 50-ml mark to absorb the sulfur dioxide liberated from the sample. The gas washing bottle had a coarse porosity 12 mm diameter fritted cylinder gas dispersion tube. A similar gas dispersion tube was used in the Nessler tube. Polyethylene tubing was used for all connections with quick-disconnect tubing connectors on both sides of the gas washing bottle. It was found convenient to run several gas trains in parallel, since samples and standards could then be run at the same time and temperature.

Reagents. As follows.

1,10-Phenanthroline solution, 0.03 mol l^{-1}. The 1,10-phenanthroline was first dissolved in 100 ml reagent alcohol before being diluted to 1 l with water. If the solution became coloured, it was discarded.

Ferric ammonium sulfate solution, 0.01 mol l^{-1}. To suppress ferric hydrolysis, 1 ml concentrated sulfuric acid per liter of solution was added. This solution was filtered to remove any insoluble matter and the amount of acid adjusted if necessary so that a mixture of one part of this solution and ten parts of the 1,10-phenanthroline solution had a pH between 5 and 6.

Ammonium bifluoride solution, 5% should be stored in polyethylene and dispensed with plastic pipette.

Hydrochloric acid solution, 6N.

Octyl alcohol and crystalline sulfamic acid. Used as obtained from J. T. Baker Chemical Company.

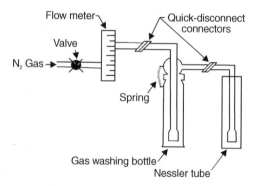

Fig. 4.9. Gas train apparatus for evolution and adsorption of sulfite (From [4.113])

Stock sulfite standard, 0.006 mol l^{-1}. This solution is unstable and must be standardized iodometrically immediately before being used to prepare the stabilized sulfite standard.

Stabilized sulfite standard. Immediately after standardizing the stock sulfite standard, 10 ml was pipetted into a 500-ml volumetric flask partially filled with 1% potassium mercuritetrachloride solution (K_2HgCl_4) and diluted to mark with this same solution. This stabilized standard was usually stable for 30 days if kept in a refrigerator, but was discarded or filtered if a precipitate began to develop at the bottom.

The absorbing solutions were prepared by adding to each 100-ml Nessler tube 5.0 ml 1,10-phenanthroline solution, 0.5 ml ferric ammonium sulfate solution, 25 ml distilled water and five drops octyl alcohol (to act as defoamer). Gas dispersion tubes were then placed into the Nessler tubes.

The gas washing bottles were prepared by adding 0.1 g sulfamic acid and 100 ml of sample (or an aliquot containing less than 120 µg SO_3^{2-} diluted to 100 ml). After adding 10 ml of hydochloric acid solution, the gas washing bottles were immediately connected in parallel to gas trains (as shown in Fig. 4.9). The nitrogen flow was adjusted to 2.0 l min^{-1} for each train and the samples purged for 60 min.

After exactly 60 min, the Nessler tubes were disconnected and the temperature of one of the absorbing solutions was measured to the nearest 0.5 °C, followed immediately by the addition of 1.0 ml ammonium bifluoride to each solution. It was found that controlling the time interval from the start of sulphur dioxide evolution to the ammonium bifluoride addition was necessary for good reproducibility of one run with another. After rinsing and removing the gas dispersion tubes from the Nessler tubes, the Nessler tubes were diluted to the 50-ml mark and mixed by rapidly moving the tubes in a circular motion. Rubber stoppers and tygon tubing caused high results and therefore were not allowed to come in contact with the absorbing solutions.

The colour complex was quite stable after the addition of ammonium bifluoride. The absorbances vs distilled water were read at 510 nm using either a 5-cm cell for a range of 0–40 µg sulphite per aliquot or a 1-cm cell for a range of 0–120 µg sulphite. A pipette was used to transfer the clear lower portion of the absorbing solution to the colorimetric cell after allowing the octyl alcohol to rise to the surface. The concentration of sulfite ion was calculated from a calibration curve obtained by analyzing a procedure blank and several stabilized sulfite standards.

Good agreement was obtained between results obtained by this procedure and those obtained by the standard iodometric procedure [4.114].

4.2.1.2 Spectrophotometric Method

Sulphite, sulphate and dithionate have been determined spectrophotometrically in the presence of each other [4.115].

4.2.1.3 Segmented Flow Analysis

This technique can be applied to the determination of sulphite in the concentration range $0-25\,\text{mg}\,l^{-1}$ (as SO_3) in water.

4.2.1.4 Continuous Flow Analysis

Continuouse flow chemiluminescence analysis has been applied to the determination of sulphite and sulphur dioxide in water [4.116].

4.2.1.5 Flow Injection Analysis

Al-Tamrah et al. [4.117] have described a flow injection chemiluminescence method for the determination of sulphite in water.

Sonne and Dasgupta [4.198] carried out a simultaneous photometric flow injection analysis determination of sulphites, sulphate, sulphide, polysulphide and thiosulphate.

4.2.1.6 Kinetic Multicomponent Analysis

Gonzalez et al. [4.206] have described a method of kinetic multicomponent analysis based on miscellar analysis for the determination of sulphite and thiocyanate. The method is based on reaction with 5, 5'-dithiobis (2 nitro benzoic acid) in aqueous cetyltrimethyl ammonium bromide miscelles. Down to $0.2-1.5 \times 10^{-4}\,\text{mol}\,l^{-1}$ sulphite can be determined with a relative error of less than 5%.

4.2.1.7 Ion-Selective Electrodes

Sulphite (and thiosulphate) can be titrated potentiometrically with mercuric ions using a sulphide ion-selective electrode as indicator electrode in 0.1 N sodium hydroxide medium. Sulphide and polysulphides interfere in this procedure [4.118].

4.2.1.8 High Performance Liquid Chromatography

A sensitive method for sulphite is based on high performance liquid chromatography followed by detection using inductively coupled plasma atomic emission spectroscopy [4.119].

4.2.1.9 Ion Chromatography

Ion chromatography has been applied to the determination of sulphite and arsenate [4.120], and sulphite and selenite [4.121] in natural waters.

4.2.2 Rainwater

4.2.2.1 Ion Chromatography

This technique has been applied to the determination of sulphite (and fluoride, chloride, bromide, nitrite, nitrate and phosphate) in rainwater [4.122].

4.2.3 Industrial Effluents

4.2.3.1 Ion Chromatography

Ion chromatography has been applied to the determination of sulphite in photographic processing effluents [4.123].

Petrie et al. [4.203] applied ion chromatography to the determination of sulphite and dithionite in mineral leachates.

4.2.4 Waste Waters

4.2.4.1 Spectrophotometric Method

Bhat et al. [4.124] have reported a spectrophotometric method for the determination of low levels of sulphite (and sulphide and sulphate) in waste waters. The method involves the reduction of the bis-2,9 dimethyl 1-10-phenanthroline copper II ion by sulphite.

4.2.4.2 Ion Chromatography

This technique has been applied [4.125] to the determination of sulphite (and chloride) in waste waters.

4.2.5 Boiler Feed Water

4.2.5.1 Ion Chromatography

This technique has been applied [4.126] to the determination of sulphite (and phosphate) in boiler feed waters.

4.2.6 Wine

4.2.6.1 Flow Injection Analysis

Flow injection analysis methods have been described for the determination of sulphite in wine and beverages [4.127]. In these methods the sample is injected into a carrier stream and mixed with sulphuric acid. The joint stream passes along a PTFE membrane in a gas diffusion cell. The dioxide formed diffuses through the membrane into a mixture of p-rosaniline and formaldehyde forming a red-purple complex which is evaluated spectrophotometrically at 585 nm.

4.2.6.2 Ultraviolet Spectroscopy

In the presence of oxygen, sulfite is oxidized by sulfite-oxidase (SO_2-OD) to sulfite:

$$SO_3^{2-} + O_2 + H_2O \xrightarrow{SO_2-OD} SO_4^{2-} + H_2O_2. \tag{4.1}$$

The hydrogen peroxide formed in this reaction is reduced by the enzyme NADH-peroxidase (NAHD-POD) in the presence of reduced nicotinamide-adenine dinucleotide (NADH):

$$H_2O_2 + NADH + H^+ \xrightarrow{NADH-POD} 2H_2O + NAD^+. \tag{4.2}$$

The amount of NADH consumed in reaction (4.2) is equivalent to the amount of sulfite. NADH is determined by UV spectroscopy by means of its absorbance at 334, 340 and 365 nm.

This method has been applied [4.128] to the determination of sulphite in wine, spirits, beer, fruit juices, solid foodstuffs, jam, potato products and spices.

4.2.7 Foodstuffs

4.2.7.1 Titration Method

The Food and Drug Administration (USA) have stated that levels of sulphite in food must be declared on the label if they exceed 10 mg kg^{-1}. This is because of reported ill effects of higher levels of sulphite on a proportion of the population, particularly asthmatics.

Tecator Ltd. [4.129] supply automated equipment for the determination of low levels of sulphite in foods and beverages. This equipment is based on forced stream distillation in the presence of acid to isolate sulphur dioxide from the sample. The sulphur dioxide is swept by a stream of nitrogen into hydrogen peroxide where it is oxidized to sulphuric acid and titrated with standard base to the bromophenol blue end-point.

4.2.7.2 Flow Injection Analysis

Alternatively, Tecator supply apparatus based on flow injection analysis [4.130]. In one of these methods the sample is injected into a carrier stream which is merged with an acid stream. Sulphur dioxide is formed at this low pH and diffuses through a gas permeable membrane into a reagent stream containing formaldehyde and *p*-rosaniline. The sulphur dioxide and the reagent form a red-purple complex which is evaluated spectrophotometrically at 560 nm. To analyse for total sulphite, a sample preparation step must be included in which bound sulphite in the sample is extracted with tetrachloromercurate or formaldehyde. These compounds form a strong complex with sulphite and, to liberate this portion, alkali must be added before acidification of the sample. This can be done on-line.

In an alternative flow injection analysis method [4.131], the injected sample is merged with a strong citric or sulphuric acid solution. The low pH (about 2) will convert the sulphite to sulphur dioxide, which diffuses through a gas permeable membrane into a stream of malachite green solution, buffered to pH 8. The sulphur dioxide decolorizes malachite green, and the decrease in absorbance is measured at 615 nm.

Both the flow injection analysis methods discussed above have a detection limit of 1 mg kg^{-1} sulphite in sample and a relative standard deviation of 1%.

4.2.7.3 Spectrophotometric Method

Smith [4.132] has carried out comparative determinations of sulphite in dried apricots, potato flakes, wine, vinegar and lemon juice by spectrophotometric methods and sulphite oxidase enzyme electrodes. Good correlation was obtained between the two techniques.

4.2.8 Pharmaceuticals

Sulphite has been determined in pharmaceutical matrices by a combination of spectrophotometry and flow injection analysis [4.133].

4.3 Thiosulphate and Polythionates

4.3.1 Natural Waters

4.3.1.1 Titration Methods

Thiosulphate (and sulphate) can be titrated potentiometrically with mercuric ions using a sulphide ion-selective electrode in 0.1 N sodium hydroxide medium [4.134]. Sulphite can be decomposed with formaldehyde and, consequently, thiosulphate can be titrated alone. Polysulphide and sulphide can be converted to thiosulphate and titrated.

Thiosulphates can be determined in natural waters by titration with mercury II solutions. Sulphides, polysulphides and sulphites interfere in this procedure [4.135].

4.3.1.2 Spectrophotometric Methods

Dithionate has been determined spectrophotometrically in the presence of sulphite and sulphate [4.136].

4.3.1.3 Anodic Stripping Voltammetry

Differential pulse anodic stripping voltammetry has been used [4.137] to detect thiosulphate, a potential substrate for sulphuric acid forming thiobacilli. The method was selective for thiosulphate in the presence of hydrogen sulphide, sulphite, polysulphide and elemental sulphur. Reproducible results were obtained for thiosulphate concentrations of $3-5$ mg l^{-1}.

4.3.1.4 Flow Injection Analysis

Sonne and Dasgupta [4.198] carried out a simultaneous photometric flow injection analysis of thiosulphate, sulphate, sulphide, polysulphide and sulphite.

4.3.1.5 Ion Chromatography

Ion chromatography has been applied to the analysis of low concentrations of polythionate in admixture with thiosulphates, and dithionites in admixture with sulphite. This is discussed further in Chap. 11 (Ion Chromatography).

Sunden et al. [4.138] determined thiosulphate in natural waters by ion chromatography.

4.3.2 Industrial Effluents

4.3.2.1 Titration Methods

Makhija and Hitchen [4.139] determined thiosulphates and polythionates in mining effluents and mill effluents by an acidimetric method. It involves sodium hydroxide titration of the acid liberated on addition of mercuric chloride (which oxidizes thiosulphate and polythionates to sulphuric acid), after adjustment of the initial pH value to either 4.3 (using methyl orange) or 8.2 (using phenophthalein). In this method a suitable proportion of the solution containing a polythionate having two, three or five sulphur atoms is acidified to either pH 4.30 or pH 8.20 with 0.005 N sulphuric acid. Sodium hydroxide and mercuric chloride is then added to release acid quantitatively. The stoichiometric equation is:

$$2S_xO^{2-} + 3HgCl_2 + 4H_2O \rightarrow HgCl_2 \cdot 2HgS + 8H^+ \\ + 4Cl^- + 4SO_4^{2-} + (2x-6)S.$$

A similar equation applies for thiosulphate:

$$2S_2O_3^{2-} + 3HgCl_2 + 2H_2O \rightarrow HgCl_2 \cdot 2HgS + 4H^+ + 4Cl^- + 2SO_4^{2-}.$$

In both cases the acid generated is titrated with standard sodium hydroxide solution to determine the sum of the concentration of the four sulphur anions:

$$S_2O_3^{2-}; S_3O_6^{2-}; S_4O_6^{2-}; S_5O_6^{2-}.$$

4.3.2.2 Atomic Absorption Spectrometry

Chakraborty and Das [4.140] have described a method for preconcentrating and determining thiosulphate in photographic waste effluents in which a stable ion-associated complex is formed between lead, thiourea, and thiosulphate in alkaline medium. This complex was extracted into n-butyl acetate:butanol (2:1) solvent mixture prior to determination in amounts down to 19 $\mu g\,l^{-1}$ by atomic absorption spectrometry.

4.3.2.3 High Performance Liquid Chromatography

Takano et al. [4.141] applied high performance liquid chromatography on an anion exchange column with differential pulse polarographic detection to the determination of thiosulphate and tri-, tetra-, penta- and hexa-thionates in trade effluents. The method is accurate to within 10% at the $0.001-1\,\text{nmol}\,l^{-1}$ concentration range.

4.4 Thiocyanate

4.4.1 Natural Waters

4.4.1.1 Spectrophotometric Methods

The spectrophotometric method for the determination of free cyanide in natural waters described by Nagashima and Ozawa [4.142] (Sect. 3.4.1) has also been applied to the determination of thiocyanate.

A spectrophotometric method has been described for the determination of thiocyanate (and simple and complex cyanides) in water in which the sample is reacted with acid cuprous chloride and the released hydrogen cyanide absorbed in sodium hydroxide solution. The alkali is then acidified and reacted with chloramine-T to produce cyanogen chloride. A pyridine- pyrazolone reagent is added and the absorbance of the resulting blue dye is evaluated at 620 nm. [4.144]

Thiocyanate, after complexation with mercury II and n-phenylbenzo hydroxamic acid, has been determined spectrophotometrically [4.145].

4.4.1.2 Ion-Selective Electrodes

Thiocyanate has been determined using a liquid membrane electrode [4.146].

4.4.1.3 Gas Chromatography

Nota et al. [4.143] have described a gas chromatographic headspace analysis technique for the determination of $0.01-100\,\text{mg}\,l^{-1}$ thiocyanate (and cyanide) in coke oven effluents. This method is discussed further in Sect. 11.4.

4.4.1.4 High Performance Liquid Chromatography

Stetzenbach and Thompson [4.147] employed high performance liquid chromatography on anion exchange columns to determine thiocyanate in admixture with chloride, bromide, iodide and nitrate in natural water. This method is discussed further under Multihalide Analysis in Sect. 2.5.

4.4.1.5 Kinetic Multicomponent Analysis

Gonzalez et al. [4.206] described the miscellar determination of mixtures of thiocyanate and sulphite. The method is based on reaction with 5,5′-dithiobis-(2-nitrobenzoic acid) in aqueous cetyltrimethyl ammonium bromide miscelles. Down to $0.5-1.5 \times 10^{-4}\,\text{mol}\,l^{-1}$ concentrations of thiocyanate can be determined with a reletive error of less than 5%.

4.4.1.6 Paper Chromatography

Thieleman [4.148] attempted the separation of thiocyanate, cyanate and cyanide ions by thin-layer and paper chromatography. Thin-layer chromatography was not successful but these three ions were successfully separated by paper chromatography with the solvent methanol:pyridine:dioxan (7:2:1) in 7–8 h. Cyanide and cyanate were identified by means of bromocresol purple and thiocyanate by 15% ferric chloride solution, as spray reagent. Carbonate ions have the same R_F value as cyanide and cyanate.

4.4.2 Industrial Effluents

4.4.2.1 Titration Methods

Atkinson et al. [4.149] have described a technique for titrating thiocyanate/cyanide mixtures in hydrometallurgical effluents. The mixtures are titrated with silver nitrate solution using an automatic potentiometric titrimeter with a silver/glass electrode pair. When thiocyanate is to be determined, cyanide is masked with formaldehyde.

A Sargent-Welch recording potentiometric titrimeter Model DG was used. It was fitted with a Radiometer P401 NH silver electrode and a Sargent-Welch S30050-15A glass electrode that functioned as the reference.

In order to determine thiocyanate in a thiocyanate-cyanide mixture, a suitable aliquot was placed in a 250-ml beaker, diluted to about 125 ml with de-ionised water and treated with 5 ml of formalin to mask the cyanide. The mixture was stirred and allowed to stand for 10 min. Nitric acid ($6\,\text{mol}\,l^{-1}$, 5 ml) was added to give a total volume of 135 ml and the sample was titrated with silver nitrate solution. For cyanide determination, a similar aliquot was taken and diluted to 135 ml with de-ionised water as before. The potentiometric titration was then carried out to the first sharp end-point. A second end-point was also observed if the titration was continued. The concentrations of thiocyanate and cyanide were calculated as follows:

$$(SCN^- \, (mg\,l^{-1})) = \{AgNO_3 \text{ molarity } (mol\,l^{-1}) \times AgNO_3 \text{ titration volume} (1) \times 58.08 \times 1000 \, (mg\,mol^{-1})\}/\text{sample volume (l)}$$

$$(CN^- \, (mg\,l^{-1})) = \{AgNO_3 \text{ molarity } (mol\,l^{-1}) \times AgNO_3 \text{ titration volume}(1) \times 2 \times 26.018 \times 1000 \, (mg\,mol^{-1})\}/\text{sample volume (l)}.$$

Table 4.18 gives data for the potentiometric determination of thiocyanate in the presence and absence of cyanide. The cyanide added is without effect on the thiocyanate determination when it is masked by formalin.

In a method [4.150] for the determination of thiocyanate in the effluent from a film processing unit, ferro and ferricyanides are first precipitated by the addition of zinc sulphate. Zinc is then precipitated from the filtrate by adding sodium carbonate and thiocyanate is determined in the final filtrate absorbtiometrically using ferric chloride. Down to $1\,mg\,l^{-1}$ thiocyanate can be determined.

4.4.2.2 Gas Chromatography

See section on cyanides under gas chromatography (Sect. 3.4.4).

Table 4.18. Potentiometric determination of thiocyanate in the presence and absence of cyanide. (From [4.149]; Copyright 1982, Royal Society of Chemistry, UK)
An initial titration volume of 135 ml was used

Thiocyanate/mg		Cyanide present/mg	Absolute error/mg	Recovery, %
Taken	Found			
25.18	25.46	–	0.28	101.1
25.18	25.50	10.0	0.32	101.3
12.50	12.65	–	0.15	101.2
12.50	12.66	10.0	0.16	101.3
9.47	9.52	–	0.05	100.5
9.47	9.58	20.0	0.11	101.2
4.74	4.76	–	0.02	100.4
4.74	4.76	100.0	0.02	100.4
1.26	1.14	–	−0.12	90.5
1.26	1.13	50.0	−0.13	89.7
0.625	0.565	–	0.06	90.4
0.625	0.578	50.0	−0.047	92.2
0.252	0.241	–	0.011	85.6
0.252	0.249	100.0	0.003	98.8
0.125	0.121	–	−0.004	96.8
0.125	0.109	100.0	−0.016	87.2

4.4.3 Waste Waters

4.4.3.1 Spectrophotometric Methods

Luthy et al. [4.151] have described a spectrophotometric method for the determination of thiocyanate and cyanide in coal gasification waste waters. They showed that the copper pyridine method with pre-extraction was applicable but that high concentrations of carbonate and sulphide must first be removed from the sample.

Botto et al. [4.152] used p-phenylenediamine as the chromogenic reagent in the spectrophotometric determination of thiocyanate and uncomplexed cyanide in waste water.

4.4.3.2 Ion-Selective Electrodes

Thiocyanate has been determined in waste water using a cyanide-selective electrode [4.153]. In this method thiocyanate is first converted to cyanogen bromide using bromine water. Excess bromine is removed by the addition of phenol. The cyanogen bromide is then converted to cyanide by the addition of sulphurous acid and then sodium hydroxide added. Cyanide is then estimated using the cyanide selective electrode:

$$SCN^- + 4Br_2 + 4H_2O \rightarrow CNBr + SO_4^{2-} + 7Br^- + 8H^+$$
$$CNBr + SO_2 + H_2O \rightarrow HCN + Br^- + SO_4^{2-} + H_2O + OH^-$$
$$HCN + OH^- \rightarrow CN^- + H_2O.$$

4.4.3.3 Paper Chromatography

Cyanide and cyanate in waste water have also been separated by paper chromatography [4.154]. Separation is achieved by use of isopropyl alcohol:ethanol:water (9:4:3) as solvent with Filtrak FN 8 paper. After drying, the paper is sprayed with bromocresol green solution. On a green background the ions appear as clear blue spots (R_F values-cyanide 0.04, cyanate 0.29). The spots may also be located by treating the paper with silver nitrate solution and exposing it to hydrogen sulphide.

4.4.4 Urine

4.4.4.1 Flow Injection Analysis

Cox et al. [4.207] have described stable modified electrodes for flow injection analysis of thiocyanate in urine. Platinum electrodes were modified by adsorption of iodine and coated with cellulose acetate. They also developed glass coated carbon electrodes, modified by anodization in a $RuCe_3$ $K_4Ru(CN)_6$ mixture. Both electrodes are stable amperometric indicators for thiocyanate in urine.

4.5 Sulphide

4.5.1 Natural Waters

4.5.1.1 Titration Methods

Wronski [4.155] determined $\mu g l^{-1}$ levels of sulphide in natural waters by thiomercurimetric titration. The procedure consists of separation of hydrogen sulphide as bis (triethyl-lead) sulphide by precipitation with triethyl-lead chloride in hexanol followed either by direct titration with o-hydroxymercuribenzoic acid in the presence of dithizone or by fluorimetric titration in hexanol-ethanol with tetramercurated fluorescein (excitation at 520 nm). The fluorimetric procedure is applicable to concentrations of hydrogen sulphide of greater than $0.01\ \mu g l^{-1}$ and the direct titration which determines thiols and sulphide is applicable to concentrations above $2\ \mu g l^{-1}$. For the direct titration, thiols are removed with acrylonitrile and hydrogen cyanide with formaldehyde. Thiols, cyanide, xanthates and dithiocarbamates do not affect the end-point of the fluorimetric titration. Neither method is affected by the common ions present in natural water samples.

Dyrssen and Wedborg [4.156] determined sulphide (polysulphides, thiosulphates, sulphites and thiols interfere) in natural water by titration with mercury II solutions. The stoichiometry of the titrations were calculated using selected values of relevant stability constants. Mathematical functions for the evaluation of the titrimetric equivalence points were derived. The presence of halide ions interfered only in the titration of sulphite by the formation of halide complexes.

Sulphide has been determined [4.157] in natural water by potentiometric or visual titration with standard solutions of caesium octacyanomolybdate ($Cs_3Mo(CN)_8$) or caesium octacyanotungstate ($Cs_3W(CN)_8$) ions. With the former titrant, the reaction is carried out in $0.05\ mol\,l^{-1}$ borate buffer at pH 9.5. With the latter titrant the reaction is carried out in $0.05\ mol\,l^{-1}$ phosphate buffer at pH 11.5. Methylene blue is used as indicator in the visual titration procedure. In the determination of 1–40 mg of sulphide the average error was $\pm 0.14\%$. Dithionite and sulphite also react but there is no interference from sulphate or thiosulphate ions.

4.5.1.2 Spectrophotometric Methods

Spectrophotometric methods for sulphide have been developed based on the use of the following chromogenic reagents; NN-dimethyl-p-phenylene diamine (methylene blue methods) [4.158, 4.159], sodium nitroprusside [4.160, 4.161], 1,10-phenanthroline-ferric chloride [4.162], bis-(2,9-dimethyl-1, 10-phenanthroline copper ion [4.163], and copper quinolin-8-olate [4.164]. The latter two methods are based on solvent extraction of the coloured produced formed with sulphide followed by spectrophotometry.

Caspieri et al. [4.161] automated the nitroprusside procedure using an AutoAnalyser. The samples are treated in a reproducible manner, and the colour is measured after a definite time from addition of the reagents. The inclusion of a distillation step increases the selectivity of the procedure and eliminates the necessity for filtration. The precision is satisfactory in the concentration range $1-10\,\mu g\,l^{-1}$ of sulphide and is slightly better than that for the manual procedure. The coefficient of variation at the $10\,mg\,l^{-1}$ level for the AutoAnalyzer and manual procedures are 4.2% and 3.6%, respectively. If high selectivity is not required, the distillation of hydrogen sulphide may be omitted. At a sampling rate of 40 per hour, about 250 samples can be analysed daily.

Bethea and Bethea [4.160] overcame some of the problems encountered in this method by lengthening the mixing coils and alternating reagent flow rates. A detection limit of $0.05\,mg\,l^{-1}$ sulphide was achieved.

In the methylene blue procedure [4.159] a mixed reagent of ferric chloride and NN-dimethylphenylene diamine is added to the sample to form methylene blue. After 20 min the extinction is measured at 670 nm. The method is applicable to waters containing between 0.03 and $32\,mg\,l^{-1}$ of sulphide. Within this range, Beer's law is obeyed and the method is free from salt effects and temperature dependence. The precision is $\pm 2\%$ at the 95% confidence level.

Grasshoff and Chau [4.158] automated the methylene blue procedure using an AutoAnalyser.

Ceba et al. [4.165] developed a method which is claimed to be a simple, rapid, sensitive and selective means of determining microgram levels of sulphide in water. The technique involves the use of chloroform solutions of copper quinolin-8-olate, which are re-extracted by aqueous solutions of sulphide, for the indirect spectrophotometric determination of sulphide. The effects of variables on the decolouration of the coloured complex by the aqueous sulphide were established, and the presence of other sulphur anions (sulphite, thiosulphate, and thiocyanate) did not interfere. Results compared favourably with those from the methylene blue method.

Reagents. All reagents and solvents were of analytical-reagent grade.

Copper quinolin-8-olate. Anhydrous copper quinolin-8-olate was obtained in the solid state by the gravimetric method using quinolin-8-ol [4.166]. From a suitable amount of this solid (about 60 mg in 1000 ml of chloroform) prepare a dilute organic solution. This solution has an absorbance value of about 1.0 at 410 nm against chloroform.

Sulphide standard solution, 1.000 gl^{-1} as S. Dissolve sodium sulphide Na$_2$S°9H$_2$O) in distilled, de-ionized water. Renew and standardise it every day by the iodimetric method. Dilute the standard solution as appropriate to produce a working solution containing 50.0 µg l^{-1} of S. Use each 50.0 µg l^{-1} working solution for only 1 h after preparation, then discard it.

Apparatus. SP 8000 and a Spectronic 100 Spectrophotometer, or equivalent equipped with 1.0-cm glass or quartz cells, a Radiometer PHM 2 and a Crison 501 pH meter, with glass–calomel electrodes and a Bibromatic Selecta 384 shaker were used.

Procedure. Place a portion of the aqueous solution containing 10–70 µg of sulphide into a 100-ml separating funnel and add sufficient dilute hydrochloric acid or sodium hydroxide solution to ensure a final pH in the range 4.5–8.0 and adjust the final volume to 50 ml with distilled water. Add 10 ml of the chloroform solution of copper quinolin-8-olate and shake the funnel mechanically for 6 min. Once the phases have separated, transfer the lower layer into a flask containing anhydrous sodium sulphate. Treat the samples in the same way, but shake with an aqueous solution without sulphate. The decolouration values are obtained as the differences between spectrophotometric readings at 410 nm of the organic phase shaken with the aqueous phase without sulphide and the organic phase shaken with the aqueous solutions of sulphide. Calibration graphs are obtained by using known amounts of sulphide treated in the same way.

A study of the effect of other ions was carried out by preparing a series of synthetic samples containing 40.0 µg of sulphide and variable amounts of foreign ions. The results obtained for binary mixtures are summarised below.

Reletive amount tolerated,	ion tested sulphide m/m
Na, K, Cl, No$_3$, SO$_4$, Ca, Mg, Ba, Sr	250–400
No$_2$, S$_2$O$_3$, CNS, SO$_3$	100–125
tartrate, citrate, PO$_4$, F, BO$_3$	25–50
Al, EDTA, CN	1–8

As can be seen, the procedure is highly selective and some results, especially for the sulphide mixtures with thiosulphate, have analytical interest, e.g. the relative amount of thiosulphate tolerated is greater than that in the methylene blue method.

The recommended method was tested by applying it to the determination of sulphide in water samples. A sample was prepared by adding a known amount of sodium sulphide to a sample of Guadiana River water. The decrease in the sulphide concentration in this sample was studied as a function of time in order to establish the evolution of a possible accidental spill. The results obtained are given in Table 4.19 together with those obtained by using the methylene blue method. These results agree for concentrations of sulphide in the range 0.5–1.0 mg l^{-1} whereas for sulphide concentrations greater than 1.0 mg l^{-1} the methylene blue method gives values higher than the method described by Ceba et al. [4.165]. To confirm the above results, two synthetic samples of water were prepared with the following composition: 5 mg l^{-1} of aluminium, 10 mg l^{-1} of sodium, 10 mg l^{-1} of potassium, 30 mg l^{-1} of magnesium, 30 mg l^{-1} of calcium, 185 mg l^{-1} of chloride, and 1.25 or 0.9 mg l^{-1} of sulphide. The results obtained as a function of time are given in Table 4.20.

4.5.1.3 Spectrofluorimetric Method

A method [4.167] [4.168] based on mercury II 2-(2-pyridyl) benzimidole is very sensitive, being capable of determining down to 0.3 µg l^{-1} sulphate. The sample is acidified with acetic acid and the hydrogen sulphide formed is swept into the reagent buffered to pH 6.2–7.3. The fluorescence produced due to sulphide is measured at 381 nm (excitation 311 nm).

Table 4.19. Evolution of the sulphide concentration in a sample of Guadiana river water. (From [4.165]; Copyright 1982, Royal Society of Chemistry, UK) Initial concentration of sulphide = 1.5 µg ml^{-1}

Time/min	Re-extraction method		Methylene blue method	
	Decloration value	S^{2-} found/mg l^{-1}	Absorbance	S^{2-} found/ mg l^{-1}
10	0.853	1.15	0.605	1.28
30	0.831	1.11	0.571	1.20
60	0.722	0.98	0.456	0.98
90	0.679	0.92	0.448	0.95
120	0.613	0.82	0.395	0.84
150	0.561	0.76	0.361	0.77
200	0.419	0.58	0.280	0.60
240	0.328	0.44	0.230	0.48

Table 4.20. Comparison of re-extraction spectrophotometric and methylene blue methods for the determination of sulphide. (From [4.165]; Copyright 1982, Royal Society of Chemistry, UK)

Time/min	Re-extraction method				Methylene blue method			
	Decloration value		S^{2-} found/mg l^{-1}		Absorbance		S^{2-} found/mg l^{-1}	
	Sample 1	Sample 2	Sample 1	Sample 2	Sample 1	Sample 2	Sample 1	Sample 2
10	0.830	0.602	1.12	0.81	0.562	0.383	1.20	0.81
30	0.828	0.600	1.12	0.81	0.568	0.381	1.20	0.81
60	0.808	0.593	1.10	0.80	0.550	0.385	1.16	0.82
90	0.781	0.575	1.05	0.78	0.525	0.376	1.10	0.80
120	0.753	0.542	1.02	0.73	0.508	0.347	1.07	0.73

4.5.1.4 Ion-Selective Electrodes

Work on sulphide sensitive electrodes was reported by Whitfield [4.169], Bock and Puff [4.170], and Shpeizer [4.172]. Al-Hitti et al. [4.171] carried out direct potentiometric titration of sulphide sulphur in natural water using standard cadmium sulphate solution as titrant and an electrode comprising silver deposited onto a carbon reed. The sensitivity of this procedure is 2 mg l^{-1} sulphide.

Guterman and Ben-Yaakov [4.173] determined total dissolved sulphide in the pH range 7.5–11.5 by ion-selective electrodes. Their instruments combine a sulfide ion activity electrode and pH glass electrode to produce a pH-independent output voltage that is proportional to total dissolved sulfide concentration.

Guleus [4.174] developed a portable monitor for measurement of total dissolved sulphide in natural water based on the glass/sulphide electrode couple.

Tse et al. [4.201] used an electrode with electrochemically deposited N, N', N'', N'''-tetramethyl tetra-3, 4-pyridimoporphydrazino cobalt for the detection of sulphide ions.

4.5.1.5 Flow Injection Analysis

Sonne and Dasgupta [4.198] have carried out a simultaneous determination of sulphide, polysulphide, sulphate, sulphite and thiosulphate.

Milosalvjevic et al. [4.204] have described a flow injection gas diffusion method incorporating an amperometric detector for the preconcentration and determination of trace amounts of sulphide in natural waters. Only cyanide interferes in this method.

Kuban et al. [4.200] used nitroprusside and methylene blue spectrophotometric methods for a silicone membrane differential flow injection determination of sulphide in water. In this method the hydrogen sulphide evolved from an acidified sample was collected in alkali prior to analysis. A detection limit of 2 µg l^{-1} was achieved by the nitroprusside procedure.

4.5.1.6 Gas Chromatography

Lech and Bagander [4.96] determined sulphide in aqueous solutions using gas chromatography with a flame detector.

4.5.1.7 Polarography

Differential pulse polarography has been used [4.175] to determine sulphide in natural waters. Jaya et al. [4.176] preconcentrated sulphide at concentrations ranging from 0 to 40 µg l^{-1} onto a silver electrode. The electrode was preconditioned at a constant negative potential of 1.2 V. The open circuit mode of the potentiostat was selected and the sample stirred for 10 min, after which the current measuring mode was selected immediately. Peak currents or charges passing were compared with those obtained from standard solutions. The procedure was precise and reliable with a similar sensitivity to, and better selectivity than, conventional cathodic-stripping volutammetry. Sulphide at $\mu g\, l^{-1}$ levels was determined in natural water samples, with a precision of 2%.

4.5.1.8 Ion Chromatography

Bond et al. [4.177] used ion chromatography to determine sulphide and cyanide in natural waters. Peschet and Tinet [4.178] used this procedure to codetermine sulphide, hypochlorite and cyanide in ground waters. These methods are discussed further in Chap. 11.

Goodwin et al. [4.179] determined traces of sulphide in turbid waters by gas dialysis ion chromatography. Sulphide quantitation was achieved in zinc preserved samples by this technique with a $1.9\,\mu g\, l^{-1}$ detection limit, a precision of $0.64\,\mu g\, l^{-1}$ and a recovery of 98–101%.

4.5.1.9 Miscellaneous Methods

Garber et al. [4.180] compared various methods for the determination of sulphide in waters. These included methylene blue spectrophotometric procedures, iodiometric titration, automated analysis and coulometric procedures.

4.5.2 Potable Water

4.5.2.1 Ion Chromatography

The ion chromatographic procedure described in Sect. 10.4.2 is applicable to the determination of sulphide in potable waters.

4.5.3 Industrial Effluents

4.5.3.1 Titration Methods

A sulphide ion-selective electrode has been used [4.181] as an indicator electrode in the potentiometric titration of sulphide in paper mill effluents with standard mercuric chloride. In 0.1 N sodium hydroxide medium, sulphide and polysulphide are titrated. As the electrode responds to mercuric ions, thiosulphate and sulphite can also be titrated at pH 7.0 to 7.5. The polysulphide plus sulphite can be converted to thiosulphate and can be titrated. Sulphite can be decomposed with formaldehyde and, consequently, thiosulphate can be titrated alone.

4.5.3.2 Gas Chromatography

Hawke et al. [4.182] described a method for the determination of sulphides in effluents using gas chromatography with a flame photometric detector. Samples were acidified and the hydrogen sulphide brought to solution vapour equilibrium. Following analysis by gas chromatography the concentration of hydrogen sulphide was determined by interpolation on a calibration graph. The precision, accuracy, bias and effect of potentially interfering substances were determined. The samples could be processed in 1 h by this method.

Knoery and Cutter [4.205] have described a method for the determination of sulphide and carbonyl sulphide in natural waters using specialised collection procedures followed by gas chromatography using flame photometric detector. The following species were determined: H_2S_{aq}, HS^-, S^{2-} and particulate metal sulphides. The sample was stripped with helium and the evolved gases analysed by gas chromatography Down to $0.2\,\rho\,mol\,l^{-1}$ of total dissolved sulphide could be determined.

4.5.3.3 Polarography

Kakiyama and Komatsu [4.183] evaluated sulphide in industrial effluents by measurement of the peak at -0.61 V vs the mercury pool anode in a sodium hydroxide base electrolyte containing EDTA as lead masking agent. Sulphite, thiosulphate, sulphate, thiocyanate, iodide, bromide or chloride do not interfere.

4.5.3.4 Ion Chromatography

Rocklin and Johnson [4.184] used ion chromatography to determine sulphide (and cyanide, iodide and bromide) in industrial effluents. This technique is discussed further in Sect. 11.4.

4.5.4 Waste Waters

4.5.4.1 Spectrophotometric Methods

Bhat et al. [4.185] reported a spectrophotometric method for the determination of low levels of sulphide (and sulphate and sulphite) in waste waters. The method involves the reduction of the bis-2,9-dimethyl 1-10 phenanthroline copper II ion by sulphide in the presence of formaldehyde.

Kuban et al. [4.200] applied the spectrophotometric methods employing nitroprusside and methylene blue, described in Sect. 4.5.1, to the determination of down to $2\,\mu g\,l^{-1}$ of sulphide in waste waters.

4.5.4.2 Ion-Selective Electrodes

Both sulphide-[4.186] and iodide-selective [4.187] electrodes have been used to determine sulphide in waste water. Papp and Havas [4.186] used a sulphide ion-selective electrode as an indicator electrode in the potentiometric titration of sulphide with mercuric chloride. Novkirishka et al. [4.187] used an iodide-selective electrode in an indirect potentiometric titration procedure to determine sulphide in industrial waste waters. Sulphide is first precipitated as cadmium sulphide by addition of cadmium acetate in an alkaline medium. The isolated cadmium sulphide is dissolved in excess standard iodine solution and excess iodine estimated by potentiometric titration using the iodide selective electrode.

4.5.5 Sewage

4.5.5.1 Ion-Selective Electrodes

Glaister et al. [4.188] studied three sulphide ion-selective electrodes in cascade flow and flow-through modes to investigate carrier stream, sample size and flow rate parameters in the analysis of sulphide in sewage effluents. Results were compared with those obtained by direct potentiometry. The electrodes were successfully used as detectors of sulphide during flow injection analysis and the presence of ascorbic acid in the standard antioxidant buffer minimized deleterious effects of hydrogen peroxide in the samples. Both the cascade flow and flowthrough modes of electrodes yielded sulphide concentrations similar to those obtained by colorimetric methods.

4.5.5.2 Draeger Tube Methods

Ballinger and Lloyd [4.189] have described a detailed procedure for the rapid determination of down to $0.06\,mg\,l^{-1}$ sulphides in sewage samples. Hydrogen sulphide is brought to solution-vapour equilibrium in a closed flask under controlled conditions. The concentration of hydrogen sulphide vapour is determined by means of Draeger tubes, and related to the concentration in solution by means of a calibration graph.

In aqueous solution, hydrogen sulphide dissociates as a weak diprotic acid:

$H_2S + H_2O \rightleftharpoons HS^- + H_2O^+$

$HS^- + H_2O \rightleftharpoons S^{2-} + H_3O^+$.

At a given pH, the degree of dissociation depends upon the temperature and activity of the sulphide species. At pH 5.0, dissolved sulphides are present almost entirely as undissociated hydrogen sulphide. It follows from Henry's law that the vapour pressure of hydrogen sulphide above its aqueous solution is a function of the mole fraction of dissolved hydrogen sulphide. In this method, hydrogen sulphide is brought to solution–vapour equilibrium in a closed flask, and the concentration of hydrogen sulphide vapour in the air space is determined by a conventional gas analyser. The pH of the solution is adjusted to 5.0 by means of a buffer solution and the ionic strengths of samples and standards are maintained at a constant level by the strong electrolytes incorporated into this solution.

Reagents. Chlorine free deionized or distilled water use throughout.

Sulphide stock solution (around $1000\,mg\,l^{-1}$ sulphide). Wash crystals of sodium sulphide nonahydrate with water and place about 8 g in a 1-l graduated flask. Dissolve and dilute to the mark with water, the pH of which has been adjusted to 9.0 with dilute sodium hydroxide. This solution should be standarized against N/80 iodine. Restandardize the sulphide solution before each batch of analyses and discard after 5 days.

Sulphide working solution (around $50\,mg\,l^{-1}$). Pipette an appropriate volume of stock sulphide solution into a 1-l graduated flask, containing water of pH 9.0. The pipette tip should dip below the water surface. Dilute to the mark with water of pH 9.0. Prepare the solution freshly as required.

Buffer solution, pH 5.0. Dissolve 140 g anhydrous disodium hydrogen phosphate (Na_2HPO_4), 106.5 g citric acid monohydrate and 80 g sodium chloride in water and dilute to 2 l in a graduated flask. This solution is stable for several months if stored in a dark glass bottle. Before analysing an unknown sample type, check the pH of the buffer sample mixture to ensure that a pH of 5.0 ± 0.2 has been achieved. If the pH deviates from these limits, increase the volume of buffer and decrease the volume of dilution water corresponding.

Apparatus. (Fig. 4.10). Draeger multi-gas detector pump-Model 21/31. Draeger hydrogen sulphide tubes $0-200\,mg\,l^{-1}$, catalogue reference 1/C (A). 1-l Pyrex conical quickfit flask, socket 34/35 (B). 35 mm silicone rubber bung, drilled in two places to give a tight fit to a 9 mm rod (C). 300 mm soda-glass tube 9 mm OD, 5 mm i.d. (D). Small rubber bung to plug (D). 60 mm PVC tube 10 mm o.d., 6 mm i.d. (E). Thread the tube through one of the holes in (C).

Procedure. Ensure that reagents and water for dilution are at the same temperature $\pm 1\,°C$. Without breaking the seal at either end of the Draeger tube, push it firmly into the gas sampling tube (E). Quickly transfer a volume (x ml) of

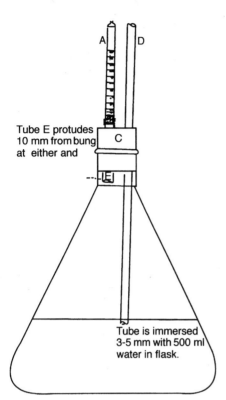

Fig. 4.10. Diagram of apparatus for determination of sulphide

sample to the flask containing 5 l water. Add samples by pipette, with the tip dipping below the surface of the dilution water. Immediately add 25 ml buffer solution and quickly push the bung into the flask. Plug tube (D) and shake the flask vigorously and reproducibly at 1 s intervals for 1 min (note 1 below). Remove the Draeger tube and soak up any liquid from tube (E) with a twist of absorbent paper. Break the seal at either end of the tube and replace it firmly in the gas sampling tube with the white band downwards. Remove plug from tube (D), expel the air from the gas detector pump and push the pump section firmly into the Draeger tube, ensuring that the pump is fully compressed immediately before release. Release the pump and support in position by hand until no more bubbles emerge from bottom of tube (C). Remove the Draeger tube and record the hydrogen sulphide concentration on the 0–200 mg l^{-1} scale (note 2 below). Ensure that no liquid has been sucked into the Draeger tube.

Note 1: use a stop-watch to time this operation. The shaking motion must cause the sample to foam and entrain air and should be carried out by flicking the wrist rather than with a vertical motion which may trap liquid in the gas sampling tube (E).

Note 2: the reading of this scale depends upon the ability of the analyst to detect the white/grey-brown boundary. Calibration and analysis should be carried out by the same analyst. Tubes should be read in the daylight and rotated to check whether the stain is distributed uniformly. It should be possible to estimate the scale reading to the nearest 5 units. The scale reading gives $mg\,l^{-1}$ hydrogen sulphide by volume.

Carry out the procedure described above using the following dilutions of sulphide working solution in water, within $\pm 1\,°C$ sample temperature.

Sulphide working solution (ml)	Sulphide free tap water (ml)
2.5	498
5.0	495
10.0	490
12.5	488

Construct a graph of $mg\,l^{-1}$ sulphide in solution against the reading on the $0.200\,mg\,l^{-1}$ Draeger scale. The concentration of sulphide in the reagents and dilution water are negligible and a blank correction is not normally necessary.

The above method determines undissociated hydrogen sulphide, and hydrosulphide and sulphide ions. To determine total sulphide, including above plus other compounds liberating hydrogen sulphide from a cold solution containing $0.5\,mol\,l^{-1}$ hydrochloric acid, follow the above procedure using 50 ml acid chloride solution instead of buffer solution.

The effect of various compounds likely to occur in sewage on results obtained by this method is shown in Table 4.21. The comparative freedom of interference from other sulphur-containing anions is of particular note. Sulphite, metabisulphite and thiosulphate do not interfere significantly at 100-fold excess; detergents depress the recovery of sulphide at 100-fold excess but do not interfere significantly at 10-fold excess.

4.5.6 Soil

Ramesh et al. [4.190] used an indirect method based on atomic absorption spectrometry for the determination of sulphide in flooded acid sulphate soils.

Hydrogen sulphide, evolved during the anaerobic metabolism of sulphate, is readily converted into insoluble metal sulphides, chiefly iron II sulphide, in flooded acid sulphate soils that are especially rich in iron.

The widely used method for determining sulphide is based on the precipitation of the sulphide in hydrogen sulphide as zinc sulphide and subsequent determination by methylene blue formation or iodine titrimetry. Ray et al. [4.191] have described an alternative simple method for determining sulphide involving the precipitation of zinc sulphide by the action of zinc on the hydrogen sulphide in flooded acid sulphate soils, and then indirect determination of

Table 4.21. Effect of Constituents and contaminants of sewage on a dissolved mg l^{-1} sulphide solution

Compound	Apparent % change in sulphide concentration		
	10 mg l^{-1}	100 mg l^{-1}	1000 mg l^{-1}
Nitrite N as sodium nitrate		<5%	−95%
Ammn.N as NH$_4$Cl		<5%	<5%
Sulphite SO$_3$ as Na$_2$SO$_3$7H$_2$O		<5%	<5%
Thiosulphate S$_2$O$_3$ as Na$_2$S$_2$O$_3$		<5%	+9.5%
Metabisulphite S$_2$O$_3$ as Na$_2$S$_2$O$_3$		<5%	−20%
Anionic detergents, as Manoxol-OT	<5%	−24%	
Acetone			<5%
Mineral oil			<5%
Methylated spirit			<5%
Chloroform			<5%
Effect of sewage organic strength (simulated by a dispersion of raw sausage meat in water)		Approx. COD 600 mg l^{-1} Oil & grease 300 mg l^{-1} SS 300 mg l^{-1}	Approx. COD 6000 mg l^{-1} Oil & grease 3000 mg l^{-1} SS 3000 mg l^{-1}
		−7%	−10%

^{a}In order to maintain pH 5.0 ± 0.2 it was necessary to add 100 ml buffer solution to overcome the alkalinity or sodium sulphite.

sulphide by determining the zinc in the precipitate and also the zinc remaining in solution, after the precipitation, by atomic absorption spectrometry.

These workers showed that about 85% of sulphide was recovered from sodium sulphide standards by both the iodometric titration method and their atomic absorption method. The latter was simple, rapid and reproducible with variations of less than 5% between replicates.

4.5.7 Environmental Samples

Jovanovic and Jovanovic [4.192] used a deposited iron wire silver-silver sulphide electrode to determine low levels of sulphide and cyanide in biochemical and environmental samples.

4.5.8 Blood

Lindell et al. [4.193] determined sulphide levels in blood using an ion selective electrode. The sulphide was first preconcentrated by trapping in sodium hydroxide solution.

4.5.9 Air

Wood and Marr [4.194] have described improvements in the ethylene blue spectrophotometric method for the determination of hydrogen sulphide in air.

4.6 Polysulphide

4.6.1 Natural Waters

Polysulphides can be determined in natural waters by titration with mercury II solutions. Sulphides, thiosulphates, sulphites and thiols interfere in this procedure [4.195].

Sonne and Dasgupta [4.198] have determined polysulphides, sulphides, sulphate, sulphite and thiosulphate by simultaneous photometric flow injection analysis.

4.6.2 Industrial Effluents

Borchardt and Easty [4.196] developed a gas chromatographic method for determining polysulphide in kraft pulping liquors. The polysulphide is decomposed to elemental sulphur in buffer solution. The elemental sulphur is derivatised with triphenylphosphine. The resulting triphenylphosphine sulphide is determined by flame ionization-gas chromatography.

4.6.3 Petroleum Fractions

Polysulphide has been determined in petroleum fractions by potentiometric analysis [4.197].

4.7 References

4.1 Ceausesin D (1968) Reta Chim 19: 676
4.2 Duvivier L (1985) Tribune du Cebedeau 38: 3
4.3 MacKellar WJ, Wiederanders RS, Tallman DE (1978) Anal Chem 50: 160
4.4 Pagenkopf GK, Brady W, Clampet J, Purcell MA (1978) Analytica Chimica Acta 98: 177
4.5 Basargin NN, Nagina AA (1969) Trudy Kem Analit Khim 17: 331
4.6 Enaki II, Nabivanets BI, Gidrobiol Zu (1972) 8: 124 Ref: Zhur Khim 19GD (4) Abstract No. 4G220 (1973)

4.7 Akos-Szabo Z, Inczedy J (1968) Magy Kem Lap 23: 528
4.8 Savvin SB, Akimova TG, Dedkova VP, Varshal GM (1969) Zhur Analit Khim 24: 1868
4.9 Duvivier B (1984) Tribune du Cebedeau 37: 175
4.10 Archer EE (1957) Analyst (London) 82: 208
4.11 Pakalns P, Farrer YJ (1979) Water Research 13: 991
4.12 Askne C (1977) Swedish Water and Air Pollution Research Laboratory Report No. B-372, 17 pp
4.13 Lambert JL, Manzo DJ (1971) Analytica Chimica Acta 54: 530
4.14 Jones PA, Stephen WI (1973) Analytica Chimica Acta 64: 85
4.15 Toei L, Miyata H, Yamawaki Y (1977) Analytica Chimica Acta 94: 485
4.16 Goguel E (1969) Analytical Chemistry 41: 1034
4.17 Bjarnborg B, Korhonen K (1985) Vatten 41: 36
4.18 Nasu T (1969) Japan Analyst 18: 1183
4.19 Henrikksen A, Paulsen IMB (1974) Vatten 30: 187
4.20 Berglund L, Henriksen A (1970) Laboratory Practice 19: 918
4.21 Reijnders HFR, Van Staden JJ, Griepink B (1979) Fresenius Zeitschrift fur Analytishe Chemie 298: 156
4.22 Davis JB, Lindstrom F (1972) Analytical Chemistry 44: 524
4.23 Fishmann MJ, Pascoe RF (1970) Prof Rap U.S Geological Survey Denver Colorado No. 700-C C222-225
4.24 Airey D, Dal Pont G, Sandars G (1984) Analytica Chimica Acta 166: 79
4.25 Coleman RL, Shults WD, Kelley MT, Dean JA (1972) Anal Chem 444: 1031
4.26 Bonda J, Michenkova E, Sponar J (1984) Vodni Hospodarstri 34: 195
4.27 Krug FJ, Filho HB, Zagatto EAG, Jorgensen SS (1977) Analyst (London) 102: 503
4.28 Slamna J, Mols JJ, Board JM, Van der Sloot HA, Van Raaphorst JG (1979) International Journal of Environmental Analytical Chemistry 7: 161
4.29 Van Raaphorst JG, Slanina J, Banger D, Lingerak WA (1977) National Bureau of Standards U.S. Special Publication 464. Proceedings of the 8th Symposium
4.30 Stephen WI (1970) Analytica Chimica Acta 50: 413
4.31 Nasu T, Kitagawa T, Mori T (1970) Japan Analyst 19: 673
4.32 Morgen EA, Vlasov NA, Tyutin VA (1969) Gidrokhim. Mater 50: 92 Ref. zhur Khim 19GD (23) Abstract No. 23G162 (1969)
4.33 Hori T, Sugiyama M, Himenko S (1970) Japan Analyst 19: 401
4.34 Kondo O, Miyata H, Toei K (1982) Anal Chim Acta 134: 353
4.35 Van Staden JF (1982) Fresenius Zeitschrift fur Analytische Chemie 310: 239
4.36 Krug FJ, Zagatto EAG, Reis BF, Bahia D, Jacintho AO, Jorgensen SS (1983) Analytica Chimica Acta 145: 179
4.37 American Public Health Association, American Water Works Association and Water Pollution Control Federation, Standard Methods for the Examination of Water and Wastewater, 14th edn., American Public Health Association, New York, 1975, p. 496
4.38 Nakashima S, Yagi M, Zenki M, Doi M, Toei K (1984) Fresenius Zeitschrift fur Analytische Chemie 317: 29
4.39 Pleskach LI, Chirkova GD, Fedorova ZV (1972) Zhur Analit Khim 27: 140
4.40 Little IP, Reeve R, Proud GM, Lulham A (1969) Journal of Science and Food Agriculture 20: 673
4.41 Siemer DD, Woodriff R, Robinson J (1977) Applied Spectroscopy 31: 168
4.42 Montiel A (1982) Trib Cebedeau 25: 292
4.43 Kokkenonen P, Palko M, Hajunen GHJ (1987) Atomic Spectroscopy 8: 98
4.44 Miles DL, Cook JM (1982) Analytica Chimica Acta 141: 207
4.45 Jasinski R, Trachtenberg J (1973) Analytical Chemistry 45: 1277
4.46 Ross JW, Fraub MS (1969) Analytical Chemistry 41: 967
4.47 Hulanicki A, Lewandowski R, Lewenstam A (1976) Analyst (London) 101: 939
4.48 Jones DL, Moody GS, Thomas JDR and Hangos M (1979) Analyst (London) 104: 973
4.49 Hara H, Horvai G, Pungor E (1988) Analyst (London) 113: 1817
4.50 Mascini M (1973) Analyst (London) 98: 325

4.51 Lukianets IG, Kulish NG, Zabolotskii VI and Meshechkov AI (1985) Soviet Journal of Water Chemistry and Technology 7: 42
4.52 Bocek P, Miedziak I, Demi M, Janak J (1987) Journal of Chromatography 137: 83
4.53 Zelinski I, Zelenska V, Kamienski D, Havasii P, Hednarova V (1984) Journal of Chromatography 299: 300
4.54 Stainton MP (1974) Limnology and Oceanography 19: 707
4.55 Markovarga G, Csiky I, Jousson JA (1984) Analytical Chemistry 56: 2066
4.56 Smee BW, Hall GEM, Koop DJ (1978) Journal of Geochemical Exploration 10: 245
4.57 Van Os MJ, Slanina J, De Ligny CL, Hammers WE, Agterdenbos J (1982) Analytica Chimica Acta 144: 73
4.58 Cronan CS (1979) Analytical Chemistry 51: 1340
4.59 Wilken RD, Kock HH (1985) Fresenius Zeitschrift fur Analytische Chemie 320: 477
4.60 Flanagan JD, Downie RA (1976) Analytical Chemistry 48: 2047
4.61 Ferus R, Torrades F (1985) Analyst (London) 110: 403
4.62 Ligon WV, Dorn SB (1985) Analytical Chemistry 57: 1993
4.63 Meehan BJ, Tariq SA (1973) Talanta 20: 1215
4.64 Department of the Environment of National Water Council Standing Committee of Analysts, HM Stationary Office, London. Methods for examination of waters and associated materials. Sulphate in waters, effluents and solids (1979) 378pp (R22A:B ENV) (1980)
4.65 Galloway JN, Likens GE, Hawley ME (1984) Science, 226: 829
4.66 Burns D, Thorburn CJS, Hayes WP, Kent DM (1974) Mikro Chimica Acta 2: 245
4.67 Suzuki K, Ohzeki K, Kamhara T (1981) Analytica Chimica Acta 136: 435
4.68 Reijnders HFR, Melis PHAA, Griepuik B (1983) Fresenuis Zertschrift fur Analytische Chemie 314: 627
4.69 Madsen BC, Murphy RJ (1981) Analytical Chemistry 53: 1924
4.70 Reijnders HFR, Van Staden JJ, Griepuik N (1980) Fresenius Zeitschrift fur Analytische Chemie 300: 273
4.71 Slanina I, Bakker F, Bruyn-Hes A, Mols JJ (1980) Analytica Chimica Acta 113: 331
4.72 Reijnders HFR, Melis PHA, Grie B (1983) Fresenius Zeitschrift fur Analytische Chemie 314: 627
4.73 Wagner GH, Steel KT (1982) American Laboratory, 12 July
4.74 Faigle W, Klockow D (1981) Fresenius Zeitschrift für Analytische Chemie 306: 190
4.75 Pyen GS, Erdmann DE (1983) Analytica Chimica Acta 149: 355
4.76 Wetzel RA, Anderson CL, Schleicher H, Crook DG (1979) Analytical Chemistry 51: 1532
4.77 Jones UK, Tartar JG (1985) International Laboratory, september.
4.78 Schwabe R, Darimont T, Mohlman T, Pabel M, Sonneborn M (1983) Journal of Environmental Analytical Chemistry 14: 169
4.79 Yamamoto M, Yamamoto H, Yamamoto Y, Zatsuchita S, Baba N, Ikeshiga T (1984) Analytical Chemistry, 56: 822
4.80 Bucholz AE, Verplough CJ, Smith JL (1982) Journal of Oceanographic Science 20: 499
4.81 Oikawa K, Saitoh H (1982) Chemosphere 11: 933
4.82 Wagner F, Valenta P, Nurnberg W (1985) Fresenius Zeitschrift fur Analytische Chemie 320: 470
4.83 Rowland AP (1986) Analytical Proceedings Chemical Society (London) 23: 308
4.84 Matsuchita S, Tada Y, Baba N, Hosako J (1983) J Chromatography 259: 459
4.85 Regnet W, Quentin KE (1981) Zeitschrift fur Wasser und Abwasser, Forschung 14: 106
4.86 Van Staden JF (1982) Fresenius Zeitschrift fur Analytische Chemie 312: 438
4.87 Siemer DD, Woodriff R, Robinson J (1977) Applied Spectroscopy 31: 168
4.88 Schwabe R, Darimont T, Mohlman T, Pabel M, Sonneborn M (1983) Journal of Environmental Analytical Chemistry 14: 169
4.89 Kotik FI (1971) Zavod Lab 37: 541
4.90 White WW, Henry MC (1972) Plating 59: 429
4.91 Petrova EI (1971) Zhur Analit Khim 26: 402

4.92 Pilipenko AT, Maksimenko TS, Terletskaya AV (1983) Soviet Journal of Water Chemistry and Technology 5: 73
4.93 Bhat SR, Eckert JM, Gibson NA (1983) Water (Australia) 10: 19
4.94 Mosko J (1984) Analytical Chemistry 56: 629
4.95 Srivastava SK, Jain CK (1985) Water Research 19: 53
4.96 Lech C, Bagander LE (1988) Analytical Chemistry 60: 1680
4.97 Brunt K (1985) Analytical Chemistry 57: 1338
4.98 Tanaka K, Ishizuka T (1982) Water Research 16: 719
4.99 Green LW, Woods JR (1981) Analytical Chemistry 53: 2187
4.100 Roberts FM, Gjerde DT, Fritz JS (1981) Analytical Chemistry 53: 1691
4.101 Tretter H, Paul G, Blum I, Schreck H (1985) Fresenins Zeitschrift fur Analytische Chemie 321: 650
4.102 Chauhan PPS, Chanhan CPS (1979) Soil Science 128: 193
4.103 Al-Ghabsha TS, Bogdanski SL, Townshend A (1980) Analytica Chemica Acta 120: 383
4.104 Hardijk CA, Capponburg TE (1985) Journal of Microbiological Methods 3: 205
4.105 Krug FJ, Zagatto EAG, Reis BF, Bahia FOO, Jacintho AO, Jargensen SS (1983) Analytica Chimica Acta 145: 179
4.106 Krug FJ, Bergamin FOH, Zagatto EAG, Jorgenson SS (1977) Analyst (London) 102: 503
4.107 Little IP, Reeve R, Proud GM, Lulhan AJ (1969) Science of Food and Agriculture 20: 673
4.108 Bradfield EG, Cooke DT (1985) Analyst (London) 110: 1409
4.109 Ogner G, Hangen A (1977) Analyst (London) 102: 453
4.110 Basargin NN, Men'shikova NL, Belova VS, Myassishcheva LG (1968) Zh Analit Khim 23: 732
4.111 Bruno P, Caselli M, DiFano A, Traini A (1979) Analytica Chimica Acta 104: 379
4.112 Dryssen D, Wedborg M (1986) Analytica Chimica Acta 180: 473
4.113 Haskins JE, Kendall H, Baird RB (1984) Water Research 18: 751
4.114 APHA Standard Methods for the Examination of Water and Wastewater, 15th edn. p. 451, APHA Washington D.C. (1980)
4.115 Badri B (1988) Analyst (London) 113: 351
4.116 Sarantonis EG, Koukli II, Calokerinos AC (1988) Analyst (London) 113: 603
4.117 Al-Tamrah SA, Townshend A, Wheatley AR (1987) Analyst (London) 112: 883
4.118 Papp J, Havas J (1970) Magy Kem Foly 76: 307
4.119 Migneault DR (1989) Analytical Chemistry 61: 272
4.120 Hansen LD, Richter BE, Rollins DK, Lamb JD, Eatough J (1979) Analytical Chemistry 51: 633
4.121 Sunden T, Lingrom M, Cedergren A (1983) Analytical Chemistry 55: 2
4.122 Oikawa K, Saitoh H (1982) Chemosphere 11: 933
4.123 McCormick MJ, Dixon LM (1985) Journal of Chromatography 322: 478
4.124 Bhat SR, Eckert JM, Gibson NA (1983) Water (Australia) 10: 19
4.125 Halcombe DJ, Meserole VT (1981) Water Quality Bulletin 6: 37
4.126 Stevens JS, Turkelson V-T (1977) Analytical Chemistry 49: 1176
4.127 Tecator Ltd. UK. Application Notes ASN 61-23/83 (1983) Determination of total sulphite in wine by flow injection analysis and gas diffusion and ASN 61-22/84 (1984) Determination of free sulphite in beverages by flow injection analysis and gas diffusion.
4.128 BohringerMaunheim, Lewes, Sussex, UK. Methods of Enzymic Food Anaysis. P 57 Determination of sulphite (1984)
4.129 Tecator Ltd, Box 70, S-26321, Hoganes, Sweden. Application Note AN 90/87. Determination of total sulphite in foods using the 1026 distilling unit (1987)
4.130 Tecator Ltd. Box 70, S-26321, Hoganes, Sweden, Application Note AN 61/83. Determination of sulphite in liquid samples by flow injection analysis and gas diffusion (1983)
4.131 Tecator Ltd, Box 70, S-26321, Hoganes, Sweden, Application Note. ASTN 44/88. Determination of sulphite in food and drinks by flow injection analysis (1988)
4.132 Smith UJ (1987) Analytical chemistry 59: 2256
4.133 Brown DS, Jenk DR (1987) Analyst (London) 112: 899
4.134 Papp J, Havas J (1970) Magy Kem Foly 76: 307
4.135 Dryssen D, Wedborg M (1986) Analytica Chimica Acta 180: 473

4.136 Badri B (1988) Analyst (London) 113: 351
4.137 Konig WA (1985) Fresenius Zeitechrift fur Analytische Chemie 322: 581
4.138 Sunden T, Lingron M, Cedergren A (1983) Analytical Chemistry 55: 2
4.139 Makhija R, Hitchen A (1978) Talanta 25: 79
4.140 Chakraborty D, Das AK (1988) Atomic Spectroscopy 9: 115
4.141 Takano B, McKibben MA, Barnes HL (1984) Analytical Chemistry 56: 1594
4.142 Nagashima S, Ozawa T (1981) International Journal of Environmental Analytical Chemistry 10: 99
4.143 Nota G, Miraglia VR (1981) Improtac Acampora, A Journal of Chromatography 207: 47
4.144 Imperial Chemical Industries Ltd. Effluent and Water Treatment Journal 17: 77 (1977)
4.145 Agrawal YK, Bhatt PM (1987) Analyst (London) 112: 1767
4.146 Hassan SSM, Elmosalamy MAAF (1987) Analyst (London) 112: 1709
4.147 Stetzenback K, Thompson GM (1983) Groundwater 21: 36
4.148 Thieleman H (1970) Mikoochim Acta 3: 645
4.149 Atkinson GF, Byerley JJ , Mitchell BJ (1982) Analyst (London), 107: 398
4.150 Abramkina TV, Kerov VA, Pigulevskii EV (1972) Trudy Leningr Inst Kinoinzhenerov 19: 158 Ref. Zhur Khim 19GD (21). Abstract No. 21G145 (1972)
4.151 Luthy RG, Bruce SG, Walters RW, Nakles DV (1979) Journal of Water Pollution Control Federation 51: 2267
4.152 Botto RI, Karchmer HH, Eastwood MW (1981) Analytical Chemistry 53: 2375
4.153 Nota G (1975) Analytical Chemistry 47: 763
4.154 Thieleman H (1970) Pharmazie 25: 271
4.155 Wronski M (1971) Anal Chem 43: 606
4.156 Dyrssen D, Wedborg M (1986) Analytica Chimica Acta 180: 473
4.157 Nasson SS, Bok LDC, Grobler SR (1974) Z Analyt Chem 268: 287
4.158 Grasshoff KM, Chau KM (1971) Analytica Chimica Acta 53: 442
4.159 Cline JD (1969) Liminology and Oceanography 14: 455
4.160 Bethea NJ, Bethea RM (1972) Analytica Chimica Acta 61: 311
4.161 Caspieri P, Scott R, Simpson EA (1969) Analytica Chimica Acta 45: 547
4.162 Rahim SA, Salim AY, Shereef S (1973) Analyst (London) 98: 851
4.163 Bhat SR, Eckert JM, Geyer R, Gibson NA (1980) Analytica Chimica Acta 114: 293
4.164 Jars FV, Munoz Leyva JA (1982) Analyst (London) 107: 781
4.165 Ceba RM, Leyva JAM, Jana FV (1982) Analyst (London) 107: 781
4.166 Erdey L (1965) Gravimetric analyses Part 2, 1st edition, Pergamon Press, Oxford
4.167 Bark LS, Dixon A (1970) Analyst (London) 95: 786
4.168 Kokkonen P, Palko M, Lajunen LHJ (1987) Atomic Spectroscopy 8: 98
4.169 Whitfield M (1971) Limnology and Oceanography 16: 829
4.170 Bock R, Puff HJ (1968) Z Analyst Chem 240: 381
4.171 Al-Hitti IK, Moody GJ, Thomas JDR (1983) Analyst (London) 108: 43
4.172 Shpeizer GM, Gidrokhim (1970) Mater 53: 97 Ref Zhur Khim 19GD (3) Abstract No. 3G190
4.173 Guterman H, Ben-Yaakov S (1983) Anal Chem 55: 1731
4.174 Gulens J (1985) Water Research 19: 201
4.175 Leppinen J, Vahlila S (1986) Talanta 33: 795
4.176 Jaya S, Rao P, Rao GP (1986) Analyst (London) 111: 717
4.177 Bond AM, Heritage ID, Wallace AG, McCormick MJ (1982) Analytical Chemistry 54: 582
4.178 Peschet JL, Tinet C (1986) Techniques Sciences Methods 81: 351
4.179 Goodwin LR, Franscom D, Urso A, Dieken FP (1988) Analytical Chemistry 60: 216
4.180 Garber WR, Nagano J, Wada FN (1970) Water Pollution Control Federation 42: R209
4.181 Papp J, Havas J (1970) Magy Kem Foly 76: 307
4.182 Hawke DJ, Lloyd A, Martinson DM, Slater PG, Excell C (1985) Analyst (London) 110: 269
4.183 Kakiyama H, Komatsu M (1970) Japan Analyst 19: 902
4.184 Rocklin RD, Johnson EL (1983) Analytical Chemistry 55: 4
4.185 Bhat SR, Eckert JM, Gibson NA (1983) Water (Australia) 10: 19
4.186 Papp J, Havas J (1970) Magy Kem Foly 76: 307

4.187 Novkirishka M, Michailov G, Christova R (1981) Fresenius Zeitschrift fur Analytische Chemie 305: 411
4.188 Glaister MG, Moody GJ, Thomas JDR (1985) Analyst (London) 110: 113
4.189 Ballinger D, Lloyd A (1981) Water Pollution Control 80: 648
4.190 Ramesh C, Ray PK, Nayar AK Misra O, Sethunathen N (1980) Analyst (London) 105: 984
4.191 Ray RC, Nayar PK, Sethunathen N (1980) Analyst (London) 105: 984
4.192 Jovanovic VM, Jovanovic MS (1988) Analyst (London) 113: 71
4.193 Lindell H, Tappinen P, Savolainen H (1988) Analyst (London) 113: 839
4.194 Wood CF, Marr IL (1988) Analyst (London) 113: 1635
4.195 Dryssen D, Wedborg M (1986) Analytica Chimica Acta 180: 473
4.196 Borchardt LT, Easty DB (1984) Journal of Chromatography 299: 471
4.197 Farroha SM, Habboush AE, Saffo FF, Abdul-Nour MH (1987) Analyst (London) 112: 1173
4.198 Sonne K, Dasgupta PK (1991) Analytical Chemistry 63: 427
4.199 Singh RP, Pambid ER, Abbas NM (1991) Analytical Chemistry 63: 1897
4.200 Kuban V, Dasgupta PK, Marx JN (1992) Analytical Chemistry 64: 36
4.201 Tse Y, Hong, Janda P, Lever ABP (1994) Analytical Chemistry 66: 384
4.202 Schneide EP, Durst RA (1977) Analytical Letters 10: 55
4.203 Petrie LM, Jakely ME, Brandwig RL, Kroenings JG (1993) Analytical Chemistry 65: 952
4.204 Milosalvjevic EB, Solujie L, Hendrix JL, Nelson JH (1988) Analytical Chemistry 60: 2791
4.205 Knoery J, Cutter GA (1993) Analytical Chemistry 65: 976
4.206 Gonzalez V, Moreno B, Sicilia D, Rubio S, Perez-Bendito D (1993) Analytical Chemistry 65: 1897
4.207 Cox JA, Gray T, Kulkarni KR (1988) Analytical Chemistry 60: 1710

5 Phosphorus Containing Anions

5.1 Phosphate

5.1.1 Natural Waters

5.1.1.1 Spectrophotometric Methods

Skjemstad and Rieve [5.1] have reported three automated procedures for the simultaneous determination of phosphate, ammonia and nitrate plus nitrite at $\mu g\ l^{-1}$ levels in natural waters. Phosphate is estimated by reaction with molybdate and reduction to molybdenum blue with ferrous ammonium sulphate. Ammonia is determined by the salicylate/dichloroisocyanurate reaction in the presence of nitroprusside after in-line distillation. Nitrate plus nitrite is estimated by using an in-line copperized cadmium reductor, diazotizing the nitrite with sulphanilamide and coupling with N-l-naphthylethylenediamine. Mercuric chloride used as a preservative does not interfere in these procedures.

Throughout the methods described below, deionized water with a conductivity of less than $1\mu s$ (micro siemen) was used for standards, reagents, and rinsing of glassware. All sample tubes and glassware were washed with "BRIJ-35" detergent each time before use and soaked in 5 N hydrochloric acid at frequent intervals. All samples were filtered through a Millipore 0.45 µm filter before analysis.

Their determination of phosphate utilized the manifold and flow sequence are illustrated in Fig. 5.1.

Reagents. As follows.

Ammonium molybdate. Dissolve 33.7 g of ammonium molybdate in 800 ml of water. Add 493 ml of concentrated sulphuric acid to 1 l of water and when cool add to the molybdate and make up to 2 l.

Ferrous ammonium sulphate. Dissolve 67.5 g of ferrous ammonium sulphate in 400 ml of boiled water. Add 10 ml of 10 N sulphuric acid and make up to 500 ml and transfer to the ferrous ammonium sulphate container illustrated in Fig. 5.1. Prepare a fresh solution each day.

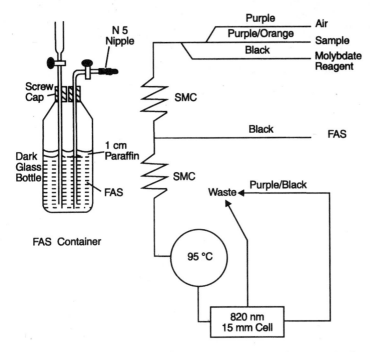

Fig. 5.1. Flow sequence for automated determination of phosphate (From [5.1])

Standard phosphate solution. Make 4.39 g KH_2PO_4 upto 1 l (1 ml = 1000 µgP). From this solution prepare a range of working standards in the range 20 – 1000 µg ml^{-1} of phosphorus.

Their determination of ammonia utilized the manifold and flow sequence given in Fig. 5.2.

Reagents. As follows.

Sodium salicylate. Dissolve 4.3 g of sodium hydroxide in approximately 180 ml of water and add 14.7 g of salicylic acid. When dissolved, add 0.12 g of sodium nitroprusside and adjust to pH 5.2 with sodium hydroxide solution or solid salicylic acid and make to 200 ml. It is advisable to prepare a fresh solution each day.

Dichloroisocyanurate. Dissolve 1 g of commercial sodium dichloroisocyanurate (62% available chlorine) in 100 ml of water and add to 100 ml of 1.2 N sodium hydroxide.

Boric acid. Make 2.06 g boric acid upto 1 l.

Sodium carbonate, 16 g l^{-1}

Standard ammonia solution. Dissolve 4.74 g of dried ammonium sulphate in 1 l of water (1 ml = 1000 µg HN_3 — N). From this solution prepare a set of working standards in the range of 20 to 500 µg l^{-1} of N.

The determination of nitrate plus nitrite utilized a manifold and flow sequence based on a combination of two published procedures. An in-line cad-

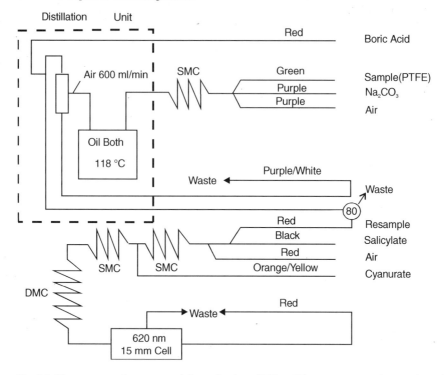

Fig. 5.2. Flow sequence for automated determination of $NH_3 - N$ ($2 \times$ range expansion, 2 min wash, 2 min sample) (From [5.1])

mium reducing column buffered to pH 8.6 as recommended by Henriksen and Selmer-Olsen [5.2] reduces nitrate to nitrite, which is then determined by a colorimetric finish with a flow sequence and reagent concentration identical to one described by Terey [5.3].

Phosphorus in natural waters can exist in a multiplicity of forms, including ortho-, pyro-, poly-, meta-, organic, colloidal, and suspended phosphorus. Each of these forms is capable of being measured either partly or wholly (depending upon the reaction conditions) as orthophosphate through hydrolysis and reaction with molybdate. Strickland and Parsons [5.4] therefore attempted to define a labile fraction of phosphorus (soluble reactive phosphorus) as the amount of phosphorus able to pass a 0.45 μm Millipore filter and reacting within 3 min with acid molybdate. This fraction is not equivalent to orthophosphate, nor dissolved inorganic phosphorus, but rather some ill-defined intermediate.

It becomes evident, therefore, that methods will measure differing proportions of the total phosphorus depending upon the reaction conditions employed and whether all or some of the above-mentioned phosphorus-containing fractions are present in any one sample.

The ability of the phosphate manifold to recover added phosphorus was tested to a maximum level of 400 µg l^{-1}. The resultant linear response indicates that the method is able to recover orthophosphate in the presence of other elements normally present which may cause interference. Adding Hg(II) showed no effect to a level of 50 mg l^{-1}. Similarly no interference was recorded from silica added as sodium silicate up to a level of 100 mg l^{-1} SiO$_2$.

The excellent precision of this manifold at low levels of phosphate phosphorus is illustrated in the statistical analysis given in Table 5.1. The overall relative standard deviation was 1% and there were no significant differences ($P = 0.01$) between days or in the sample × day interactions. There is no doubt that this method is very precise even at low levels of phosphate phosphorus.

Rickford and Willett [5.5] showed the filtration of water samples through Gelman membranes which contained wetting agents prior to analysis for phosphate caused interferences (low phosphate results) in molybdenum blue spectrophotometric methods for phosphate. This effect was due to the release of some substance from the membrane. It is recommended that low extractable membranes are used when the filtrate is required for the determination of phosphate or inorganic phosphorus.

Pakalns and Steman [5.6] have discussed the effect of surfactants on the spectrophotometric determination of phosphate by direct and solvent extraction procedures. For direct methods cationic detergent caused considerable interference, but there was little interference by biodegradable anionic detergents. The ascorbic acid method is recommended for routine determination of phosphate in most waters. For the extraction procedure the levels of cationic and non-ionic detergents must not exceed 2 and 10 mg l^{-1} respectively. The maximal level of anionic detergents should be 8 mg l^{-1}.

Stainton [5.7] has discussed errors in molybdenum blue methods for determining orthophosphate in natural waters. He suspected that the release of

Table 5.1. A Precision of PO$_4$-P determinations. (From [5.1]; Copyright 1978, American Society of Agronomy Inc.)

Sample	Day 1 µg l^{-1}	Day 2 (mean of 10)	Day 3	Sample mean
1	14.2	14.6	14.3	14.4
2	263.5	263.5	264.3	263.7
3	9.4	9.6	9.4	9.5
4	21.5	21.5	21.1	21.4
5	100.4	100.8	100.7	100.6
Day mean	81.8	82.0	82.0	81.9
L.S.D. between			$P = 0.05$	$P = 0.01$
sample means			0.4	0.6
day means			0.3	0.5
Any pair of sample x day means			0.8	1.0
Relative SD				1.0%

orthophosphate from colloidal phosphorus was the source of such errors as had been shown by previous workers, and he therefore studied the effect of a variety of the more popular molybdenum blue methods on the colloidal phosphorus fraction. All variations in the method caused a significant loss of phosphorus from the fraction. Phosphorus originating from the colloidal fraction tested as orthophosphate gave an overestimate of the true orthophosphate present.

Downes [5.8] gives details of an automated method for determination of reactive phosphorus in concentrations lower than 0.5 mg l^{-1} in natural waters which also contained arsenic, silicon and mercuric chloride preserving agent. The addition of thiosulphate in acid solution removes various interferences.

Sulphide interference in the determination of phosphate by molybdenum blue methods has been discussed [5.9].

Lei et al [5.10] give details of a procedure for the determination of traces of phosphorus in natural waters by absorption spectrometry, using a long capillary cell to lower the limit of detection.

Problems with standard methods for the determination of phosphate by standard molybdenum blue methods has prompted development of a method based on the use of hydrazine – molybdic acid reagent [5.11].

Goulden and Brooksbank [5.12] have described an automated method for eliminating arsenate interference in the determination of phosphate in natural waters by the molybdenum blue method. The arsenate is reduced to arsenite by thiosulphate in an acidic medium before the colour-producing reagents are added.

Harmsen [5.13] used derivative spectrophotometry to determine phosphate in turbid water samples using the conventional molybdenum blue spectrophotometric procedure. Turbidities less than 0.8 did not interfere with adsorbance. Precautions were needed for phosphorus concentrations higher than 3 mg l^{-1}. Nevertheless recovery tests were accurate to within 4%.

Potman and Lijkema [5.14] have discussed the following determination of down to 0.4 µg l^{-1} phosphate in small samples of borehole and interstitial water.

Reagents. As follows.

Molybdate solution. Dissolve 1.157 g $(NH_4)_6 MO_7 4H_2O$ in 250 ml of water.

Hydrochloric acid, concentrations 3 N, 2 N and 0.4 N. Sodium hydroxide, 0.1 N.

Diethyl ether: isobutanol (5:1); (A.R.) freshly distilled and stored in quartz bottle.

Ammonium thiocyanate solution, 5 g ammonium thiocyanate in 100 ml water.

Copper sulfate. Dissolve 0.04 g $CuSO_4 \cdot 5H_2O$ and add 1 ml of sulphuric acid to 1000 ml water.

Phosphate standards. Prepared from potassium dihydrogen phosphate dried at 110 °C

Apparatus. Spectrophotometer PMQ II, Zeiss, or equivalent. 100 ml separation funnels with Teflon stopcock, polyethylene stopper and a short delivery tube cut off at a 6° angle. Glassware steamed and rinsed with distilled water prior to use.

Procedure. Mix in a separation funnel 5 ml 2 N hydrochloric acid, 6 ml molybdate solution and at most 25 ml of the sample. If necessary fill up with water to 25 ml. After 10 min add 25 ml of the ether:isobutanol solvent and shake for 1 min. Remove the aqueous phase carefully by swinging the separation funnel after removal of the bulk of the water layer in order to drain away drops adhering to the wall and stopper. Wash the organic solution twice with 20 ml 0.4 N hydrochloric acid for 1 min and remove the aqueous phase carefully. Shake the ether solution for 1 min with 10 ml 0.1 N sodium hydroxide, separate the alkaline solution completely (see step 3) and repeat with 5 ml 0.1 N sodium hydroxide, shaking for about 30 s.

Mix the combined alkaline solutions with 11 ml of the copper sulphate solution, 8.5 ml concentrated hydrochloric acid, 7 ml acetone, 10 ml of the 5% ammonium thicyanate solution and fill up with 3 N hydrochloric acid to 50 ml. Mix thoroughly. Measure the absorbance after 45 min at 470 nm in a 1 cm cell, comparing with standards after correction for the blank.

Iron and silicate interfere in the procedure but only at higher concentrations (Table 5.2), i.e. a 1000-fold excess of iron and a 100-fold excess of silica do not interfere.

The only alternative spectrophotometric methods for the determination of phosphate to those based on molybdenum blue appear to be those which utilize malachite green as the chromogenic reagent [5.15,5.16]. Fernandez et al. [5.15] have discussed an automated version of the malachite green micromethod for determining inorganic phosphate in natural waters. The molar absorptivity of the technique was six times more sensitive than that of molybdenum blue methods. Reproducibility was very good and interferences from silicate and arsenate were easily removed.

Motomizu et al. [5.17] have described a spectrophotometric method for the determination of phosphate in river waters based on the measurement at 650 nm of the colour produced with molybdate and malachite green. The limit of detection is in the μg l^{-1} range. In this method the sample solution was

Table 5.2. Interference of iron and silica in phosphate determination. (From [5.13]; Copyright 1984, Elsevier Science Publishers BV, Netherlands)

μgP in sample	0.34	0.34	0.34	0.34	0.34
μgFe in sample	0	170	255	340	510
Abs. (corrected)	0.157	0.158	0.159	0.182	0.189
μgP recovered	0.34	0.34	0.35	0.40	0.41
μgP in sample	0.34	0.34	0.34	0.34	0.34
μg Si in sample	0	17	25.5	34	51
Abs. (corrected)	0.163	0.168	0.161	0.168	0.205
μgP recovered	0.34	0.35	0.34	0.35	0.43

acidified with sulphuric acid and heated in a water bath above 90°C for 40 min and subsequently it was coloured with molybdate and malachite green. The colour was stabilised by adding poly (vinyl alcohol).

Reagents. As follows.

Malachite green solution, 2×10^{-3} mol l^{-1}. Dissolve commercially available malachite green (oxalate), 0.91 g in distilled water to give 1 l of solution.

Molybdate solution, 0.68 mol l^{-1}. Dissolve ammonium molybdate, $(NH_4)(Mo_7O_{24}) \cdot 4H_2O$ (Nakarai Chemicals) (120 g) in distilled water to give 1 l of solution.

Standard phosphate solution. Dry potassium dihydrogen orthophosphate at reduced pressure (about 5 mm Hg) at 60°C to constant weight. Dissolve the dried compound (0.2722 g) in distilled water to give 1 l of solution (2×10^{-3} mol l^{-1}). For calibration purposes, dilute the solution to 4×10^{-5} mol l^{-1} with distilled water before use; the diluted solution contains 1.24 µg ml^{-1} of phosphorus and 0–4 ml of this solution were used.

Polyvinyl alcohol solution. Dissolve commercially available polyvinyl alcohol (number average degree of polymerisation 500) (1 g) in 100 ml of hot water and use after filtration through filter-paper.

Sulphuric acid. Commercially available concentrated sulphuric acid (97%, 18.2 mol l^{-1}).

Prepare the reagent solution by mixing 300 ml of 0.68 mol l^{-1} molybdate solution, 47 ml of concentrated sulphuric acid and 250 ml of 2×10^{-3} mol l^{-1} malachite green solution. About 30 min after mixing, filter the solution through a membrane filter (pore size 0.45 µm). Use the filtrate as the reagent solution.

The procedure for orthophosphate is as follows. After sampling a test water in a glass bottle, acidify it with sulphuric acid to pH 2–3 and, if necessary, filter it through a membrane filter (pore size 0.45 µm). Transfer up to 20 ml of the acidified sample solution, containing up to 5 µg of phosphorus, into a 25-ml calibrated flask and dilute to 20 ml with distilled water. Add 1 ml of 7.5 mol l^{-1} sulphuric acid. Add 3 ml of the reagent solution. Within 2 min of mixing the sample and reagent solution, add 0.5 ml of PVA solution, dilute to the mark with water and mix. Measure the absorbance at 650 nm within 2 h.

For total inorganic phosphate, the procedure is as above except that after adding 1 ml 7.5 mol l^{-1} sulphuric acid the solution is heated above 90°C for 40 min then cooled to room temperature.

The maximum absorption wavelength is 650 nm and at this wavelength the calibration graph was linear for amounts of phosphorus from 0 to 5 µg. The molar absorbtivity was 7.8×10^4 mol^{-1} cm^{-1}.

The precision of the method was evaluated by the above procedure without pretreatment, by determining the same amounts (1.55 µg) of phosphorus in ten experiments. The results obtained are shown in Table 5.3. The standard deviation of the absorbance of the reagent blank is less than 0.002 absorbance unit, which corresponds to 0.009 µg of phosphorus. The standard deviation and relative standard deviation for 1.55 µg phosphorus are less than 0.005 absorbance unit and 1.1% respectively.

Table 5.3. Standard deviations (SD) and relative standard deviations (RSD) for the reagent blank and for the determination of phosphorus. (From [5.17]; Copyright 1983, Royal Society of Chemistry, UK)

Sample	Parameter	Time after preparation of reagent solution			
		1 h	1 day	2 days	4 days
Reagent blank	$\bar{x}^a$	0.022 ± 0.004	0.023 ± 0.002	0.022 ± 0.004	0.021 ± 0.001
	SD	0.0023	0.0012	0.0017	0.0008
	RSD, %	10.4	5.1	7.6	3.7
Phosphorus (1.55 μg)	$\bar{x}^b$	0.330 ± 0.009	0.327 ± 0.006	0.327 ± 0.006	0.320 ± 0.005
	SD	0.0036	0.0036	0.0027	0.0026
	RSD, %	1.1	1.1	0.8	0.8

[a] Average of the absorbances of the reagent blanks in 10 experiments. Reference: water.
[b] Average of the absorbances of the phosphorus (1.55 μg) in 10 experiments. Reference: reagent blank.

Table 5.4. Standard deviations (SD) and relative standard deviations (RSD) for the determination of phosphorus in waters. Phosphorus was determined with pre-treatment. (From [5.17]; Copyright 1983, Royal Society of Chemistry, UK)

	Water sample		
	Tap water[a]	Asahi River[b]	Zasu River[c]
Volume taken/ml	20	20	10
Average absorbance[c]	0.030 ± 0.002	0.052 ± 0.5	0.389 ± 0.006
Average phosphorus content, p.p.b.[d]	9.5 ± 0.4	12.6 ± 0.5	188.3 ± 2.3
SD	0.35	0.42	4.6
RSD, %	3.7	3.3	2.4

[a] Sampled on August 24th, 1982.
[b] Sampled on August 21st, 1982.
[c] Reference: reagent blank.
[d] From ten experiments.

Table 5.4 gives the results for the standard deviation and relative standard deviation in ten experiments on real water samples. The relative standard deviations are less than 4%. The results of recovery tests in Table 5.5 show that the recoveries ranged between 95.0 and 101.3%.

The interference of diverse ions was examined and the results obtained are shown in Table 5.6. The amounts of ions, except silicate, generally present in river water are much smaller than those listed in Table 5.6. Silicate ion at concentrations above 5×10^{-5} mol l^{-1} reacts gradually with molybdate to form heteropolyacid and this results in positive errors, but large amounts of silicate ion were eliminated by acidifying the sample solution and filtering in through a membrane filter (pore size 0.45 μm).

Recently, eutrophication in lakes and reservoirs has become an important problem. Phosphorus is a critical nutrient aiding the eutrophication process. The principal sources of phosphorus in environmental waters are agricultural effluents, municipal waste waters, industrial waste waters, etc.

Table 5.5. Recovery test. Phosphorus was determined with pre-treatment. (From [5.17]; Copyright 1983, Royal Society of Chemistry, UK)

	Water sample		
	Tap water[a]	Asahi River	Zasu River
Volume taken/ml	17	17	10
Phosphorus added/µg[a]	0.078	0.078	0.78
Absorbance[b]	0.192 ± 0.002	0.205 ± 0.002	0.530 ± 0.003
Recovery, %[c]	100.3	101.2	95.0

[a] Potassium dihydrogen orthophosphate was added.
[b] Reference: reagent blank.
[c] Average of five experiments.

Table 5.6. Effect of diverse ions. (From [5.17]; Copyright 1983, Royal Society of Chemistry, UK)
Phosphorus taken: 3.72 µg

Ion	Tolerance limit[a]
HCO_3^-	6 mg
Al^{3+}	3 mg
$Fe^{2+}, CO^{2+}, Ni^{2+}, Cu^{2+}, Zn^{2+}, Ca^{2+}, K^+, NO_3^-$	1 mg
Mg^{2+}, Na^+, NH_4^+	0.5 mg
Si(IV)	15 µg
ClO_4^-	5 µg
W(VI)	4 µg
As(V)	2 µg
V(V)	1 µg

[a] The tolerance limit is defined as the concentration level at which the interference causes an error of not more than ±2%.

As the eutrophication of lakes, streams and reservoirs has increased in recent years, sanitary engineers have become increasingly aware of the widespread use of phosphorus in the environment and the important role it has in the eutrophication process. Phosphates have been applied as plant fertilizers for many centuries in one form or another but the development of synthetic fertilizers has greatly increased the amount applied. Metaphosphates are often used to treat water used in cooling systems and boilers to control corrosion. With the banning of alkyl benzene sulfonate for household detergents in the mid 1960s, phosphate builders were used for synthetic detergents and this greatly increased the phosphorus content of domestic wastewater.

Phosphates occur naturally in bodies of water and littoral sediments but their increased use and subsequent discharge into the environment has greatly increased the concentrations in the environment in recent years. Since phosphorus is an essential nutrient for the growth of aquatic plants and organisms, the result in many instances had been the overgrowth of algae and aquatic plants to the point that the usefulness of the body of water for other purposes has been greatly impaired.

The importance of phosphorus in the eutrophication of surface waters is well known. Extensive monitoring is needed to provide data on this nutrient for effective water quality management. An indispensable key to the accomplishment of this objective are methods for analyzing phosphates that are accurate, specific, and reliable.

Most colorimetric determination procedures in common use are based on the formation of 12-molybdophosphoric acid and its reduction to a blue heteropoly compound. There are quite a few reductants that can be used for the formation of phosphomolybdenum blue. A few are used for the high range determinations, such as vanadomolybdophosphoric acid and amino acid. Stannous chloride has been most widely used for low range determinations.

To and Randall [5.18] have reported on the use of ascorbic acid as a reductant and compared its reliability to that of a method using stannous chloride. They report that, using ascorbic acid, the colours are more stable than occurs when stannous chloride is used. Consequently, the ascorbic acid method is more amenable to automation, although there are some limitations to the method. In the preparation of the combined reagent, care should be taken that all reagents reach room temperature and that they are mixed in order as prescribed. Otherwise turbidity and sometimes even a precipitate may form.

The ascorbic acid solution and the combined reagent are very unstable. They have to be prepared weekly and stored in the refrigerator when not in use. Otherwise the reagents will turn yellow and become useless.

There are interferences from some ions such as arsenates, nitrites and hexavalent chromium.

Boyd and Tucker [5.19] found that Millipore membrane filters (0.45 μm), Gelman glass-fibre filters and Whatman No. 42 and No. 1 filter papers were suitable for preparing pond water samples for orthophosphate determinations. The stannous chloride method was considered as accurate and precise as the ascorbic acid method for orthophosphate, but more sensitive. Nevertheless, the greater stability of the coloured sol by the latter method made it more suitable for analysing pond waters for filtrate orthophosphates. Concentration of filtrable orthophosphate changed with time of storage at 25 °C in polyethylene bottles for up to 48h; samples should ideally be analysed within 1 h of collection, although storage for 2–8h does not greatly affect the concentration.

Studies by Tarapchak [5.20] have indicated that methods using molybdenum blue to determine orthophosphate in surface waters may result in an over-estimation of the amount of available phosphorus. Evidence was obtained that molybdenum accelerates hydrolysis of organic phosphorus in the presence of acid and also either causes hydrolysis or forms complexes with organic phosphorus before acidification. Possible ways for limiting these effects are discussed.

Tarapchak et al. [5.21] have shown that when the molybdenum blue method was applied to samples of Lake Michigan water, differences in the period of exposure of samples to acid molybdate and differences in molybdate concentration can affect the accuracy of the results.

Nurnberg [5.22] found that iron and hydrogen sulphide, often present in anoxic lake waters, interfere with the determination of soluble reactive phosphorus. Methods for reducing the interference are indicated. They propose an alternative more complicated procedures for soluble reactive phosphorus in anoxic waters for determination of total (unfiltered) reactive phosphorus. Results are on average only 2% lower, and the method can be automated if required.

Hydrogen sulfide concentrations higher than 1 mg l^{-1} and ferrous iron concentration above 0.20 mg l^{-1} produce interferences in the soluble reactive phosphate analysis (e.g. 80% underestimation of soluble reactive phosphorus).

Ferrous iron is not problematic if the sample is kept anoxic before and during filtration. On the other hand vigorous aeration is obligatory if hydrogen sulfide is present.

Soluble reactive phosphorus is defined as the phosphorus which passes a 0.45 µm filter and which reacts with the molybdate blue reagent within a short period of time (Strickland and Parsons [5.23]).

The chemical analysis of the phosphorus fractions was based on the molybdate blue reagent (Murphy and Riley [5.24]). For the standard soluble reactive phosphorus analysis the sample was filtered through a 0.45 µm cellulose acetate filter, which was prerinsed with 250 ml distilled and 5 ml sample water, at a vacuum pressure of 34.5 Pa (5 psi). Mixed reagents were added at once to 30 ml of the filtrate and the absorbance was read in a 10-cm cell after 10 ± 1 min. The timing was kept short and constant to minimise hydrolysis of organic phosphates.

When high levels of iron were present, the anaerobic sample was transferred by a 50-ml glass syringe from the bottle to the filter manifold. The filter had been prerinsed with nitrogen bubbled water to expel any air, and the filtration was carried out under a permanent nitrogen atmosphere. The handling time of the sample from the time of opening the anoxic bottle until adding the phosphorus reagents after filtration did not exceed 5 min.

When high levels of hydrogen sulfide were present, the sample was aerated by an aquarium pump for 1 or 2 h and then the standard soluble reactive phosphorus analysis (see above) was carried out.

Purging the sample with air to sweep out hydrogen sulphide prior to developing the molybdenum blue colour has a beneficial effect on the determination of soluble reactive phosphorus. After 30 min of vigorous aeration the turbidity blank did not show interference and soluble reactive phosphorus had increased to a constant value. The length of time needed to drive off the gas depends on the concentration of hydrogen sulphide, temperature, sample volume and the extent of aeration. To determine simply and quickly if aeration has been sufficient, the blank reagent can be added. The samples will turn yellow in the presence of hydrogen sulfide and produce a high absorbance at 330 and 885 nm.

Nurnberg [5.22] concludes that the correlation between iron concentration and the removal rate of soluble reactive phosphorus and the induction of phosphorus precipitation by iron addition to waters which do not naturally precipitate phosphorus all suggest that ferrous iron controls the amount of soluble phosphorus when the water is aerated. If no precautions are taken to avoid oxidation of ferrous iron, adsorption onto iron hydroxides or precipitation of ferric phosphate can occur; this lowers the apparent soluble reaction phosphorus concentration after the filtration step. This underestimation of soluble reactive phosphorus can be as high as 40% within 2 h of aeration, depending on iron concentration and oxygen tension. However in purely buffered softwater, ferric iron phosphorus flocculants smaller than 0.45 μm are formed. Such waters shown no interference in the standard oxic analysis of soluble reactive phosphorus even though aeration of iron to the ferric state and loss of free orthophosphate do occur.

Waters with low iron concentration can contain high hydrogen sulfide levels. Hydrogen sulfide grossly interferes with the phosphorus analysis due to a reaction with the acid antimony molybdate reagent and can lead to an 80% underestimate of soluble reactive phosphorus. Sufficient aeration of a high hydrogen sulphide containing sample is required to obtain a valid estimate of reactive phosphorus.

Rather than incorrectly estimating soluble reactive phosphorus (if no precautions against the possible interferences are taken) or carrying out a relatively complicated soluble reactive phosphorus analysis (if precautions are taken), sampling anoxic water without any precautions to maintain anoxia and then, if necessary, aerating the aliquot before a total reactive phosphorus analysis is suggested. In this procedure it is essential that a blank reagent (described in the method) for each sample is prepared to correct for the turbidity and colour. This blank also indicates whether the sample has been sufficiently aerated to reduce the interference of hydrogen sulfide. Once the aeration time is determined, the total reactive phosphorus procedure can easily be carried out.

Total reactive phosphorus represents the unfiltered fraction of phosphorus which reacts with molybdate reagent. The initial collection and handling of anaerobic samples for total reactive phosphorus analysis was identical to the soluble reactive phosphorus procedure. Because of the lack of filtration, all total reactive phosphorus absorbances have to be corrected for turbidity. The turbidity blank is obtained by adding a blank reagent in which double distilled water replaced the ascorbic acid reductant normally used and reading after approx. 10 min at the same wave length.

Stauffer [5.25] has carried out studies to optimize the determination of phosphate in groundwater using reduced molybdenum blue. High levels of silica may be responsible for phosphorus determinations being in error by three orders of magnitude.

Matsubara et al. [5.26] have described a rapid method for the determination of trace amounts of phosphate (and arsenate) in water by spectrophotometric detection of their heteropoly acid- malachite green aggregates following preconcentration by membrane filtration.

5.1.1.2 Segmented Flow Analysis

This technique, described in Sect. 1.3.2, can be applied to the determination of orthophosphate in the concentration ranges $0-200$ mg l^{-1} and $0-0.1$ mg l^{-1} in water. The technique can also be applied to the determination of total phosphates in the range $0-25$ mg l^{-1} (as P).

5.1.1.3 Flow Injection Analysis

Fogg and Bselsu [5.27] developed a sequential flow injection voltammetric determination at a glassy electrode of phosphate (and nitrite) by injection of reagents into a sample stream. Good precision was obtained in the range $5-500$ μmol l^{-1} phosphate.

Motomizu et al. [5.28] determined phosphate in river water by flow injection analysis. The technique used malachite green in acidic medium as chromogenic reagent and spectrophotometric measurements are made at 650 nm. The procedure is very sensitive and can be used at a high sampling rate.

Reagents. As follows. Standard phosphate solution. Potassium dihydrogen phosphate is dissolved in distilled water 2×10^{-3} mol l^{-1}. Dilute accurately before use.

Reagent solution. Dissolve 19.4 g of ammonium heptamolybdate tetrahydrate, 0.092 g of Malachite Green (oxalate) in about 500 ml of distilled water, add 250 ml of ethanol and 70 ml of concentrated sulphuric acid and dilute to 1000 ml with distilled water. This solution is 0.11 mol l^{-1} in molybdenum and 2.2×10^{-4} mol l^{-1} in Malachite Green. Filter through a 0.45 μm membrane filter before use. This solution is stable for at least one month if stored in the dark.

Carrier solution. Sulphuric acid (0.35 mol l^{-1}).

Apparatus. Shimadzu UV-140-02 double-beam spectrophotometer or equivalent, with a 10-mm micro flow-cell (18 μl) and recorded with a Toa Dempa FBR-251A recorder. A double plunger micropump (Kyowa Seimitsu KHU-W-104) was used for the reagent solution and the carrier solution. The sample was injected by a 6-way injection valve (Kyowa Seimitsu KMH-6V) into the carrier stream (sample volume 240 μl for use of reagent A). The flow lines were made from Teflon tubing. The reaction coils were 7.5 m long (1 mm bore) for use of reagent A, and were wound on glass rods. To cancel the pulse from the reciprocal pump, 10 m damping coils (0.5 mm bore) and an air damper (volume 24 ml) were used. The backpressure tubing (0.25 mm bore) was 3 m long to prevent formation of air bubbles. A diagram of the flow system (with dimensions) is shown in Fig. 5.3.

Procedure. The optimal analytical conditions require a flow rate of 2 ml min^{-1} of Reagent A solution. (see Reagents). The flow system in Fig. 5.3 (reaction coil 7.5 ml long and 1 mm bore) is used for preparing the calibration graph. Inject

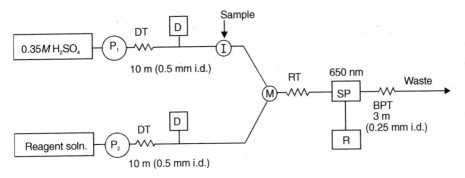

Fig. 5.3. Schematic flow diagram. P_1 and P_2, pumps; DT, damping coil (0.5 mm × 10 m); D, air damper; I, sample injector; M, mixing joint; RT, reaction coil (1 mm × 7.5 m or 0.5 mm × 5 m); SP, spectrophotometer with flow cell; R, recorder; BPT, back pressure coil (0.25 cm × 3 m) (From [5.28])

the sample solution from a sample loop of 30 cm of 1 mm bore Teflon tubing (sample volume about 240 µl). The calibration graphs are linear up to 120 and 800 µg l^{-1} (at different recorder sensitivities).

In Table 5.7 tolerable levels of co-existing ions likely to occur in river water samples are shown.

Only orthophosphate can be determined by the above. However, the condensed phosphates can easily be hydrolysed to orthophosphate by heating the acidified solution for 30 min in a water bath at about 90 °C. In this work the water samples were made 0.35 mol l^{-1} in sulphuric acid in stoppered test-tubes, heated for about 45 min in a water bath at about 95 °C cooled to room temperature and injected into the carrier stream.

Janse et al. [5.29] have optimized conditions for the determination of phosphate by flow injection analysis. The flow rates of the water carrier stream and the reagent streams, the injection volume and the lengths of coils are the parameters on which the procedure was optimized.

Table 5.7. Tolerable concentrations of co-existing ions. Phosphorus taken: 120 ng/ml. (From [5.28]; Copyright 1983, Pergamon Publishing Co., UK)

Ion	Maximum tolerable concentration, mol l^{-1}
Na$^+$, Cl$^-$	0.1
Mg^{2+}, Ca^{2+}	0.01
Al^{3+}, K$^+$, Fe^{3+}, NO$_3^-$, HCO$_3^-$	10^{-3}ᵃ
SiO$_3^{2-}$	10^{-4}
WO$_4^{2-}$, VO$_3^-$	10^{-5}
AsO$_4^{3-}$	2 × 10^{-6}

ᵃMaximum tested.

5.1.1.4 Atomic Absorption Spectrometry

Phosphate (and arsenate and arsenite) have been determined in natural waters by a technique based on flotation spectrometry and extraction – indirect atomic absorption spectrometry using malachite green as an ion – pairing reagent [5.30].

5.1.1.5 High Performance Liquid Chromatography

Sakurai et al. [5.31] have described a high performance liquid chromatographic procedure for the determination of down to 0.5 mg l^{-1} phosphate in waste water and river waters. The method is based on the solvent extraction of molybdoheteropoly yellow with methyl propionate. The corresponding silicon compound is not extracted into this solvent. Thus the interference by silicon is excluded, even at concentrations as high as 10 mg l^{-1}.

Apparatus. Hitachi Model 635 high speed liquid chromatograph equipped with a UV detector and a Hitachi Model 100 – 50 double beam spectrophotometer with a 5-cm glass cell.

Reagents. As follows. Standard phosphate stock solution (100 mg l^{-1} P). Prepare by dissolving 0.4394 g of potassium dihydrogen phosphate in 1 l of water.
Ammonium molybdate solution (3%). Prepare by dissolving ammonium molybdate $(NH_4)_6MO_7O_{24} \cdot 4 H_2O$ in hot water at about 60 °C.
Nitric acid, 4.4N.
Methyl propionate, Analar.
Nitric acid, Aristar.
Perchloric acid, Aristar.

Procedure. For determination of dissolved inorganic phosphate, take 20 ml of the sample solution in a 50 ml beaker and add 1 ml of 4.4 N nitric acid. Then heat the solution in a water bath at about 50 – 60 °C, add 1 ml of 3% ammonium molybdate solution and allow the mixture to stand for a further 5 min in the water bath with occasional stirring. After standing for a further 5 min in running water, transfer the solution into a separatory funnel (50-ml) with 10 ml of water containing 3 ml of methyl propionate. Selectively extract the complex into the organic phase by shaking for 2 min. Centrifuge for 2 min (3000 rpm) to remove water suspended in the organic phase. Determine the molybdoheteropoly yellow in the organic extract by a high speed liquid chromatograph with UV spectrophotometric detector at nm under the working conditions shown in Table 5.8.

For determination of total phosphorus take 20 ml of the sample solution in a 50- ml beaker and heat with 5 ml each of concentrated nitric and perchloric acids on the hot plate until white smoke is produced. After cooling to ambient

Table 5.8. Condition for the determination of phosphorus molybdohetcropoly yellow with HPLC. (From [5.31]; Copyright 1983, Springer Verlag GmbH, Germany)

Packing material	Lichrosorb RP-18 (mean particle size: 5 μm)
Column size	4 mm$\varnothing$ × 150 mm
Eluent	30 % H_2O in CH_3CN
Flow rate	0.9 ml/min (70 kg cm^{-2})
Temperature	ambient
Detector	UV 251 nm
Range	0.04 AUFS
Chart Speed	5 mm min^{-1}
Injecting sample solution	5 μl

temperature, add 10 ml of water and 1 ml of 4·4 N nitric acid to the beaker and dissolve the residue by warming. Heat the solution obtained in a water bath at about 50–60 °C then continue as above.

A wavelength of 251 nm at the UV detector gave an absorption maximum for phosphorus molybdoheteropoly yellow in methyl propionate.

Morgan and Dani [5.32] have described the development of an enzymatic method of determining phosphate eluted from a reverse phase high performance liquid chromatographic column. The method is applicable to sewage effluents and natural water samples. It is based on the nucleoside phosphorylase catalysed conversion of inosine and orthophosphate to hypoxanthine. This method is claimed to give better results than standard molybdenum blue spectrophotometric methods:

$$\text{Phosphate} + \text{inosine} \xrightarrow[\text{Mg}^{2+}]{\text{NP}} \text{ribose-1-phosphate} + \text{hypoxanthine}.$$

The inosine and hypoxanthine were separated by reversed phase high performance liquid chromatography and the amount of hypoxanthine produced was related to the phosphate concentration. Quantitation of the hypoxanthine peak was found to be linear with orthophosphate up to about 30 mg l^{-1}. A detection limit of 0.75 mg l^{-1} could be obtained after dialysis of the commercial enzyme. Interference studies showed that the enzymatic assay, unlike the colorimetric molybdate blue technique, was essentially unaffected by complex matrices such as polyphosphates and phosphoesters.

Reagents. As follows triply distilled water. Tris-(hydroxymethyl) aminomethane (Tris). Aldrich (Milwaukee, WI USA).

M-Aminophenol. Eastman Kodak (Rochester, NY, USA); recrystallized from diethylether before use.

Inosine.

Reagent solution of 20.7 m Mol l^{-1} inosine, 6.06 m mol l^{-1} magnesium sulfate, 0.4 mg ml^{-1} nucleoside phosphorylase and 0.4 mg ml^{-1} m-aminophenol in 0.1 mol l^{-1} Tris buffer, pH 7.4.

Apparatus. HPLC system-high pressure Milton Royl Model 396 pump (Laboratory Data Control, Riviera Beach, FL., USA) was modified with a tee configuration pulse dampener consisting of a 52 cm × 4.6 mm stainless steel tube. The injector, equipped with a 130 µl sample loop, was a Rheodyne Model 7010 with a loop filler port Model 7011 (Rheodyne, Berkeley, CA. USA). The separations were carried out with a 5 cm × 4.6 mm i.d. column and a 15 cm × 4.6 mm i.d. working column. Both stainless steel columns were packed with 10 µm LiChrosorb RP-18 (E. Merck, Darmstadt, G.F.R) using a Model 10-600-30 pneumatic amplifier pump (SC Hydraulic Engineering, Los Angeles, C.A. USA) and a high pressure slurry packer (Alltech, Deerfield, IL, USA). An Altex Model 153 UV detector, constant wavelength (254 nm) (Altex, Berkeley, CA, U.S.A.) was used to monitor the column effluent. Peaks were recorded with a Fisher Recorder Model 5000 (Houston Instruments, Austin, TX, U.S.A.). In addition, the chromatograms were digitized by an Apple 2$^+$ minicomputer (Apple, Cupertino, CA. U.S.A.) equipped with a 12 bit analogue to digital converter (Inter-active Microware, State College, PA U.S.A.).

Spectrophotometric work was performed on a Hewlett Packard Model 8450 UV-VIS spectrophotometer equipped with a Model 9872 B x-y plotter (Hewlett Packard, Palo Alto, CA. U.S.A.).

Procedure. Add 0.7 ml of inosine magnesium sulphate nucleotide phosphorylase m-aminophenol reagent to a 2-ml volumetric flask and fill to the mark with the sample. After mixing by inversion of the flask, allow the solution to react for 30 min in a constant temperature bath at 20°C. Stop the reaction at the end of 30 min by placing the volumetric flask into boiling water for 2 min. After cooling, chromatograph the solution using a mobile phase of 0.1 mol l^{-1} Tris buffer, pH 7.4 at a flow rate of 1.8 ml min^{-1}. This effectively separates inosine from hypoxanthine on the C_{18} column. To help compensate for any loss in column efficiency and maintain good accuracy, use an internal standard, m-aminophenol which elutes between hypoxanthine and inosine in the assay procedure.

Figure 5.4 shows a typical chromatogram for a standard 42 µg l^{-1} phosphate solution and hypoxanthine peaks resulting from various phosphate samples after reaction with the enzyme. The calibration curve had a slope of 0.043 ± 0.002 and an intercept of 0.124 ± 0.033 with a correlation coefficient of 0.9981. Linearity up to 30 mg l^{-1} was observed. Relative standard deviation of triplicate runs was 10% or less. The detection limit, twice the signal of the blank, was determined to be 1.5 mg l^{-1}. Dialysis of the commercial enzyme in 0.1 mol l^{-1}. Tris buffer, pH 7.4, for 24 h was found to reduce the blank and therefore the detection limit by at least a factor of 2.

A plot of phosphate concentration given by the enzymatic method vs that found by the molybdate blue method had a slope of 0.999 ± 0.046 and an intercept of 0.421 ± 1.222 with a correlation coefficient of 0.996.

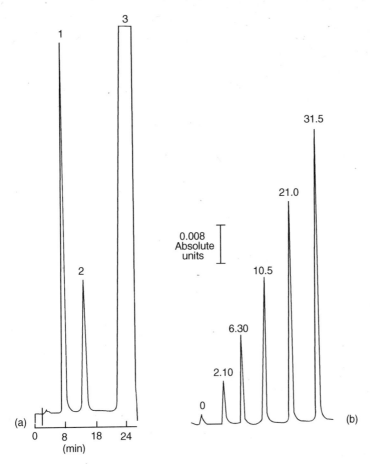

Fig. 5.4. a Sample chromatogram of a 42 ppm phosphate sample. Absorbance units full scale, au.f.s. = 0.08. Peaks: 1 = hypoxanthine; 2 = m-aminophenol; 3 = inosine. **b** Hypoxanthine peak for various phosphate concentrations in mg l^{-1} (indicated above each Peak) (From [5.32])

To compare the selectivity of the enzymatic method with that of the molybdate blue method, interference studies were conducted for both methods. The first study was the effect of tripolyphosphate in the sample. The molybdate blue method showed a positive deviation with increasing amounts of tripolyphosphate. The reason for this positive deviation can be attributed to the acid hydrolysis of tripolyphosphate to orthophosphate. The enzymatic method also showed a slight positive deviation, probably due to orthophosphate impurity in the tripolyphosphate used.

Braungart and Russel [5.33] applied high performance liquid chromatography to the determination in natural water of molybdoheteropoly acids of phosphorus, silicon, arsenic and germanium as ion associates. The method is based on solvent extraction of molybdoheteropoly compounds as ion associates by dichloromethane with the addition of amine to improve the stability of the associates. Separation of the heteropoly compounds without decomposition was possible using reversed phase high pressure liquid chromatography systems.

Moss and Stephen [5.34] have described the determination of phosphate (and chloride, bromide and iodide) in natural waters by high performance liquid chromatography.

5.1.1.6 Column Coupling Isotachophoresis

This technique offers very similar advantages to ion chromatography in the determination of anions in water, namely multiple ion analysis, little or no sample pretreatment, speed, sensitivity and automation. Very similar advantages are offered by capillary isotachophoresis. Zelinski et al. [5.35] applied the technique to the determination of $0.02-0.1$ mg l^{-1} quantities of phosphate, chloride, fluoride, nitrate, nitrite and sulphate in river waters. This technique is discussed further in Chap. 11.

5.1.1.7 Ion Chromatography

Phosphate has been codetermined with chloride, bromide, fluoride, nitrate, nitrite and sulphate in natural waters [5.36] and codetermined with fluoride and bromide in river waters [5.37]. See also Sect. 11.2–11.5.

5.1.1.8 Differential Pulse Polarography

Hight et al. [5.38] give details of a procedure for the determination of phosphate in aqueous samples by differential pulse polarography, based on determination of molybdenum in 2-phosphomolybdic acid. Phosphate is first converted to the phosphomolybdate complex which is then extracted into isobutyl acetate. High sensitivity is achieved by measuring the polarographic wave due to the catalytic reduction of perchlorate or nitrate in the presence of molybdenum (VI). The method is suitable for samples as small as 3.5 ml which contain as little as 9 µg l^{-1} of phosphorus. The average relative deviation is 3.0% at the 0.045 mg l^{-1} phosphorus level and 1.6% at the 1.2 mg l^{-1} level.

Isobutyl acetate is then removed by evaporation and the residue dissolved in dilute sodium hydroxide, followed by acidification with perchloric acid. Phosphate is then determined indirectly by measurement of the catalytic current of perchlorate or nitrate in association with molybdenum VI.

Reagents. All chemicals are either analytical or primary standard grade. No further purification is necessary. Use demineralized water to make all solutions.

Stock molybdenum solution. Prepare monthly by dissolving 17.66 g of ammonium molybdate tetrahydrate $(NH_4)_6Mo_7O_{24} \cdot 4H_2O$ in 1 l of demineralized water. Store in a polyethylene container. Prepare working solutions of lower molybdenum concentration daily by dilution.

Stock phosphate solution. Prepare monthly by dissolving 1.3609 g of oven-dried (2 h at 100 °C) primary-standard grade potassium dihydrogen phosphate in 1 l of demineralized water. Store in a polyethylene container. Prepare working solutions of lower phosphate concentrations daily by dilution.

Glassware. Wash all glassware with 6 mol l^{-1} hydrochloric acid and rinse thoroughly with demineralized water. Wash polyethylene stoppers in 6 mol l^{-1} hydrochloric acid, rinse with demineralized water, then wash in 8 mol l^{-1} nitric acid and again rinse with demineralized water.

Apparatus. Record differential pulse polarograms on a Princeton Applied Research Electrochemical System Model 170 equipped with potentiostatic control and a mechanical drop-detachment device. A dropping mercury electrode working electrode, a saturated calomel (SCE) reference electrode and a platinum-wire counter-electrode are used. The polarographic cell, drop-detachment device and mercury column with its reservoir are placed inside a grounded aluminium cage to prevent pickup of extraneous electrical signals from the laboratory surroundings. Reaerate solutions with nitrogen purified by passage through vanadium (II) chloride solution and keep a blanket of the nitrogen over the solutions during the recording of polarograms.

Procedures. Standardization of molybdenum solution is achieved by activating a 6×1 cm column of freshly amalgamated zinc by passing 20 ml of 1.44 mol l^{-1} hydrochloric acid (12 ml of concentrated acid diluted to 100 ml) through it at a rate of 1 drop sec^{-1}. Prepare a 0.01 mol l^{-1} molybdenum solution in 1.44 mol l^{-1} hydrochloric acid by diluting 10.00 ml of stock molybdenum solution with 12.0 ml of concentrated hydrochloric acid to 100.0 ml with demineralized water. Dilute 10 ml of iron (III) solution (prepared by dissolving 15.06 g of ferric ammonium sulphate $(FeNH_4(SO_4)_2-12H_2O)$ in 5 vol.% sulphuric acid) to about 100 ml with demineralized water and place in the receiving beaker. Cover this beaker with a sheet of "Parafilm". Place the tip of the Jones reductor below the "Parafilm" cover and the continuously flush the receiving beaker with nitrogen during the reduction and titration.

Pass exactly 15.00 ml of the acidic 0.01 mol^{-1} molybdenum solution through the Jones reductor at a rate of 15 drops/min and rinse the reductor with 10.0 ml of demineralized water. Collect the reduced solution and the rinsings in the receiving beaker and titrate the iron (II) produced with the 0.1N permanganate. Treat a blank solution of 15.00 ml of 1.44 mol^{-1} hydrochloric acid similarly.

Standardization of molybdenum solutions by differential pulse polarography is done by pipetting exactly 5.00 ml portions of 4.0 mol l^{-1} perchloric acid and various known microlitre amounts of standard $0.0143 \text{ mol l}^{-1}$ molybdenum solution into a series of 10.00 ml volumetric flasks and diluting to volume with 0.125 mol l^{-1} sodium hydroxide. Analyse the solutions by differential pulse polarography, after deaeration with nitrogen for 5 min immediately before polarography. Scan the voltage from + 0.20 V to − 0.60 V vs SCE. Maintain the solutions at 22 ± 0.5 °C.

Samples are prepared by placing them in an acid-washed nalgene container and immediately cooling in ice. Filter the sample through a 0.45 µm membrane filter.

Determination of phosphate in low concentration range is done as follows. Prepare standard solutions for phosphate concentrations in the range $3 \times 10^{-7} - 3 \times 10^{-6}$ mol l^{-1} (9 - 90 µg l^{-1}) by pipetting 60 - 600 µl of standard 2.52×10^{-4} mol l^{-1} phosphate solution into a series of 50-ml standard flasks and diluting to the mark with demineralized water.

Pipette 3.50 ml of a solution to be analysed into a 1.5 × 15 cm Pyrex testtube fitted with a high-density polyethylene or glass stopper. Add 500 µl of 4.24% ammonium molybdate solution, followed by 1.00 ml of 3.33 mol l^{-1} hydrochloric acid (27.8 ml of concentrated acid diluted to 100 ml). After 2 min add 5.00 ml of isobutyl acetate, stopper the test-tube and shake the contents for 1 min. Transfer 4.00 ml of the organic (upper) phase (containing the phosphomolybdic acid) to a 100 ml beaker and evaporate the solvent by placing the beaker in a laboratory hood providing a strong draught. Dissolve the residue in 5.00 ml of 0.125 mol l^{-1} sodium hydroxide. Acidify this solution with 5.00 ml of 4.0 mol l^{-1} perchloric acid. Transfer to the polarographic cell and analyse by differential pulse polarography. Analyse similarly a blank solution containing no phosphate.

Determination of phosphate in high concentrations is done as follows. Prepare standard solutions for high phosphate concentrations in the range from 1×10^{-6} to 6×10^{-3} mol l^{-1} by pipetting 10 - 500 µl of 0.01 mol l^{-1} standard phosphate solution into 50-ml standard flasks and diluting to the mark with demineralized water. Place 1 ml of the diluted phosphate solution in the testtube followed by 0.50 ml of 0.1 mol l^{-1} molybdate and 1.0 ml of 1.65 mol l^{-1} hydrochloric acid. After 2 minutes add 2.0 ml of isobutyl acetate and shake the tube vigorously. Transfer exactly 1 ml of the organic phase into a beaker. Evaporate the isobutyl acetate and dissolve the residue in 5.0 ml of 0.125 mol l^{-1} sodium hydroxide. Acidify this solution with 5.0 ml of 4 mol l^{-1} perchloric acid or 1.0 ml of 2 mol l^{-1} sulphuric acid, followed by addition of 4.0 ml of 5.0 mol l^{-1} sodium nitrate. Transfer the resulting solution into a polarographic cell and analyse it for molybdenum content by differential polarography. Measure the magnitude of the catalytic peak and compare it with that for the blank.

Determine the amount of phosphate in aqueous samples by extraction of 12-phosphomolybdic acid and measure the catalytic perchlorate current in the presence of molybdenum (VI).

Hight et al. [5.38] evaluated this procedure by measuring peak currents for molybdenum (VI) alone in 2.0 mol l^{-1} perchlorate or 2.0 mol l^{-1} nitrate media and comparing them with the values obtained for known phosphate solutions by extraction and polarographic analysis of 12-phosphomolybdic acid. The results are summarized in Table 5.9. Analysis results for the Environmental Protection Agency quality-control sample with very low concentration of phosphate and for natural water samples are presented in Table 5.10.

Conversion of orthophosphate into 12-phosphomolybdic acid proceeds by the reaction

$$PO_4^{3-} + 12MoO_4^{2-} + 27H^+ \rightarrow H_3PO_4(MoO_3)_{12} + 12H_2O$$

and depends not only on the concentration of acid but also on the ratio of molybdenum(VI) to phosphate. The hydrochloric acid concentration in the aqueous phase is kept to 0.66 mol l^{-1}, which ensures not only rapid formation of 12-phosphomolybdic acid but also enables selective extraction of this acid, thus avoiding the possible interference of silicate and arsenate, which remain in the aqueous phase. In this step, the concentration of molybdate must not be less than 0.02 mol l^{-1}. This provides a molybdate to phosphate ratio of 400 or more. Under these conditions, formation of 12-phosphomolybdic acid is rather fast and it can be extracted after 2 min. However, if the molybdate concentration is reduced to 0.01 mol l^{-1}, the extraction efficiency falls to below 50%. Also, if the aqueous phase phosphate concentration is low, less than the expected amount of molybdenum is recovered in the polarographic cell, and the extraction of phos-

Table 5.9. Current as a function of phosphate concentration[a]. (From [5.38]; Copyright 1982, Pergamon Publishing Co., UK)

Amount of phosphate in aqueous phase ng	Peak current nA	Molybdenum(VI) in polagraphic cell	
		found	expected
In 2.0 mol l^{-1} perchlorate			
0.00	54 ± 5		
155	200 ± 7	2.5	3.00
310	390 ± 10	5.1	6.00
620	760 ± 15	10.0	12.00
1240	1500 ± 20	20.6	24.00
In 2.0 mol l^{-1} nitrate			
15.5	100 ± 6	0.25	0.30
38.7	100 ± 8	0.70	0.75
77.5	282 ± 10	1.40	1.5
155	570 ± 15	2.75	3.00
310	1060 ± 20	5.40	6.00

[a]Samples are prepared by taking 1.00 ml of the secondary standard and diluting it with 0.50 ml of 0.1 mol l^{-1} Mo(VI) and 1.0 ml of 1.65 mol l^{-1} HCL. Heteropoly acid is extracted into 2.00 ml of isobutyl acetate and 1.00 ml of the organic phase is analysed. Molybdenum (VI) concentration is obtained by comparison of the peak currents with those for aqueous molybdate obtained by multiplying the 12-phosphomolybdate concentration by 12 assuming conversion and extraction to be complete.

Table 5.10. Orthophosphate determination in EPA quality-control sample and in surface water[a]. (From [5.38]; Copyright 1982, Pergamon Publishing Co., UK)

Determination Standards for EPD sample analysis	Phosphorus in aqueous phase $\mu g\,l^{-1}$	Peak current nA[b]	Phosphorus $\mu g\,l^{-1}$
	9.4	206 ± 8	–
	31.2	495 ± 15	–
	40.6	616 ± 15	–
	46.8	700 ± 20	–
	93.6	1201 ± 30	–
EPA sample unspiked	–	699 ± 20	45
spiked[c]	–	1016 ± 25	76
Standards for surface water	626	1560 ± 20	
	1250	2700 ± 25	
	1870	4000 ± 40	
Skokie Lagoon unspiked	–	2630 ± 25	1220
Spiked[c]	–	3320 ± 35	1540

[a] Sample analysed by taking a 3.50 ml aliquot, 0.50 ml of 0.24 mol l^{-1} Mo (VI) and 1.00 ml of 3.33 mol l^{-1} HCl. The 12 phosphomolybdic acid is extracted with 5.0 ml of isobutyl acetate and 4.00 ml of the organic phase are analysed.
[b] Polarographic settings: scan-rate 5 mV/s, pulse-amplitude 25 mV, drop time 1 s/drop, sampling time 5 ms. Peak currents are the average values of three measurements.
[c] EPA quality-sample water is spiked with 0.100 ml of 2.4×10^{-4} mol l^{-1} phosphate per 25 ml of sample and surface water is spiked with 0.010 ml of 0.0101 mol l^{-1} phosphate per ml of sample.

phomolybdic acid is incomplete. Nevertheless linear calibration graphs can be obtained even when the aqueous phase phosphate concentration is as low as 3×10^{-7} mol l^{-1}. Since the amount of phosphomolybdic acid extracted depends on the aqueous phase phosphate concentration, it is necessary to use standard phosphate solutions that bracket the concentration of the sample. Because of the small amount of free molybdate extracted in the blank, it is also necessary to analyse a blank (for the whole procedure) along with the standard phosphate solutions.

Matsunaga et al. [5.39] have also discussed the differential pulse auodic voltammetry of phosphates.

5.1.1.9 Enzymic Methods

Enzymic methods have been developed for the determination of phosphates in natural waters. Stevens [5.40] gives details of an automated enzymic method for determination of orthophosphate. Results of analysis of water samples by this method and by the soluble reactive phosphorus method indicated that a form of phosphorus was present which was determined as soluble reactive phosphorus but not as enzymatic orthophosphate; the presence of a hydrous ferric oxide-orthophosphate complex is postulated.

Steinberg et al. [5.41] described an automated enzymic method for phosphate based on the property of orthophosphate of inhibiting the hydrolysis of certain fluorogenic substrates by means of the alkaline phosphatase of *Esch. coli*. A particular substrate is 4-methylumbelliferyl phosphate which has a maximal excitation frequency of 360 nm. Results obtained are compared with values given by the molybdenum blue method, which point to discrepancies in the latter. A detection limit of 0.1 $\mu g\, l^{-1}$ phosphate phosphorus is reported.

Chrost et al. [5.42] described a method for determining enzymically hydrolysable phosphate based on determination of inorganic phosphate released after the hydrolysis of organophosphoric esters by free, dissolved phosphohydrolases (mainly phosphatase) produced by biota. The method gave higher values in hypereutrophic waters than the molybdenum blue procedures and similar values for less eutrophic waters. The method was recommended for eutrophic waters where the orthophosphate concentration exceeded 25 $\mu g\, l^{-1}$ phosphorus.

5.1.1.10 Miscellaneous Methods

Various other methods for the determination of phosphate in water exist, including atomic absorption inhibition, neutron activation analysis of the tungstophosphate [5.43–5.45], X-ray fluorescence spectrometry following adsorption of phosphate onto silica gel modified with N-substituted diamine functional groups [5.46], and radio-activation analysis as the ^{185}W tungsto molybdophosphoric acid complex [5.47].

Ryden et al. [5.48] have studied the adsorption of inorganic phosphates by laboratory ware, including glass, polypropylene and polycarbonate. They conclude that the use of acid washed glassware should be discouraged and a polycarbonate which has been previously "phosphated" by treatment with hot hydrochloric acid followed by potassium dihydrogen phosphate is the most suitable sample container material.

Nelson and Romkous [5.49] have shown that, after freezing of water samples followed by storage at $-20\,°C$ for 3 days, the soluble phosphate concentration decreased by 2–27%. Rapid freezing with liquid nitrogen produced about half the effect of slow freezing. No decrease in the phosphate concentration was found when samples were stored at 2 °C (for 3 days) or when they were frozen slowly after removal of sediments by centerfuging.

Lindquist and Cox [5.50] determined phosphate in lakewater by cathodic stripping chronopotentiometry at a copper electrode. Good agreement was obtained with results obtained by spectrophotometric procedures.

Shapiro [5.51] fixed lakewater samples for phosphate determination by adding aqueous ammonium molybdate to the sample immediately after collection. As soon as convenient the samples were extracted with isobutyl alcohol and the organic layers stored for up to two weeks before analysis. During this period the determined phosphate content does not decrease by more than 20%.

5.1.2 Sea and Estuary Water

5.1.2.1 Spectrophotometric Methods

Johnson and Pilson [5.52] have described a spectrophotometric molybdenum blue method for the determination of phosphate and arsenate and arsenite in estuary water and sea water. A reducing reagent is used to lower the oxidation state of any arsenic present to $+3$, which eliminates any absorbance caused by molybdoarsenate, since arsenite will not form the molybdenum complex. This results in an absorbance value for phosphate only.

A commonly used procedure for the determination of phosphate in sea water and estuarine waters involves the formation of the molybdenum blue complex at $35-40\,°C$ in an Autoanalyser and spectrophotometric evaluation of the colour produced. Unfortunately when applied to sea water samples, depending on the chloride content of the sample, peak distortion or even negative peaks occur which make it impossible to obtain reliable phosphate values. This effect can be overcome by the replacement of the distilled water used in such methods by a solution of sodium chloride of an appropriate concentration related to the chloride concentration of the sample. The chloride content of the wash solution need not be exactly equal to that of the sample. For chloride contents in the sample up to $18\,000$ mg l^{-1} (i.e. sea water) the chloride concentration in the wash should be within $\pm 15\%$ of that in the sample (Table 5.11). The use of saline standards is optional but the use of saline control solutions is mandatory. Using good equipment, down to 0.02 mg l^{-1} phosphate can be determined by such procedures. For chloride contents above $18\,000$ mg l^{-1} the chloride content of the wash should be within $\pm 5\%$ of that in the sample.

Airey et al. [5.53] have described a method for the removal of sulphide prior to the determination of phosphate in anoxic estuarine waters. Mercury II chloride was used to precipitate free sulphide from samples of anoxic water. The sulphide-free supernatent liquid was used to estimate sulphide by measuring the

Table 5.11. Dependence of chloride content of wash liquid on chloride. (From [5.52]; Copyright 1972, Elsevier Science Publishers BV, Netherlands)

Content of sample	
Chloride content of wash mg l^{-1}	Chloride content of sample mg l^{-1}
0	0–1800
4850	1800–7200
10000	7200–12600
15200	12600–18000

concentration of unreacted mercury II, as well as to determine phosphate by a spectrophotometric method in which sulphide interferes. The detection limit for phosphate was $1\,\mu g\,l^{-1}$.

5.1.2.2 Ion Chromatography

Tyree and Bynum [5.54] used ion chromatography to determine low levels of phosphate in sea water.

5.1.3 Rainwater

5.1.3.1 Continuous Flow Analysis

Continuous flow analysis has been used to determine phosphate (and chloride, nitrite, nitrate and sulphate) in rainwater [5.55].

5.1.3.2 Flow Injection Analysis

A computer controlled multichannel continuous flow analysis system has been applied to the measurement of phosphate (and nitrite, nitrate, chloride and sulphate) in rainwater [5.56, 5.57].

5.1.3.3 Gas Chromatography

Faigle and Klockow [5.58] applied gas chromatography to the determination of traces of phosphate (and nitrate and sulphate) in rainwater. The salts were freeze dried and converted to the corresponding silver salts. These were then converted to n-butyl esters with the aid of n-butyl iodide, the n-butyl esters being determined by direct injection into a gas chromatographic column comprising 3% OV-17 on Chromosorb G.

5.1.3.4 Ion Chromatography

Applications of ion chromatography to the determination of phosphate and other anions in rainwater are summarised in Table 5.12. See also Sect. 11.2.

Table 5.12. Determination of phosphate in rainwater by ion chromatography

Type of water sample	Codetermined anions	References
Rain	Cl, F, Br, NO_2, BrO_2	5.59
Rain	NO_3, SO_4	5.60
Rain	Cl, NO_3, SO_4	5.61
Rain	Cl, NO_2, NO_3, SO_4	5.62
Rain	Cl, I, Br, NO_2, NO_3 SO_4, SO_3	5.63

5.1.4 Potable Water

5.1.4.1 Ion Chromatography

This technique has been applied to the determination of phosphate (and chloride, fluoride, nitrate and sulphate) in potable water and rainwater [5.59]. See also Sect. 11.3.

5.1.5 Industrial Effluents

5.1.5.1 Ion Chromatography

This technique has been applied to the determination of phosphate (and chloride, sulphate, nitrite, nitrate, fluoride and bromide) in treated waters [5.64]. See also Sect. 11.4.

5.1.6 Waste Waters

5.1.6.1 Spectrophotometric Method

Buchanan and Easty [5.65] showed that the official methods [5.66, 5.67] for determining orthophosphate and total phosphorus (i.e. phosphate after digestion with persulphate) in waste water as reduced molybdophosphate are subject to interference when applied to waste samples. Thus sulfides interfere in the determination of orthophosphate but not in that of total phosphorus because they are oxidised by the persulphate. Lignin sulphonates interfere in the determination of total phosphorus and somewhat less in the determination of orthophosphate. Addition of excess of aqueous bromine to the sample prevents interference from sulphide and greatly reduces that from ligninsulphonates. In the determination of orthophosphate the unconsumed bromine is removed by a stream of air before further steps in the determination. It is also shown that the precipitation of alkali lignin and other substances, which occur when the solution oxidation is neutralized after the persulphate according to the official procedure, can be eliminated by filtering the solution while it is still acidic, the amount of acid used in the reagent solution being reduced to compensate for that already present.

5.1.6.2 Ultraviolet Spectroscopy

The automated ultraviolet spectrophotometric method [5.68] for the determination of ortho- and ortho- plus hydrolysable or total phosphate in waste water involves the use of two independent flow systems, which are flow-charted, concurrently allowing these phosphates to be measured colorimetrically as orthophosphate. Features of the method include compensation of the bias effect of turbidity during orthophosphate determination, and the elimination of the need to neutralize acidified samples in hydrolysable and total phosphate determinations prior to colour development. The method and equipment both allow continuous operation for up to 24 h.

Standard deviations over the ranges $0-3$ mg l^{-1} phosphate and $3-30$ mg l^{-1} phosphate were 0.03 and 0.2 respectively. Up to 100 mg l^{-1} of silicate, iron or nitrate did not interfere in the determination of 30 mg l^{-1} phosphate.

5.1.6.3 Atomic Absorption Spectrometry

In the atomic absorption inhibition titration method [5.69] the phosphate in the sample is treated with incremental amounts of magnesium solution and magnesium determined by atomic absorption spectrometry. The end-point is indicated by a sharp rise in the magnesium absorption; features of the titration curve after the end-point are related to species formed in the flame. Some observations on polyphosphate titrations are presented, and the method was applied to the determination of phosphate in surface and waste waters.

5.1.6.4 High Performance Liquid Chromatography

The methods developed by Sakurai et al. [5.70] and Morgan and Danielson [5.71] are also applicable to the determination of respectively, phosphates in waste waters and in sewage effluents.

5.1.6.5 Ion Exclusion Chromatography

Tanaka and Ishizuka [5.72] have investigated the possibility of determining orthophosphate in waste waters by ion exclusion chromatography on a cation exchange resin in the H^+ form by elution with acetone:water. They discuss optimal conditions for the separation of phosphate from chloride, sulphate, carbonate etc. which are always present in waste water and sewage samples.

A Spectra Physics flow coulometer detector β-liquid chromatograph equipped with a Hitachi 630 and a conductiometric detector (Yanagi moto C-202) was used for isocratic elution with acetone:water. In the flow coulometric detector the following electrochemical reaction was used to detect H^+ ion from H_3PO_4 eluted.

p-benzoquinone $+ 2H + 2e^- \rightarrow$ hydroquinone ($+0.45$ V vs Ag/AgI).

Details of the flow coulmetric detective have already been described by Tanaka and Muto [5.73].

A glass jacketed column packed with a Hitachi 2613 strongly acidic cation exchange resin in the H^+ form was used at 30°C. The flow rate of eluent was 1 ml min^{-1}. Aqueous sample solutions containing phosphate and other anions (0.5 – 2 ml) were introduced onto the column with a variable loop injector and chromatographed by ion-exclusion as the corresponding acids.

Figure 5.5 shows the chromatogram of a mixture of 10 mg l^{-1} of chloride, sulphate, nitrate, phosphate and carbonate obtained with the flow columetric detector by elution with 60% acetone:water. As can be seen from Fig. 5.5, phosphate could be separated from the strong acid anions and carbonate. The R_s value between the strong acid anions and phosphate was about 1.7. This R_s value suffices for the quantitation of phosphate by the peak area measurement with a computing integrator.

In Fig. 5.5 an unknown peak was observed between phosphate and carbonate. This peak was due to water in the sample solution introduced into the chromatograph. However, the peak did not interfere with the quantitation of phosphate.

5.1.6.6 Ion Chromatography

Green and Woods [5.74] have described the codetermination of phosphate, fluoride, chloride and sulphate in waste water by ion chromatography. See also Sect. 11.4

5.1.6.7 Sample Preservation

Rossin and Lester [5.75] evaluated some of the methods available for the preservation of condensed phosphates in samples of domestic waste waters, i.e. the prevention of the hydrolysis of condensed phosphate to orthophosphate. They

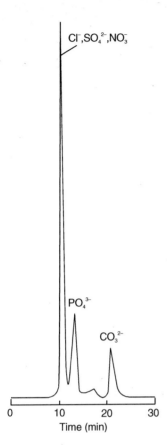

Fig. 5.5. Ion exclusion chromatogram of mixture of 10 mg l^{-1} each of Cl$^-$, SO$_4^{2-}$, NO$_3^-$, PO$_4^{3-}$, and CO$_3^{2-}$ - obtained by elution with 60% acetone water. Detected with FCD (100 mV full scale) (From [5.72])

showed that mercuric chloride at 40 mg l^{-1} at normal pH is the most effective preservative examined for samples stored at 20°C for 6 days (Table 5.13). Samples should be filtered as soon as possible after they have been taken. Mercuric chloride at high pH was not a satisfactory preservative. Addition of sodium hydroxide to raise the pH increases phosphatase activity but lowers the rate of spontaneous hydrolysis and, on balance, is not a satisfactory treatment. Formaldehyde at elevated pH was less efficient than mercuric chloride at normal pH.

The samples treated with mercuric chloride at normal pH presented the smallest difference from the mean with a range between −0.07 and +0.06. When formaldehyde was used at high pH, the range was between −0.16 and 0.00. The samples containing mercuric chloride at pH 10 showed a range between −0.15 and +0.21. A variation between −0.21 and +0.35 was found when pH was increased in the filtered samples. The use of formaldehyde at normal pH resulted in a variation between −0.25 and +0.34.

Table 5.13. Concentration of orthophosphate in settled sewage samples, filtered and unfiltered, with different preservatives, analysed at different times. (From [5.75]; Copyright 1980, Environment Technology Letters)

Sample		Orthophosphate as P mg l^{-1} after storage				
		0	1 day	3 days	6 days	Mean
Settled	unfiltered	5.26	5.96	6.33	6.63	—
Sewage	filtered	4.28	4.20	4.48	4.74	4.43
	D.M.[a]	−0.15	−0.23	+0.05	+0.31	
Settled	unfiltered	4.19	5.35	6.04	6.44	—
Sewage	filtered	4.19	4.03	4.14	4.59	4.24
pH 10	D.M.[a]	−0.05	−0.21	−0.10	+0.35	
Settled	unfiltered	4.89	4.77	5.34	5.85	—
Sewage	filtered	4.09	4.16	4.14	4.24	4.16
HgCl$_2$	D.M.[a]	−0.07	0.00	−0.2	+0.06	
Settled	unfiltered	5.11	5.24	5.49	5.53	—
Sewage	filtered	3.56	3.54	3.76	3.90	3.69
HgCl$_2$ + pH 10	D.M.[a]	−0.13	−0.15	+0.07	+0.21	
Settled	unfiltered	5.00	5.52	6.20	6.48	—
Sewage	filtered	4.15	4.16	4.54	4.74	4.40
Formaldehyde	D.M.[a]	−0.25	−0.24	+0.14	+0.34	
Settled	unfiltered	5.11	5.43	5.43	5.46	—
Sewage	filtered	3.84	4.09	4.04	4.00	4.00
Formaldehyde						
pH 10	D.M.[a]	−0.16	+0.09	+0.04	0.00	

[a]D.M. = Deviation from the mean concentration in filtered samples.

5.1.7 Sewage

5.1.7.1 Spectrophotometric Method

The Department of the Environment, UK [5.76] has issued details of spectrophotometric methods for the determination of orthophosphate in sewage effluents.

These methods are based on reaction with acid molybdate reagent to form a phosphomolybdenum blue complex which is determined at 882 nm. The first method has a range of 0–0.40 mg l^{-1} (after dilution of sample if required) and the second is mainly designed for oligotrophic waters with phosphorus contents in the range 0–25 µg l^{-1}. In addition, the report discusses various methods for converting other forms of phosphorus to orthophosphate, and the elimination of interference due to arsenic, based on the reduction of arsenate to arsenite.

Bretscher [5.77] discusses reduction reagents for the spectrophotometric determination of phosphate in sewage as the phosphomolybdenum blue complex. He points out that the disadvantage of aqueous stannous chloride reduction reagent is that it oxidizes rapidly and a fresh solution must prepared daily.

Solutions of stannous chloride in glycerol were found to be stable for at least six months.

5.1.7.2 High Performance Liquid Chromatography

The high performance liquid chromatographic method described in Sect. 5.1.1 for the determination of phosphate in natural waters is also applicable to sewage effluents.

5.1.8 Boiler Feed Water

5.1.8.1 Ion Chromatography

This technique has been used for the determination of phosphate and sulphate in boiler feed water [5.78].

5.1.9 Soil and Plant Extracts

5.1.9.1 Spectrophotometric Method

Spectrophotometric evaluation at 880 nm of the phosphomolybdate complex has been used to determine phosphates in sodium bicarbonate extracts (pH 8.5) of soil [5.79].

5.1.9.2 Ion Chromatography

Bradfield and Cooke [5.80] have described an ion chromatographic method using a UV detector for the determination of chloride, nitrate, sulphate and phosphate in water extracts of plants and soils. Plant materials are heated for 30 min at 70 °C with water to extract anions. Soils are leached with water and Dowex 50-X4 ion exchange resin added to the aqueous extract which is then passed through a Sep-Pak C18 cartridge and the eluate then passed through the ion chromatographic column. The best separation of these anions was obtained using a 5×10^{-4} mol l^{-1} potassium hydrogen phthalate solution in 20% methanol at pH 4.9. A reverse phase system was employed. Detection times were 5.5, 7.9, 12.6 and 18 min for chloride, nitrate, phosphate and sulphate respectively. Recoveries ranged from 84 to 108% with a mean of 97%. This technique is discussed further in Chap. 11 (Ion Chromatography).

5.1.10 Blood and Serum

5.1.10.1 Flow Injection Analysis

This technique using a glassy carbon electrode as a voltammetric detector, has been employed for the determination of phosphate in water, blood, eluent, serum and hydroponic water [5.81]. In this method the liquid sample is injected into an eluent which is 2% vol.% in ammonium molybdate and 0.6 vol.% in concentrated sulphuric acid. Molydbophosphate, which is determined by reduction at the glassy carbon electrode, is fully formed when a 3 m delay coil is incorporated before the detector and a flow rate of 4 ml min^{-1} is used. The method is applicable to samples containing more than 10^{-5} mol l^{-1} phosphate.

5.2 Condensed Phosphates

Condensed phosphate is a generic title for all phosphates which are formed by the removal of one or more water molecules from orthophosphoric acid, which may be represented as $3H_2O \cdot P_2O_5$. These include metaphosphoric acid ($H_2O \cdot P_2O_5$ i.e. HPO_3), pyrophosphoric acid ($2H_2O \cdot P_2O_5$ i.e. $H_4P_2O_7$), ultraphosphoric acid (general formula $mH_2O \cdot P_2O_5$ where $0 \leqslant m < 1$), and polyphosphoric acid (general formula $mH_3PO_4 - (m-1)H_2O$ where $1 \leqslant m < \infty$). The most common condensed phosphates are pyrophosphate and tripolyphosphate which have dissociable protons and behave as typical polyprotic acids. Both exist in three forms.

Pyrophosphoric acid is represented as $H_4P_2O_7$ ($2H_2OP_2O_5$). The three forms of pyrophosphate are

$H_3P_2O_7^{1-}$, $H_2P_2O_7^{2-}$ and $H P_2 O_7^{3-}$.

Tripolyphosphoric acid is represented as $H_5P_3O_{10}$ ($3H_3PO_4 - 2H_2O$). The three forms of tripolyphosphate are

$H_3P_3O_{10}^{2-}$, $H_2P_3O_{10}^{3-}$ and $HP_3O_{10}^{4-}$.

A number of factors affect the rate at which condensed and organic phosphates undergo hydrolytic degradation in aqueous solution. The major environmental factors in decreasing order of effectiveness are listed in Table 5.14. Consideration of Table 5.14 indicates that pH is one of the factors which exerts a considerable influence on the rate of hydrolysis.

In domestic waste water, all these factors may influence the hydrolysis of condensed phosphate to orthophosphate. Thus if the sample is not analysed immediately, changes in the concentrations of condensed and orthophosphates will occur. If the forms of phosphate are to be determined rather than total phosphate, hydrolysis must be prevented.

Table 5.14. Major environmental factors affecting the hydrolytic degradation of condensed phosphates

Factor	Approximate effect on rate
Temperature	10^5–10^6 faster from freezing to boiling
pH	10^3–10^4 slower from strong acid to base
Enzymes	As much as 10^5–10^6 faster
Colloidal gels	As much as 10^4–10^5 faster
Complexing cations	Very many-fold faster in most cases
Concentration	Roughly proportional
Ionic environment in the solution	Several-fold change

5.2.1 Ion Exclusion Chromatography

Ion exclusion chromatography provides a convenient way to separate molecular acids from highly ionized substances. The separation column is packed with a cation exchange resin in the H^+ form so that salts are converted to the corresponding acid. Ionized acids pass rapidly through the column while molecular acids are held up to varying degrees. A conductivity detector is commonly used.

Tanaka et al. [5.82–5.84] have reported that the separation of phosphate from chloride, sulphate and several condensated phosphates $P_2O_7^{4-}$, (pyrophosphate) and $P_3O_{10}^{2-}$ and $P_3O_{10}^{3-}$ (tripolyphosphates) could be achieved by ion exclusion chromatography on a cation exchange resin in the H^+ form by elution with an acetone:water and dioxan:water mixture. As the separation mechanism of the ion exclusion chromatography by elution with water alone for numerous anions or their respective acids is based on the Donnan membrane equilibrium principle (ion exclusion effect), it is a highly useful technique for the separation of non-electrolytes such as carbonic acid from electrolytes such as hydrochloric acid and sulphuric acid [5.85]. Ion exclusion chromatography has also been coupled to ion chromatography to determine simultaneously both weak and strong acids [5.86]. The ion exclusion chromatography separation of orthophosphate from the strong acid anions by elution with organic solvent:water described above is based on this ion exclusion effect and/or the partition effect between the cation exchange resin phase (water rich) and the mobile phase (organic solvent rich) owing to the hydration of the resin [5.82] and phosphate has been monitored as the corresponding acid (H_3PO_4) with a flow coulometric detector for the detection of H^+ ion and a conductometric detector.

5.2.2 High Performance Liquid Chromatography

High performance liquid chromatography has also been used [5.87] to determine orthophosphate, pyrophosphate and tripolyphosphate.

Speciation and quantitative analysis of orthophosphate, pyrophosphate, and tripolyphosphate are performed by using high performance liquid chromatography with inductively coupled argon plasma emission spectrometric detection. High performance liquid chromatography is used to separate mixtures of phosphates on an anion exchange column using tartrate magnesium buffer. The ICP is used as a selective detector by observing P II emissions at 214.9 nm. The detection limit is 0.5, 1 and 3 µg respectively for ortho-, pyro- and tripolyphosphate respectively.

In this procedure a Waters Associates liquid chromatograph was connected to anion exchange columns. Column packings used were µ-Bondapak-NH$_2$ and Nagel SA. The outlet of the liquid chromatograph was connected to cross flow nebulizer of the ICP spectrometer (Jarrell-Ash-Atom-Comp 750) with Teflon tubing (1/16 in × 5 in). Eluant flow rate was set at 1 ml min^{-1}.

Phosphorus was monitored at 214.9 nm (second order). The analog signal was taken out through the profile mode of a computer program. In order to filter noise, a 50 kΩ resistor and 20 µF condenser were installed before the recording, giving an approximate time constant of 1 s.

The ICP operational parameters were RF power 1.3 kw, coolant gas 18 l min^{-1}, sample gas 0.5 l min^{-1}, plasma gas 0.1 min^{-1}, and observation height in plasma 15 mm.

The results in Table 5.15 indicate that under these conditions the detection response is substantially independent of the chemical form of the phosphorus compounds.

Optimum chromatographic conditions are summarized in Table 5.16.

Among the eluants examined, oxalate plus magnesium gave the best resolution, tartrate plus magnesium gave the second best but formate plus magnesium and acetate plus magnesium gave poor resolutions. Oxalate plus magnesium buffer should be freshly prepared as it is subject to decomposition. The column should be rinsed after usage.

Calibration curves prepared by injecting an aliquot of different concentrations showed a linear relationship with a dynamic range of $10^{-4} - 10^{-2}$ mol l^{-1}. The practical detection limit was estimated for the chromatogram taking the

Table 5.15. ICP response factor for various phosphorus compounds. (From [5.87]; Copyright 1981, American Chemical Society)

Compounds		Response factor	Purity
KH_2PO_4	Orthophosphate	1.00	1.00
$Na_4P_2O_7 \cdot 10H_2O$	Pyrophosphate	1.01	1.02
$Na_5P_3O_{10}$	tripolyphosphate	1.00	1.00
$(NaPO_3)_n$	metaphosphate	1.02	0.93
$C_{10}H_{12}N_5O_7PNa \cdot 6H_2O$	–	0.98	1.14
$C_{10}H_{12}N_5O_{10}P_2$	–	1.03	0.98
$C_{10}H_{14}N_5O_{13}P_3Na_3 \cdot 3H_2O$	–	1.02	0.97

Table 5.16. HPLC variables. (From [5.87]; Copyright 1981, American Chemical Society)

Column	μ-Bondapak-NH$_2$ 1/8 in. × 1 ft
	Nucleosil N(CH$_3$)$_3$ 1/8 in. × 1 ft
Eluant	Tartaric acid (0.1 mol l^{-1}) + MgSO$_4$ (0.01 mol^{-1}) + NaOH (0.1 mol^{-1})
Buffers	Oxalic acid (0.1 mol^{-1}) + MgSO$_4$ (0.01 mol^{-1}) + NaOH (0.1 mol^{-1})
	Ammonium formate (0.2 mol^{-1}) + MgSO$_4$ (0.01 mol^{-1}) + HCl (0.1 mol^{-1})
	Acetic acid (0.2 mol^{-1}) + NaOH (0.1 mol^{-1})
	Tris(hydroxymethyl)amine (0.3 mol^{-1}) + H$_2$SO$_4$ (0.075 mol^{-1})

peak height to noise level ratio at 2. It was 0.5 μg of P, 1 μg of P and 3 μg of P for mono-, di- and triphosphate.

With the use of a basic eluant such as ammonium or Tris buffer, it was noticed that triphosphate was rapidly decomposed to orthophosphate and pyrophosphate during chromatography. Figure. 5.6 shows the chromatogram of orthophosphate, pyrophosphate, triphosphate and metaphosphate using Tris buffer as an eluant. It may reasonably be expected that the triphosphate peak would appear later than pyrophosphate. However, no peak appeared in such a later retention time range and, instead, two peaks which had the same retention time as orthophosphate and diphosphate appeared. The areas of the first and second peaks had the ratio of 1:2 indicating the mole ratio of 1:1. This suggests that the third P — O bond of triphosphate is readily cleaved at the column head giving 1 mol of phosphate and 1 mol of diphosphate. Diphosphate is stable enough to give a single peak during chromatography. When triphosphate aqueous solution was kept at room temperature for 1 week, significant deterioration was observed. Basic condition and the presence of μ-Bondapak-NH$_2$ might accelerate decomposition rate.

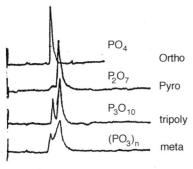

Fig. 5.6. Chromatogram of orthophosphate, pyrophosphate, tripolyphosphate and trimetaphosphate. Eluant was Tris-sulfate buffer (pH 8.2) (From [5.87])

Table 5.17. A Elution recovery of orthophosphate from HPLC column at different sample concentrations. (From [5.87]; Copyright 1981, American Chemical Society)

Concn, mol l^{-1}	Emission intens integrated	Recovery, %
10^{-3}	114	100
10^{-4}	125	109
10^{-5}	108	95
10^{-6}	100	87

When the ionic strength of the water sample is not high, phosphate can be trapped at the column top and be eluted with eluant buffer. By use of this concentration technique, dilute solutions can be analysed. Orthophosphate aqueous solution (10^{-4}, 10^{-5} and 10^{-6} mol l^{-1} was pumped into the column at the rate of 1.5, 1.5 and 2.0 ml min^{-1} respectively. Since too much time is required to load the water sample, a faster loading rate is chosen for the most dilute sample (10^{-6} mol l^{-1}). Recovery from the column is listed in Table 5.17.

In Fig. 5.7 a typical chromatogram obtained by this technique for a mixture of orthophosphate, pyrophosphate and tripolyphosphate ($P_3O_{10}^{5-}$) is shown.

Yaza et al. [5.88] used high performance liquid chromatography with a photodiode array detector to analyse mixtures of polyphosphates, orthophosphates and monoflurophosphates.

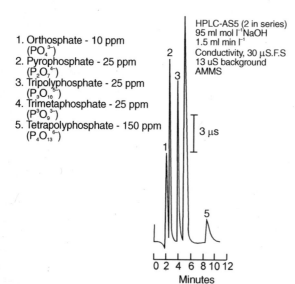

Fig. 5.7. Chromatogram of orthophosphate, pyrophosphate tripolyphosphate, trimetaphosphate and tetrapolyphosphate Column is μ-Bondapak. NH$_2$ and the eluant was oxalate-Mg buffer at the flow rate of 1.0 ml min^{-1}, 50 μl of mixed solution containing 10^{-3} mol l^{-1} of each phosphate (From [5.87])

5.3 References

5.1 Skjemstad JO, Rieve R (1978) Environmental Quality 7: 137
5.2 Henrickslon A, Selmer-Olsen AR (1970) Analyst (London) 95: 514
5.3 Tery DR (1966) Anal Chim Acta 34: 41
5.4 Strickland JOH, Parsons TR (1960) A manual of seawater analysis Bulletin of the Fisheries Research Board, Ottawa, Canada 125: 1
5.5 Rickford GP, Willett IR (1981) Water Res 15: 512
5.6 Pakalns P, Steman HT (1976) Water Research 10: 437
5.7 Stainton MP (1980) Canadian Journal of Fisheries and Aquatic Sciences 37: 472
5.8 Downes MT (1978) Water Research 12: 743
5.9 de Jonge VN, Villerius LA (1980) Marine Chemistry 9: 191
5.10 Lei W, Fujiwara F, Fuwa K (1983) Anal Chem 55: 951
5.11 Beyer A (1977) Gas-u-Wasserfach (Wasser and Abwasser) 118: 327
5.12 Goulden PD, Brooksbank P (1974) Limnology and Oceanography 19: 705
5.13 Harmsen J (1984) Anal Chim Acta 156: 339
5.14 Potman W, Lijkema L (1979) Water Research 13: 801
5.15 Fernandez JA, Niell F, Lucena J (1985) J Limnology and Oceanography 30: 227
5.16 Altmann HJ, Feverstenau E, Gielewski A, Scholz L (1971) Z Analyt Chem 256: 274
5.17 Motomizu S, Wakimoto T, Toei K (1983) Analyst (London) 108: 361
5.18 To YS, Randall CW (1977) Journal of Water Pollution Control Federation 49: 689
5.19 Boyd CE, Tucker L (1980) Transactions of American Fisheries Society 109: 314
5.20 Tarapchak SJ (1983) Journal of Environmental Quality 12: 105
5.21 Tarapchak SJ, Bigelow SM, Rubitschun C (1982) Canadian Journal of Fisheries and Aquatic Sciences 39: 296
5.22 Nurnberg G (1984) Water Research 18: 296
5.23 Strickland JDH, Parsons TR (1968) in "A Practical Handbook of Seawater Analysis", Fisheries Research Board of Canada 167: 311pp
5.24 Murphy J, Riley JP (1962) Anal Chim Acta 27: 31
5.25 Stauffer RE (1983) Anal Chem 55: 1205
5.26 Matsubara C, Yamamoto Y, Takamura K (1987) Analyst (London) 112: 1257
5.27 Fogg AG, Bselsu NK (1984) Analyst (London) 109: 19
5.28 Motomizu S, Wackimoto T, Toei K (1983) Talanta 30: 333
5.29 Janse TAHM, Van der Weil PFA, Kateman G (1983) Anal Chim Acta 155: 89
5.30 Nasu T, Kan M (1988) Analyst (London) 113: 1683
5.31 Sakurai N, Kadohata K, Ichinose N (1983) Fresenius Zeit schrift fur Analytische Chemie 314: 634
5.32 Morgan DK, Danielson ND (1983) Journal of Chromatography 262: 265
5.33 Braungart M, Russel H (1985) Chromatographia 19: 185
5.34 Moss PE, Stephen MI (1985) Analytical Proceedings, Chemical Society (London) 22: 5
5.35 Zelinski I, Zelinska V, Kamenski D, Havassa P, Lednarova V (1984) Journal of Chromatography 294: 317
5.36 Smee BW, Hall GEM, Koop DJ (1978) Journal of Geochemical Exploration 10: 245
5.37 Dits JS, Smeenk GNM (1985) H_2O 18: 7
5.38 Hight SC, Bet-pere F, Jaseskis B (1982) Talanta 29: 721
5.39 Matsunaga K, Kudo I, Yanagur M (1986) Anal Chim Acta 185: 355
5.40 Stevens RJ (1979) Water Research 13: 763
5.41 Steinberg C, Kuhl K, Schrimpf A (1982) Zeitschrift fur Analytische Chemie 15: 26
5.42 Chrost RJ, Siuda W, Albrecht D, Overbeck T (1986) ominology and Oceanography 31: 662
5.43 Lin CI, Huber CO (1972) Anal Chem 44: 2200
5.44 Crawford WE, Lin CI, Huber CO (1973) Anal Chim Acta 64: 387
5.45 Allen HE, Hahn RB (1969) Environmental Science and Technology 3: 844
5.46 Leyden DE, Nonidez WK, Carr PW (1975) Anal Chem 47: 1449

5.47 Hahn RB, Schmitt TM (1969) Anal Chem 41: 359
5.48 Ryden JC, Syers JK, Harris RF (1972) Analyst (London) 97: 903
5.49 Nelson DW, Romkous MJM (1972) Journal of Environmental Quality 1: 323
5.50 Lindquist GL, Cox JA (1974) Anal Chem 46: 360
5.51 Shapiro J (1973) Limnology and Oceanography 18: 143
5.52 Johnson DL, Pilson EQ (1972) Anal Chim Acta 58: 289
5.53 Airey D, Dal Pont G, Sanders G (1984) Analytica Chimica Acta 166: 79
5.54 Tyree SY Bynum MAO (1984) Limnology and Oceanography 29: 1337
5.55 Reijuders HFR, Melis PHAA, Griepuik B (1983) Freseinius Zeitschrift für Analytische Chemie 314: 627
5.56 Slanina I, Bakker F, Bruya-Hes A, Mols JJ (1980) Analytica Chimica Acta 113: 331
5.57 Reijuders HFR, Melis PHA, Griepuik B (1983) Zeitschrift für Analytische Chemie 314: 627
5.58 Faigle W, Klockow D (1981) Fresenius Zeitschrift für Analytische Chemie 306: 190
5.59 Schwabe R, Darimont T, Mohlman T, Pabel M, Sonneborn M (1983) Journal of Environmental Analytical Chemistry 14: 169
5.60 Pyen GS, Erdmann DE (1983) Analytica Chimica Acta 149: 355
5.61 Wetzel RA, Anderson CL, Schleicher H, Crook DG (1979) Analytical Chemistry 51: 1532
5.62 Matzuchita S, Baba N, Ikeshiga T (1984) Analytical Chemistry 56: 822
5.63 Oikawa K, Saitoh H (1982) Chemosphere 11: 933
5.64 Mosko J (1984) Analytical Chemistry 56: 629
5.65 Buchanan MA, Easty DB (1973) Tapp 56: 127
5.66 "Methods for Chemical Analysis of Water and Wastes", Environmental Protection Agency, Cincinnati, Ohio (1971)
5.67 Murphey A, Riley JP (1963) Analytical Abstracts 10: 1219
5.68 Osburn QW, Lemmel DE, Downey RL (1974) Environmental Science and Technology 8: 363
5.69 Crawford WE, Lin CI, Huber CO (1973) Anal Chim Acta 64: 387
5.70 Sakurai N, Kadohata K, Ichinose N (1983) Freseinius Zeitschrift fur Analytische Chemie 314: 634
5.71 Morgan DK, Danilson ND (1983) Journal of Chromatography 262: 265
5.72 Tanaka K, Ishizuka T (1982) Water Research 16: 719
5.73 Tanaka Y, Muto G (1973) Anal Chem 45: 1864
5.74 Green LW, Woods JR (1981) Analytical Chemistry 53: 2187
5.75 Rossin AC, Lester JN (1980) Environmental Technology Letters 1: 9
5.76 Department of the Environment/National Water Council Standing Committee of Analysts. H.M. Stationary Office, London. Method for the examination of water and associated materials. Phosphorus in waters, effluents and sewages 1980 (1981)
5.77 Bretscher U, Gas U (1976) Abwasserfach (Wasser, Abwasser) 117: 31
5.78 Stevens TS, Turkelson VT (1977) Analytical Chemistry 49: 1176
5.79 Murphy J, Riley JP (1962) Analytica Chimica Acta 27: 31
5.80 Bradfield EG, Cooke DJ (1985) Analyst (London) 110: 1409
5.81 Fogg AC, Bsebsu NT (1982) Analyst (London) 107: 566
5.82 Tanaka F, Ishizaka T (1980) Journal of Chromatography 190: 7
5.83 Tanaka K, Sunahara H (1978) Bunseki Kagaku 27: 95
5.84 Tanaka K, Nakajima K, Sunahara M (1977) Bunseki Kagaku 26: 102
5.85 Tanaka K, Ishizuka T, Sunahara M (1979) Journal of Chromatography 174: 153
5.86 Dionex Application Notes. No. 19 Ion Chromatography systems Analysis of strong and weak acids in coffee extracts, Dionex Corporation (1979)
5.87 Morita A (1981) Anal Chem 53: 1997
5.88 Yaza N, Nakashima S, Nakazato VT, Uida N, Kodama H, Tateda A (1992) Analytical Chemistry 64: 1499

6 Silicon-Containing Anions

6.1 Silicate

6.1.1 Natural Waters

6.1.1.1 Spectrophotometric Methods

The National Water Council (UK) [6.1] has published two spectrophotometric methods for the determination of soluble molybdate-reactive silicon (mainly monomeric and dimeric silicic acids and silicate). The first is based on the use of ascorbic acid for the reduction step, and the second is based on the use of 1-amino-2-naphthol-4-sulphonic acid to effect the reduction when analysing clean waters, especially those of low silicon content. In addition, three alternative pretreatments are described for converting other forms to soluble molybdate-reactive silicon. The method with ascorbic acid as the reducing agent is particularly suited to a wide range of silicon concentrations up to 100 mg l^{-1}, measurements being made in 10-mm cells at 700 nm, but it is claimed to be easily modified for higher or low sensitivity.

Pakalns [6.2] studied the effects of surfactants on the spectrophotometric molybdenum blue method for the determination of silicon. Three molybdenum blue methods using different reducing agents were used; 1-amino-2-naphthol-4-sulphonic acid [6.3], sodium sulphite [6.4] and ascorbic acid [6.5]. Pakalns [6.2] studied interference by anionic, cationic and non-ionic detergents, sodium tripholyphosphate, and sodium pyrophosphate. Interferences were very large for the cationic and non-ionic detergents, but negligible for the biodegradable linear alkyl sulphate type detergents. The amino-naphthol-sulphonic acid method is recommended (Table 6.1).

Defosse [6.6] described a rapid spectrophotometric method for determining silicon based on reaction with a reagent comprising quinine sulphate and ammonium molybdate acidified with nitric and sulphuric acids. The column was evaluated at 420 nm against a reagent blank. The method is applicable to samples containing up to 5 mg l^{-1} phosphorus. Other reagents used for the reduction of β-silicomolybdic acid to methylene blue include sodium sulphite-4-amino-3-hydroxynaphthalene-1-sulphonic acid [6.7, 6.8], sodium sulphite-metal [6.11], stannous chloride-glycerol [6.9], and stannic chloride-ascorbic acid-oxalic acid [6.10].

Table 6.1. The effect of standard addition on the recovery of silicon with and without linear alkyl sulphate (LAS) addition. (From [6.2]; Copyright 1976, Pergamon Publishing Co., UK)

Detergent	Amount taken ($mg\ l^{-1}$)	Method	Recovery (%)			
			No LAS added		LAS added	
			$1.0\ mg\ l^{-1ac}$ $7.5\ mg\ l^{-1b}$	$2.0\ mg\ l^{-1ac}$ $15\ mg\ l^{-1b}$	$1.0\ mg\ l^{-1ac}$ $7.5\ mg\ l^{-1b}$	$2.0\ mg\ l^{-1ac}$ $15\ mg\ l^{-1b}$
Cationic	20	A	90	90	100[d]	100[d]
	40	A	90	80	100[d]	100[d]
	75	A			100[d]	100[d]
	100	B	100	100		
	150	B	108	103	100[e]	100[e]
	200	B			100[e]	100[e]
	15	C	92	92		
	40	C	52	81	100[f]	100[f]
	100	C			100[f]	100[f]
Anionic	2500	A	115	115		
	750	B	114	107		
	2500	C	118	118		
Nonionic	50	A	100	100		
	100	A	100	92	100[g]	100[g]
	150	A	96	86	100[g]	100[g]
	150	B	125	110	100[h]	100[h]
	200	B			100[h]	100[h]
	10	C	104	104		
	20	C	114	110	100[i]	100[i]
	30	C	127	119	j	j

[a] Concentration of silica in Method A (1 amino-Z-naphthol-4-Sulphonic acid).
[b] Concentration of Silica in Method B (Sodium Sulphite).
[c] Concentration of Silica in Method C (ascorbic acid).
[d] Six-fold Concentration of LAS added.
[e] Five-fold Concentration of LAS added.
[f] Ten-fold Concentration of LAS added.
[g] 15 mg of LAS added.
[h] Five-fold Concentration of LAS added.
[i] Twelve-fold Concentration of LAS added.
[j] Heavy turbidity.

Pavlova et al. [6.9] extract the silicomolybdic acid into butanol or isoamyl alcohol prior to reduction to methylene blue in order to improve the sensitivity of the method.

Kasahara et al. [6.13] collect the methylene blue on a solvent-soluble membrane filter which is then dissolved in N,N-dimethylformide and determined spectrophotometrically. The detection limit of this method is 0.5 µg $SiO_2\ l^{-1}$.

Yoshimura et al. [6.12] determined silicic acid in natural water at the µg l^{-1} level or lower by gel-phase colorimetry based on the direct measurement of absorbance of molybdenum blue adsorbed on sephadex gel.

6.1.1.2 Gas Phase Chemiluminescence Analysis

Fujiwara et al. [6.24] have described the application of this technique to the determination of silicate in natural waters. Silicate is converted to silane using lithium aluminium hydride in a Teflon tube heated to 200 °C. The silane is collected and mixed with ozone and analysed by chemiluminescence analyses. The detection limit is 0.5 µg silicon (absolute) Phosphorus and arsenic interfere.

6.1.1.3 Flow Injection Analysis

This technique has been applied to the determination of silicate [6.14].

6.1.1.4 Segmented Flow Analysis

This technique, discussed in Sect. 1.3.2, can be applied to the determination of silicate in the concentration ranges 0–5 and 0–100 mg l^{-1} (as Si) in water.

6.1.1.5 Ion Chromatography

Okada and Kuwamoto [6.15] have described an ion chromatographic method for the determination of silicate in natural waters. This method is discussed more fully in Sect. 11.1.

6.1.2 Potable Water

6.1.2.1 Polarographic Method

Iyer et al. [6.16] converted trace amounts of silica in potable water to silicomolybdate by addition of ammonium molybdate, followed by methylethylketone to ensure that only the beta form of silicomolybdate was present. Buffering with citrate masked any phosphate or excess molybdate. The initial potential of the hanging mercury drop electrode was kept at -0.2 V for 2 min and silicomolybdate accumulated by adsorption in the interface. Its bulk concentration was determined by measuring the peak height at -0.36 V.

Reagents. As below.
  Standard silicate solution (1 g Si l^{-1}). Fuze 1.000 g of quartz powder (99.99% purity) with 5 g of anhydrous sodium carbonate in a platinum crucible. Dissolve the melt in water, make up to 1000 ml and store in a polyethylene container. Prepare fresh standard aliquots containing 0.1 mg l^{-1} Si.

Ammonium molybdate (1%). Dissolve 1 g of ammonium heptamolybdate in 100 ml of water.

Filtrate buffer. Dissolve 21 g of citric acid monohydrate in 200 ml of 1 mol l^{-1} sodium hydroxide, dilute to 950 ml, and adjust the pH to 2.5 with 1 mol l^{-1} hydrochloric acid and make up to 1000 ml.

Methylethylketone, E. Merck, Darmstadt, FRG.

Store all solutions in polethylene flasks, except the ketone.

Apparatus. Voltammetric measurements performed using a Polarographic analyzer, model 170, in a thermostated Metrohm cell using a three electrode system. The potentiostat and the incorporated electronic integrating circuit make the recording of current potential and charge potential curves possible. The working electrode is a Metrohm hanging mercury drop electrode (HMDE), type E 410. The reference electrode is a saturated Ag/AgCl electrode. The auxiliary electrode is a coiled platinum wire in a tube separated from the cell by a porous glass plug. The temperature of the cell was maintained at $25 \pm 0.05\,°C$ using a thermostat. The solution is deaerated by passing a stream of pure nitrogen (99.999%). The nitrogen atmosphere is maintained over the solution during measurement.

Procedure. Pipette an aliquot of sample solution containing 0.15 to 1.0 µg silicon into a polyethylene wide-mouthed bottle. Add 1 ml ammonium molybdate and dilute to 15 ml with water. Adjust the pH to 1.6 ± 0.1 mol l^{-1} hydrochloric acid. After 30 min add 0.2 ml of 1 mol l^{-1} hydrochloric acid, 5 ml methylethylketone and make up to 50 ml with citrate buffer. Take the voltammogram after a reaction period of 30 min and subsequent deaeration for 15 min at open circuit. Keep the hanging drop mercury electrode initially at a potential of -0.2 V for a waiting time t_a of exactly 2 min and scan in the range from -0.2 V to -0.8 V at a scan rate of 100 mV/s. Measure the peak height at the potential of -0.36 V. Make two standard additions for the evaluation of the silica content in the sample.

A typical voltammogram of silicomolybdate at the hanging drop mercury electrode obtained by following the standard procedure is given in Fig. 6.1. The reduction of the silicomolybdate gives a well defined peak at -0.36 V, followed by reduction of the molybdate ion at -0.52 V. With increasing concentration of silica only the first peak at -0.36 V increased linearly unlike the second peak.

Results obtained on applying the procedure to samples of potable water are summarized in Table 6.2. For comparison the samples were analysed by the spectrophotometric molybdenum blue method. The values obtained by voltammetry agree satisfactory with this method, confirming the accuracy of the voltammetric determination.

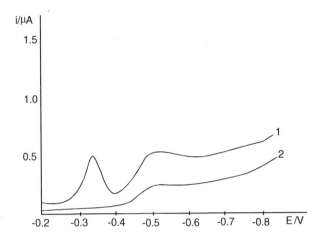

Fig. 6.1. Voltammogram of silicomolybdate at the HMDE in the presence of methylethylketone, 0.2 mol l^{-1} citrate buffer, pH 2.5, 3×10^{-7} mol l^{-1} SiO$_2$, 10% ketone. MMDE surface area 2.8×10^{-2} cm^2, scan rate v 100 mV s^{-1}, adsorption time t 120 s at the starting potential $E_2 - 0.2$ V. Curve 1 – supporting electrolyte, curve 2-silico- molybdate (From [6.16])

Table 6.2. Determination of silica (SiO$_2$) in potable water. (From [6.16]; Copyright 1981, Marcel Dekker Inc., USA)

Nature of water samples		Voltammetric method µg l^{-1} Si O$_2$	Spectrophotometric method
Drinking water	1	140	130
	2	150	155
	3	190	180
	4	110	130
	5	130	140

6.1.3 Industrial Effluents

6.1.3.1 Spectrophotometric Method

A modification of the standard molybdenum blue spectrophotometric method has been used to determine silicate in industrial effluents [6.17].

6.1.3.2 Flameless Atomic Absorption Spectrometry

This technique has been shown to be capable of determining down to 25 µg l^{-1} silica in industrial process waters [6.18].

6.1.4 Waste Waters

6.1.4.1 Atomic Absorption Spectrometry

Looyenga and Huber [6.19] used atomic absorption inhibition titration to determine silicate in waste water. The method is based on the strong inhibition by silicate of the absorption of magnesium ions which is measured continuously during titration. The sample solution (pH 3 to 4) is first passed through Amberlite IR-120 resin (H^+ form) to remove magnesium and other interfering cations and the percolate is then titrated with a standard solution (100–200 mg l^{-1}); the titrated solution is simultaneously aspirated into the hydrogen air flame for measurement of the absorption at 285.2 nm. The end-point is detected by a sharp increase in the magnesium signal. The detection limit is 0.1 mg l^{-1} and there is no interference from phosphate or sulphate.

6.1.5 Boiler Feed Water

6.1.5.1 Spectrophotometric Methods

Spectrophotometric methods have been described for the determination of traces of silicate in high pressure boiler feed water [6.20, 6.21] and deionized water [6.22]. silicomolybdate procedures employing 4-amino-3-hydroxynaphthalene-1-sulphonic acid have been employed [6.20, 6.21]. Spectrophotometric evaluation of the reduced complex is carried out at 805–810 nm in these methods.

It is essential, in order to avoid silica contamination, to use precleaned polyethylene sample bottles and apparatus. Ramachandran and Gupta [6.22] performed studies on the effects of different reducing agents on silicate determination by the molybdenum blue method. Best results were obtained using ascorbic acid and antimony. The antimony did not participate in the complex but enhanced its absorbance. The recommended procedure is described. It was relatively rapid and sensitive. Beer's law was obeyed for silicate concentrations of 20–1000 µg l^{-1}.

Pilipenko et al. [6.23] compared spectrophotometric and chemiluminescent methods of determining silicate in deionized water. They studied different reducing agents in the molybdenum blue spectrophotometric method, namely tungstosilicic acid, stannous chloride, 1-amino-2-naphthol-sulphonic acid and mixtures of metal and sodium sulphite. They also studied the extraction of the blue tungstosilicic acid complex by isobutanol and the formation of complexes of tungstosilicic acid with crystal violet and methylene blue. The chemiluminescent method was based on the oxidation of luminol by molybdosilicic acid in alkaline medium. These workers concluded that the best method, from the points of view of detection limit, reproducibility, simplicity and rapidity, was one based on the reduction of molybdosilicic acid by 1-amino-2-naphthol-4-sulphonic acid.

6.2 References

6.1 Department of the Environment of National Water Council Standing Committee of Analysts, H.M. Stationary Office, London (1981) Methods for the examination of water and associated materials (RP22 BCENV) Silicon in water and effluents
6.2 Pakalns P (1976) Water Res 10: 1083
6.3 American Public Health Association (1971) Standard Methods for the examination of water and wastewater APHA, AWWA, WPFC 13th ediction APHA Washington
6.4 Brown E, Skongstad MW, Fishman MJ (1970) Methods for the collection and analysis of water samples for dissolved minerals and gases, U.S. Government Printing Office
6.5 Pakalns P (1971) Anal Chim Acta 54: 281
6.6 Defosse C (1972) Trib Cebedeau 25: 267
6.7 Morrison IR, Wilson IR (1969) Analyst (London) 94: 54
6.8 Fanning KA, Pilson MI (1973) Anal Chem 45: 136
6.9 Pavlova MV, Podal'skaya BI, Shafran IG (1972) Trudy uses Nauchno-issled. Inst Khim Reakt osabo Inst Khim Veskchestv, 34: 185. Ref: Zhur Khim 19GD (11) Abstract No. 11G200
6.10 Smith JD, Milne PE (1981) Anal Chim Acta 123: 263
6.11 Strickland PC, Parsons P (1968) Bulletin Fisheries of the Research Board of Canada, Ottawa, Canada 65: 110
6.12 Yoshimura K, Motomura M, Tarutini T (1984) Anal Chem 56: 2342
6.13 Kasahara I, Ferai R, Murai Y, Hata N, Toguchi S, Goto K (1987) Anal Chem 59: 787
6.14 Fogg AG, Bschsu NK (1981) Analyst (London) 106: 1288
6.15 Okada T, Kuwamoto T (1985) Analytical Chemistry 57: 258
6.16 Iyer CSP, Valanta P, Nurnberg HW (1981) Analytical Letters 14: 921
6.17 Duce FA, Yamamura SS (1970) Talanta 17: 143
6.18 Rawn JA, Henn EL (1979) Anal Chem 51: 452
6.19 Looyenga RW, Huber CO (1971) Anal Chem 43: 498
6.20 Morrison IR, Wilson AL (1969) Analyst (London) 94: 54
6.21 Baker PM, Farrant BR (1968) Analyst (London) 93: 732
6.22 Ramachandran R, Gupta PK (1985) Anal Chim Acta 172: 307
6.23 Pilipenko AT, Ryabushko VO, Terletskaya AV (1983) Soviet Journal of Water Chemistry and Technology, 5: 45
6.24 Fujiwara K, Uchida M, Chen M, Kuwamoto Y, Kumamaru T (1993) Analytical Chemistry 65: 1814

7 Boron-Containing Anions

7.1 Borate

7.1.1 Natural Waters

7.1.1.1 Spectrophotometric Methods

The most popular spectroscopic method for the determination of borate in water is based on the use of curcumin as chromogenic reagent [7.1 – 7.4] which reacts with borate to produce rosocyanine, a pink coloured dye.

Choi and Chen [7.2] have developed a curcumin method using a 20 vol.% 2-ethyl-1, 3-hexanediol in methyl isobutyl ketone as an extracting solvent. A 5-cm cylindrical quartz cell is used to measure absorbance against 100% ethyl alcohol at a wavelength of 550 nm. This method eliminates inconvenient time limits and the interference of fluorides or nitrates. Using a 5-cm cylindrical cell, the detectable range of boron is approximately $0.25 – 3.0\,mg\,l^{-1}$ with less than 5% error.

Reagents. As follows.
    Hydrochloric acid, 1 N.
    2-ethyl 1,3-hexanediol in methyl isobutyl ketone, 20 vol.%.
    Curcumin, 0.375 g curcumin 100 ml^{-1} glacial acetic acid.
    Sulphuric acid, concentrated.

Procedure. Transfer 1 ml of solution into a 10-ml polypropylene tube (17 × 100 mm) with a friction-fit closure. Add 2 ml hydrochloric acid and mix gently. Then add 3 ml of 20 vol.% 2-ethyl-1,3-hexanediol in methyl isobutyl ketone and shake vigorously for 2 min. Leave the solution to stand until the two phases are clearly separated. Transfer 1-ml aliquot from the upper, organic solvent layer of the extract into a 50-ml polypropylene centrifuge tube with a screw cap. Add 2 ml of curcumin reagent and then 0.5 ml of concentrated sulphuric acid. Close the centrifuge tube immediately after the addition of concentrated sulphuric acid and swirl the bottle until the yellowish colour turns to a dark purple colour. The reaction may be terminated at any time under 23 h by adding 50 ml of 100% ethyl alcohol. Measure the absorbance of the solution against 100% ethyl alcohol at a wavelength 550 nm with a 5-cm cylindrical quartz cell (glass and quartz function equally well in the visible portion of the spectrum). The absorbance of the solution can be measured 60 m after the addition of alcohol. The time span between the addition of alcohol and the measurement of absorbance can be arranged according to the number of samples, but better accuracy is achieved when the intervals are relatively close.

The results obtained from the study of interference from fluoride, magnesium, calcium and nitrate ions are listed in Tables 7.1–7.3. These data clearly show that there is no significant interference from any of these ions.

The detection limit of this method is approximately $0.2 \, mg \, l^{-1}$ borate.

Aznarez et al. [7.4] have described a more sensitive spectrophotometric and fluorimetric method based on, respectively, the curcumin complex and molecular fluorescence with dibenzoylmethane. In both cases the boron is extracted into isobutyl methyl ketone with methylpentanediol.

Apparatus. A Pye Unicam SP8 100 spectrophotometer or equivalent with special equipment for fluorescence measurements, a thermostatic bath and a mechanical shaker. Glass materials must be avoided in order to eliminate boron contamination. PTFE, polyethylene or platinum materials are recommended.

Reagents. As follows.

Boron stock standard solution, $1000 \, \mu g \, ml^{-1}$. Dissolve 2.860 g of dried boric acid in 500 ml of water.

Boron working standard solutions. Dilute the stock standard solution just before use.

Extraction solution, 20 vol.% methylpentanediol in isobutyl methyl ketone, stored in a polethylene bottle.

Curcumin solution, 0.1% w/v in glacial acetic acid. Prepared just before use.

Dibenzoylmethane solution, 0.1% w/v in isobutyl methyl ketone.

Quinine sulphate solution, 0.05% w/v in $0.1 \, mol \, l^{-1}$ sulphuric acid.

Procedure. Dilute 50 ml of natural water with hydrochloric acid (1 + 1) to 100 ml in a calibrated flask. Place a measured volume of sample solution (less than 50 ml) containing 10–100 µg of boron (spectrophotometric method) or 0.5–5 µg of boron (fluorometric method) in a separating funnel. Extract three times with 10-ml volumes of isobutyl methyl ketone in order to eliminate any iron interference. Add 10 ml of extraction solution and shake for about 5 min with a mechanical shaker. Dry the organic phase with about 1 g of anhydrous sodium sulphate.

Spectrophotometric Procedure. Pipette 3 ml of the organic phase into a polyethylene test tube with a hermetic cap and add 2 ml of curcumin solution and 2 ml of concentrated phosphoric acid. Shake the sealed test tube for 2 min and heat at $70 \pm 3 \,°C$ for 1 h in a thermostated bath. After rapid external cooling to room

Table 7.1. Magnesium effect on boron determination using improved curcumin method. (From [7.2]; Copyright 1979, American Water Works Association)

Solution	mg l^{-1}	Absorbance difference between solution and blank after time elapsed							Mean absorbance difference	Standard deviation	Coeffic of variation %
		10 min	50 min	100 min	150 min	200 min	250 min				
Boron 0.5	Magnesium 0	0.184	0.184	0.187	0.186	0.189	0.180		0.1887	0.0038	2.04
	0.5	0.214	0.193	0.192	0.195	0.197	0.190				
	5.0	0.214	0.184	0.195	0.190	0.199	0.187				
	25.0	0.190	0.179	0.190	0.185	0.192	0.181				
	50.0	0.188	0.182	0.194	0.187	0.187	0.183				
	250.0	0.206	0.185	0.192	0.192	0.197	0.186				
2.5	0	1.061	1.037	1.034	1.036	1.047	1.031		1.0376	0.0043	0.44
	25.0	1.056	1.040	1.044	1.038	1.049	1.041				
	250.0	1.038	1.031	1.032	1.032	1.041	1.031				

Data obtained at 10 minutes are not included in the statistical analysis.

Table 7.2. Calcium effect on boron determination using improved curcumin method. (From [7.2]; Copyright 1979, American Water Works Association)

Solution	mg l^{-1}	Absorbance difference between solution and blank after time elapsed						Mean absorbance difference	Standard deviation	Coeffic of variation %
		10 min	50 min	100 min	150 min	200 min	250 min			
Boron	Calcium									
0.5	0	0.199	0.188	0.187	0.194	0.192	0.196	0.1976	0.0045	2.28
	10	0.201	0.196	0.193	0.198	0.199	0.200			
	100	0.215	0.202	0.203	0.200	0.200	0.202			
	500	0.189	0.180	0.179	0.179	0.176	0.184			
	1000	0.211	0.199	0.201	0.201	0.201	0.200			
2.5	0	1.075	1.069	1.063	1.065	1.070	1.070	1.0722	0.0092	0.86
	10	1.073	1.050	1.063	1.061	1.058	1.066			
	100	1.076	1.068	1.071	1.067	1.068	1.066			
	500	1.081	1.072	1.077	1.079	1.076	1.078			
	1000	1.071	1.072	1.079	1.081	1.076	1.072			
	2000	1.097	1.078	1.087	1.091	1.086	1.088			

These data are excluded in the statistical analysis Data obtained at 10 min and excluded data are not included.

Table 7.3. Nitrate effect on boron determination using improved curcumin method. (From [7.2]; Copyright 1979, American Water Works Association)

Solution	mg l^{-1}	Absorbance difference between solution and blank after time elapsed						Mean absorbance difference	Standard deviation	Coeffic of variation %
		10 min	50 min	100 min	150 min	200 min	250 min			
Boron	Nitrate									
0.5	0	0.199	0.188	0.182	0.184	0.187	0.180			
	5	0.212	0.192	0.190	0.192	0.188	0.197			
	25	0.208	0.197	0.194	0.197	0.197	0.199	0.1899	0.0048	2.54
	50	0.197	0.184	0.186	0.184	0.191	0.189			
	100	0.214	0.194	0.192	0.189	0.184	0.190			
2.5	0	1.066	1.045	1.034	1.040	1.044	1.048			
	5	1.082	1.040	1.041	1.044	1.048	1.044			
	25	1.084	1.053	1.048	1.046	1.042	1.046	1.0438	0.0023	0.22
	50	1.078	1.042	1.038	1.044	1.046	1.036			
	100	1.070	1.046	1.048	1.042	1.048	1.042			

temperature, measure the absorbance of the solution at 510 nm against a reagent blank solution within 45 min.

Prepare a calibration graph as follows. To different volumes of standard solution containing 10–100 µg of boron add an equal volume of hydrochloric acid (1 + 1) and extract with 10 ml of extraction solution according to the above extraction procedure. Pipette 3 ml of the organic phase and carry out the same spectrophotometric procedure.

Fluorimetric Procedure. Pipette 3 ml of the organic extraction phase into a polyethylene test tube with a hermetic cap and add 2 ml of dibenzoylmethane solution and 2 ml of concentrated phosphoric acid. Shake the sealed test tube for 2 min and heat at $80 \pm 3\,°C$ for 30 min in a thermostated bath. After rapid external cooling to room temperature, measure the relative fluorescence intensity of the solution at 400 nm or by using a Kodak 2B cut-off filter, with excitation at 390 nm and quinine sulphate solution as reference, within 45 min.

Prepare a calibration graph as follows. To different volumes of standard solution containing 0.5–5 µg of boron, add an equal volume of hydrochloric acid (1 + 1) and extract with 10 ml of extraction solution according to the above extraction procedure. Pipette 3 ml of the organic phase and carry out the same fluorimetric procedure.

The absorption spectrum of the boron–curcumin compound in isobutyl methyl ketone obtained following the above procedure exhibits maximum absorbance at 510 nm when measured against a reagent blank solution.

The fluorescent excitation spectrum of the boric acid dibenzoylmethane compound in isobutyl methyl ketone against quinine sulphate solution is shown in Fig. 7.1. The wavelength of the maximum excitation radiation was 390 nm. The maximum relative fluorescence intensity was measured at 400 nm or by using a Kodak 2B cut-off filter (400 nm cut-off).

The calibration graph at 510 nm obtained in the curcumin method is a straight line and Beer's law is obeyed from 0.5 to $5\,\mu g\,ml^{-1}$ of boron in the final measured solution (correspondent to 10–110 µg of boron in the aquous phase). The molar absorptivity, calculated from the slope of the statistical working calibration graph at 510 nm, was $2905\,l\,mol^{-1}\,cm^{-1}$. The precision of the method for ten replicate determinations was 0.6%. The absorbance of the reagent blank solution at 510 nm was 0.010 ± 0.003 for ten replicate determinations. Therefore, the detection limit was $40\,\mu g\,l^{-1}$ (as boron) in the final measured solution. The corresponding detection limit in the fluorimetric method was $1\,\mu g\,l^{-1}$ (as boron). Interference effects in both methods are illustrated in Table 7.4. The interference of iron at concentrations greater than $7 \times 10^{-5}\,mol\,l^{-1}$ can be eliminated by removing iron as the chloro complex by extraction with isobutyl methyl ketone. The total elimination of iron was not necessary as the phosphoric acid masked the residual iron in the boric acid–curcumin reaction.

The results for boron determinations in some natural waters are shown in Table 7.5

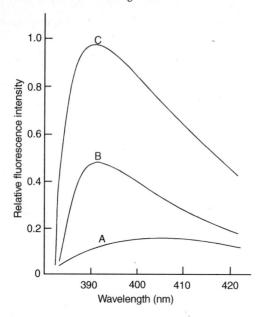

Fig. 7.1. Fluorescence excitation spectra against quinine sulphate solution as reference: *curve A*, reagent blank solution; *curve B*, boron - DBM, 50 µg l^{-1} of boron; *curve C*, boron - DBM, 100 µg l^{-1} of boron (From [7.4])

Table 7.4. Effect of foreign ions on boron determination. (From [7.4]; Copyright 1983, Royal Society of Chemistry, UK)

Foreign ion	Maximum concentration tested without giving interference mol l^{-1}	
	Spectrophotometric method[a]	Fluorimetric method[b]
Cl^-	7.2	7.2
SO_4^{2-}	2.5	2.5
NH_4^+	1.0	0.8
Na^+	0.2	0.2
K^+, Ca^{2+}, Al^{3-}	0.5	0.4
NO_3^-, NO_3^-, Cr^{3+}	0.2	0.2
$Mg^{2+}, HCO_3^-, CO_3^{2-}$	0.05	0.05
Mn^{2+}	0.05	0.01
F^-	0.02	0.01
$Cu^{2+}, Zn^{2+}, Ni^{2+}$	4×10^{-3}	4×10^{-3}
$Sr^{2+}, PO_4^{3-}, SIO_3^{2-}, Br^-$	3×10^{-3}	3×10^{-3}
Fe^{3+}	7×10^{-5c}	7×10^{-5c}
Fe^{3+} (by elimination with three 10-ml IBMK washes)	0.1	0.01

[a] Determination of 58.3 µg of boron by the spectrophotometric method.
[b] Determination of 2.38 µg of boron by the fluorimetric method.
[c] Tolerance limit (M) as the concentration level at which the interferent causes an error of not more than ±2% (spectrophotometric method) or ±3% (fluorimetric method).

Table 7.5. Determination of boron in natural waters by the fluorimetric method. (From [7.4]; Copyright 1983, Royal Society of Chemistry, UK)

Mean boron content – standard deviation [a]/µg l^{-1}	Mean recovery %[b]
200 ± 3	98.2
182 ± 3	97.5
132 ± 2	99.3
143 ± 2	99.7
124 ± 2	100.2
151 ± 2	100.8
253 ± 5	101.0
237 ± 3	99.9
186 ± 4	99.8
157 ± 2	190.6
255 ± 3	98.6
152 ± 2	97.7
208 ± 3	97.8
249 ± 2	99.9
201 ± 3	99.6

[a]Ten determinations.
[b]Boron added to spiked samples, 2.0 µg; three determinations.

A further chromogenic reagent that has been used for the determination of borate is Azomethine [7.3, 7.5, 7.6] which forms a yellow complex with borate. Randow [7.6] evaluated the Azomethine H method for several natural water samples and also for model solutions containing humic acids with standard amounts of boron added to the sample. The effects of illumination, temperature and time on the colour development were examined, together with other factors such as the initial colour of the sample and the container materials. Over the temperature range 12–24 °C the absorbance maximum was reached within 1 h and remained stable for 1 h afterwards. The sample colour intensity necessitated a proportionate correction to the observed value. Recoveries of added boron ranged from 86 to 104% for concentrations between 20 and 200 µg l^{-1}. The limit of detection was 16 µg l^{-1}. Glass vessels were only suitable for brief contact during operations such as pipetting or dilution; all other operations should be performed in polythene beakers or flasks.

Bilikova [7.3] executed microgram level determinations of boron in natural waters by two methods, namely the modified curcumin method with rosocyanin, and the azomethane method. Both methods allowed determination of boron in water at between 0.1 and 1.6 µg in 50- and 100-ml volumes respectively.

The Department of the Environment UK [7.7] have discussed four alternative methods for the determination of boron comprising the curcumin method, automated and manual methods based on azomethine H, with measurements of absorbance of the complexes formed, and a mannitol titration method suitable for a wide range of boron concentrations from 0.5 to 1000 mg l^{-1}. Interference by nitrate is discussed.

Yoshimura et al. [7.8] determined boron in natural waters after specific adsorption on Sephadex gel.

Other chromogenic reagents that have been employed for the spectrophotometric determination of borate include crystal violet which forms a complex with boron, having an adsorption maximum of 600 nm [7.9], ferroin [7.10], carmine [7.11], chromotropic acid [7.12, 7.13], and various diketones [7.14]. Spectrofluorimetric reagents that have been used include 4'-chloro-2-hydroxy-4-methoxybenzophenone [7.15, 7.16], 1,8-dihydroxynaphthalene-4-sulphonic acid [7.17], dianthrinamide [7.18], and azomethine [7.7].

7.1.1.2 Segmented Flow Analysis

This technique, discussed in Sect. 1.3.2, has been applied to the determination of $0-2$ mg l^{-1} borate in water.

7.1.1.3 Ion-Selective Electrodes

These are available commercially for the determination of borate in water.

7.1.1.4 Flow Injection Analysis

Lussier et al. [7.26] determined borate in light and heavy water by flow injection analysis with indirect UV-visible spectrophotometric detection.

7.1.1.5 Ion Chromatography

Hill and Lash [7.19] have discussed the determination of borate in natural water, see Sect. 11.1 for further detail.

7.1.2 Industrial Effluents

7.1.2.1 Ion Chromatography

The technique mentioned above [7.19] has also been applied to industrial effluents.

7.1.3 Sewage

7.1.3.1 Spectrophotometric Methods

Borate has been determined [7.20] in amounts down to 0.02 mg l^{-1} in sewage effluents spectrophotometrically by reaction with phenol and carminic acid in concentrated sulphuric acid to produce a coloured compound with an absorption maximum at 610 nm.

The standard curcumin method [7.21, 7.22] has been found to be suitable for the determination of borate in industrial effluents.

7.1.4 Soil

7.1.4.1 Spectrophotometric Methods

A method has been described [7.23] for the determination of borate in soils based on conversion of borate in a hot water extract to fluoroborate by the action of orthophosphoric acid and sodium fluoride.

The concentration of fluoroborate is measured spectrophotometrically as the blue complex formed with methylene blue which is extracted into 1, 2 dichloroethane. Nitrates and nitrites interfere but these can be removed by reduction with zinc powder and orthophosphoric acid.

7.1.5 Plants

Ogner [7.24] has described an automated analyser method for the determination of boron-containing anions in plants. This is based on the formation of a fluorescent complex between these anions and carminic acid at pH7. The plant tissues are ashed at 550 °C and the residue dissolved in 0.5 N hydrochloric acid prior to adjustment to pH 6 – 7 with sodium carbonate solution. The solution is excited at 470 nm and fluorescence intensities measured at 585 nm. Interferences by the reaction of some cations with carminic acid is overcome by passing the solution through an ion exchange column to exchange the cations for sodium ions. Analytical recoveries of boron anions were in the range 98 – 104%. The detection limit of the method was 5 µg l^{-1} boron.

7.1.6 Plant Extracts

Lopez Gracia et al. [7.25] have described a rapid and sensitive spectrophotometric method for the determination of boron complex anions in plant extracts and waters which is based on the formation of a blue complex at pH 1–2 between the anionic complex of boric acid with 2,6-dihydroxybenzoic acid and crystal violet. The colour is stabilized with polyvinyl alcohol. At 600 nm the calibration graph is linear in the range 0.3–4.5 µg boron per 25 ml of final solution with a relative standard deviation of $\pm 2.6\%$ for $\mu g\ l^{-1}$ of boron. In this procedure to determine borate in plant tissues, the dried tissue is treated with calcium hydroxide, then ashed at 400 °C. The ash is digested with 1 N sulphuric acid and heated to 80 °C, neutralized with cadmium hydroxide then treated with acidic 2,6-dihydroxy-benzoic acid and crystal violet and the colour evaluated spectrophotometrically at 600 nm. Most of the ions present in natural waters or plant extracts do not interfere in the determination of boron complex anions by this procedure. Recoveries of boron from water samples and plant extracts were in the range of 97–102%.

7.2 References

7.1 Dyrssen DW, Novikov YP, Uppstroem LF (1972) Anal Chim Acta 60: 139
7.2 Choi WW, Chen K (1979) American Water Works Association 71: 153
7.3 Bilikova A (1982) Vodni Hospodarstvi Series B 32: 266
7.4 Aznarez J, Bonilla A, Vidal JC (1983) Analyst (London) 108: 368
7.5 Spencer RR, Erdmann DE (1979) Environmental Science and Technology 13: 954
7.6 Randow FFE (1985) Zeitschrift fur Wasser and Abwasser, Forschung 18: 290
7.7 Department of the Environment of National Water Council Standing Committee of Analysts, HM Stationary Office, London, Methods for the examination of waters and associated materials Boron in waters, effluents, sewage and some solids (1981)
7.8 Yoshimura K, Karija R, Tarutani T (1979) Anal Chim Acta 109: 115
7.9 Garcia IL, Cordoba MH, Sanchez-Pedreno C (1985) Analyst (London) 110: 1259
7.10 Bassett J, Matthews PJ (1974) Analyst (London) 99: 1
7.11 Malyugu DP (1969) Zavod Lab 35: 279
7.12 Lapid J, Farhz S, Karesh Y (1971) Analytical Letters 9: 355
7.13 Korenaga T, Motomizum S, Toei K (1978) Analyst (London) 103: 745
7.14 Lambert JL, Paurkstelis JV, Bruckdorfoxn RA (1978) Anal Chem 50: 820
7.15 Liebich B, Monnier D, Marcantonatos M (1970) Anal Chim Acta 52: 305
7.16 Monnier D, Marcantonatos M (1982) Mitt Geb Lebensmitteluntersm U Hyg 63: 212
7.17 Motomizu S, Toei K (1980) Analyst (London) 105: 955
7.18 Levinson AA (1971) Water Research 5: 41
7.19 Hill CJ, Lash RP (1980) Analytical Chemistry 52: 24
7.20 Lionnel LJ (1970) Analyst (London) 95: 194
7.21 Burton NG, Tait BH (1969) Journal of the American Water Works Association 61: 357

7.22 American Public Health Authority (APHA) Standard methods for the Analysis of waters and wastewaters, 12th edition, New York (1965)
7.23 Ducret L (1957) Analytica Chimica Acta 17: 213
7.24 Ogner G (1980) Analyst (London) 105: 916
7.25 Lopez Garcia I, Cordolia MN, Sanchez-Pedreno C (1985) Analyst (London) 110: 1259
7.26 Lussler T, Gilbert R, Hubert J (1992) Analytical Chemistry 64: 2201

8 Carbonate, Bicarbonate and Total Alkalinity

8.1 Total Alkalinity

8.1.1 Natural Waters

8.1.1.1 Titration Methods

Legrand et al. [8.1] carried out, alkalinity and acidity titrations on polar snow. The method is accurate to $\pm\,0.2\,\mu$ equiv l^{-1} acidity or alkalinity. Contamination free sampling and sample handling techniques are described.

The National Water Council (U.K.) [8.2] has described three methods for determining alkalinity in river water. These comprise the determination of alkalinities in the range 20 – 100 mg l^{-1} as calcium carbonate, the determination of alkalinity at low levels, i.e. below 20 µg l^{-1} as calcium carbonate, and the use of a continuous flow air-segmented system for automated determinations in the range 20 – 300 mg l^{-1} as calcium carbonate in polluted waters and certain types of trade waste waters. Each of the methods described is empirical in nature but correct adherence to the stated procedure will enable reproducible results to be obtained. In the preamble, a number of alternative systems commonly used to express alkalinity values are compared and some examples of the amounts of alkalinity contributed by the sodium salts of weak acids are tabulated.

Pearson [8.3] has proposed a method for determining alkalinity (and acidity) employing fixed endpoints and involving less manipulation of samples than existing methods of analysis. It involves calculation of alkalinity and acidity values to equivalence end-points from results of titrations to pH 4.5 and pH 8.3. The validity of the technique was confirmed experimentally within a limited range of conditions, although some anomalous results may be obtained in some waters such as those of high ionic strength.

The measurement of alkalinity in natural water has been discussed by several other workers [8.4 – 8.9].

8.1.1.2 Spectrophotometric Methods

Hillbom et al. [8.10] describe a probe which allows potentiometric titrations to be automatically carried out in situ in natural waters.

Automated methods of measuring alkalinity are typically based on the change of pH of a buffer containing methyl orange. The methyl orange is buffered at a pH at the acid end of the transition range of methyl orange. Additions of alkalinity cause the colour of the methyl orange to change gradually from orange to yellow as the pH increases within the transition range.

Two instruments that are used to automate the measurement of alkalinity are the Technicon AutoAnalyzer and the Coulter Kem-O-Lab Industrial Model, made by Coulter Electronics. Coulter Electronics has published methods that are useful for concentrations of alkalinity from 10 to 500 mg l^{-1} as $CaCO_3$, whereas published methods for the Technicon AutoAnalyser are useful only in the range of 100 – 500 mg l^{-1}. Willis and Mullings [8.12] modified this apparatus to make it suitable for the measurement of concentrations of less than 50 mg l^{-1}, a concentration commonly encountered in natural water samples.

Reagents. As follows.

Methyl orange Stock Solution, methyl orange 0.05% in water.

pH 3.1 buffer. Dilute 5.1 g of potassium acid phthalate with 87.5 ml of 0.1 N hydrochloric acid to 500 ml with distilled water and adjust the pH to 3.1 with 1 N hydrochloric acid.

Methyl orange reagent. Mix varying amounts of methyl orange stock solution and buffer (Table 8.1) and dilute with distilled water. Prepare this reagent fresh daily. If the reagent is to be used with the Technicon Auto analyser then also add 2 ml of Brij-35 (obtained from Technicon Instrument).

Alkalinity Standards. Prepare alkalinity standards of approximately 1000 mg l^{-1} as $CaCO_3$ by dissolving 1.060 g of sodium carbonate in 1000 ml of distilled water. Determine the exact concentration by titrating with 0.02 N hydrochloric acid.

Apparatus. The Technicon AutoAnalyzer II, Industrial Model, was used with the same manifold as described by Technicon for the measurement of alkalinity. A 15- × 2.0-mm flow cell and a 550-nm filter were used. The electronic gain on the colorimeters can be varied from 0 to 100. For the purpose of comparison the gain was set at 10.0. The Coulter Kem-o-Lab (IKL) was used following manufacturers recommended procedure for measuring alkalinity. The IKL automatically takes a standard volume of sample and mixes it with 0.5 ml of distilled water and 1.0 ml of methyl orange reagent. Measure the absorbance with a 2-cm flow cell and a wavelength of 540 nm 12 min after the reagent is added.

Procedure. Perform the titration with a Brinkman pH meter, Model 506, to monitor the pH and a Dosimat E535 (Brinkman) to deliver the titrant. Titrate a 25-ml sample to a preliminary end-point of pH 5.6 with 0.02 N sulphuric acid. With this volume of titrant as a measure of the approximate alkalinity, the sample was further titrated to a final end-point based on the relationship of equivalence point pH to total alkalinity as described by Thomas and Lynch [8.11].

Willis and Mullings [8.12] concluded that three reagents would be sufficient to measure alkalinity at concentrations ranging from 5 to 700 mg l^{-1} $CaCO_3$ on the Technicon AutoAnalyser. Solutions with concentration from 5 to 550 mg l^{-1} $CaCO_3$ will be within the linear range of at least one reagent. The reagents used are listed in Table 8.1 along with the range of linear and usable concentrations. The reagents that were used with the Coulter Instrument are the same as those described in their manual and are included in Table 8.1 for completeness. The calibration curves have a linear portion in the middle and a curved portion on each end. The usable range includes the linear portion as well as some of the curved portion at each end.

In Table 8.2 are shown results obtained in alkalinity determinations on natural water samples by the manual titration method and the two automated methods discusssed above.

The titration method has greater reproducibility than either of the two automated methods, but it is much slower. About 8 h are required for a set of 80 samples, whereas for both the automated methods analysis takes less than 2 h for a set of 80. Also, once the standardization is completed, the automated instruments require almost no attention. Both instruments were equally reproducible, i.e. the coefficient of variation was 2% for four or more samples analyzed on separate days. Special precautions were needed in the automated methods to avoid carryover effects between one sample and the next.

Van Staden and Van Vleit [8.13] have described a simple, automated procedure for determining total alkalinity in surface water, groundwater and potable water. Up to 120 samples per hour can be analysed with a coefficient of variation of better than 1.40%.

8.1.1.3 Segmented Flow Analysis

This technique, discussed in Sect.1.3.2, can be applied to the determination of total alkalinity in the concentration range 0–500 mg l^{-1} (as $CaCO_3$) in water.

8.1.1.4 Flow Injection Analysis

Basson and Van Staden [8.14] have described a single point titration system for the determination of total alkalinity in surface waters, ground and domestic waters, based on flow injection analysis. Water samples are injected automatically into the flow system using a flow injection sampling unit. A sampling cycle of 30 s allows 120 samples per hour to be analysed. The sample is reacted with an acid linear response buffer solution and the pH of the resulting solution is measured with a glass electrode in a flow through assembly.

Reagents. Unless otherwise specified all reagents are of analytical grade quality.

Carrier stream, 0.1 mol l^{-1} sodium chloride solution. Dissolve 5.844 g sodium chloride in 1 l distilled water.

Table 8.1. Reagents and instruments used for different alkalinity determination. (From [8.12]; Copyright 1983, American Chemical Society)

Reagent[a]	Instrument	ml of buffer	ml of methyl orange	ml of Brij-35	Range, mg l^{-1}	
					Linear	Usable
A	TAA[b]	10	85	2	10–90	4–140
B	TTA[b]	60	85	2	50–250	10–600
C	TAA[b]	125	50	2	200–650	10–900
D	IKL[c]	43.5	32.6	0	10–90	4–150
E	IKL[c]	392	32.6	0	160–900	50–1000

[a] Each reagent also contains enough water to make a total volume of 500 ml.
[b] TAA-Technicon AutoAnalyser.
[c] IKL-Coulter Kem-O-Lab Industrial Model IKL.

Table 8.2. Comparison of results of alkalinity tests by titration. (From [8.12]; Copyright 1983, American Chemical Society)

Sample no.	Titration, mg l^{-1}		Coulter kem-o-Lab mg l^{-1}		Technicon Autoanalyser mg l^{-1}	
	Mean	std dev	Mean	Std dev[a]	Mean	Std dev[a]
1103	26.1	0.9	26.5	0.5	25.4	0.2
1111	44.6	0.3	44.1	0.5	44.3	0.1
1112	14.9	0.7	15.2	0.9	13.9	1.2
1113	13.4	0.7	13.8	1.3	13.4	2.3
1114	20.3	0.7	10.6	0.6	18.7	1.5
1116	56.6	0.3	57.3	0.3	57.8	0.9
1118	147	0.3	144	3.5	141	1.4
1119	104	0.3	102	3.6	104	2.9
1120	226	1.6	259	2.4	264	1.2
Ave		0.6		1.4		1.3

[a]Based on five measurements for each determination of mean or standard deviation.

Buffer solution. Prepare the following buffer component solutions.

a. Add 6.01 g diethylmalonate to 7 ml concentrated hydrochloric acid solution and dilute to 500 ml with distilled water.
b. Weigh 7.88 g citric acid and dilute to 500 ml with distilled water.
c. Weigh 5.22 g p-nitrophenol and dilute to 500 ml with distilled water.
d. Weigh 3.87 g diethylbarbituric acid and dilute to 500 ml with distilled water.
e. Weigh 2.32 g boric acid, dissolve and dilute to 500 ml with distilled water.
f. Weigh 3.53 g phenol, dissolve and dilute to 500 ml with distilled water.

Prepare a stock buffer solution by mixing 50 ml of solution (a), 50 ml of solution (b), 59 ml of solution (c), 150 ml of solution (d), 50 ml of solution (e) and 50 ml of solution (f). This solution is made 0.10 mol l^{-1} with respect to sodium chloride.

Apparatus. Peristaltic pump turning at 30 rpm, Scientific flow-injection sampler (Breda), Orion model 601 pH meter, Beckman micro blood electrode consisting of a micro sensor assembly (Beckman Part No 580621) and a pH-electrode (Beckman Part No. 39045), Hitachi model QPD 53 potentiometric recorder.

Procedure. A schematic flow diagram for the system is shown in Fig. 8.1. Tygon tubing of 0.50 mm diameter was used and all connectors were made from Perspex. The tube lengths used are as indicated in Fig. 8.1.

Automatically inject water samples into the flow system by using a flow injection sampling unit. Use a sampling cycle of 30 s between successive samples giving a capacity of 120 samples per hour. Actuate the valve system on a time basis which is correlated with the sampler unit. The sampling valve actuates every 28 s after movement of the sampler to the next sample.

Figure 8.2 shows the appearance of recorder tracing as well as calibration curves when analysing a series of standards in the range 20–300 mg l^{-1} bicarbonate solution. Although the calibration curve is not linear over the complete concentration range, experimental results indicate linearity in the range 30–75 mg l^{-1}.

In Table 8.3 a comparison is made of results obtained by the above flow injection method, by an electrometric titration method, and by an automated bromocresol green indicator method.

Flow injection results compare favourably with a standard electrometric method and in some instances better results were obtained when compared to the automated bromocresol green method. This is probably due to colour interferences from these water specific samples. Student's t-test was applied in comparing both sets of results with the manual electrometric method. Table 8.3 reflects these calculations. Both procedures give results that differ to a statistically insignificant extent when compared to the electrometric procedure. However, the difference between the procedures is smaller than the difference between the continuous flow procedure and the electrometric procedure.

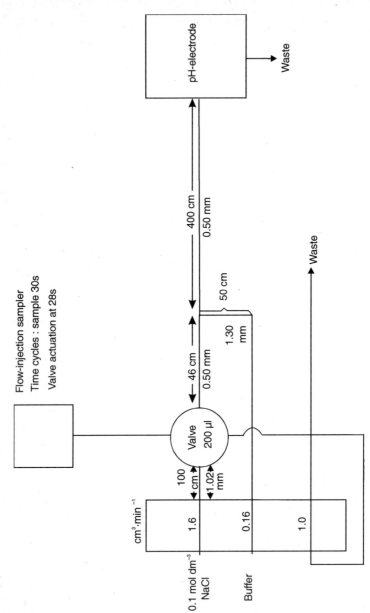

Fig. 8.1. Flow system for the determination of total alkalinity. Sampling rate 120 samples h^{-1}. Tube length and diameters are given in cm and mm (From [8.14])

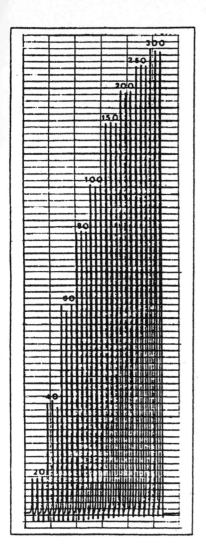
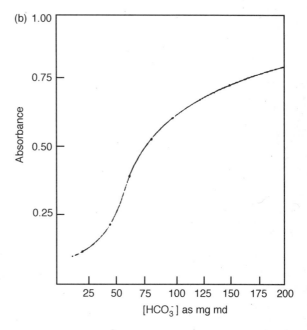

Fig. 8.2. **a** Calibration peaks obtained with 10 cm² buffer solution diluted to 100 cm³ with distilled water. The values on top of the peaks represent mgl^{-1} bicarbonate ions. **b** Corresponding calibration graph From [8.14]

8.1.1.5 Alkalinity, pH and Total Ionic Concentration Flow Injection Analysis

Canate et al. [8.15] have described a flow injection analysis system for the simultaneous determination of the alkalinity, pH and total ionic concentration of potable water. The pH measurements are performed by means of a glass calomel microelectrode inserted in to the water stream. The alkalinity and total ionic concentration are determined by flow injection analysis titrations, acid base reactions and spectrophotometric detection. In addition, the determination of the total ionic concentration requires the incorporation of an ion exchange minicolumn. The results obtained are in agreement with those obtained by conventional methods for the determination of these parameters and are obtained with a higher degrees of automation, which results in a smaller reagent consumption, greater convenience and higher reproducibility and sampling fre-

Table 8.3. Comparison of results obtained by the FIA-method the electrometric method and the automated segmented bromo cresol green method[a]. (From [8.14]; Copyright 1980, United Trades Press, UK)

Sample	Automated segmented method [HCO_3^-] in mg l^{-1}	Electrometric method [HCO_3^-] in mg l^{-1}	FIA-method [HCO_3^-] in mg l^{-1}	Coefficient of variation %
1	226	268	248	0.99
2	485	546	532	0.46
3	266	298	305	0.98
4	266	306	313	0.56
5	213	232	236	0.71
6	250	276	288	0.66
7	120	142	135	0.98
8	160	259	287	0.48
9	267	291	324	0.31
10	480	317	332	0.56
11	107	120	110	0.26
12	117	123	110	0.27
13	119	123	112	0.24
14	120	125	113	0.31
15	113	109	110	0.42
16	113	117	108	0.55
17	96	98	95	0.30
18	138	142	132	0.22
19	178	201	187	0.28
20	183	204	199	0.35
21	174	194	193	0.21
22	174	196	186	0.48
23	167	186	178	0.21
24	65	68	66	0.30
25	30	28	26	1.10

[a] 15 tests were carried out on each sample.

quency. A microcomputer is installed for on-line data collection and the treatment and display of results.

Reagents. As follows. Aqueous solution of methyl red in an acidic medium. Phenolphthalein in a basic medium and standard buffer solutions (Merck) of pH 4.00 and 7.00.

Apparatus. Pye Unicam SP6-500 spectrophotometer connected to a Radiometer REC-80 recorder, a FIA tron 721 flow cell with a glass calomel microelectrode connected to a Beckman 3500 pH meter for continuous monitoring of the pH, a 150 × 3 mm Omnifit minicolumn packed with Dowex 50-X8 cationic resin in proton form, a Gilson Minipuls-2 peristaltic pump, a Hellman 178.10QS flow cell, two Tecator L100-1 injection valves, a Rheodyne 5301 three-way valve and a Hewlett Packard HP–85 microcomputer with HP 3478A analogue to digital converter and an HP-1B 82937 interface.

Manifold. Figure 8.3 shows the reversed FIA configuration used. It involves the use of two detectors; a potentiometric detector ME (a glass microelectrode) located before the confluence of the sample with any reagent for the continuous monitoring of pH, and a spectrophotometric detector for the monitoring of the acid base reactions. Two serial injection valves are placed along the water stream. The first of these, S′, is aimed at incorporating the ion exchange microcolumn (IE) into the stream in order to determine the total ionic concentration. This valve is synchronised with the selecting valve S and in the filling position of valve S′, selects channel 1 (methyl red in an acidic medium) which fills the loop of V_i (ordinary injection valve 1126 μl). The alkalinity of the water sample is determined in this position. In the emptying position, S′ incorporates the minicolumn IE, S selects channel 2 (phenolphthalein in a basic medium) and the determination of the total ionic concentration is carried out. The flow rate adopted was 0.75 ml min^{-1} and the reactor lengths $L_1 = 2.40$ cm and $L_2 = 110$ cm.

The microcomputer collects the data from both detectors and, by means of the computer program described below, processes the data and gives the pH, alkalinity and total ionic concentration for each sample.

Computer Program. The microcomputer collects the pH and absorbance values (measurement frequency, 0.2 s) which are introduced into a BASIC program. The data acquisition process starts at a time from the reagent injection equal to the travel time, so that as little memory as possible is used. Data treatment is carried out according to the scheme in Fig. 8.4. The width of each peak (in seconds) at heights pre-fixed by the user for the sample, standards and blanks, is collected. The program features a subroutine for the optimisation of each system. The program also allows data to be discarded from anomalous peaks or standards that would give a low correlation coefficient for the linear range of the calibration graph. This corresponds to the change in the width (Δt) at a height h as a function of log C (C = analyte concentration).

Alkalinity is measured by titration with hydrochloric acid to the methyl red end-point.

Similar to a conventional titration, the reaction is monitored after the equivalence point so that the blank signal (distilled water) shows a higher peak

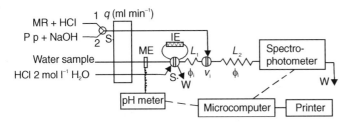

Fig. 8.3. Manifold for the simultaneous determination of pH, alkalinity and total ionic concentration, MR = methylred, pp = phenol phthalein (for details see text) (From [8.15])

width. The greater the alkalinity, the smaller is the peak width. The analytical signal was $\Delta t_{blank} - \Delta t_{sample}$, so that the slopes of the calibration graphs were positive. To determine total ionic concentration the salts present in the water sample are transformed into their corresponding acids on passing through the cationic resin in the H$^+$ form. The acids are then titrated with sodium hydroxide and phenolphthalein. In this method the minicolumn is placed in the loop of valve S' (Fig. 8.3) which is inserted into the circulating stream in the emptying position.

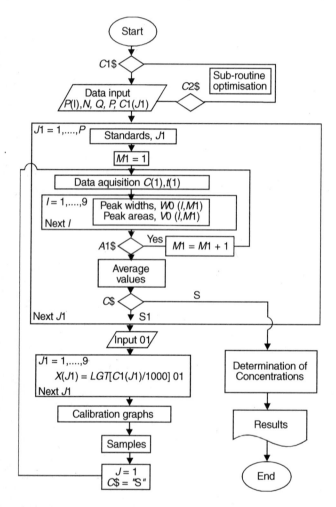

Fig. 8.4. Flow diagram of the computer program, C1% = inquiries to the program; P(I) = measurement height; N = number of absorbance–time data points; Q = interval between two consecutive data points; P = number of standards; C1 (J1) = concentration of standards; C(I) = concentration of analyte; t(I) = time; WO = peak width; VO = peak area; 01 = molecular mass; S = sample; S1 = standard; and J1, M1 and 1 = counters (From [8.15])

In the filling position, the minicolumn is regenerated with a sodium hydroxide stream and washed with distilled water, thus allowing continuous use for 4–5 h.

The reaction is monitored at 552 nm (after the equivalence point in the conventional titration) so that, as in the above system, the peak width is wider for the blank than for the samples.

As the insertion of a glass calomel microelectrode into the flow injection analysis configuration introduced no species that disturb the system, pH monitoring can be readily carried out by inserting the sensor prior to the reagent injection (Fig. 8.3) and calibrating it with standard buffer solutions at the working flow rate.

Table 8.4 summarizes the features of the simultaneous determination of alkalinity, pH and total ionic concentration.

Canate et al. [8.15] tested the validity of their procedure by analysing a series of synthetic samples, the composition and results of which are shown in Table 8.5. In general, the errors in the determination of alkalinity are higher than those of the total ionic concentration.

The method was also applied to the determination of these parameters in potable water. The results (Table 8.6) have been compared with those of conventional manual methods (e.g. for alkalinity titration with hydrochloric acid and methyl orange as indicator and for total ionic concentration, passage of the water sample through a cationic ion exchange column in acidic form and titration of the eluate with sodium hydroxide and phenolphthalein as indicator). The results obtained were in good agreement with these conventional methods.

8.1.2 Potable Waters

8.1.2.1 Spectrophotometric Method

The method discussed by Van Staden and Van Vleit [8.13], and discussed under Natural Waters in Sect. 8.1.1.2 has been applied to the determination of total alkalinity in potable waters.

8.1.2.2 Flow Injection Analysis

The flow injection analysis method described by Basson and Van Staden [8.14] has been applied to the determination of total alkalinity in potable waters.

Table 8.4. Features of the simultaneous determination of pH, alkalinity and total ionic concentration. (From [8.15]; Copyright 1987, Royal Society of Chemistry)

Parameter	Detection method	Range of determination[a]	Equation of calibration graph[b]	Regression coefficient	R.S.d.,% (concentration)
pH	Potentiometric (continuous contol)	3–11			
Alkalinity	Spectrophotometric	Δt: 5.0–600 µg ml^{-1} CaCO$_3$ ΔS: 10.0–600 µm ml^{-1} CaCO$_3$	$\Delta t = 31.28 \log C + 151.5$ $\Delta S = 3.48 \log C + 17.3$	0.9952 0.9965	± 1.5 (11.9 µg ml^{-1}) ± 1.7 (11.9 µg ml^{-1})
Concentration	Spectrophotometric	0.7 – 2.8 mequiv. l^{-1} Cl$^-$ 2.8–27 mequiv. l^{-1} Cl$^-$	$\Delta S = 3.09 \log C + 9.4$ $\Delta t = 34.43 \log C + 109.3$ $\Delta S = 2.53 \log C + 7.4$	0.9998 0.9994 0.9790	± 3.4 (7.04 mequivl^{-1}) ± 2.1 (7.04 mequiv. l^{-1}) ± 3.5 (7.04 mequiv. l^{-1})

[a] Δt = peak width/s; ΔS = peak area covered by the base line and the measured height.
[b] C = analyte concentration.

Table 8.5. Determination of pH, alkalinity and total ionic concentration in natural and potable water by the Canate and the conventional methods. (From [8.15]; Copyright 1987, Royal Society of Chemistry, UK)

Sample	pH	Alkalinity/ μg ml^{-1} CaCO$_3$		Total ionic concentration/mequiv. l^{-1}	
		FIA	Conventional[a]	FIA	Conventional[a]
Bottled waters –					
Font-bella	7.99	194.79	195.47	4.856	4.760
Cabras	8.58	481.11	453.21	12.373	12.444
Unbottled waters –					
Córdoba (urban)	7.95	134.64	137.62	4.689	4.781
La Rambla (urban)	7.96	400.68	393.81	17.933	17.808
La Rambla (well)	7.51	366.79	366.29	13.145	13.340

[a]See text.

Table 8.6. Determination of pH, alkalinity and total ionic concentration in synthetic samples. (From [8.15]; Copyright 1987, Royal Society of Chemistry, UK)

Sample composition, p.p.m.	pH	Alkalinity/ $\mu g\ ml^{-1}$ $CaCO_3$	Error, %	Total ionic concentration/ $mequiv°l^{-1}$	Error, %
Cl^-, 5; SO_4^{2-}, 5; NO_3^-, 5; HCO_3^-, 10	6.74	15.84	−3.4	0.666	1.9
Cl^-, 25; SO_4^{2-}, 30; NO_3^-, 10; HCO_3^-, 30	7.34	46.59	−5.3	2.519	1.9
Cl^-, 60; SO_4^{2-}, 60; NO_3^-, 20; HCO_3^-, 60	7.65	99.95	1.5	5.363	2.6
Cl^-, 125; SO_4^{2-}, 150; NO_3^-, 40; HCO_3^-, 150	8.24	252.9	2.7	12.010	−1.5
Cl^-, 250; SO_4^{2-}, 250; NO_3^-, 50; HCO_3^-, 175	8.30	301.4	4.9	18.637	−0.7
Cl^-, 350; SO_4^{2-}, 350; NO_3^-, 50; HCO_3^-, 250	8.38	420.3	2.4	26.564	1.63

8.2 Carbonate

8.2.1 Natural Waters

8.2.1.1 Ion Chromatography

This technique has been applied to the codetermination of carbonate and bicarbonate in natural and potable waters [8.16].

8.2.2 Waste Waters

8.2.2.1 Ion Exchange Chromatography

Ion exchange chromatography has been applied to the determination of carbonate [8.17] (and chloride, sulphate and phosphate) in waste waters. This technique is discussed in more detail in Sect. 5.1.6 (Phosphate).

8.2.3 High Purity Waters

8.2.3.1 Ion Chromatography

This technique has been applied to the determination of carbonate in high purity water [8.18].

8.2.4 Soil

Collins [8.19] has described a gasometric method for the determination of carbonate in soil based on reaction with hydrochloric acid and subsequent measurement of the volume of carbon dioxide produced.

8.3 Bicarbonate

8.3.1 Natural Waters

8.3.1.1 Titration Method

Feij and Smeenk [8.20] and Colin [8.21] discuss the titrimetric determination of bicarbonate.

8.3.1.2 Segmented Flow Analysis

This technique, discussed in Sect. 1.3.2, can be applied to the determination of bicarbonate in the concentration range $0-0.5\,\text{mg}\,l^{-1}$ as (CO_2) in water.

8.3.1.3 Ion Chromatography

This technique has been applied to the determination of bicarbonate in natural waters [8.22]. The method is discussed in further detail in Chap.11.

8.3.1.4 Combustion Method

Harmsen et al. [8.23] have described a rapid method for the determination of bicarbonate in ground water and surface water. Using the inorganic channel of the Beckmann total organic carbon analyser these workers determined the TOC content of a 2.0 µl water sample. Since the total inorganic carbon is dependent on the pH of the water, use of a determined pH value allows determination of the total bicarbonate based on the equilibrium equations of carbonic acid in water. Calculation in terms of mg l^{-1} is also possible. These workers tabulate the percentages of carbon dioxide, bicarbonate ion and carbonate ion and total organic carbon in water at 20 °C as a function of pH over the range pH 7.0 to 8.0 at 0.1 pH unit intervals. The method is rapid (2–3 min) and has an accuracy of better than 2%, is insensitive to other weak acids and requires only small sample volumes.

8.3.2 Potable Waters

8.3.2.1 Ion Exclusion Chromatography

Tanaka and Fritz [8.22] have described a procedure for the determination of bicarbonate and carbon dioxide in water by ion exclusion chromatography using water as the eluent and an electrical conductivity detector. The sensitivity of detection is improved approximately ten-fold by the use of two ion exchange enhancement columns inserted in series between the separating column and the detector. The first enhancement column converts carbonic acid to potassium bicarbonate and the second enhancement column converts the potassium bicarbonate to potassium hydroxide.

Apparatus. The chromatographic system consisted of an eluent reservoir, pump, pulse dampener, precolumn for removing carbon dioxide gas from the eluent, a 100-μl sample loop, separation column, first and second enhancement ion exchange columns, a Wescan 213 conductivity detector and a strip chart recorder. A flow rate of 1.0 ml min^{-1} was used.

The plastic separating column was 7.5 × 100 mm and was packed with a cation exchange resin in the H$^+$ form TSK SCX, 5 μm (polystyrene divinyl benzene copolymer-based material with a high cation exchange capacity 4.2 mg equiv g^{-1} available from TSK, Tokyo, Japan). The first enhancement column was constructed of plastic (4.6 × 50 mm) and packed with a cation exchange resin in the K$^+$ form (TSK SCX, t μm; TSK IC-Cation for cation chromatographic use, 10 μm silicon based material with low cation exchange capacity 0.45 mequiv g^{-1}); TSK SP-5 PW for HPLC use, 10 μm (hydroxylated, polyether based material with a low cation exchange capacity 0.3 m equiv g^{-1} or TSK IC-Cation SW for cation chromatography use, 5 μm (polystyrene divinyl benzene copolymer with low cation exchange capacity 30 μ equiv g^{-1}. The second enhancement column was constructed of plastic (4.6 × 50 mm) and packed with an anion exchange resin in the hydroxy form TSK SAX (5 μm).

The precolumn was constructed of plastic (7.5 × 100 mm) and packed with an anion exchange resin in the hydroxy form (TSK SAX, 5 μm).

All samples were filtered with a 0.45 μm polytetrafluoroethylene membrane filter before injection into the column.

Tanaka and Fritz [8.22] applied this procedure to the determination of bicarbonate or carbon dioxide in some potable water samples. Figure 8.5 shows the ion exclusion chromatograms obtained before and after a softening treatment. The results indicated that the method is useful in the field of water quality control for water treatment facilities.

Environmental waters such as river, lake and well waters were analyzed for carbon dioxide and bicarbonate by the three column ion exclusion method without any difficulty. It may be possible to use the first peak to estimate the total salts of strong acids (such as chloride and sulfate) in water samples.

The detection limit for bicarbonate was about 2 µmol l^{-1} and the relative standard deviation about 1%.

Strong acids and salts of strong acids, which are converted to acids by the separation column, pass rapidly through ion exchange columns and do not interfere with the carbon dioxide/bicarbonate peak which elutes later. The effect of weak acids or salts of weak acids was determined by adding various carboxylic acids to samples containing a known amount of sodium bicarbonate.

8.3.2.2 Ion Chromatography

This technique has been applied to the determination of bicarbonate in potable waters. See Sect. 11.3.

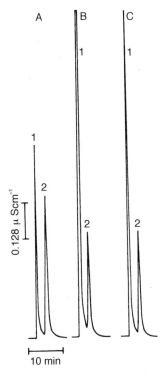

Fig. 8.5. Ion exclusion chromatograms of HCO^{3-} in some tap waters; (A) raw tap water of City of Ames, 1A, after 10-fold dilution (0.653 m mol l^{-1} × 10); (B) potable water after softening treatment (0.446 m mol l^{-1}; (C) potable water (La Salle, IL) after 20-fold dilution. (0.470 m mol × 10). Peak 1 is strong acid anions and peak 2 is HCO^{3-} (From [8.22])

8.4 References

8.1 Legrand MR, Aristarain AJ, Delmas RJ (1982) Anal Chem 54: 1336
8.2 Department of the Environment – National Water Council, HM Standing Committee of Analysts, Stationary Office (London) (1982) The determination of alkalinity and acidity in water.
8.3 Pearson J (1981) J Water Pollution Control Federation 53: 1243
8.4 Bartenshaw MP (1977) Anal Chim Acta 91: 339
8.5 Rosopulo A, Quentin KE (1971) Z Analyt chem 253: 27
8.6 Feij LAC, Smeenk JGMM (1981) H_2O 14: 131
8.7 Willis RB, Mullings GL (1983) Anal Chem 55: 1173
8.8 Hillbom E, Liden J, Pettersson S (1983) Anal Chem 55: 1180
8.9 Colin F, Tribune du Cebedeau (1985) 38: 3
8.10 Hillbom E, Liden J, Pettersson S (1983) Analytical Chemistry 55: 1180
8.11 Thomas JFJ, Lynch JJ (1980) Journal of the American Water Works Association 52: 259
8.12 Willis RB, Mullings GL (1983) Analytical Chemistry 55: 1173
8.13 Van Staden JF, Van Vleit HR, Water SA (1984) 10: 168
8.14 Basson WD, Van Staden JF (1980) Laboratory Practice 29: 632
8.15 Canate F, Rios A, Lunque de Castro MD, Valcarcal M (1987) Analyst (London) 112: 263
8.16 Brandt G, Kettrip A (1985) Fresenuis Zertschrift fur Analytische Chemie 320: 485
8.17 Tanabe K, Ishizuki T (1982) Water Research 16: 719
8.18 Brandt G, Matysehek G, Kettrup A (1985) Fresemius zeitschrift fur Analytische Chemie 321: 653
8.19 Collins SHJ (1906) Society of Chemical Industries (London) 25: 518
8.20 Feij LAC, Smeenk JGMM (1981) H_2O 14: 131
8.21 Colin F (1985) Tribune de Cebedeau 3: 3
8.22 Tanaka K, Fritz JS (1987) Analytical Chemistry 59: 708
8.23 Harmsen J, Van Drumpt H, Muylaert JM (1979) H_2O 12: 585

9 Metal-Containing Anions

9.1 Arsenite

9.1.1 Natural Waters

9.1.1.1 Atomic Absorption Spectrometry

Arsenite (and arsenate and phosphate) have been determined in natural waters by a technique based on flotation spectrophotometry and extraction–indirect atomic absorption spectrometry using malachite green as an ion-pair reagent [9.1].

9.1.1.2 Ion Chromatography

Hoover and Jager [9.2] have discussed the determination of arsenite together with other anions (selenite and selenate) in potable and ground waters. This method is discussed in further detail in Sect. 11.1.2 (Ion Chromatography).

9.2 Arsenate

9.2.1 Natural Waters

9.2.1.1 Spectrophotometric Methods

Matsuhara et al. [9.3] have described a rapid method for the determination of trace amounts of arsenate (and phosphate) in water by spectrophotometric detection of their heteropolyacid-malachite green aggregates following preconcentration by membrane filtration.

9.2.1.2 Flow Injection Analysis

This technique has been applied to the determination of arsenate in natural waters [9.4].

9.2.1.3 Atomic Absorption Spectrometry

Arsenate (and arsenite and phosphate) have been determined in natural waters by a technique based on flotation spectrophotometry and extraction–indirect atomic absorption spectrometry using malachite green as an ion-pair reagent [9.1].

9.2.1.4 Ion Chromatography

This technique has been applied to the determination of arsenate (and sulphite) in natural waters [9.5].

9.2.1.5 Polarography

An indirect polarographic method has been used to determine traces of arsenate in mineral waters [9.6]. In this method arsenate is converted to uranyl arsenate which is filtered off and excess uranyl ion is determined by polarography at -3.60 V. Phosphates interfere in this procedure.

9.2.2 Potable Water

9.2.2.1 Ion Chromatography

The ion chromatographic technique for the determination of arsenate described in Sect. 11.3.3 to applicable to the determination of arsenate in potable water.

9.3 Arsenite/Arsenate

9.3.1 Natural Waters

9.3.1.1 Spectrophotometric Methods

In the method of Johnson and Pilson [9.7], in one aliquot of sample arsenite is oxidized to arsenate, in a further aliquot arsenate is reduced to arsenite and a third aliquot is untreated. Each aliquot is submitted to a molybdenum blue procedure. The first gives a value for phosphate plus arsenate plus arsenite, the second for phosphate and the third a value for phosphate plus arsenate. The content of each anion is calculated from these results. The method is applicable at the $3\ \mu mol\, l^{-1}$ level.

Johnson [9.8] has also described a spectrophotometric method for the determination of arsenate and phosphate in natural water in which total arsenate plus phosphate are determined by classical molybdenum blue procedure [9.9]. Arsenate is then reduced to arsenite which does not form a coloured complex with molybdic acid. A difference technique then enables the separate determination of phosphate and arsenate.

9.3.1.2 Ion Chromatography

Urasa and Ferede [9.10] used direct current plasma as an element-selective detector for the simultaneous ion chromatographic determination of arsenic III and arsenic V in the presence of other common anions. Matrix effects were eliminated, and a wide range of element compositions can be used without experiencing detector limitations.

9.3.1.3 Polarography

Arsenate has been determined [9.11] by differential pulse polarography on acidic aqueous solutions containing polyhydroxy compounds such as d-mannitol. Peak heights measured at -0.55 V in a medium containing $2\,\text{mol}\,l^{-1}$ aqueous perchloric acid and 4.5 g mannitol in 50 ml of solution gave linear calibration curves over the range $20-160\,\mu g\,l^{-1}$. Arsenite could be similarly determined at -0.34 V or -0.42 V with or without mannitol respectively, the method of measurement at -0.42 V being used where arsenite occurred in the presence of arsenate. Detection limits were approximately $10\,\mu g$ arsenic l^{-1} for both forms. Examples of results obtained for arsenite and arsenate recovery in a river water sample are given.

9.4 Selenite

9.4.1 Natural Waters

9.4.1.1 Spectrophotometric Methods

Bodini and Alzamora [9.12] used reaction of 4, 5, 6-triaminopyridine with selenite in acidic medium as the basis of a spectrophotometric method for estimating traces of the latter in natural water. The reaction product is a piazselenol with an absorption maximum at 362 nm. The precipitate is dissolved in $3.5\,\text{mol}\,l^{-1}$ phosphoric acid in order to avoid interference from iron and chloride or tin(II). The calibration graph is linear up to $10\,\text{mg}\,l^{-1}$ selenium and the

9.4.1.2 Spectrofluorimetry

Takayanagi and Wong [9.13] give details of a fluorimetric procedure for the determination of tetravalent selenium in natural waters. The optimal conditions for fluorimetric determination of total selenium were also investigated.

A method has been described [9.14] for the determination of selenite in natural waters based on the formation of piazselenols by reaction with substituted 1,2-diaminobenzenes. The derivatives were extracted into solvents prior to analysis by fluorimetry or gas chromatography with electron capture detection, according to published methods. The use of 1,2-diamino-3,5-dibromobenzene and 4-nitro-1,2-diaminobenzene enabled detection limits of $1\,ng\,l^{-1}$ and $6\,ng\,l^{-1}$ of selenium in water to be achieved, respectively. Using diaminonaphthalene as a fluorimetric reagent, a detection limit of $50\,ng\,l^{-1}$ could be obtained, although at the risk of severe interference from ferric iron, nitrite and nitrate. All organoselenium compounds could be decomposed by ultraviolet irradiation, with formation of selenite.

9.4.1.3 Ion Chromatography

Hoover and Jager [9.2] have discussed the determination of selenite together with other anions (selenate and arsenite) in potable and ground waters by ion chromatography. This method is discussed in further detail in Sect. 11.3.

9.4.1.4 Polarography

Lin Ahmed et al. [9.15] determined selenite by a polarographic technique involving the use of an acidic supporting electrolyte. Generally, hydrochloric or perchloric acids are used.

Christian et al. [9.16] reported that the polarographic reduction of selenite in hydrochloric acid exhibits two main waves corresponding to the following reactions:

$$SeIV + 6e^- + 2H^+ \rightarrow H_2Se$$

at $E_{1/2} = -0.01$ V vs SCE, followed by depolarisation of mercury by the reduction product:

$$H_2Se + Hg \rightarrow HgSe + 2H^+ + 2e^-.$$

The second wave at $E_{1/2} = -0.54$ V vs SCE corresponds to the further reduction of the HgSe to hydrogen selenide:

$$HgSe + 2H^+ + 2e^- \rightarrow H_2Se + Hg.$$

The only elements and anions likely to interfere in the determination of selenite are arsenite, copper II, iron III, lead II and zinc II. Arsenite and copper II strongly enhance the selenium peak, whilst iron III has a suppressant effect.

9.4.1.5 Miscellaneous Methods

Brimmer et al. [9.17] have carried out an investigation of the quantitative reduction of selenate to selenite in aqueous solution.

9.4.2 Potable Water

The ion chromatography procedure for the determination of selenite described in Sect. 11.3.3 is applicable to the analysis of potable waters.

9.4.3 Soil Extracts

9.4.3.1 Ion Chromatography

Karlson and Frankenberger [9.18] developed a single column ion chromatographic method for the determination of selenite in soil extracts with the simultaneous determination of chloride, nitrite, nitrate and phosphate. Separation of the anions was conducted on a low capacity anion exchange column and anions were quantified by a conductiometric detection.

The element stream consisted of 1.5 m mol l^{-1} phthalic acid adjusted to pH 2.7 with formic acid. The method requires minimal sample treatment, allowing for precise measurements of trace levels of selenite in the presence of high background levels of chloride, nitrate and nitrite.

Interfering chloride anions were removed by reaction with a silver saturated cation exchange resin. The detection limit of selenite was 3 µg l^{-1} with a concentrator column. The relative standard deviation using a 500 µl loop was 2.0% with standards and 6.7% in soil extracts (0.5 mg l^{-1}). Selenite levels found in soil extracts ranged from 0.8 to 99.6 µg l^{-1}.

9.4.4 Milk

9.4.4.1 Gas Chromatography

Shimoishi [9.19] has described a method for the determination of selenite in milk, milk products and albumin, based on the conversion of selenite to

5-nitropiazselenol following reaction with 1,2-diamino-4-nitrobenzene. The final estimation is performed by gas chromatography with a detection limit of 0.005 µg. About 60% of the total selenium content of milk was found to exist as selenite. Between 0.003 and 0.058 mg l^{-1} selenite was found in milk.

In this procedure the sample is heated to 140 °C with concentrated nitric acid. The cooled digest is then diluted and extracted with toluene. Recoveries were in the range 97–104% and the detection limit was 0.005 µg selenium.

9.4.5 Urine

9.4.5.1 Ion Exchange Chromatography

Blotcky et al. [9.20] determined selenite in urine by a method based on anion exchange chromatography followed by molecular neutron activation analysis.

The method takes advantage of the high sensitivity of neutron activation analysis for ^{77}Se after quantitative separation of selenite ions. Recoveries are 100% with a limit of detection of 10 µg l^{-1} selenite (as selenium).

9.4.6 Biological Samples

9.4.6.1 Gas Chromatography

Uchida et al. [9.21] have described a method for the determination of selenite in biological samples such as blood and liver, and also plant materials such as spinach and tomato leaves.

Selenite reacts with 1,2-diamino-3,5-dibromobenzene to form 4,6-dibromopiazselenol which is detected by means of a gas chromatograph equipped with an electron capture detector.

The sample digest is treated with lanthanum nitrate and ammonia and the precipitate formed separated by centrifuging. The supernatant liquid is drawn off and rejected and the precipitate dissolved by the addition of 1,2-diamino-3,5-dibromobenzene. The 4,6-dibromopiazselenol complex is extracted into toluene and this extract gas chromatographed.

A relative standard deviation of 3.9% at the 0.076 mg kg^{-1} selenite level was obtained by this procedure. Between 0.03 and 1.1 mg kg^{-1} selenite was found in various biological and plant materials by this method.

9.4.6.2 Anodic Stripping Voltammetry

Ahmed et al. [9.22] determined selenium IV (selenite) in biological samples by using differential pulse stripping voltammetry.

9.5 Selenate

9.5.1 Natural Waters

9.5.1.1 Ion Chromatography

Hoover and Jager [9.2] have discussed the determination of selenate together with other anions (selenite and arsenite) in potable and ground waters. This method is discussed in further detail in Sect. 11.3.3.

9.5.2 Potable Water

9.5.2.1 Spectrofluorimetry

Selenate has been determined in potable water fluorimetrically using napthalene-2,3-diamine [9.23]. The complex is extracted into cyclohexane and evaluated at 521 nm (excitation 389 nm). The calibration graph is rectilinear in the range 0.02 – 1.0 µg of selenium. The coefficient of variation for samples containing 0.5 and 1.0 ng of selenium was 6.3% and a 1000 – fold excess of nitrate, barium, cadmium, calcium, chromium III, cobalt, copper II, iron III, magnesium, manganese II, nickel III or zinc did not interfere.

9.6 Selenate/Selenite

9.6.1 Sea Water

9.6.1.1 Spectrofluorimetry

Itoh et al. [9.24] determined selenium IV in estuarine waters by an anion exchange resin modified with bismuthiol 11 and diaminonaphthalene fluorophotometry. An Amberlite IRA-400 anion exchange resin was modified by mixing with an aqueous solution of bismuthiol-11 to give 0.2 mmol bismuthiol per resin. Sample solution was eluted through a column packed with the modified resin. Selenium IV adsorbed as selenotrisulphate was then eluted from the column with either 0.1 mol l^{-1} penicillamine or 0.1 mol l^{-1} cysteine. The eluate was then subjected to an acid digestion procedure to reduce selenium to the tetravalent state with diaminonaphthalene for fluorimetric determination. Approximate agreement with the tellurium coprecipitation method was obtained. The application of both methods to the analysis of estuarine waters permitted the separate determination of both selenium IV and selenium VI since the tellurium coprecipitation method did not differentiate between the two species.

9.6.1.2 Atomic Absorption Spectrometry

The selective hydride generation procedure has been applied [9.25] as a basis for the differential determination of dissolved selenite, selenate, total selenium and organoselenium compounds in natural waters.

Lansford et al. [9.26] have described a method for determination of selenium in water by atomic absorption spectrophotometry after reduction to the selenite state. The presence of mercury and arsenic can be tolerated in concentrations up to $25\,\mu g\,l^{-1}$; interference by arsenic in excess of this level may be controlled by limited addition of stannous chloride.

Cheam et al. [9.27, 9.28] have described an automated hydride generation system for the determination of selenate and selenite arsenite.

Ohta and Suzuki [9.29] using a molybdenum micro-tube atomizer have developed an atomization technique for the determination of selenium in water samples. The highest absorption peak was achieved in pure oxygen, as was the lowest atomization temperature; the analytical wavelength was 196.00 nm. Extraction of selenium with diethyldithiocarbamate eliminated matrix interference and copper addition suppressed interference from co-extracted elements. No selenium VI was extractable from citrate buffered solution containing EDTA, so that use of these reagents allowed separate determination of selenium IV and selenium VI. Recoveries of 90–106% were achieved in analysis of water samples containing $30-130\,\mu g\,l^{-1}$ selenium.

Apparatus. Atomic absorption measurements were made using a Nippon Jarell Ash 0.5 m Eberttype monochromator coupled to an R 106 photomultiplier tube (Hamamatsu TV Co.) and a laboratory-built fast-response amplifier. The output signal from the amplifier was monitored on a Memoriscope (Iwatsu MS-5021) with a time constant of $0.1\,\mu s$. The light source was a selenium electrodeless discharge lamp operated by a 27-MHz microwave generator (Hamamatsu C977 power supply). The analytical wavelength was 196.0 nm.

The molybdenum micro-tube (25 mm long and 1.5 mm i.d.) was used as atomizer. The micro-tube was mounted on two copper supports so that there was no localized variation in tube temperature. The micro-tube atomizer was enclosed in a Pyrex chamber (300 ml volume) which had two silica end-windows to allow transmission of the light beam. The chamber was purged with argon at a flow rate of $500\,ml\,min^{-1}$. The power for heating the atomizer was applied by a step-down transformer.

The tube temperature was measured with a photodiode (Hamamatsu TV Co. S641). The signal from the photodiode was calibrated with an optical pyrometer. The signal from the photodiode and the absorption signal were recorded simultaneously on Memoriscope. All sample solutions were injected with a glass micropipette into the micro-tube through a 0.3-mm hole which was drilled at the mid-point of the tube.

Reagents. As follows.
> Nitric acid, concentrated.
> Hydrochloric acid, 5 N.
> Ethylene diamine tetracetic acid, disodiumsalt, 1% w/v aqueous.
> Cupric nitrate, 0.1% w/v aqueous.
> Ammonia, concentrated.
> Sodium diethyldithiocarbamate, 0.2% w/v aqueous.
> Carbon tetrachloride.
> Selenium stock solution. Prepare by dissolving selenium dioxide in water.
> Selenium VI solution. Prepare from potassium selenate.
> Arsenic solution. Prepare from arsenic trioxide.
> Antimony solution. Prepare from antimony III potassium tartrate.
> Tellurium solution. Prepare from its metal.
> Citrate buffer (0.7 mol l^{-1}). Prepare by neutralizing citric acid solution with ammonia.

Procedure. Transfer 25–50 ml of water sample into a Teflon beaker, add 2 ml of nitric acid and evaporate to dryness on a water bath. Add 10 ml of hydrochloric acid (5 N) to the residue and heat the solution in a boiling water bath to reduce selenium VI. Evaporate the solution to 0.5 ml. After adding 1 ml of EDTA solution (1%), 2 ml of citrate buffer solution and 0.5 ml of copper II nitrate solution (0.1%), adjust the pH of the solution to 4.5 with ammonia. Transfer the resultant solution into a separatory funnel, dilute to 5 ml and add 1 ml of sodium diethyldithiocarbamate solution (0.2%). Extract the selenium complex with 5 ml of carbon tetrachloride. Transfer 1 µl of the organic phase into the metal micro-tube atomizer and dehydrate at 100 °C, followed by ashing at 200–300 °C. Then atomize selenium by heating to a final temperature of 2000 °C. For the determination of selenium IV omit the reduction step of selenium VI with hydrochloric acid, and extract the selenium complex from buffered sample solution after the addition of copper and EDTA. Prepare a calibration curve by atomizing organic extracts obtained from standard solutions as above.

Most cations and anions have interferences on atomization of selenium. Figure 9.1 shows the effect of sodium chloride on selenium. The effects of 100-fold amounts of lead, arsenic and tellurium on atomization of selenium are shown in Fig. 9.2. Samples were checked for background effects and reagent blank. Bismuth and antimony also depressed the selenium absorption.

Arsenic lowered somewhat the peak temperature in atomization profile for selenium. Copper tends to suppress the interferences of diverse elements on atomization of selenium. However, the interferences from large concentrations of diverse elements and matrices were not improved even in the presence of copper at the atomization step. Therefore, the separation of selenium from matrices was recommended.

Selenium is extracted as diethyldithiocarbamate complex from the solution containing citrate and EDTA [9.30]. Ohta and Suzuki [9.29] found that only a

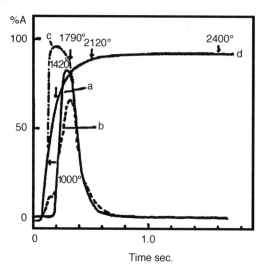

Fig. 9.1. Memoriscope traces for effect of sodium chloride on atomization of selenium: (a) 0.73 ng Se (IV); (b) 0.63 ng Se (IV) and 62.5 ng NaCl; (c) 0.63 ng Se (IV) and 6.25 µg NaCl; (d) temperature increase. Gas flow rate 500 ml Ar min^{-1} (From [9.29])

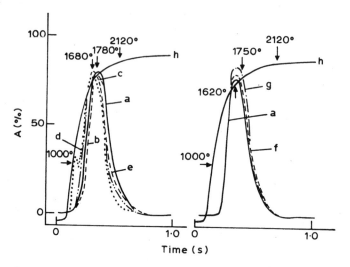

Fig. 9.2. Memoriscope traces for effect of diverse elements: (a) 0.63 ng Se (IV); (b) 0.63 ng Se (IV) and 63 ng Pb; (c) 0.63 ng Se (IV), 63 ng Pb and 63 ng Cu; (d) 0.63 ng Se (IV) and 63 ng As; (e) 0.63 ng Se (IV), 63 ng As and 63 ng Cu; (f) 0.63 ng Se (IV) and 63 ng Te; (g) 0.63 ng Se (IV), 63 ng Te and 63 ng Cu; (h) temperature increase. Gas flow rate 500 ml Ar min^{-1} (From [9.29])

few elements, such as copper, bismuth, arsenic, antimony and tellurium, are also extracted together with selenium. They examined this for effects of 100-fold amounts of elements co-extracted with the selenium diethyldithiocarbamate complex. An appreciable improvement of interferences from diverse elements was observed in the presence of copper. Silver depressed the selenium absorption in the case of atomization of diethyldithiocarbamate complex, but the interference of silver was suppressed in the presence of copper. The atomization profile from diethyldithiocarbamate complex was identical with that from selenide.

No selenium VI is extractable from citrate buffered solution containing EDTA and copper. Therefore, tetra- and hexavalent selenium could be determined separately.

Some water samples were analyzed for selenium and the results are given in Table 9.1 with the recovery of selenium added to samples. The recovery was satisfactory. The forms of selenium in waters are known to be selenite and selenate [9.27]. Selenium occurs in natural water at concentrations ranging from less than $0.0002\ \mu g\,l^{-1}$ to greater than $50\ \mu g\,l^{-1}$. Therefore, a large sample size is necessary for analysis at lower concentration level.

9.6.2 Natural Waters

9.6.2.1 Gas Chromatography

Uchida et al. [9.31] have described a method for determination of selenium in rivers and sea water by gas chromatography with electron-capture detection. The specific reaction of 1,2-diamino-3,5-dibromobenzene with selenium IV is used, the product (4,6-dibromopiazselenol) being extracted into toluene and determined from its peak height in the gas chromatogram. Selenium (-II, 0) is oxidized to selenium IV by bromine solution and selenium VI is reduced to the same state by bromine/bromide redox buffer solution and determined as the piazselenol as above.

Apparatus. A Shimdazu Model GC-5A gas chromatograph, equipped with a ^{63}Ni electron capture detector. Pack a glass column (1 m × 3 mm i.d.) with 15%

Table 9.1. Determination of selenium in water samples. (From [9.29]; Copyright 1980, Springer-Verlag GmbH, Germany)

Sample	Selenium $\mu g\,l^{-1}$ Added Se (IV)	Se(VI)	Found	Recovery %
A	–	–	0.4	
	130	–	130	100
	130	–	120	92
B	–	–	0.4	
	63	–	63	100
	–	63	57	90
	31	31	31[a]	100
	31	–	33[a]	106
	31	31	61	98

[a] No reduction procedure with hydrochloric acid was applied.

SE-30 on 60–80 mesh Chromosorb W. The column and detector temperatures were maintained at 200 and 280 °C respectively. The nitrogen flow rate was 28 ml min^{-1}. A Shimadzu Model 101 recorder was used at a chart speed of 5 mm min^{-1}.

Reagents. As follows.
    Hydrochloric acid, concentrated.
    Toluene, redistilled.
    Perchloric acid, aqueous 2:1.
    Bromine aqueous, 3%.
    Hydroxylamine hydrochloride, aqueous, 1 mol l^{-1}.
    Hydrobromic acid, 47%.
    1,2-Diamino-3,5-ditromobenzene monochloride.
    The synthesis of this reagent has been reported [9.32]. Dissolve the reagent (0.6 g) in 500 ml of concentrated hydrochloric acid and wash with toluene (25 ml) to remove toluene-soluble matter. Store the solution in a brown glass bottle. It can be used safely for 1 month.

    Se 0 stock solution (1 mg of Se in 1 ml of CS$_2$). Dissolve elemental selenium (500 mg) in 500 ml of carbon disulfide. Prepare working solutions by appropriate dilution. Determine the concentration of the working solution from the calibration graph by gas chromatography after the conversion of Se 0 to Se IV by bromine-bromide redox buffer [9.33]. The solution could be used safely for 1 week.

    Se IV stock solution (0.975 mg of Se ml^{-1}). Dissolve selenium dioxide (1.433 g) in 1000 ml of distilled water and standardize gravimetrically. The solution is stable for 6 months.

    Se VI stock solution (1.178 mg of Se ml^{-1}). Dissolve selenic acid monohydrate (220 mg) in 100 ml of distilled water and prepare working solutions by appropriate dilution. Determine the concentration by gas chromatography, after Se VI had been reduced to Se 0 and then oxidized to Se IV [9.33].

Procedure. Pretreatment of sample waters, add concentrated hydrochloric acid to the sample immediately after sampling (1 ml l^{-1}). Filter by a membrane filter (pore size 0.45 µm) to remove suspended matter as soon as possible.

    Se IV determination (procedure 1) is carried out as follows; to a 500 ml separating funnel add 500 ml of the pretreated sample water and 20 ml of concentrated hydrochloric acid and mix well. Shake the solution vigorously with 25 ml of toluene to saturate with toluene and to remove toluene soluble matter. Transfer the aqueous phase into another 500-ml separating funnel after phase separation. To the solution add 0.12% 1,2-diamino-3,5-dibromobenzene solution (10 ml), allow the solution to stand for 2 h, add 1 ml of toluene, shake the solution vigorously for 5 min and extract the 4,6-dibromopiazselenol. Wash the extract twice with 3 ml of perchloric acid (2 + 1), inject 2 µl of the extract into the gas chromatograph and measure the peak height of 4,6-dibromopiazselenol.

    Se(-II, 0, IV) determination (procedure 2) is carried out as follows. Add to the pretreated sample solution (500 ml) 25 ml of concentrated hydrochloric acid and 2 ml of bromine solution (3%) and mix for 5 min at room temperature. By this treatment Se(-II,0) is oxidized completely to the quadrivalent state but Se VI is not reduced. Add 2 ml of hydroxylamine hydrochloride (1 mol l^{-1}) to the solution to reduce the excess bromine to bromide. Wash the solution with 25 ml of toluene as mentioned in procedure 1. Determine as for procedure 1. Calculate selenium (-II,0) from the difference between the peak heights of the piazselenol by procedures 1 and 2.

Se(-II,0,IV,VI) determination (procedure 3) is carried out as follows. To the pretreated sample water (500 ml) add concentrated hydrochloric acid (50 ml), 47% hydrobromic acid (25 ml) and 3% bromine solution (0.5 ml) and boil gently for 15 min. By this treatment Se(-II,0) is oxidized to Se IV and Se VI is reduced to Se IV. Thus all selenium is now in the quadrivalent state. After the solution is cooled to room temperature, add 2 ml 1 mol l^{-1} hydroxylamine hydrochloride to reduce the excess bromine. Wash the solution with 25 ml of toluene and determine the total selenium by the same method as procedure 1. Calculate selenium VI by deducting Se(-II,0,IV) from Se(-II,0,IV,VI).

Distilled water containing 4.6 ng of Se 0, 4.8 ng of Se IV and 5.9 ng of Se VI were analyzed by procedures 1, 2 and 3 and good recoveries were found (Table 9.2).

As an example, the gas chromatograms of selenium in river water by procedures 2 and 3 are shown in Fig. 9.3 a,b respectively. Selenium VI is calculated from the difference between the peak heights of 4,6-dibromopiazselenol in these chromatograms. The amounts and oxidation states of selenium in river water in Japan are shown in Table 9.3. The total amount of selenium in river water varied greatly, e.g. from 16 to 230 ng l^{-1}.

9.6.2.2 High Performance Liquid Chromatography

High performance ion exchange chromatography has been used [9.34] to separate selenite and selenate which were determined by post column derivatization using 2,3-diaminonaphthalene. Peak broadening was minimized by an air-segmented continuous flow autoanalyser. 2,3-diaminonaphthalene reacted with

Table 9.2. Analysis of simulated mixtures of 4.6 ng of Se (0), 4.8 ng of Se(IV), and 5.9 ng of Se (VI). (From [9.31]; Copyright 1980, American Chemical Society)

	Distilled water (500 ml)		
	Peak height	Amount recovered ng	Recovery %
Procedure 1	16.3	4.9	102
(Se (IV))	15.0	4.5	94
Procedure 2	31.2	9.5	101
(Se (0) + Se (IV))	30.5	9.3	99
Procedure 3	51.9	15.9	104
(total Se)	52.2	16.0	105

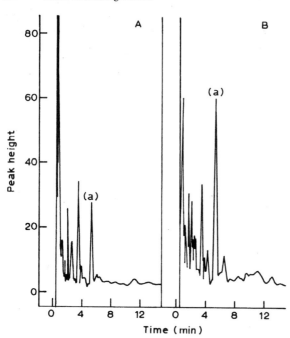

Fig. 9.3. Recorder tracings of selenium analysis of a river water sample: **A** analysis by procedure 2, **B** analysis by procedure 3 (see text). (From [9.31])

Table 9.3. Determination of selenium in river water. (From [9.31]; Copyright 1980, American Chemical Society)

Sample[a]	ng l^{-1}			
	Se(−II, 0)	Se(IV)	Se(VI)	total Se
Yoshii River, Okayama Pref., Feb 5, 1979	12	16	202	231
Takahashi River, Okayama Pref., Feb 5, 1979	5	4	7	16
Asahi River, Okayama Pref., Feb 5, 1979	11	<2	9	20
Asahi River, Okayama Pref., Feb 17, 1979	11	2	3	16

[a]Samples contained 1 ml^{-1} of concentrated hydrochloric acid.

selenium IV but not with selenium VI. Selenium VI was determined after on-line reduction with hydrogen bromide (1.8 N) at 100 °C. Selenium IV was detected in potable water at 0.17 µg l^{-1}. Injection of larger volumes improved the detection limits. Samples of river or lake water were concentrated by column head techniques.

Cartoni and Coccioli [9.35] characterized mineral waters using high performance liquid chromatography. Filtered mineral water was pumped through a column packed with LiChrosorb RP-18. Traces of organic material dissolved in the water were adsorbed onto the first part of the column and were then desorbed using an acetonitrile gradient. The eluate was monitored with an ultraviolet variable wavelength detector.

9.6.2.3 Polarography

AC polarography and differential pulse polarography [9.36] and anodic stripping voltammetry [9.37] have been used to determine selenium.

9.6.2.4 X-Ray Fluorescence Spectrometry

Robberecht and Van Grieken [9.38] determined $\mu g\ l^{-1}$ levels of total dissolved selenium and selenite in natural waters by X-ray fluorescence spectrometry. Detection limits for tetravalent selenium and total selenium are 50 ng l^{-1} and 60 ng l^{-1} respectively. Coefficients of variation are 10% for tetravalent and hexavalent selenium at the $0.5-1\ \mu g\ L^{-1}$ selenium level.

9.6.2.5 Sample Preservation

Cheam and Agemian [9.39] studied the preservation and stability of inorganic selenium compounds at $\mu g\ l^{-1}$ levels in natural water samples. They adopted the use of sulphuric acid as the preserving medium as this was found not to interfere with the hydride generation method they adopted for the determination of arsenite and arsenate. They found that the use of polyethylene sample containers and pH adjustment to 1.5 (with 0.2 vol.% sulphuric acid) provided optimal conditions for the preservation of samples for periods of up to 125 days. Storage at 4 °C also preserves samples effectively over the same time span.

9.6.3 Potable Waters

9.6.3.1 Ion Chromatography

The ion chromatographic method discussed in Sect. 11.3.3 has been applied to the determination of selenite and selenate (and arsenite) in potable waters.

9.6.4 Industrial Effluent

9.6.4.1 Spectrofluorimetry

A fluorimetric procedure based on the use of 2,3-diaminonaphthalene has been used to determine mμg amounts of total selenate and selenite in industrial effluents [9.40]. The fluorescent intensity of the complex is measured at 522 nm (excitation at 366 nm).

9.6.4.2 Polarography

Batley [9.41] has given details of a procedure for determination of trivalent selenium in effluents in the concentration range 2–100 µg l^{-1} by using the sensitive adsorption controlled peak obtained by differential pulse polarography in dilute acid solution. Heavy metals which could interfere were removed on Chelex 100 resin. Hexavalent selenium could be determined by the same method after photolytic reduction in the absence of oxygen. For selenium concentrations below 2 µg l^{-1} preconcentration by anion exchange was necessary.

9.7 Tellurate

9.7.1 Natural Waters

9.7.1.1 Spectrophotometric Methods

A spectrophotometric method utilizing tetramethyl thiourea as chromogenic reagent has been described for the determination of tellurate in water [9.42].

9.8 Chromate/Dichromate

9.8.1 Natural Waters

9.8.1.1 Spectrophotometric Methods

Aoyama et al. [4.3, 4.4] have described a rapid method for the determination of down to 3 µg l^{-1} of chromate in which hexavalent chromium is complexed with diphenylcarbazide, and this complex is floated with a sodium lauryl sulphate anionic surfactant. After dilution the subsided foam is measured spectrophotometrically. The continuous flotation procedure is as follows.

Place the water (1 l), previously acidified to 0.1 mol l^{-1} with sulphuric acid, in the separation tube, pass nitrogen into the tube at about 105 ml min^{-1}, and pump in the 1% sodium lauryl sulphate solution at 0.39 ml min^{-1}. Continuously supply the sample pH and the 1% diphenylcarbazide solution to the separation tube at a rate of 2 l h^{-1} and 0.230 ml min^{-1} respectively. A steady state flotation was achieved 1h after flotation had started. Mix the sample and diphenyl carbazide solutions in the mixing chambers before entering the tube. Collect the foam subsided in the collector and intermittently transfer to flasks and dilute to 50 ml with water, and measure absorbances as described above.

The effect of diverse ions on the flotation of chromium VI were investigated. Only copper II and iron III in 10-fold amounts, and vanadium V in 20-fold amounts caused fading of the colour. Even in thoses cases, however, chromium VI could be measured by a standard addition method.

Diphenylcarbazone and diphenylcarbazide have been widely used for the spectrophotometric determination of chromium [9.45]. Only relatively recently, however, has the nature of the complexation reactions been elucidated. Cr III reacts with diphenylcarbazone where Cr VI reacts (probably via a redox reaction combined with complexation) with diphenylcarbazide [9.46]. Although speciation would seem a likely prospect with such reactions, commercial diphenylcarbazone is a complex mixture of several components, including diphenylcarbazide, diphenylcarbazone, phenylsemicarbazide, and diphenylcarbadiazone with no stoichiometric relationship between the diphenylcarbazone and diphenylcarbazide [9.47]. As a consequence, use of diphenylcarbazone to chelate Cr III selectively also results in the sequestration of some Cr VI. Total chromium can be determined with diphenylcarbazone following reduction of all chromium to Cr III.

9.8.1.2 Segmented Flow Analysis

The technique discussed in Sect. 1.3.2 has been applied to the determination of chromate in water in the 0–2 mg l^{-1} concentration range.

9.8.1.3 Flow Injection Analysis

This technique has been used to determine chromate in water.

Marshall and Mottola [9.48] evaluated a method for determining the ionic forms of chromium in water:

Chromium VI

$H_2CrO_4 \rightleftharpoons HCrO_4^-$

$HCrO_4^- \rightleftharpoons CrO_4^{2-} + H^+$

$CR_2O_7^{2-} + H_2O^- \rightleftharpoons 2HCrO_4^-$

Chromium III

$Cr^{3+} + H_2O \rightleftharpoons Cr(OH)^{2+} + H^+$

$Cr(OH)^{2+} + H_2O \rightleftharpoons Cr(OH)_2^+ + H^+$

$Cr^{3+} + 3OH^- = \rightleftharpoons Cr(OH)_3$

$Cr^{3+} + 4OH^- \rightleftharpoons Cr(OH)_4^-$

Ruz et al. [9.49] speciated different oxidation states of chromium. They were able to obtain a concentration profile for the chromium III and chromium VI species, namely $HCrO_4^-$, $Cr_2O_7^{2-}$ and CrO_4^{2-}. Ruz et al. [9.50] also used reverse flow injection analysis for the simultaneous and sequential determination of chromate and chromium III by unsegmented flow methods. A photometric detector was employed.

Ion exchanger phase absorbtiometry coupled to flow injection analysis has been used to determine trace levels of dichromate in water [9.51].

9.8.1.4 Atomic Absorption Spectrometry

Subramanian [9.52] studied the factors affecting the determination of trivalent and hexavalent chromium (chromate) by direct complexation with ammonium pyrrolidinedithiocarbamate, extraction of the complex into methyl isobutyl ketone, and determination by graphite furnace atomic absorption spectrometry. Factors studied included the pH of the solution, concentration of reagents, period required for complete extraction, and the solubility of the chelate in the organic phase. Based on the results, procedures were developed for selective determination of trivalent and hexavalent chromium without the need to convert the trivalent to the hexavalent state. For both states the detection limit was $0.2 \mu g l^{-1}$.

A method has been developed for differentiating hexavalent from trivalent chromium [9.53]. The metal is electrodeposited with mercury on pyrolytic graphite-coated tubular furnaces in the temperature range 1000–3000 °C, using a flow-through assembly. Both the hexa- and trivalent forms are deposited as the metal at pH 4.7 and a potential of -1.8 V against the standard calomel electrode, while at pH 4.7 at and -0.3 V the hexavalent form is selectively reduced to the trivalent form and accumulated by adsorption. This method was applied to the analysis of chromium species in samples of different salinity, in conjunction with atomic absorption spectrophotometry. The limit of detection was $0.05 \mu g$ chromium l^{-1} and relative standard deviation from replicate measurements of $0.4 \mu g$ chromium (VI) was 13%. Matrix interference was largely overcome in this procedure.

Reagents. Use Merck Suprapure grade reagents.
   Acetate buffer, pH 4.7, $2 \, mol \, l^{-1}$.
   Mercuric nitrate, aqueous, $1.4 \times 10^{-2} \, mol \, l^{-1}$.
   Tracer solutions of chromic chloride (^{51}Cr) and sodium chromate (^{51}Cr), supplied by the Isotope Division, Australian Atomic Energy Commission.
   Water samples were collected in acid-washed high density polythene bottles. After filtration through a $0.45 \mu m$ membrane filter samples were stored samples at $4°C$ until analyzed.

Apparatus. Elecrodepositions were carried out using a Princeton Applied Research model 174 polarographic analyzer. The cell, constructed from Perspex, consisted of a threaded top section containing the three electrodes and gas inlet tubes and a detachable base section capable of holding 25 ml of solution. Pyrolytic graphite-coated tubes (9 mm × 3 mm i.d.) supplied by Ultra Carbon Corp., Bay City, Mich., were used as replaceable cathodes. For experiments on deposit distributions, tubes supplied by Varian Techtron were also used. A tube was held by a polythene screw in the recessed end of the Teflon tube. Pressure contact was made with the end of the graphite electrode, by a platinum wire inserted through a separate hold in the Teflon tube. The pyrolytic graphite coating was scraped from the upper end of the electrode to ensure good electrical contact. During electrolysis, the graphite should be fully covered by solution. The anode consisted of a coil of platinum wire. A ceramic plug salt bridge was used to connect a Beckman No. 39178 fibre-junction, saturated calomel electrode (SCE). Nitrogen flow could be regulated either above or under the solution via Pyrex glass inlet tubes.

Solution was circulated through the cathode and returned to the cell at 2.2 ml s^{-1} using a Masterflex pump drive (model 7544–80) and head (model 7015–20).

Atomic absorption analyses were performed on a Varian Techtron AA-5 spectrometer using the CRA-90 furnace accessory. A Varian Techtron chromium hollow cathode lamp was operated at 8 mA, and the 357.9 nm line monitored, using the power supply in the atomization mode with a programmed ramp at 500 °C s^{-1}, to 2200 °C with a 2 s hold. Peak heights were recorded on a Mace FBQ100 chart recorder. A fast detection system (time constant 2 ms) was used to record signal profiles. The signal from the Photomultiplier was led directly to the input of a Biomation model 805 waveform recorder which was triggered by the furnace power supply when it commenced the atomization stage. Recorded signals were displayed on a Tektronix 7504 oscilloscope and plotted using the slow analogue output of the waveform recorder.

Furnace temperatures were measured using an Ircon series 6000 radiation thermometer (Ircon Inc Skokie, III) with the temperature range of 1000 to 3000 °C. Radiation from the central inner area of the furnace was monitored and the output of the radiation thermometer was fed directly into the waveform recorder.

Procedure. Clean the cell and associated circulation system before use with dilute nitric acid, followed by distilled water. To a 25-ml volumetric flask, add 0.2 ml of 2 mol l^{-1} acetate buffer pH 4.7 and 0.3 ml of 1.4×10^{-2} mol l^{-1} mercuric nitrate and make up to volume with sample. Transfer the solution to the cell base. With Teflon-tipped tweezers place a graphite tube in the cathode holder and tighten the holding screw using an acid-cleaned Perspex screwdriver. Position the cathode holder, activate the pumping system, and deaerate the solution with nitrogen for 5 min. Preset the deposition potential and with gas now passing over the solution, commence the electrodeposition for the required time. At the completion of this period, unscrew the cell base, allowing the pump tubing to empty, and then quickly hold in position a second cell base filled with distilled water. Allow several seconds for water to be drawn through the cathode tube, remove the cell base, and withdraw the cathode holder. Remove excess water from the outside face of the graphite tube with a filter paper. Unscrew the polythene screw, remove the graphite tube using the Teflon-coated tweezers, and turn off the applied voltage and pumping system. Position the tube on a watch glass under an infrared lamp to dry for several minutes and then place in a pill-pack until ready to measure. For measurement, transfer the tube to the graphite furnace atomizer unit of the atomic absorption spectrometer.

The practical utility of this electrodeposition technique was tested by the analysis of chromium species in a number of saline water samples. Absorbances for chromium deposited at -0.3 V and -1.8 V from samples buffered to pH 4.7 were calibrated by the method of standard additions. This permitted calculation of the concentrations of labile chromium VI and chromium III respectively. Implicit in these calculations is the assumption that the concentration of labile chromium VI are equal at the two deposition potentials. This is not unreasonable since the existence of any complexed chromium VI species is unlikely. The only species likely to be in a bound form will be the small fraction adsorbed on collodial matter, which is assumed to behave similarly at both potentials.

The determination of total chromium requires a more vigorous treatment than the boiling of an acidified sample used for other bound heavy metals since the dissociation of inert chromium III complexes is slow. Experiments with ultraviolet irradiation in the presence of hydrogen peroxide showed that after a 6 h irradiation of an acidified sample, all chromium species had been reduced to chromium III with all organic matter being fully decomposed. The completeness of this reduction was confirmed by measurements at -0.3 V.

Results for the analysis of some saline waters are shown in Table 9.4.

In a further method hexavalent chromium (i.e. chromate and dichromate) is reduced by diethyldithiocarbamate to trivalent chromium [9.54, 9.55] with which it forms an isobutyl methyl ketone soluble complex. Preconcentrated chromate is then determined in the solvent extract by atomic absorption spectrometry at 357.9 nm.

Parkow et al. [9.56] have described a procedure for the differential analysis of traces of chromium VI (chromate) and chromium III in natural waters. The

Table 9.4. Chromium species in saline waters, $\mu g\ l^{-1}$. (From [9.53]; Copyright 1980, American Chemical Society)

Source of sample	Salinity %	Labile Cr(VI)	Labile Cr(III)	Total Cr
Cronulla Point, Australia	35	0.13	0.08	0.25
Woronora River, Australia	21	0.20	0.03	0.31
Georges River, Australia	35	0.68	0.22	1.25

sample is filtered, acidified, and divided into three portions, one of which is left untreated while the others are passed through a cation exchange resin and an anion exchange resin, respectively. The three aliquots are then treated with nitric acid, evaporated, and analysed by atomic absorption to give the concentrations of cationic, anionic, and non-ionic chromium. The concentration of chromium III is probably closely related to the sum of the cationic and non-ionic fractions and the concentration of chromium VI corresponds to the anion portion.

Reagents. Potassium dichromate and all other chemicals and buffer solutions used were reagent grade.

Apparatus. Analyses for chromium were performed on a Perkin Elmer 303 Atomic Absorption spectrophotometer, with a Honeywell Electronix 194 recorder. Anion exchange resin, AG1-X4 (Cl-form, 100–200 mesh) and cation exchange resin 50W-X4 (Na form, 100–200 mesh): analytical grade, Bio-Rad Laboratories. Glassfibre filter papers (diameter 70 mm S&S No. 25) used to remove suspended matter from river water samples: Schleicher and Schuell Corporation. Membrane fiters: (0.45 µm pore size), Millipore Corporation.

Deionized water, distilled from an all Pyrex still and used for dilution and for the rinsing of (1 mol l^{-1} HNO_3 washed) glassware.

Procedure. Collect samples in (1 mol l^{-1} nitric acid washed, doubly deionized water rinsed) polyethylene containers. Rinse the containers (4 l volume) several times with the water to be analysed. Filter a 3-l water sample through glass fibre filters. After filtration, divide the 3-l sample into equal portions of 1 l each. Pass one aliquot through a 4.0 × 0.5 cm column of anion exchange resin, one through a similar column of cation exchange resin, and leave one untreated. Add 10 ml of 1.0 mol l^{-1} nitric acid to each 1 l aliquot and then, slowly (over 4 h) evaporate (without boiling) to a volume of 10 ml. Analyse by atomic absorption spectrometry using the Cr line at 357.9 nm and the method of standard additions using a reducing flame (yellow) to maximize the chromium signal. The concentrations of total, anionic, cationic, and non-ionic chromium in the original 3-l sample are obtained by difference. The precisions of the analyses for total, anionic, cationic, and non-ionic chromium were respectively ± 14, ± 20, ± 20, and $\pm 25\%$ relative.

9.8.1.5 High Performance Liquid Chromatography

Reverse phase high performance liquid chromatography has been used to carry out the simultaneous determination of Cr III and Cr VI in water [9.57]. The sample is buffered to pH 5.8 and allowed to react overnight at room temperature with 5% sodium diethyl dithiocarbamate. It is then extracted with chloroform and analysed by reversed phase high performance liquid chromatography after the residue from evaporation of the chloroform has been dissolved in acetonitrile, using methonal: water (65:35) as the mobile phase. Detection was by UV at 254 nm. Corrections were made for interferences between chromium III and chromium VI and concentrations of 2–10 mg l^{-1} of each ion were determined with accuracies of the order of $\pm$ 5%.

Apparatus. HP 1084 liquid chromatograph. Column for reversed phase, 250 mm × 4.6 mm i.d., 2 µm first pore, filled with LiChrosorb PP-8 particle size 10 µm (Merck, Darmstadt, FRG).UV detector, 254 nm.

Procedure. Transfer the sample (1 ml) to a 25-ml separating funnel. Add 5 ml 0.2 mol l^{-1} acetate buffer solution (pH = 4.0 or pH = 5.8). Then add 1 ml 5% sodium diethyldithiocarbamate and allow the reaction mixture to stand overnight. Extract the reaction mixture twice with 5 ml chloroform. Evaporate the chloroform under reduced pressure.

For analysis, dissolve the residue in 1 ml acetonitrile and chromatograph under the following conditions:
    Column   RP-8
    Mobile phase Methanol:water (65:35)
    Flow rate   3 ml mm^{-1}
    Detection   254 nm

In this method chromium III chelates at pH 5.8 but not at pH 4.0. Evaluation of the peaks shown in Fig. 9.4 produced with chromium III and chromium VI standard solutions enables the concentration of these two species in unknown water samples to be deduced.

9.8.1.6 Polarography

Differential pulse voltammetry has been used to determine chromate and chromium III [9.58, 9.59].

Crosmun and Mueller [9.59] achieved a detection limit of 10 µg l^{-1} chromate in natural waters.

9.8.1.7 Miscellaneous Methods

Rajendrababu and Nanda Kummar [9.60] have described a simple paper chromatographic method for the rapid detection of chromate in natural water in amounts down to 2 µg. Chemiluminescence with luminol peroxide [9.61] and with lopine [9.62] have been used to determine low concentrations of chromate in water. The former technique achieved a detection limit of 0.05 mg l^{-1} and the latter 0.3 µg l^{-1} without preconcentration and 0.015 µg l^{-1} with preconcentration.

Cranston and Murray [9.63] have reviewed methods for the determination of chromium species in natural waters.

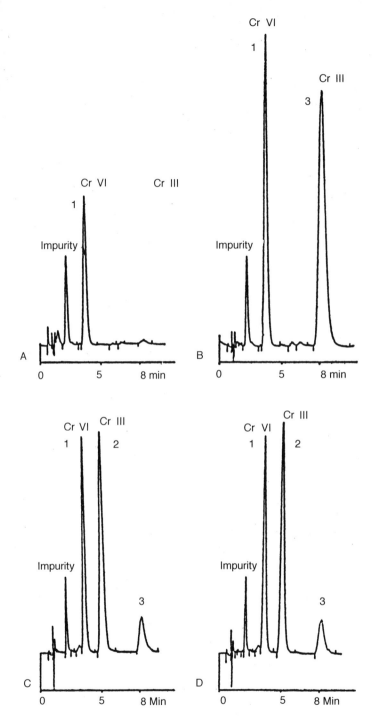

Fig. 9.4. A–D Chromatograms of reaction mixtures of Cr (III) and Cr (VI) Conditions for HPLC: column RP-8, mobile phase: methanol-water (65:35); flow rate, 3 cm^3 min^{-1} detection at 254nm: **A** 10.0 µg Cr III, pH 4.0; **B** 10.0 µg Cr III, pH 5.8; **C** 10.0 µg Cr VI, pH 4.0; **D** 10.0 µg Cr VI, +10.0 µg Cr III pH 4.0. 1, unidentified disulphide; 2, Cr(S$_2$CH (C$_2$H$_5$)$_2$)$_2$(OS$_2$ CN(C$_2$H$_5$)$_2$), (Cr VI); 3, Cr(S$_2$CN(C$_2$H$_5$)$_2$)$_3$, (CRIII) (From [9.57])

9.8.1.8 Sample Preservation

Stollenwerk and Grove [9.64] investigated the effect on the concentration of hexavalent chromium of sample preservation techniques. Acidification of samples could reduce hexavalent chromium to the trivalent form. The rate of such reduction was increased by impurities such as nitric oxide present in nitric acid. Organic matter could also accelerate reduction of hexavalent chromium in acidified samples, and could also reduce the chromium even in unacidified samples. The best method for preserving hexavalent chromium involved filtration and cooling of the sample. Storing the sample at $4\,^\circ C$ might preserve the concentration of hexavalent chromium for some time, depending on the reactivity of other constituents in the sample, but should not be relied on. Measurement of total chromium is recommended as a method of checking on chromium reduction.

9.8.2 Waste Waters

9.8.2.1 Spectrophotometric Methods

Wei et al. [9.65] described a spectrophotometric method for the determination of chromium VI in electroplating waste waters and total chromium III and VI in industrial waste waters using 2-(5-bromo-2-pyridylazo)-5-diethylaminophenol. This was found to be the most sensitive reagent for chromium. The complex formed was very stable.

Reagents. As follows.
Standard chromium – 1.00 mg ml^{-1} of chromium. Prepare by dissolving 282.9 mg of potassium dichromate in 100 ml of distilled water. Prepare solutions of lower concentrations by diluting aliquots of the stock solution.
    2-(5-bromo-2-pyridylazo)-5-diethylaminophenol.
    Acetate buffer, pH 4.7.
    Hydroxylamine, 10% w/v.
    Ethanol.
    Ferric chloride, 0.1% w/v.
    Sodium hydroxide, 0.1% w/v.
    Chloroform.

Apparatus. Spectrophotometer used is a Zeiss Specord UV-VIS, double beam, automatic recording instrument or equivalent with matched 1 cm silica cells.

Procedures. Transfer a sample containing not more than 15 µg of chromium into a 100-ml flask. Add, with stirring, 2 ml of acetate buffer (pH 4.7), 1 ml of 10% hydroxylamine hydrochloric solution, 2 ml of 2(5-bromo-2-pyridylazo)-5-diethylaminophenol solution and rinse the walls of the flask with 5 ml of water. The flask is fitted with a stopper which carries a funnel. Place the flask on an electric plate having a surface temperature of 100–105 °C. Heat the contents of the flask until the volume of the solution is reduced to 3–5 ml. Rinse the walls of the flask with 5 ml of ethanol and continue to reflux for 40 min. Remove the flask from the hot plate, and after cooling to room temperature add 5 ml of ethanol to dissolve the reagent and reaction product. Transfer the solution to a 25-ml volumetric flask and make up to the mark with distilled water. Measure the absorbance of the coloured solution at 600 nm, using a reagent blank as reference.

Determination of chromium VI in electroplating waste water is carried out as follows. To 50 ml of the water sample, add 0.1 ml of 0.1% ferric chloride solution, heat to boiling, and add dropwise 1% sodium hydroxide solution until the solution becomes alkaline (pH 8–9). Continue to heat gently until the ferric hydroxide precipitate is completely coagulated (to separate and remove chromium III, cobalt III, nickel II and copper II, and other metal ions). Remove the precipitate by filtration and wash it several times with hot water. Transfer the filtrate together with washings to a 100-ml volumetric flask and make up to volume with distilled water. Pipette an aliquot of this solution and carry out the determination by the procedure described above.

Determination of total chromium III, VI in industrial waste water is carried out as follows. To a water sample (5–20 ml) add 2 ml of acetate buffer (pH 4.7), 2 ml of 2(5-bromo-2-pyridylazo)-5-diethylaminophenol solution and mix thoroughly. Allow the solution to stand for 30 min, extract with several 10-ml portions of chloroform until the organic phase shows an orange colour. After separation of the two phases, discard the chloroform layer. Wash the aqueous layer once with chloroform and separate the chloroform as completely as possible. Transfer the aqueous layer to a 100-ml flask and carry out the determination by the procedure previously described.

The effect of foreign ions on the determination of 10 µg of chromium III was examined. The following ions (when present in the amounts in mg shown in brackets) do not interfere:

$Na^+, K^+, NH_4^+, Cl^-, Br^-, I^-, NO_3^-, SO_4^{2-}, Ac^-$ (10);

$Li^+, Ca^{2+}, Mg^{2+}, Sr^{2+}, Al^{3+}, Mn^{2+}, Cd^{2+}, SCN^-, SiO_3^{2-}, PO_4^{3-}$ (1);

$Be^{2+}, Pb^{2+}, As^{3+}, La^{2+}, La^{3+}, Ce^{3+}, Y^{3+}, Ge^{4+}, Sc^{4+}, UO_2^{2+}$ (0.1);

Zn^{2+}, Th^{4+}, F^- (0.05) and $Ag^+, Au^{3+}, Hg^{2+}, Sc^{3+}, WO_4^{2-}, VO_3^-$ (0.01).

Equal amounts of $Fe^{3+}, Cu^{2+}, Co^{2+}, Ni^{2+}, Ga^{3+}, Tn^{3+}$ and Tl^{3+} produce positive errors.

Table 9.5 shows the results of determination of chromium by the above method in comparison with the diphenylcarbazide method. o-Toluidine, [9.66] o-nitrophenylfluorone acid [9.67], and 2-(5-bromo-2-pyridylazo)-5-diethylaminophenol have been used as chromogenic reagents in the determination of hexavalent chromium (chromate) in waste waters. Qi and Zhu [9.67] reacted

Table 9.5. Results of analysis by present method in comparison with diphenylcarbazide method. (From [9.65]; Copyright 1982, Springer-Verlag GmbH, Germany)

Sample No.	Water taken ml	Chromium found by present method µg	Chromium content in water sample mg l^{-1}	Chromium found by diphenylcarbazide method mg l^{-1}
1	20.0	2.30	0.115	0.110
2	10.0	6.00	0.600	0.550
3	2.0	5.56	2.78	2.65
4	4.0	7.40	1.82	1.73
5	2.5	9.1	3.64	3.76

chromium VI with o-nitrophenylfluorene in the presence of acetyltrimethyl ammonium bromide to form a purple red complex at pH 4.7 – 6.6 by heating at 50 °C for 10 min. The wavelength of maximal absorbance was 582 nm. Beer's law is obeyed up to 0.2 µg l^{-1} chromium VI. Interference due to copper II, iron III and aluminium III was eliminated by the addition of a masking agent containing potassium fluoride, trans-1, 2-diaminocyclohexane tetraacetic acid and potassium sodium tartrate.

9.8.2.2 Atomic Absorption Spectrometry

Martin and Riley [9.68] used atomic absorption spectrometry to determine hexavalent chromium in waste waters.

9.8.2.3 Polarography

Hanzdorf and Januer [9.69] applied differential pulse polarography to the determination of chromate in waste water samples and investigated interferences by other cations.

9.8.2.4 Miscellaneous Methods

Pavel et al. [9.70] examined the behaviour and persistence of trace amounts of chromium VI in waste water samples as a function of pH, container material, initial concentration and temperature. Based on these results they devised a method for preserving the hexavalent form of chromium. Following adjustment to pH 7–8 addition of 10 vol.% of a solution containing 0.15 mol l^{-1} sodium bicarbonate and 0.1 mol l^{-1} EDTA (disodium salt) reduced losses of chromium VI to less than 20% after 7 days, compared with a 60–80% loss in unprotected samples.

9.8.3 Potable Water

9.8.3.1 Atomic Absorption Spectrometry

Posta et al. [9.87] have described a high performance flow flame atomic absorption spectrometric method for the determination of tri- and hexavalent chromium and the preconcentration of hexavalent chromium.

A high performance liquid chromatographic integrator at the output of the flame atomic absorption spectrometer makes it possible to process simultaneously the signals of both oxidation states of chromium. Detection limits for tri- and hexavalent chromium ore, respectively, 0.03 and 0.02 $\mu g\ l^{-1}$.

9.8.4 Oxides, Silicate Glasses and Cements

9.8.4.1 X-Ray Microprobe Analysis

Bajt et al. [9.88] applied synchotron X-ray microprobe analysis to the determination of the chromate content. X-ray absorption near edge structure analysis was used. Down to 10 mg l^{-1} chromate could be determined.

9.9 Germanate

9.9.1 Natural Waters

9.9.1.1 Flow Injection Analysis

Fogg and Bsetsu [9.71] have discussed the determination of germanate in natural waters by flow injection analysis.

9.10 Molybdate

9.10.1 Natural Waters

9.10.1.1 Atomic Absorption Spectrometry

Vazquez-Gonzalez et al. [9.72] have described a method for preconcentrating and determining molybdenum by electrothermal atomization atomic absorption spectrometry after preconcentration by means of anion exchange using Amberlite IRA-400 in resin citrate form. The optimal analytical parameters were established by drying, carbonization, charring, atomization, and cleaning in a graphite furnace. The precision and accuracy of the method were investigated. Less than 0.2 µg l^{-1} molybdenum could be determined by this procedure

Samchuk [9.73] developed an atomic absorption method for the determination of molybdenum in natural waters, with preconcentration on carbon modified by complex-forming organic reagents and on chelated sorbents. The methods for preparing the modified carbons and chelated sorbents are described. The work was done on a two-beam atomic absorption spectrometer with a graphite atomizer and deuterium background corrector.

9.10.2 Sea Water

9.10.2.1 Atomic Absorption Spectrometry

A limited amount of work has been carried out on the determination of molybdenum in sea water by atomic absorption spectrometry [9.74 – 9.76] and graphite furnace atomic absorption spectrometry [9.77]. In a recommended procedure [9.77] a 50-ml sample of sea water at pH 2.5 is preconcentrated on a column of 0.5 g p-aminobenzylcellulose, then the column is left in contact with 1 mol l^{-1} ammonium carbonate for 3h, after which three 5-ml fractions are collected. Finally, molybdenum is determined by atomic absorption at 313.2 nm with use of the hot-graphite-rod-technique. At the 10 mg l^{-1} level the standard deviation was 0.13 µg.

9.11 Tungstate

9.11.1 Natural Waters

9.11.1.1 Atomic Absorption Spectrometry

This technique has been used to determine tungstate as the benzoin oxime complex in methyl isobutyl ketone extracts of natural waters [9.78].

9.11.1.2 Inductively Coupled Plasma Atomic Emission Spectrometry

Hall et al. [9.79] preconcentrated tungsten and molybdenum from natural spring water onto activated charcoal prior to analysis by inductively coupled plasma atomic emission spectrometry and inductively coupled plasma mass spectrometry. The detection limit for both elements by ICPMS was 0.06 µg l^{-1}, and by

ICPAES the detection limits were 1.2 and 0.4 µg l^{-1}, respectively, for tungsten and molybdenum.

9.11.2 Tissues

9.11.2.1 Atomic Absorption Spectrometry

Indirect atomic absorption spectrometry has been used for the indirect determination of tungstate in rat tissue [9.80].

9.12 Uranate

9.12.1 Natural Waters

9.12.1.1 Spectrophotometric Method

Kuroda et al. [9.81] observed that uranium is strongly absorbed from acidified saline solutions by a strongly basic ion exchange resin in the presence of azide ions. The distribution coefficient of uranium with 0.5 mol l^{-1} sodium chloride increased rapidly with an increase in azide concentration, and was much higher than the coefficient obtained with hydrazoic acid alone. The sorbed uranium was easily eluted with 1 mol l^{-1} hydrocholoric acid, and was determined spectrophotometrically. Recoveries of 98.3 – 99.6% uranium were obtained from artificial sea water containing 3.4 µg l^{-1} uranium.

9.13 Vanadate

9.13.1 Natural Waters

9.13.1.1 Spectrophotometric Methods

Murthy et al. [9.82] determined vanadium IV and vanadium V in water by solvent extraction of the complexes with 2-hydroxy-acetophenone oxime followed by spectrophotometry.

9.13.2 Soil and Plant Extracts

Abbasi [9.83] determined metavanadate in solution by a method based on the formation of a violet colour with vanadium V on addition of a chloroform solution of N-(p-NN dimethylanilino-3-methoxy-2-naphtho)-hydroxamic acid to the acidified (4 – 6 mol l^{-1} hydrochloric acid) sample. This solution was evaluated spectrophotometrically at 570 nm. The detection limit was 0.05 µg vanadium at a dilution ratio of 1:10^7. Very few interferences occur in this procedure.

The method was also applied to extracts of soils, plants and geological samples.

9.14 Titanate

9.14.1 Natural Waters

9.14.1.1 Spectrophotometric Methods

Titanates have been determined spectrophotometrically in water and soil extracts by a method based on reaction with N-p-methoxyphenyl-2 furohydroxamic acid which forms a golden yellow colour with an absorption maximum at 385 nm with titanium V. In this method [9.84] the titanate is reduced to titanium III with stannous chloride and then extracted with N-p-methoxy-phenyl-1-2-furohydroxamic acid in chloroform. The chloroform extract was then evaluated spectrophotometrically. Beer's law is obeyed over the range 0.5 – 10 mg l^{-1} titanate and 1.7 mg l^{-1} of titanium could be detected.

The following ions did not interfere when present at 50 times the concentration of titanate: $Cu^{2+}, Co^{2+}, Zn^{2+}, Hg^{2+}, Ag^+, Cd^{2+}, UO_2^{2+}, V^{4+}, Fe^{2+}, Fe^{3+}, Al^{3+}, La^{3+}, Sc^{3+}, Ga^{3+}, Th^{4+}$, halides, acetate, carbonate, nitrate, oxalate.

Plant and soil samples were brought into solution by ashing (plants), alkali fusion (soils) and acid treatment.

9.15 Ferrocyanide

9.15.1 Industrial Effluents

9.15.1.1 Spectrophotometric Method

Ferrocyanides in coke plant effluents have been determined spectrophotometrically using 2,2'-bipyridyl or 1,10-phenanthroline [9.85]. The ferrocyanide is first broken down by digestion with formaldehyde prior to the determination of uncomplexed iron.

9.16 Metal Cyanide Complexes

9.16.1 Natural Waters

9.16.1.1 Ion Chromatography

The cyanide ion in inorganic cyanides can be present as both complexed and free cyanide. In order to study the chromatography of metal cyanides, Rocklin and Johnson [9.86] prepared and assayed solutions of cadmium, zinc, copper, nickel, gold, iron and cobalt cyanides. Table 9.6 lists the percentage of total cyanide detected.

The results suggest that the complex cyanides can be grouped into three catogories depending on the cumulative formation constant and stability of the complex.

Category 1 includes the weakly complexed and labile cyanides $Cd(CN)_4^{2-}$ (log $B_4 = 18.78$) and $Zn(CN)_4^{2-}$ (log $B_4 = 16.7$). These complexes completely dissociate under the chromatographic conditions used, the cyanide being indistinguishable from free cyanide.

Category 2 includes the moderately strong cyanide complexes $Ni(CN)_4^{2-}$ (log $B_4 = 31.3$) and $Cu(CN)_4^{3-}$ (log $B_4 = 30.3$). Although these complexes are labile, they are retained on the column and slowly dissociate during the chromatography. This slow dissociation produces tailing which lasts for several minutes as the free cyanide elutes and is detected. As the results presented in Table 9.6 demonstrate, the tailing and the non-quantitative recovery of cyanide preclude the use of direct injection to determine total cyanide in samples containing copper and nickel. These samples may be analyzed after acid distillation and caustic trapping. The cyanide in the caustic solution can then be determined by ion chromatography with electrochemical detection.

Category 3 includes those cyanides which are inert and therefore totally undissociated, such as $Au(CN)_2^-$ (log $B_2 = 38.3$), $Fe(CN)_6^{3-}$ (log $B_6 = 42$) and

Table 9.6. Percentage of total cyanide in metal complexes determined as "free" cyanide. (From [9.86]; Copyright 1983, American Chemical Society)

Metal complex	log β_r	%
$Cd(CN)_4^{2-}$	18.8	102
$Zn(CN)_4^{2-}$	16.7	102
$Ni(CN)_4^{3-}$	31.3	81
$Cu(CN)_4^{3-}$	30.3	52
$Cu(CN)_3^{2-}$	28.6	42
$Cu(CN)_2^-$	24.0	38
$Au(CN)_2^-$	38.3	0
$Fe(CN)_6^{3-}$	42	0
$Co(CN)_4^{3-}$	64	0

Co(CN)$_6^{3-}$ (log B$_6$=64). No free cyanide was detected for these complexes. Although these complexes do not elute under the chromatographic conditions used, they can be eluted and determined by using different chromatographic conditions and conductivity detection.

Samples containing both free cyanide (or weakly complexed cyanide) and strongly complexed cyanide can be analyzed for free cyanide by direct injection. The determination of total cyanide (both free and strongly complexed) requires distillation of the sample with caustic trapping.

9.17 References

9.1 Nas UT, Kan M (1988) Analyst (London) 113: 1683
9.2 Hoover TB, Jager GD (1984) Anal Chem 56: 221
9.3 Matsunhara C, Yamamoto Y, Takamura K (1987) Analyst (London) 112: 1257
9.4 Fogg AG, Bselsn NK (1981) Analyst (London) 106: 1288
9.5 Hansen D, Richter BE, Rollins DK, Lamb JD, Eatough J (1979) Anal Chem 51: 633
9.6 Rozanski L (1971) Chemica Analit 16: 793
9.7 Johnson DL, Pilson ME (1972) Anal Chim Acta 58: 289
9.8 Johnson DL (1971) Environmental Science and Technology 5: 411
9.9 Murphy E, Riley R (1963) Analytical Abstracts 10: 1219
9.10 Urasa IT, Ferede F (1987) Analytical Chemistry, 59: 1563
9.11 Chakraborti D, Nickols L, Irgolic KJ (1984) Fresenius Zeitschrift fur Analytische Chemie 319: 248
9.12 Bodini ME, Alzamora O (1983) Talanta 30: 409
9.13 Takayanagi K, Wong GTF (1983) Anal Chim Acta 148: 263
9.14 Harper DJ, PhD. Thesis (1984) University of Liverpool 15: 251
9.15 Lin Ahmed R, Hill JD, Magee RJ (1983) Analyst (London) 108: 835
9.16 Christian GD, Knoblock EC, Purdy WC (1967) Analytical Chemistry 35: 1128
9.17 Brimmer SP, Fawcett WR, Kulhavy KA (1987) Analytical Chemistry 59: 1470
9.18 Karlson U, Frankenberger WT (1986) Analytical Chemistry 58: 2704
9.19 Shimoishi Y (1976) Analyst (London) 101: 298
9.20 Blotcky AJ, Hansen GT, Borkar N, Ebrahim A, Rack EP (1987) Analytical Chemistry 59: 2063
9.21 Uchida H, Shimoishi Y, Toei K (1981) Analyst (London) 106: 757
9.22 Ahmed RB, Hill JO, Magee RJ (1983) Analyst (London) 108: 835
9.23 Rankin JM (1973) Environmental Science and Technology 7: 823
9.24 Itoh K, Nakayama M, Chikuma M, Tanaka H (1985) Fresenius Zeitschrift fur Analytische Chemie 321: 56
9.25 Cutter GA (1986) Electric Power Storage Institute, Palo Alto, California, Report EPRI EA-4641, Vol.1
9.26 Lansford M, McPherson EM, Fishman MJ (1974) Atomic Absorption Newsletter 13: 103
9.27 Cheam V, Agemian H (1980) Analytica Chemica Acta 113: 237
9.28 Cheam V, Murdock A, Sly PG, Lum-Shue-Chan K (1976) Journal of Great Lakes Research 2: 272
9.29 Ohta K, Suzuki M (1980) Fresenius Zeitschrift fur Analytishe Chemie 302: 177
9.30 Bode H (1954) Fresenius Zeitschrift fur Analytishe Chemie 143: 182
9.31 Uchida H, Shimoishi Y, Toei K (1980) Environmental Science and Technology 14: 541
9.32 Shimoishi Y, Toei K (1978) Anal Chim Acta 100: 65
9.33 Shimoishi Y (1980) Talanta 17: 74
9.34 Shilata Y, Morita M, Fuwa K (1985) Analyst (London) 110: 1269

9.35 Cartoni GP, Coccioli F (1986) Journal of Chromatography 360: 225
9.36 Jones A (1976) Anal Chim Acta 87: 429
9.37 Andrews RW, Johnson DC (1976) Anal Chem 48: 1056
9.38 Robberecht HJ, Van Grieken RE (1980) Anal Chem 52: 449
9.39 Cheam V, Agemian H (1980) Anal Chim Acta 113: 237
9.40 Raihle JA (1972) Environmental Science and Technology 6: 621
9.41 Batley GE (1986) Anal Chim Acta 187: 109
9.42 Ferpinski EA (1988) Analyst (London) 113: 1473
9.43 Aoyama M, Habo T, Suzuki S (1981) Analytica Chimica Acta 129: 237
9.44 Aoyama M, Habo T, Suzuki S (1981) Analytica Chimica Acta 129: 239
9.45 Sandell EB, (1959) In "Colorimetric Determination of Traces of Metals" Interscience, New York
9.46 Marchant H (1984) Analytica Chimica Acta 230: 11
9.47 Willems GJ, de Router CJ (1974) Analytica Chimica Acta 68: 111
9.48 Marshall MA, Mottola HA (1985) Analytical Chemistry 57: 729
9.49 Ruz J, Torres A, Rios A, Luque de Castro MD, Valcarcel MJ (1986) Atomatic Chemistry 8: 70
9.50 Ruz J, Rios A, Luque de Castro MD, Valcarcel M (1985) Fresenuis Zeitschrift fur Analytishe Chemie 322: 499
9.51 Yoshimura K (1988) Analyst (London) 113: 471
9.52 Subramanian KS (1988) Analytical Chemistry 60: 11
9.53 Batley GE, Matousek JP (1980) Anal Chem 52: 1570
9.54 Fukamachi K, Morimoto M, Yanagawa M (1972) Japan Analyst 21: 26
9.55 Yanagisawa M, Suzuki M, Takequichi T (1973) Mikrochimica Acta 3: 475
9.56 Parkow JF, Lieta DD, Lin JW, Ohl SE, Shum WD, Januer GE (1977) Science of the Total Environment 7: 17
9.57 Tande T, Pettersen JE, Torgrismsen T (1980) Chromatographia 13: 607
9.58 Golimowski J, Valenta P, Nurberg HW (1985) Fresenias Zeitschrift fur Analytische Chemie 322: 314
9.59 Crosmun ST, Mueller TR (1975) Analytica Chimica Acta 75: 199
9.60 Rajendrababu A, Nanda Kummar O-J (1982) Association of Optical Analytical Chemistry 65: 137
9.61 Bowling JL, Dean JA, Goldstein G, Dale JM (1975) Anal Chim Acta 76: 47
9.62 Marino DF, Ingle JD (1981) Anal Chem 53: 294
9.63 Cranston RE, Murray JW (1978) Anal Chim Acta 99: 275
9.64 Stollenwerk KG, Grove DB (1985) Journal of Environmental Quality 14: 396
9.65 Wei FS, Zhu YR, Yin PHQF, Shen NK (1982) Mikrochimica Acta 2: 67
9.66 Florea K (1973) Revue Roum Chim 18: 1993
9.67 Qi WB, Zhu LZ (1986) Talanta 33: 694
9.68 Martin TD, Riley JK (1982) Atomic Spectroscopy 3: 174
9.69 Hanzdorf C, Janner G (1984) Analytica Chimica Acta 165: 201
9.70 Pavel J, Kliment J, Stoerk S, Suter O (1985) Fresenius Zeitschrift fur Analytische Chemia 321: 587
9.71 Fogg AG, Bsetsu NK (1981) Analyst (London) 106: 1288
9.72 Vazquez-Gonzalez JF, Bermejo-Barrera P, Bermejo-Martinez F (1987) Atomic Spectroscopy 8: 159
9.73 Samchuk AI (1987) Soviet Journal of Water Chemistry and Technology 9: 57
9.74 Nakahara T, Chakrabarti CL (1979) Analytica Chimica Acta 104: 99
9.75 Van den Sloot HA, Wals GD, Das HA (1977) Progress in Water Technology, 8: 193
9.76 Van den sloot HA, Das HA (1977) Progress in Water Technology 8: 193
9.77 Muzzarelli RAA, Rochetti R (1973) Analytica Chimica Acta 64: 371
9.78 Karrey JS, Goulden PD (1975) Atomic Absorption Newsletter 14: 33
9.79 Hall GEM, Jefferson CW, Michel FA (1988) Journal of Geochemical Exploration 30: 63
9.80 Chakraborty D, Das AK (1989) Analyst (London) 114: 67
9.81 Kuroda R, Oguma K, Mukai N, Twamoto D, Talanta (1987) 34: 433

9.82 Murthy GVR, Reddy TS, Rao SB (1989) Analyst (London) 114: 493
9.83 Abbasi SA (1981) International Journal of Environmental Studies 18: 51
9.84 Abbasi SA (1982) International Journal of Environmental Analytical Chemistry 11: 1
9.85 Vail EI, Borisenko A, Orlova VZ, Koks Khim (1970) 5: 46. Ref: Zhur. Khim. (1970) 19GD (19) Abstract No. 19G152
9.86 Rocklin RD, Johnson EL (1983) Analytical Chemistry 55: 4
9.87 Posta J, Berndt H, Luo SK, Scholdach G (1993) Analytical Chemistry 65: 2590
9.88 Bajt S Clark SB, Sutton SR, Rivers ML, Smith JV (1993) Analytical Chemistry 65: 1800

10 Organic Anions

Methods for determining various organic anions are discussed below.

10.1 Formate and Acetate

10.1.1 Natural Waters

10.1.1.1 High Performance Liquid Chromatography

Kieber et al. [10.1] determined formate in natural waters including seawater by a coupled enzymatic/high performance liquid chromatographic technique. The precision is approximately $\pm 5\%$ relative standard deviation. Intercalibration with an anion chromatographic technique showed an agreement of 98%. Down to 0.5 µmol l^{-1} absolute of formate could be determined. Kieber et al. [10.1] found 0.2–0.8 mol l^{-1} formate in sea water and 0.4–10 µmol l^{-1} in rain water.

The procedure involves precolumn oxidation of formate with formate dehydrogenase which is accompanied by a corresponding reduction of β-nicotinamideadenine dinucleotide (β (NAD)$^+$) to reduced β-nicotinamide dinucleotide (i.e. β NADH). The latter is quantified by high performance liquid chromatography.

10.1.1.2 Ion-Selective Electrodes

A potentiometric gas sensing probe has been used for the selective determination of acetate ion in water [10.29].

10.2 Citrate

10.2.1 Foodstuffs and Beverages

10.2.1.1 Ultraviolet Spectroscopy

A method for the enzymic determination of citrate is based on the enzyme citrate lysase (CL) catalysed reaction of citrate to produce oxaloacetate and acetate [10.2].

$$\text{citrate} \xrightarrow{\text{CL}} \text{oxaloacetate} + \text{acetate}. \tag{10.1}$$

In the presence of enzymes malate dehydrogenase (MDH) and lactate dehydrogenase (LDH), oxaloacetate and its decarboxylation product pyruvate are reduced to L-malate and L-lactate, respectively, by reduced nicotinamide-adenine dinucleotide (NADH):

$$\text{oxaloacetate} + \text{NADH} + \text{H}^+ \xrightarrow{\text{MDH}} \text{L-malate} + \text{NAD}^+ \tag{10.2}$$

$$\text{pyruvate} + \text{NADH} + \text{H}^+ \xrightarrow{\text{LDH}} \text{L-lactate} + \text{NAD}^+. \tag{10.3}$$

The amounts of NADH oxidized in reaction (10.2) and (10.3) are stoichiometric with the amounts of citrate. NADH is measured spectrophotometrically at 344, 340 and 365 nm.

This method has been applied to the determination of citrate in liquid and solid foodstuffs, fruit juices, wine, beer, bread, meat, cheese, vegetables, fruit, margarine, edible oil and food dressings.

Indyk and Kurmann [10.3] have determined citrate in milk powders by a spectrophotometric method.

10.3 Isocitrate

10.3.1 Foodstuffs

10.3.1.1 Ultraviolet Spectroscopy

D-Isocitric acid (D-isocitrate) is oxidatively decarboxylised by nicotinamideadenine dinucleotide phosphate (NADP) in the presence of the enzyme isocitrate dehydrogenase (ICDH) [10.4].

$$\text{disocitrate} + \text{NADP}^+ \xrightarrow{\text{ICDH}} \text{2-oxoglutarate} + CO_2 + \text{NADPH} + H^+ \quad (10.4)$$

The amount of NADPH formed in reaction (10.4) is stoichiometric with the amount of D-isocitrate. NADPH is determined by ultraviolet spectroscopy by means of its absorbance at 334, 340 and 365 nm.

The bound D-isocitric acid (esters, lactones) is determined after alkaline hydrolysis according to the same principle [10.2, 10.3].

$$\text{D-isocitric acid ester} + H_2O \xrightarrow{\text{pH 9-10}} \text{D-isocitrate} + \text{alcohol} \quad (10.5)$$

$$\text{D-isocitric acid lactone} + H_2O \xrightarrow{\text{pH 9-10}} \text{D-isocitrate}. \quad (10.6)$$

This method has been applied to the determination of D-isocitrate in a variety of food products.

10.4 Oxalate

10.4.1 Foodstuffs and Beverages

10.4.1.1 Ultraviolet Spectroscopy

Oxalate decarboxylase cleaves oxalic acid (oxalate) to formic acid (formate) and CO_2 at pH 5.0 [10.5]:

$$\text{oxalate} \xrightarrow{\text{oxalate decarboxylase}} \text{formate} + CO_2. \quad (10.7)$$

Formate formed is quantitatively oxidized to bicarbonate by nicotinamide-adenine dinucleotide (NAD) at pH 7.5 in the presence of the enzyme formate dehydrogenase reaction (10.8). The increase of NADH is determined by ultraviolet spectroscopy by means of its absorbance at 334, 340 and 365 nm.

$$\text{formate} + \text{NAD}^+ + H_2O \xrightarrow{\text{formate dehydrogenase}} HCO_3^- + \text{NADH} + H^+. \quad (10.8)$$

The amount of NADH formed during reaction (10.8) is stoichiometric with the amount of oxalic acid.

This method has been applied to the determination of oxalate in fruit juices, vegetable juices, beer, solid foodstuffs, chocolate products, fruit and vegetables.

10.5 Malate

10.5.1 Foodstuffs and Beverages

10.5.1.1 Ultraviolet Spectroscopy

In the presence of L-malate dehydrogenase (L-MDH), L-malic acid (L-malate) is oxidized by nicotinamide-adenine dinucleotide (NAD) to oxaloacetate [10.6]. The equilibrium of this reaction lies far on the side of malate. Removal of oxaloacetate from the reaction system causes displacement of the equilibrium in favour of oxaloacetate. In the reaction catalyzed by the enzyme glutamate-oxaloacetate transminase (GOT), oxaloacetate is converted to L-aspartate in the presence of L-glutamate reaction (10.10):

$$\text{L-malate} + \text{NAD}^+ \xrightleftharpoons{\text{L-MADH}} \text{oxaloacetate} + \text{NADH} + \text{H}^+ \quad (10.9)$$

$$\text{oxaloacetate} + \text{L-glutamate} \xrightleftharpoons{\text{GOT}} \text{L-aspartate} + \alpha\text{-ketoglutarate}. \quad (10.10)$$

The amount of NADH formed is stoichiometric with the concentration of L-malate. It is NADH which is measured by ultraviolet spectroscopy at a wavelength of 334, 340 or 360 nm.

This method has been applied to the determination of L-malate in liquid foodstuffs, concentrates and beverages, wine and solid foodstuffs.

10.6 Lactate

10.6.1 Silage

Wilson and Ferry [10.7] have discussed sources of interferences in the determination of lactate in silage.

10.6.2 Foodstuffs and Beverages

10.6.2.1 Ultraviolet Spectroscopy

In the presence of L-lactate dehydrogenase, L-lactate is oxidized by nicotinamideadenine dinucleotide (NAD) to pyruvate [10.8]:

$$\text{L-lactate} + \text{NAD}^+ \xrightleftharpoons{\text{L-LDH}} \text{pyruvate} + \text{H}^+. \quad (10.11)$$

The equilibrium of this reaction lies almost completely on the side of lactate. However, by trapping the pyruvate in a subsequent reaction catalyzed by the enzyme glutamate-pyruvate transaminase (GPT) in the presence of L-glutamate, the equilibrium can be displaced in favour of pyruvate and NADH reaction (10.12):

$$\text{pyruvate} + \text{L-glutamate} \overset{\text{GPT}}{\rightleftharpoons} \text{L-alanine} + \alpha\text{-oxoglutarate}. \tag{10.12}$$

The amount to NADH formed in the above reaction is stoichiometric with the concentration of L-lactic acid. The increase in NADH is determined by ultraviolet spectroscopy by means of its absorbance at 334, 340 or 365 nm.

This method has been applied to the determination of L-lactate in fruit, vegetable juices, wine, beer, vinegar-containing liquids, sauerkraut juice, yoghurt, cheese and meat products.

An enzyme sensor has been used to determine lactate [10.9].

10.7 Salicylate

10.7.1 Urine

10.7.1.1 Spectrofluorimetry

Munoz de la Pena et al. [10.10] have shown that the overlapping fluorescence spectra of salicylate and salicylurate are resolved by first derivative synchronous fluorimetry. The method was applied to urine analysis.

10.7.2 Pharmaceuticals

A liquid ion exchange membrane has been used for the selective determination of salicylate in pharmaceutical preparations [10.11].

10.8 Ascorbate

10.8.1 Natural Waters

10.8.1.1 Spectrophotometric Method

Norwitz and Keliher [10.12] have discussed the determination by ascorbic and isoascorbic acids by the spectrophotometric diazotization coupling technique.

10.8.1.2 Spectrofluorimetric Method

Ascorbate at the nmol l^{-1} level has been determined by a kinetic fluorimetric method [10.13].

10.8.1.3 High Performance Liquid Chromatography

Kishida et al. [10.30] derivatized ascorbic acid with 2, 4-dinitrophenyl hydrazine to ascorbic acid bis (dinitrophenyl) hydrazone (osazone) prior to separation from other sample constituents by high performance liquid chromatography.

10.8.2 Foodstuffs

10.8.2.1 Spectrophotometric Method

Spectrophotometry of the L-ascorbate-2-oximinocyclohexanone thiosemicarbazone-iron II complex has been used to determine L-ascorbate in pharmaceutical preparations, food and urine [10.14].

Karayannis and Farasoglou [10.15] have described a kinetic spectrophotometric method for the determination of ascorbate in orange juice, parsley and potatoes.

10.8.2.2 Amperometric Method

Matsumoto et al. [10.38] have described a simultaneous kinetic based determination of ascorbate (and fructose) with a rotating bioreactor and stationary platinum ring amperometric detector. The method was applied to food samples. This determination takes advantage of the fast chemical reaction of ascorbate by hexacyanoferrate III ions and the subsequent slower production of hexacyanoferrate II in the D-fructose-5-dehydrogenase catalysed reaction between D-fructose and hexacyanoferrate III as acceptor. Sample processing is accomplished by programmed continuous flow stopped/continuous flow operation.

10.8.3 Pharmaceuticals

10.8.3.1 Ultraviolet Spectroscopy

Direct ultraviolet spectroscopy has been used to determine ascorbate in pharmaceuticals [10.16].

10.8.4 Crustacea

10.8.4.1 Solid Phase Extraction

Nels et al. [10.31] used solid phase extraction to separate ascorbic acid-2-sulphate from the cystis of the brine shrimp *Artemia franciscona*.

10.9 Dehydroascorbate

10.9.1 Foodstuffs

Dehydroascorbic acid (DHA) is reduced at pH 7.5 in the presence of dithiothreitol (DTT) to L-ascorbic acid [10.17]:

$$\text{dehydroascorbate} + \text{DTT}_{red} \xrightarrow{\text{pH 7.5}} \text{ascorbate} + \text{DTT}_{ox}. \tag{10.13}$$

l-Ascorbic acid (ascorbate) and some other reducing substances (x-H_2) reduce the tetrazolium salt MTT (3-(4,5-dimethylthiazolyl-2)-2,5-diphenyltetrazolium bromide) in the presence of the electron carrier PMS (5-methylphenazinium methyl sulfate) at pH 3.5 to a formazan. The total of all the reducing substances is measured in the sample, reaction (10.14):

$$\text{ascorbate(x-H}_2\text{)} + \text{MTT} \xrightarrow{\text{PMS}} \text{dehydroascorbate(x)}$$

$$+ \text{MTT-formazan}^- + \text{H}^+. \tag{10.14}$$

For specific determination, the ascorbate fraction of these reducing substances is removed in a blank determination using ascorbate oxidase (AAO) in the presence of oxygen (reaction (10.15). The dehydroascorbate formed does not react with MTT/PMS:

$$\text{ascorbate} + \tfrac{1}{2}O_2 \xrightarrow{\text{AAO}} \text{dehydroascorbate} + H_2O. \tag{10.15}$$

The absorbance difference of the sample minus the absorbance difference of the sample blank is equivalent to the total ascorbic acid quantity in the sample, i.e. the sum of the dehydroascorbic acid and ascorbate. The MTT-formazan is the measuring parameter and is determined spectrophotometrically by means of its absorbance in the visible range at 578 nm. In parallel with the total ascorbic acid measurement, the free ascorbic acid in the sample is measured directly according to reactions (10.14) and (10.15). For this, too, a blank determination must be carried out.

The difference between the total ascorbic acid and free ascorbic acid gives the dehydroascorbic acid.

This method has been applied to the determination of dehydroascorbate and ascorbic acid in carrots, lettuce and other solid and liquid foodstuffs.

10.10 Pyruvate

10.10.1 Milk

10.10.1.1 Ultraviolet Spectroscopy

In the presence of the enzyme L-lactate dehydrogenase (LDH), pyruvate (pyruvic acid) is reduced to L-lactate by reduced nicotinamide-adenine dinucleotide (NADH):

$$\text{pyruvate} + \text{NADH} + \text{H} \xrightarrow{\text{L-LDH}} \text{L-lactate} + \text{NAD}^+ \tag{10.16}$$

The amount of NADH oxidized during the reaction is stoichiometric with the amount of pyruvate. The decrease in NADH is determined by ultraviolet spectroscopy by means of its absorbance at 334, 340 or 365 nm.

This method has been applied to the determination of pyruvate in milk [10.18].

10.10.1.2 Flow Injection Analysis

Flow injection analysis has been applied to the determination of pyruvate in milk [10.19]. In this method pyruvate is reduced to lactate in the presence of lactate-dehydrogenase (LDH) by reduced nicotinamide-adenine dinucleotide (NADH). The amount of nicotinamideadenine dinucleotide is measured by means of its fluoresence at 365 nm, with excitation at 440 nm:

$$\text{pyruvate} + \text{NADH} + \text{H}^+ \xrightarrow{\text{LDH}} \text{L-lactate} + \text{NAD}^+. \tag{10.17}$$

10.11 Glutamate

10.11.1 Foodstuffs

10.11.1.1 Spectrophotometric Method

In the presence of the enzyme glutamate dehydrogenase (GIDH), L-glutamic acid (L-glutamate) is deaminated oxidatively by nicotinamide-adenosine dinucleotide (NAD) to α-ketoglutarate (reaction 10.18). In the reaction catalyzed by diaphorase the NADH formed converts iodonitro tetrazolium chloride (INT) to a formazan which is measured spectrophotometrically in the visible range at 492 nm [10.20]:

$$\text{L-glutamate} + \text{NAD}^+ + \text{H}_2\text{O} \xrightleftharpoons{\text{GlDH}} \alpha\text{-ketoglutarate} + \text{NADH} + \text{NH}_4^+. \quad (10.18)$$

$$\text{NADH} + \text{INT} + \text{H}^+ \xrightarrow{\text{diaphorase}} \text{NAD}^+ + \text{formazan}. \quad (10.19)$$

The equilibrium of reaction (10.18) lies far on the side of glutamate. By trapping the NADH formed with INT (reaction 10.19), the equilibrium is displaced in favour of α-ketoglutarate.

This method was applied to the determination of L-glutamate in solid and liquid foodstuffs, meat extracts, soup, cheese, sausages, vegetables and fruit products.

10.11.1.2 Potentiometric Method

Nikolalis has described a kinetic potentiometric method for the determination of glutamate in soups [10.21].

10.12 Amino Acids

10.12.1 Natural Waters

10.12.1.1 Gas Chromatography

Gardner and Lea [10.22] preconcentrated free and combined amino acid from lake water on an ion exchange column, prior to desorption with a small volume of acid followed by conversion to N-trifluoroacetyl methyl esters and gas chromatography.

10.12.1.2 Electrophoresis

Yu and Dovichi [10.23] used capillary zone electrophoresis with thermooptical absorbance detection to determine attomole concentrations of 18 amino acids.

10.13 Nitroloacetate and Ethylene Diamine Tetraacetates

10.13.1 Natural Waters

10.13.1.1 Ultraviolet Spectroscopy

Longbottom [10.24] preconcentrated nitriloacetic acid on a strong anion exchange resin, then desorbed it with sodium tetraborate at pH 9 prior to estimation by ultraviolet spectroscopy.

10.13.1.2 Polarography

Nitriloacetic acid has been preconcentrated from water at pH 3 by passage through an anion exchange column and subsequent elution with sodium chloride [10.25] prior to estimation by polarography.

10.13.1.3 Column Exchange Chromatography

Nitriloacetate and ethylene diamine tetraacetate anions have been detected in the eluate of a liquid chromatographic column using a carbon paste amperometric detector [10.26]. With no sample preparation or preconcentration, $0.1\ mg\,l^{-1}$ nitriloacetate or $1.5\ mg\,l^{-1}$ ethylene diamine tetraacetate can be determined in natural and waste waters at a relative standard deviation of less than 7%.

10.13.2 Waste Waters

10.13.2.1 Liquid Chromatography

Dai and Helz [10.39] carried out a liquid chromatographic determination of nitriloacetate and ethylene diaminetetracetate anions using an amperometric detector. Down to $0.1\ mg\,l^{-1}$ nitriloacetate and $0.15\ mg\,l^{-1}$ ethylene diaminetetracetate can be determined in waste water by this method.

10.14 Carboxylates

10.14.1 Natural Waters

10.14.1.1 Column Chromatography

Liquid chromatography separation and indirect detection using iron II^{-1}, 1,10-phenanthroline as a mobile phase additive have been used for the separation of carboxylates (and sulphonic acids) [10.27]. Detection limits approach 20 ng.

10.14.1.2 Ion Exclusion Chromatography

Okada [10.28] has described a redox suppression for the ion exclusion chromatography of carboxylic acids with conductiometric detection. The reaction between hydriodic acid (the eluant) and hydrogen peroxide (the precolumn reagent) is used as the redox suppressor for ion exclusion chromatography of carboxylic acids. The suppressor is useful with highly acidic eluents and reduces background conductance more effectively than a conventional ion exchange suppressor.

10.14.1.3 Ion Chromatography

Some typical separations of anions achieved with this system are shown in Fig. 10.1. Figure 10.1a,b illustrates ion chromatographic separations of mono and diprotic organic acids by ion exchange using anodic AMMS and CMMS micromembrane suppressors.

A further development is the Dionex HPIC AS5A-5µ analytical anion separator column. This offers separation efficiency previously unattainable in ion chromatography. When combined with a gradient pump and an anion micromembrane suppressor the AS5A-5µ provides an impressive profile of inorganic ions and organic acid anions from a single injection of sample (Fig. 10.2). Note that phosphate and citrate, strongly retained trivalent ions, are efficiently eluted in the same run that also gives baseline resolution of the weakly retained monovalents fluoride, acetate, formate and pyruvate. Quantitation of all the analytes shown in Fig. 10.2 using conventional columns would require at least three injections under different eluent conditions.

Another benefit of using the AS5A-5µ gradient pump combination is the ability to change easily the order of elution of ions with different valencies simply by changing the gradient profile. For example, if nitrate were present in high enough concentration to interfere with a malate peak, the malate peak could be moved ahead of the nitrate peak by using a slightly different gradient (Fig. 10.2b, c).

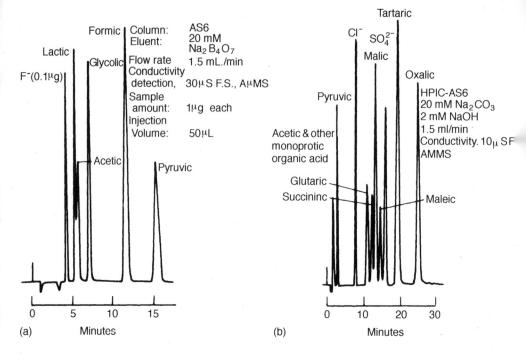

Fig. 10.1a,b. Ion chromatograms obtained with Dionex instrument using (anodic) AMMS and CMMS micromembrane suppression: **a** monoprotic organic acids by anion exchange; **b** diprotic organic acids by anion exchange (From [10.28])

10.15 Sulphonates and Chlorolignosulphonates

10.15.1 Natural Waters

10.15.1.1 Column Chromatography

This technique has been applied to the determination of sulphonates. See Sect. 10.14.1 [10.27].

10.15.2 Potable Waters

Van Loon et al. [10.40] have described a method of quantitative determination of macromolecular chlorolignosulphonates in potable water, river water and sulphite pulp mill effluents. The method is based on pyrolysis–gas chromatogra-

1. F⁻ (1.5 ppm)
2. α-Hydroxybutyrate
3. Acetate
4. Glycolate
5. Butyrate
6. Gluconate
7. α-Hydroxyvalerate
8. Formate (5 ppm)
9. Valerate
10. Pyruvate
11. Monochloroacetate
12. BrO_3^-
13. Cl⁻ (3 ppm)
14. Galacturonate
15. NO_2^- (5 ppm)
16. Glucuronate
17. Dichloroacetate
18. Trifluoroacetate
19. HPO_3^{2-}
20. SeO_3^{2-}
21. Br⁻
22. NO_3^-
23. SO_4^{2-}
24. Oxalate
25. SeO_4^{2-}
26. α-Ketoglutarate
27. Fumarate
28. Phthalate
29. Oxalacetate
30. PO_4^{3-}
31. AsO_4^{3-}
32. CrO_4^{2-}
33. Citrate
34. Isocitrate
35. *cis*-Aconitate
36. *trans*-Aconitate

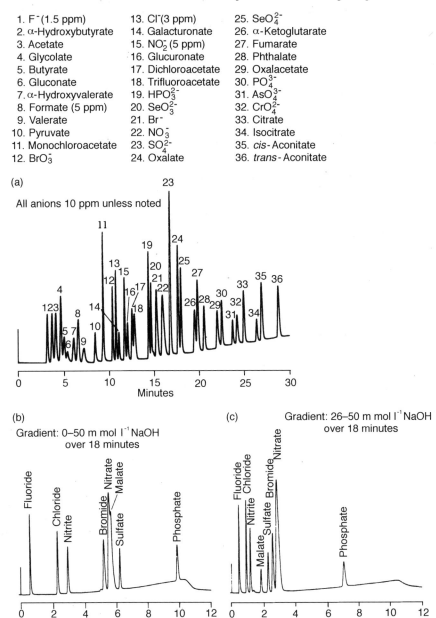

Fig. 10.2a–c. Multi-component analysis by ion chromatography (Dionex)

phy-mass spectrometry. Down to 16–32 $\mu g\, l^{-1}$ chlorolignosulphonates can be determined, 18–310 $\mu g\, l^{-1}$ having been found in river water and 60 $\mu g\, l^{-1}$ in potable water.

10.16 Acetoacetate

10.16.1 In Body Fluids

Lai and Pardue [10.32] carried out a kinetic study of the reaction of acetoacetate in body fluids with glycine and sodium nitroprusside. Acetoacetate reacts with glycine to produce an imine intermediate which tautomerizes to an enamine. Nitroprusside reacts with the imine intermediate to produce an unstable product with an adsorption maximum near to 540 nm. This product decays slowly to produce a stable product with an adsorption maximum Near 393 nm:

$$\text{aceto acetate} + \text{glycine} \underset{k_{-1}}{\overset{k_1}{\rightleftharpoons}} \text{imine} \underset{k_{-1}}{\overset{k_1}{\rightleftharpoons}} \text{enamine}$$

$$+ \text{Nitroprusside} \xrightarrow{k_2} \text{Coloured product} \xrightarrow{k_3} \text{decay product} \quad (10.20)$$

10.17 Mixed Organic Anions

10.17.1 Natural Waters

Hirajama and Fuwamoto [10.33] studied the elution behaviour of benzoate anions in ion chromatography. Anions studied included benzoate, chlorobenzoate, various anisates, dimethylbenzoates, salicylate, hydroxybenzoate, resorcylate, protocatechuate and gallate.

Okada [10.34] discussed redox suppression for the ion exclusion chromatography of carboxylates using conductiometric detection. Anions studied included tartrate, malate, malonate, citrate, glycollate, formate and fumarate.

Saari Nordhaus and Anderson [10.35] studied the separation by ion chromatography of inorganic anions and carboxylates on a mixed stationary phase. Anions studied included chloride, nitrite, nitrate, bromide, phosphate, phosphite, selenite, sulphate, acetate, lactate, propionate, butyrate and isobutyrate. By careful selection of eluent pH and ionic strength a simultaneous determination of the above anions could be achieved.

Rigas and Pietrzyk [10.36] studied ruthenium II complexes as ion interaction reagents for the liquid chromatographic separation and indirect fluorometric detection of a range of organic and inorganic anions. Ruthenium II-1, 10-

phenanthroline and ruthenium II-2,2'-bipyridine were used as mobile phase additives for the liquid analytes including (organic) glycollate, acetate, lactate, propionate, acetoacetate, hydroxybutyrate, chloracetate, isobutyrate, butyrate and (inorganic) fluoride, perchlorate, chloride, nitrite, phosphite, arsenate, nitrate, chlorate, sulphate, chromate and borofluoride.

Berglund et al. [10.37] carried out two dimensional conductiometric detection in ion chromatography sequential suppressed and single column detection of mixtures of glycollates, malonate, fluoride, borate, chloride, arsenate and nitrate.

Olsen et al. [10.41] determined carboxylic acids by isotope dilution gas chromatography Fourier transform infrared spectroscopy. The carboxylic acids were first converted to their methyl esters. Carboxylates studied include malonate, hexabenzene carboxylate, octanoiate and octandioate.

Nimura et al. [10.42] used 1-pyrenyldiazomethane as a fluorescent labelling reagent for the liquid chromatographic determination of carboxylates. Excitation was carried out at 395 nm and fluorescence measurements made at 340 nm. Down to 20-30 Fmol of palmitate, pyrenylpalmitate, lactate propionate and formate were determined.

10.17.2 Sea Water

Nanomolar quantities of individual low molecular weight carboxylates have been determined in sea water [10.43]. Diffusion of the acids across a hydrophobic membrane concentrated and separated the acids from inorganic salts and most other dissolved organic compounds including amines, acetate, propionate, butyrate, (1 and 2), valerate (1 and 2), pyruvate, acrylate, benzoate. These were all found in measurable amounts in seawater samples.

10.18 References

10.1 Kieber DJ, Vaughan GM, Mopper K (1988) Anal Chem 60: 1654
10.2 Boehringer Mannheim, Lewes, East Sussex, UK (1984) Methods of enzymic food analysis, p13, Determination of citric acid
10.3 Indyk HE, Kurmann A (1987) Analyst (London) 112: 1173
10.4 Boehringer Mannheim, Lewes, East Sussex, UK (1984) Methods of enzymic food analyses, p 31, Determination of isocitric acid
10.5 Boehringer Mannheim, Lewes, East Sussex, UK (1984) Methods of enzymic food analysis, p 41, Determination of oxalic acid
10.6 Boehringer Mannheim, Lewes, East Sussex, UK (1984) Methods of enzymic food analysis, p 39, Determination of L-malic acid
10.7 Wilson RF, Ferry RA (1977) Analyst (London) 102: 218

10.8 Boehringer Mannheim, Lewes, East Sussex, UK (1984) Methods of enzymic food analysis, p 33, Determination of L-lactic acid
10.9 Weigett D, Schubert F, Scheller F (1987) Analyst (London) 112: 1155
10.10 Munoz de la Pena A, Salinas F, Meras ID (1988) Analytical Chemistry 60: 2493
10.11 Hassan SSM, Hamanda MA (1988) Analyst (London) 113: 1709
10.12 Norwitz G, Keliher PN (1987) Analyst (London) 112: 903
10.13 Peinado J, Toribo F, Perez, Bendito D (1987) Analyst (London) 112: 775
10.14 Salinas F, Diaz TG (1988) Analyst (London) 113: 1657
10.15 Karayannis M, Farasoglou DI (1987) Analyst (London) 112: 767
10.16 Lau OW: Luk SF, Wong KS (1987) Analyst (London) 112: 1023
10.17 Boehringer Mannheim, Lewes, East Sussex, UK (1984) Methods of enzymic food analysis, p15. Determination of L-ascorbic acid
10.18 Boehringer Mannheim, Lewes, East Sussex, UK (1984) Methods of enzymic food analysis, p 43, Determination of pyruvate in Milk
10.19 Tecator Ltd, UK (1984) Application Short Note ASTN 1984 Determination of pyruvate in milk by flow injection analysis.
10.20 Boehringer Mannheim, Lewes, East Sussex, UK (1984) methods of enzymic food analysis, p 27, Determination of glutamic acid
10.21 Nikolalis DP (1987) Analyst (London) 112: 763
10.22 Gardner WS, Lea GE (1973) Journal of Environmental Science and Technology 7: 719
10.23 Yu M, Dovichi N (1989) Analytical Chemistry 61: 37
10.24 Longbottom JE (1972) Analytical Chemistry 44: 418
10.25 Habermann JP (1971) Analytical Chemistry 43: 63
10.26 Dai J, Helz CR (1988) Analytical Chemistry 60: 30
10.27 Rigas PG, Pietrzyk DJ (1987) Analytical Chemistry 59: 1388
10.28 Okada T (1988) Analytical Chemistry 60: 1666
10.29 Hasson SSM, Ahmed MA, Mageed A (1994) Analytical Chemistry 66: 492
10.30 Kishida E, Nishimoto Y, Kojo S (1992) Analytical Chemistry 64: 1505
10.31 Nels HJ, Merchie G, Lavans D, Songcloos P, De Leenheer AD (1994) Analytical Chemistry 66: 1330
10.32 Lau OS, Pardue HL (1993) Analytical Chemistry 65: 1903
10.33 Hirajama N, Fuwamoto T (1993) Analytical Chemistry 65: 141
10.34 Okada T (1988) Analytical Chemistry 60: 1666
10.35 Saari Nordhaus R, Anderson JM (1992) Analytical Chemistry 64: 2283
10.36 Rigas PG, Pietrzyk DJ (1988) Analytical Chemistry 60: 1650
10.37 Berglund I, Dasgupta PK, Lopez JL, Nara O (1993) Analytical Chemistry 65: 1192
10.38 Matsumoto K, Baza Beeza JJ, Mottola HA (1993) Analytical Chemistry 65: 1658
10.39 Dai J, Helz GR (1988) Analytical Chemistry 60: 301
10.40 Van Loon WMGM, Bron JS, de Groot B (1993) Analytical Chemistry 65: 1726
10.41 Olsen ES, Diehl JW, Froelich ML (1988) Analytical Chemistry 60: 1920
10.42 Nimura N, Kinoshita T, Yoshida T, Uetake A, Nakai C (1988) Analytical Chemistry. 60: 2067
10.43 Xiao-Hua, Yang, Lee C, Scranton MI (1993) Analytical Chemistry 65: 857

11 Applications of Ion Chromatography

Ion chromatography as originally developed by Small and co-workers in 1975 provided a method for the separation and determination of inorganic anions and cations [11.1]. This original method used two columns attached in series packed with ion exchange resins to separate the ions of interest and suppress the conductance of the eluant, leaving only the species of interest as the major conducting species in the solution. Once the ions were separated and the eluant suppressed, the solution entered a conductivity cell where the species of interest were detected. Since its introduction, ion chromatography has advanced considerably and the technique is now routinely used for the analysis of organic and inorganic anions and cations and substances, including organic acids and amines, carbohydrates and alcohols. Ion chromatography is not restricted to the separate analysis of only anions or cations. With the proper selection of eluant and separation columns, the technique can be used for the simultaneous analysis of both anions and cations.

The simultaneous determinations of anions and cations using single injection ion chromatography was introduced by Yamamoto and co-workers in 1984 [11.2]. This technique determined cations and anions simultaneously using a complexing agent, ethylenedinitrilotetraacetic acid, to complex the divalent metals. These divalent metals were later separated and detected as anions along with the uncomplexed inorganic anions.

The technique has progressed rapidly since 1975 and in 1978 a book was published on ion chromatograpic analysis of environmental pollutants [11.3]. Other early papers on the application of ion chromatography cover the determination of selected ions in geothermal well water [11.4], the determination of anions in potable water [11.5,11.6], and the separation of metal ions and anions and anions and cations [11.7].

11.1 Natural Waters

11.1.1 Mixtures of Chloride, Bromide, Fluoride, Nitrite, Nitrate, Sulphate, Phosphate and Bicarbonate

Ion Chromatography has been used for the measurement of background concentrations of fluoride, chloride, nitrate and sulphate in natural waters [11.8].

Shown in Fig. 11.1 is a schematic diagram of the Dionex ion-exchange liquid chromatograph used by Smee et al. [11.8]. The instrumentation consists essentially of a low capacity anion exchange column, the 'separator', a high capacity cation exchange column, the 'suppressor' and the detection system, a high sensitivity conductivity meter and recorder. Following the injection of a small volume (100 µl) of sample, rapid exchange and separation of the anions is accomplished with a specially prepared low capacity resin (containing quaternary ammonium exchange sites existing as a thin film on the surface to facilitate fast equilibrium).

The otherwise highly conducting background of the eluant (0.003 mol l^{-1} sodium bicarbonate/0.0024 mol l^{-1} sodium carbonate) is virtually eliminated by passage through the high capacity suppressor column whereupon carbonic acid is formed together with the acid forms of the anions of interest. The conductivity meter used as a detector then only sees a small residual background due to the weakly dissociated carbonic acid and the conductivity of the acids of interest separated in time. Additional to this basic instrumentation, there are four reservoirs (2 eluant, 1 regenerant, 1 water), a sample injection valve with a 100-µl loop, two Milton Roy fluid pumps with adjustable flow rates and an automatic timer for controlling the regeneration cycle of the cation exchange resin.

Approximately 1 ml of untreated sample was injected into the entrance port by means of a hypodermic syringe; 100 µl of sample was used to fill the sample

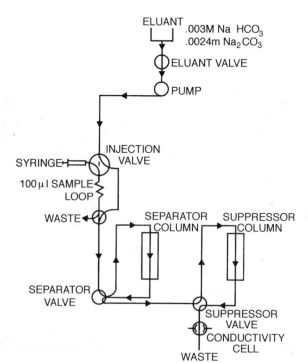

Fig. 11.1. Ion chromatographic flow diagram. Eluant conditions used for anion analysis (From [11.8])

loop, the remainder passing to waste. The determination began with a freshly regenerated separator column and proceeded sequentially through the sample series. Retention times of peaks from the resulting chromatograms were compared with known synthetic standards analysed under the same eluant strength and flow conditions. Measurements of anion concentrations in each sample were obtained by comparing peak heights with a standard curve generated from peak heights of mixed synthetic standard.

Two separator columns were studied by Smee et al. [11.8] – one 250 mm and one 500 mm in length. Examples of chromatograms obtained from standards injected into a 250-mm column are shown in Fig. 11.2. Standard 7b used in the 250 mm long column (Fig. 11.2) does not contain nitrite and bromide but concentrations of fluoride, chloride, phosphate, nitrate and sulphate are the same as for the 500-mm column. Both columns exhibit advantages and disadvantages. Better resolution of chromatographic peaks, particularly those involving nitrate, nitrite, phosphate, and bromide is obtained with the 500-mm column. A depression of the baseline created by elution of the water portion of the sample, thus lowering the background conductivity, occurs between the fluoride and chloride peaks and can result in a depression of the chloride response. This depression is negligible with the 500-mm column. Disadvantages in using the 500-mm column include a longer elution time for each sample, a decrease in precision at moderate to high concentrations created by broadening of chromatographic peaks (this may be compensated for by integrating peak areas rather than measuring peak heights), and a progressive deterioration of efficiency of the suppressor column caused by an increased volume of eluant required for the longer elution time. This latter problem may be solved by regenerating the column more frequently or by an increased use of standards with samples.

Typical chromatograms obtained from a sample of northern Ontario stream water are shown in Fig. 11.3 (250-mm column). No advantage is gained by using

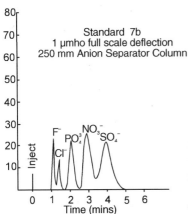

Fig. 11.2. Calibration chromatogram for a 250-mm anion separator column using standard 7b containing 0.1 μg ml^{-1} F, 0.1 μg ml^{-1} Cl, 0.1 μg ml^{-1} PO$_4^{3-}$, 1.0 μg ml^{-1} NO$_3^-$ and 1.0 μg ml^{-1} SO$_4^{2-}$ (From [11.8])

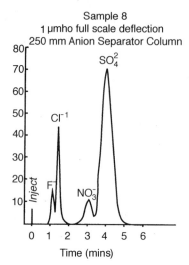

Fig. 11.3. Chromatogram of sample 8 using same conditions as for the calibration in Fig. 11.2 (From [11.8])

the 500-mm column rather than the 250-mm column. The time for analysis is more than doubled and peak heights have decreased. The increased separation of anion chromatographic peaks is not required for this sample type.

The relationship between fluoride determination by ion-selective electrode and ion chromatography is good but a larger spread of values is evident with the 500-mm column, especially with samples containing above average amounts of fluoride. Values reported by ion chromatography are usually higher than those obtained by the selective ion technique. The minimum reported concentration for fluoride by ion chromatography is 3 $\mu g\,l^{-1}$ and for selective ion electrodes 20 $\mu g\,l^{-1}$.

The detection limit of the 2-aminoperimidine method for sulphate is 5–10 $mg\,l^{-1}$. Sulphate contents reported by both columns usually correlate well with one another. The lower limit of detection for sulphate reported for the 250-mm column is 0.03 $mg\,l^{-1}$ and for the 500-ml column 0.05 $mg\,l^{-1}$, two orders of magnitude below the limit reported for the 2-aminoperimidine method.

A comparison between nitrate determination by ion chromatography and by the colorimetric technique shows that there is a correlation between the two sets of results but nitrate values obtained by the chromatographic method are consistently higher than those obtained by colorimetry. Furthermore, the 500-mm column yields higher values than the 250-mm column. Values above 0.2 $mg\,l^{-1}$ generally agree well. The limit of detection (0.02 $mg\,l^{-1}$) reported for the 250-mm column is below that reported for the 500-mm column (0.05 $mg\,l^{-1}$) because of the sharper peaks obtained with the shorter column.

Chloride contents of the natural waters varied from 0.5 to 312 $mg\,l^{-1}$ by the spectrophotometric method. A comparison of the chloride determinations reveal an excellent correlation between the chloride values obtained from the 500-mm column and the expected values. The 250-mm column exhibits a gradational increase in the difference between the ion chromatograph and the expected results with increasing chloride concentration. This difference is probably caused by a progressive overlap of the chloride peak with the fluoride peak. Minimum reported concentrations for chloride using the 250-mm column is 0.02 mg l^{-1} and for the 500-mm column, 0.01 mg l^{-1}, against 0.5 mg l^{-1} for spectrophotometric methods.

Estimates of precision based on five analyses of the replicate natural water samples are presented in Table 11.1. The 250-mm column is more precise than the 500-mm column for all anions analysed, mainly because the peak height of a response was measured rather than peak area, the shorter column producing the more narrow peaks. Fluoride determinations using the 500-mm column exhibit a progressive upward increase in fluoride content caused by the aging of the column during the analytical run. This apparent increase in fluoride content may occur at the expense of the chloride ion, the next anion through the column, as there appears to be a gradual decrease in chloride content with time. No data on nitrate was obtained from the replicate sample. Precision for sulphate through both columns is excellent, although the 250-mm column consistently produces the higher analytical values.

Van Os et al. [11.9] achieved complete separations in 6 min of 1–30 mg l^{-1} concentrations of bromide, chloride, nitrite, nitrate and sulphate using a Zipax. SAX separation column, with eluent suppression and electrical conductivity detection. The necessary high pressure packing techniques for packing the separation column is described in detail. With sample preconcentration, detection

Table 11.1. Dionex replicate samples, natural waters. (From [11.8]; Copyright 1978, Elsevier Science Publishers BV, Netherlands)

Sample No.	F (250 mm) (µg l^{-1})	F (500 mm) (µg l^{-1})	Cl (250 mm) (mg l^{-1})	Cl (500 mm) (mg l^{-1})
7	852	860	20.0	14.5
18	852	860	20.0	14.7
22	870	880	19.7	14.5
26	852	920	20.0	13.5
30	852	960	20.0	13.5
Mean	855.6	896.0	19.9	14.1
S.D.	8.05	43.4	0.14	0.59
R.S.D.	0.9 %	4.8 %	0.7 %	4.2 %

NO$_3$ (250 mm) (mg l^{-1})	NO$_3$ (500 mm) (mg l^{-1})	SO$_4$ (250 mm) (mg l^{-1})	SO$_4$ (500 mm) (mg l^{-1})
< 0.02	< 0.05	27.3	25.0
< 0.02	< 0.05	27.5	25.0
< 0.02	< 0.05	27.3	24.2
< 0.02	< 0.05	27.3	24.6
< 0.02	< 0.05	27.5	24.2
0.02	0.05	27.4	24.6
–	–	0.11	0.4
–	–	0.5 %	1.6 %

limits were reduced to about $5\,\mu g\,l^{-1}$ but calibration graphs for chloride and nitrate were not linear.

Reagents. As follows.
Adipate and succinate solutions were prepared by adjusting the corresponding acid solutions to the desired pH with sodium hydroxide.

Apparatus. Separator column Zipax SAX beads (particle diameter 25–37 µm, Du Pont de Nemours, Den Bosch, The Netherlands). (Zipax SAX is a strong pellicular anion exchanger with a capacity of about 12 µeq of quaternary ammonium groups per gram of dry material). Suppressor column resin Ag-50W-X12 ($>$ 400 mesh: Bio-Rad Laboratories). This is a strong cation exchanger prepared from a styrene-divinyl-benzene copolymer with 12% cross links. It has a capacity of 5 meq of SO_3H groups per gram of dry resin.

Ion chromatographic apparatus, Fig. 11.4, is equipped with a reciprocating piston pump (Kipp Analytica, Emmen, The Netherlands). The pulse dampener system consists of an Orlita (Giessen/Lahn, West Germany) pulse dampener and a stainless steel resistance capillary(length 3 m, i.d. 0.25 mm). The stainless steel eluent filter (Chrompack, Middelburg, The Netherlands) has a pore diameter of 2 µm. The six port sample injection valve and the eight port valve were supplied by Valco (Houston, Texas). The former is equipped with a 60-µl sample loop. Both are actuated pneumatically. The home made conductivity cell consists of two platinum electrodes in a perspex housing and has a cell volume of 6 µl. The digital conductivity detector has been described elsewhere [11.10]. The columns (precision bore polished stainless steel, i.d. 4.5 mm) and the detector cell are surrounded by water jackets and thermostated by a thermostat to within 0.01 °C. The connecting tubing (i.d. 0.25 mm) is kept as short as possible. Low dead

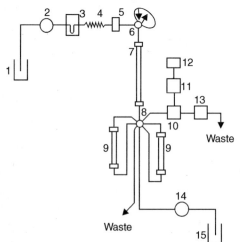

Fig. 11.4. Schematic diagram of the ion chromatograph: 1, eluent reservoir; 2, pump; 3, pulse dampener; 4, capillary; 5, filter; 6, 6-port injection valve; 7, analytical column; 8, 8-port valve; 9, suppressor columns; 10, conductivity cell; 11, conductivity detector; 12, recorder; 13, syphon counter; 14, regeneration pump; 15, regeneration liquid (nitric acid or water) (From [11.9])

volume connectors are used to minimize extracolumn peak broadening. The eluent flow rate is monitored continuously with a calibrated siphon counter.

The column packing apparatus is outlined in Fig. 11.5. The pump is of the reciprocating type and should attain a pressure of at least 400 bar when columns have to be packed with 10 µm particles. A Milton Roy pump (Chicago, Illinois) was used. Pulse dampening is achieved with a flow through Bourdon gauge, a resistance capillary, an Orlita pulse dampener and another capillary in series. The stainless steel capillaries (length 3 m, i.d. 0.25 mm) ensure a nearly pulse free flow. The slurry reservoir is a thick-walled stainless steel tube equipped with teflon sealed stainless steel nuts. The lower nut is part of the polished precolumn (length 10 cm). The internal diameter of the precolumn, connector and column should be equal. The packed bed in the column is kept in place by means of a thin plug of wool in a low dead volume connector.

The columns are packed according to the viscous slurry method [11.11]. The Zipax SAX (5 g) is suspended in 25 ml of ethyleneglycol and 2.5 g of AG-50W-X12 in 25ml of isopropanol. The slurries are degassed and homogenised by sonication. The column and precolumn are filled with tetrachloromethane in order to prevent inclusion of air bubbles in the packed bed. The slurry reservoir is filled with the slurry and thereafter methanol is added to remove all air up to the shut off valve. When the pump and dampers have been filled with methanol, the slurry reservoir is connected to the pump (with valve 7 closed). The apparatus is pressurized with methanol up to 400 bar. Thereafter, valve 7 is opened and the slurry is forced into the column. The packed bed is settled by flushing 200 ml of methanol and 200 ml of water through the column at maximum pressure.

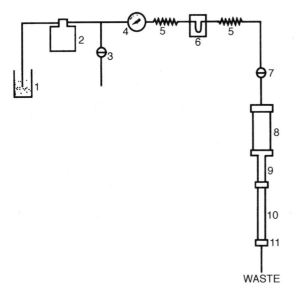

Fig. 11.5. Scheme of the column packing apparatus. 1, reservoir (methanol or water); 2, pump; 3, valve; 4, manometer; 5, capillary; 6, pulse dampener; 7, valve; 8, slurry reservoir; 9, precolumn; 10, column; 11, plug of wool in low dead volume connector (From [11.9])

Glass columns can also be slurry packed by this method when the pressure across the column is kept below 60 bar. A short glass column (100 × 4.0 mm) packed with Zipax SAX was used as a concentrator column.

The void volume of the separator column (V_m) is estimated from the retention volume of water. The capacity ratio k is calculated from the equation $k = (V_R - V_M)/V_m$, where V_R and V_M are the retention volumes of the solute and water respectively in the separator–suppressor column combination. The k values are independent of sample size (50–200 µl of solutions of 10 mg l^{-1} nitrate and 10 mg l^{-1} sulfate), only slightly dependent on the eluent flow rate (1.5 – 4.5 ml min^{-1}), and reproducible within 3%.

In Fig. 11.6 are shown the separations achieved on this system when (A) 2×10^{-3} mol l^{-1} disodium adipate and (B) 1.4×10^{-3} mol l^{-1} disodium succinate are used as eluents, both at pH 7.

On Zipax SAX the ions are eluted in the order F$^-$ < Cl$^-$ < NO$_2^-$ ≃ H$_2$PO$_4^-$ < Br$^-$ < NO$_3^-$ < SO$_4^{2-}$. Fluoride is eluted together with the negative water peak. On Zipax SAX the optimal eluent temperature for obtaining good resolution in the minimal time is about 25 °C.

Detection limits (three times the background noise) for chloride, nitrate and sulfate were determined for a 200 µl loop and for a concentrator column. These are given in Table 11.2.

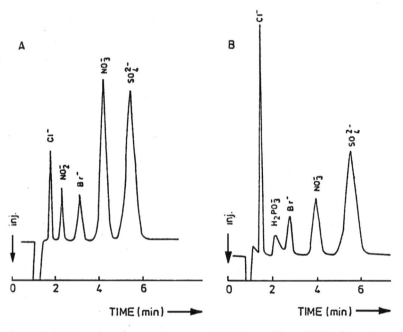

Fig. 11.6.A,B. Separation of trace anions on a 200 × 4.5 mm Zipax SAXC column: A separation of 2 mg l^{-1} Cl, 5 mg l^{-1} NO$_2^-$, 5 mg l^{-1} Br$^-$, 20 mg l^{-1} NO$_3^-$ and 20 mg l^{-1} SO$_4^{2-}$ with 2×10^{-3} mol l^{-1} Na$_2$ adipate as eluent; B separation of 5 mg l^{-1} Cl$^-$, 5 mg l^{-1} H$_2$PO$_4^-$, 5 mg l^{-1} Br$^-$, 7.5 mg l^{-1} NO$_3^-$ and 15 mg l^{-1} SO$_4^{2-}$ with 1.4×10^{-3} mol l^{-1} Na$_2$ succinate as eluent. Other conditions: 50 µl loop, eluent flow rate 2.5 ml min^{-1}, 25°C (From [11.9])

Table 11.2. Detection limits of chloride, nitrate and sulfate on the Zipax-SAX column with 1.4×10^{-3} mol l^{-1} Na$_2$-succinate as eluent. (From [11.9]; Copyright 1982, Elsevier Science Publishers BV, Netherlands)

Ion	Detection limit ($\mu g\,l^{-1}$)	
	200-μl loop	Concentrator column
Chloride	10	2
Nitrate	30	4
Sulfate	30	10

The lifetime of a Zipax SAX slurry packed separator column was at least 400 h. Deterioration of the packing resulting in peak broadening did not appear and almost no reduction of capacity was observed during that period. The lifetime of the suppressor column is at least three months.

Determination of anions in natural water is disturbed by the inevitable presence of humic substances which are strongly adsorbed on the ion chromatographic column.

Marko Varga et al. [11.12] found that a cleanup column prior to the separation column, packed with a chemically bonded amine material (Nucleosil 5 NH$_2$) was found to be effective in removing interfering humic substances. No influence was found from humic substances in concentrations up to 45 $\mu g\,l^{-1}$ on ion chromatographic analysis of nitrate and sulfate (10–100 mg l^{-1}) after passage through the cleanup column.

A short column packed with an amine bonded polymer (weak anion exchange material) connected prior to the anion separation column was found to be very effective for the selective removal of humic substances prior to anion separation by ion chromatography. About 500 sample injections could be performed before breakthrough of humic substances occurred as indicated by ultraviolet measurements made on the eluate at 225 nm.

Apparatus. Chromatographic system consisting of a pump (constametric III, LDC, Riviera Beach, Florida) a loop (200- or, 500-μl an injector) Model 7000 Rheodyne, Cotati, California) a column (see below), two detectors in series (a conductometric detector "Conductomonitor", LDC and a variable wavelength UV detector, "Spectromonitor III" LDS), and a dual channel recorder (Model 2066, LKB, Bromma, Sweden).

For the adsorption of humic substances a stainless steel column (50 mm × 4.9 mm i.d.) was used, filled with 0.8 g of 5-μm amine bonded polymer (Nucleosil 5 NH$_2$). Anions were separated by an anion exchange resin (Dionex, Sunnyvale, California) packed into a glass column (5 mm i.d. bed length 50 mm) (Pharmacia Fine Chemicals: Uppsala Sweden) which permits the length of the resin bed to be easily adjusted.

Potassium hydrogen phthalate, 5×10^{-4} mol l^{-1}, pH 5.5 was used as mobile phase. The flow rate was 2 ml min^{-1} and the conductometric detector was used in the "absolute" mode (10 μS) full scale deflection. The conductance of the eluent was electronically offset.

Under the chosen conditions, nitrate and sulfate were well resolved from other anions expected to be found in natural waters. Typical k' values were as follows: bicarbonate, H_2PO_4, chloride and nitrite < 0.5, sulphate 7.2, nitrate 1.4.

Procedure. Inject a 500-μl water sample onto the adsorption column and collect the eluate in a fraction collector. Evaporate the fraction (around 2.5 ml) which contained the anions, as sensed by the conductometric detector, to dryness (80 °C overnight) and redissolve in 250 μl of water. Analyze 100-μl aliquots on the ion chromatographic system using the ion exchange separation column and conductance detection as described above.

Marko Varga et al. [11.12] prepared a series of model solutions with 10, 30, 50, 70 and 100 mg l^{-1} of both nitrate and sulphate additionally containing 0.00, 0.09, 0.90, 4.5, 27 or 45 mg l^{-1} of reference humic acid. By use of the procedure described above, nitrate and sulfate were determined in the solutions. The calibration curve was constructed from the solutions without added humic acid.

A separate analysis of 45 mg l^{-1} reference humic acid (without anions added) gave the result: nitrate 3.7 mg l^{-1} and sulphate 5.2 mg l^{-1}. This is considered as the background level of these ions in the reference humic acid. Thus, significant amounts of nitrate and sulphate are added to the model solutions together with the humic acid.

Considering these background levels, the percent recovery (concentration found/expected concentration) was calculated for all model solutions analyzed and is presented in Table 11.3. The relative standard deviation within each set of six values was typically 2–3% both for nitrate and sulphate.

Most of the recoveries in Table 11.3 are not significantly different from 100%. For sulfate, recoveries tend to be about 10% low at high concentrations

Table 11.3. Percent recovery of nitrate and sulfate from solutions containing humic acid. (From [11.12]; Copyright 1984, American Chemical Society)

% recovery for concn of humic acid added (mg l^{-1})						
Ion concn, mg l^{-1}	0.09	0.9	4.5	9	27	45
NO_3^-						
10	102	98	105	94	103	94
30	101	109	99	108	103	105
50	107	104	103	105	99	106
70	101	106	106	108	106	102
100	106	101	108	104	102	108
SO_4^{2-}						
10	96	94	94	90	106	110
30	95	112	101	93	93	103
50	109	102	100	94	90	93
70	101	98	99	91	90	93
100	101	97	97	94	90	93

of humic acid while an opposite tendency can be noted for nitrate. One reason for this behaviour might be the competition between sulfate and nitrate and other ions present in the humic acid for the binding sites of the humic substances.

In all chromatograms an unknown peak appears. This is probably due to a disturbance in the phthalate protonation equilibrium created by the injection of aqueous samples. The retention time of the peak is independent of pH and the concentration of the eluent, but the intensity is affected by these factors. It is possible to adjust the eluent composition so the peaks of interest are separated from the "ghost peak".

Ultraviolet vs Conductivity Detectors. Cochrane and Miller [11.13] state the sensitivity of an ultraviolet detector is about ten times greater than that of an electrical conductivity detector. They used a chromatographic system consisting of the following parts in series: (1) Milton Roy mini pump set at a flow rate of 1.5 ml min^{-1}; (2) a Rheodyne 7120 sample injection valve fitted with a 100-μl sample loop; (3) Vydac 302 column, 25 cm × 4.6 mm i.d. (Separations Group); (4) LDC Model 701 conducto monitor used in the differential mode at the most sensitive, setting (× 10); (5) Schoeffel SF770 variable wavelength UV detector set at 308 nm detection wavelength and the zero suppression at −2 (maximum-5); (6) Servoscribe dual pen recorder, set at 10 mV output and 1 cm min chart speed.

Potassium hydrogen phthalate solution (5-10^{-3} mol l^{-1}) at pH 4.6 filtered through a Millipore 10.65 μm filter (type DA) was used as eluent.

Sample solutions were prepared from analytical grade reagents dissolved in the eluent.

UV detection of anions is not generally applicable except at very low wavelengths such as 215 nm. An alternative method of detection of non-UV absorbing species is to add a low level of UV absorbing substances to the eluent. The emergence of a component is then shown by a negative detector response.

The effect is easily achieved in the ion chromatography system where the preferred buffer solution (potassium hydrogen phthalate) has a strong UV response with λ_{max} at 280 nm. At this wavelength the excessively strong absorbance will not allow adequate zero suppression and the wavelength was therefore increased to a much less absorbing region (308 nm) where the background response could be backed off sufficiently for the detector to be used at maximum sensitivity. A comparison of the conductivity and UV detectors showed the following.

(1) The response is greater for UV detection by a factor of 5–30 (see Fig. 11.7) and Table 11.4. A relatively stronger, negative UV response occurs for weakly acidic anions because there is no dependence on their ionisation.
(2) Linearity ranges using UV detection are greater than those for conductance (Table 11.5).
(3) Conductance detection may give positive or negative peaks depending on eluent concentration and pH. UV detection of non-UV absorbing ions gives a response in one direction only.

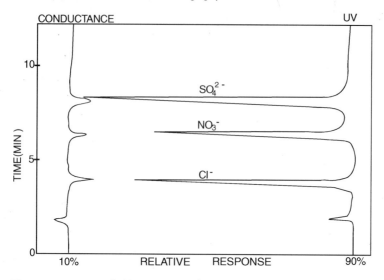

Fig. 11.7. Separation of chloride (10 mg l^{-1}), nitrate (25 mg l^{-1}) and sulphate (25 mg l^{-1}) with UV and conductance detection (From [11.13])

Table 11.4. Limits of detection for a number of anions using the conductance and UV detectors. (From [11.13]; Copyright 1982, Elsevier Science Publishers BV, Netherlands)

Anion	Limits of detection (mg l^{-1})[a]	
	Conductance	UV
Chloride	1	0.1
Bromide	1	0.2
Nitrite	1	0.2
Chlorate	2	0.2
Nitrate	3	0.3
Sulphate	3	0.3
Iodide	3	0.5
Phosphate	4	0.3
Bromate	5	0.2
Acetate	9	0.4
Formate	10	0.5
Iodate	12	0.4

[a] This is equivalent to twice the noise level at the most sensitive setting for both detectors.

A number of water extracts were analysed using the two detectors in series. The ultraviolet detection system has extended the range of application of single column ion chromatography by overcoming one of the main limitations of the single column system, i.e. the level of sensitivity for trace analysis.

Table 11.5. Linearity ranges for some anions. (From [11.13]; Copyright 1982, Elsevier Science Publishers BV, Netherlands)

Anions	Linearity range (mg l^{-1}) (upper limit)	
	Conductance	UV
Chloride	25	50
Nitrate	60	60
Sulphate	80	100

Ultraviolet detectors are used in ion chromatography. This method of detection in series with a conductivity detector has been used in a method [11.14] for the determination of anions in natural waters. This combination of detectors greatly increases the amount of information that can be collected on a given sample. The application of UV detection has the following advantages: (1) aid in the identification of unknown peaks; (2) use in resolving overlapping peaks; (3) help in eliminating problems associated with the carbonate dip; (4) reduction of problems associated with ion exclusion in the suppressor column; (5) ability to detect anions not normally detected by the conductivity detector, e.g. sulphide and arsenite.

Method. Dionex Model 14 ion chromatograph (Sunnyvale, California). Record chromatograph on a Honeywell (Fort Washington, Pennsylvania) dual pen strip chart recorder. The chromatographic conditions are summarized in Table 11.6. All of the columns were obtained from Dionex. A 100 µl-sample loop for all injections. Use a Laboratory Data Control (Riviera Beach, Florida) Spectromonitor II variable wavelength UV-Vis detector along with the normal conductivity detector.

For maximum sensitivity, the preferred position of the UV detector is after the suppressor column. The suppressor column also decreases the background absorbance of the other common ion chromatographic eluents such as 0.0002 mol l^{-1} Na$_2$CO$_3$/0.002 mol l^{-1} NaOH, 0.006 mol l^{-1} Na$_2$CO$_3$ and 0.005 mol l^{-1} Na$_2$B$_4$O$_7$ 10 H$_2$O.

Table 11.6. Chromatographic conditions. (From [11.14]; Copyright 1985, American Chemical Society)

	A	B	C	D
Precolumn	4 × 50 mm anion	3 × 150 mm L-20		
Separator column	4 × 250 mm anion	3 × 250 mm L-20	6 × 250 mm anion[a]	9 × 250 mm
Suppressor column	6 × 250 mm anion[a]	6 × 250 mm anion[a]		4 × 140 mm[b]
Eluent	0.003 mol l^{-1} NaHCO$_3$ 0.0024 mol l^{-1} Na$_2$CO$_3$	0.006 mol l^{-1} Na$_2$CO$_3$	H$_2$O	4 × 140 mm[c] 0.01 mol l^{-1} HCl
Flow rate	2.3 ml min^{-1}	2.3 ml min^{-1}	0.9 ml min^{-1}	0.9 ml min^{-1}

[a] Anion suppressor column consists of a high capacity cation-exchange column in H$^+$ form.
[b] Halide suppressor column.
[c] Postsuppressor column.

Figure 11.8 shows a separation of seven common anions monitored using the conductivity detector. Figure 11.8b was obtained with the UV detector after the suppressor column. As is illustrated in Fig. 11.8, nitrite, bromide, and nitrate absorb strongly in the UV, while fluoride, phosphate and sulfate do not show appreciable absorption above 190 nm. Chloride absorbs weakly in the UV region below 200 nm. Note the non-absorbing anions are sometimes observed by the UV detector as negative peaks, as is shown in Fig. 11.8b for phosphate. Table 11.7 contains a more extensive list of UV active anions and also lists the detection limits for many of the anions. Detection limits are generally in the sub ppm range. No attempt was made to optimize the chromatographic conditions for a particular anion. Out of the 27 anions listed in Table 11.7, all but 8 absorb in the UV above 190 nm. This demonstrates that the commonly held assumption regarding the lack of suitable chromophores by inorganic ions for UV detection is incorrect. This misconception has probably come about from the fact that, although many anions absorb in the 190–220 nm region, few of them show an absorption maximum above 190 nm. This has limited the analytical usefulness of direct UV absorbance measurement for the determination of inorganic anions. However, the lack of an absorption maximum does not severely limit the

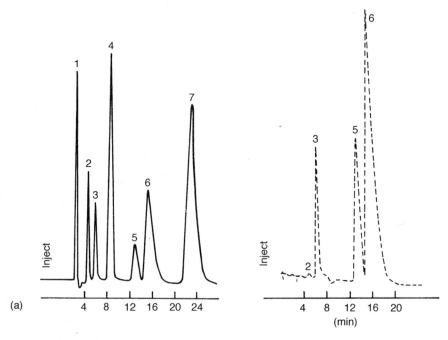

Fig. 11.8 a,b. Ion chromatogram of seven common anions: **a** conductivity detectors (—) peak $1 = 3\,\text{mg}\,l^{-1}$ F^-, peak $2 = 4\,\text{mg}\,l^{-1}$ Cl^-, peak $3 = 10\,\text{mg}\,l^{-1}$ NO_2^-, peak $4 = 50\,\text{mg}\,l^{-1}$, PO_4^{3-}, peak $5 = 10\,\text{mg}\,l^{-1}$ Br^-, peak $6 = 30\,\text{mg}\,l^{-1}$ NO_3^-, peak $7 = 50\,\text{mg}\,l^{-1}$ SO_4^{2-}; **b** UV detector (----) at 192 nm, position 2^- peak $2 = 4\,\text{mg}\,l^{-1}$ Cl^-, peak $3 = 10\,\text{mg}\,l^{-1}$ NO_2^-, peak $5 = 10\,\text{mg}\,l^{-1}$ Br^-, peak $6 = 30\,\text{mg}\,l^-$ NO_3^- (From [11.14])

Natural Waters

Table 11.7. List of UV absorbing and nonabsorbing inorganic anions. (From [11.14]; Copyright 1985, American Chemical Society)

Anion	Retention time, min	UV active	UV detection limit,[e] mg l^{-1}	UV detector position	Wavelength nm	Chromatographic condition Table 11.6
—	2.4	N[a]		2[c]	195	A
—	2.4	Y[b]		1[d]	205	A
O_3^-	2.6	Y	0.08	2	195	A
IO_2^{2-}	3.0	Y	0.2	2	195	A
$H_2PO_2^-$	3.0	N		2	195	A
H_2SO_3	3.8	N		2	195	A
BrO_3^-	4.3	Y	0.16	2	195	A
Cl^-	4.6	Y	2	2	192	A
$^{2-}$	5.4	Y		1	195	C
NO_2^-	5.9	Y	0.1	2	195	A
BF_4^-	6.6	N		2	195	B
HPO_3^{2-}	6.7	N		2	195	A
SeO_3^{2-}	6.9	Y	0.5	2	195	A
I^-	7.8	Y	0.15	2	195	B
PO_4^{3-}	8.4	N		2	195	A
SCN^-	10.8	Y	0.2	2	195	B
AsO_4^{3-}	10.9	Y	1.5	1	200	D
AsO_4^{3-}	12.2	Y	2	2	195	A
$S_2O_3^{2-}$	12.6	Y		2	195	B
Br^-	13.0	Y	0.1	2	195	A
AsO_3^{3-}	13.3	Y	1.2	1	200	D
SO_3^{2-}	13.4	Y		2	195	A
N_3^-	14.1	Y	0.3	2	195	A
NO_3^-	15.2	Y	0.1	2	195	A
ClO_3^-	15.2	Y	4	2	195	A
$SeCN^-$	18.8	Y	0.4	2	195	B
ClO_4^-	19.6	N		2	195	B
S^{2-}	21.2	Y	0.4	1	200	D
SO_4^{2-}	22.8	N		2	195	A
SeO_4^{2-}	27.2	Y	15	2	195	A

[a] N, not UV active.
[b] Y, UV active.
[c] 2, after suppressor.
[d] 1, before suppressor.
[e] 100 μL injection, peak height 2 × noise level.

analytical usefulness when the UV spectrometer is combined with a separation technique such as ion chromatography.

The UV detector, in combination with the conductivity detector, offers a number of advantages in ion chromatography. Figure 11.9 shows the separation of sub-ppm iodate and bromate monitored using both detectors. The iodate and bromate peaks monitored with the conductivity detector (solid line chromatogram) were distorted by the so-called carbonate dip. The size of the carbonate dip is controlled by the eluent (both its concentration and composition), the size of the injection loop, and the size of the suppressor column. The peak distortion

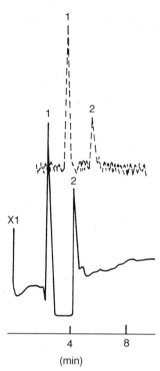

Fig. 11.9. Separation of iodate and bromate by ion chromatography: *UV detector* (------) at 195 nm, position 2; *conductivity* detector (———) peak 1, 0.8 mg l^{-1} IO$_3^-$; peak 2, 0.8 mg l^{-1} BrO$_3^-$ (From [11.14])

caused by the carbonate dip makes both qualitative and quantitative analysis more difficult. Little or no carbonate dip is observed in Fig. 11.9 (broken line chromatogram) with the UV detector after the suppressor column.

Figure 11.10a shows the UV and conductivity peaks from a 30 mg l^{-1} nitrate injection while Fig. 11.10b shows the UV and conductivity peaks from a 30 mg l^{-1} chlorate injection. Both species have the same retention time and peak shape (tailing) and with just the information from the conductivity detector it would be impossible to identify which species is present. With both detectors, it is possible to obtain a UV/conductivity peak height ratio and this information can be used as an aid in identifying unknown peaks. Additional UV/conductivity peak height ratios can be obtained by varying the wavelength. Identification is then made by running a standard solution of the suspected species (same retention time) and comparing the peak height ratios at the same wavelengths as the unknown species.

The ion chromatographic determination of weak acid anions is complicated by ion exclusion in the suppressor column, resulting in faster elution and sharper peaks, directly proportional to the degree of exhaustion of the suppressor column, A 10 mg l^{-1} nitrite standard showed a 37% increase in peak height over an 8 h period when monitored with the conductivity detector while only a minor 2% increase in peak height was observed over the same time period by using the UV detector after the separator column.

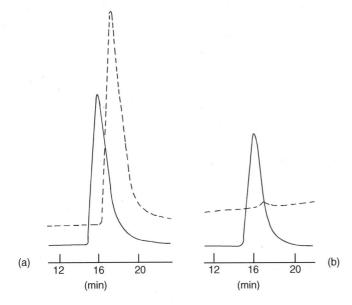

Fig. 11.10a,b. Separation of nitrate and chlorate by ion chromatography: **a** conductivity detector (———) 30 mg l^{-1} NO$_3^-$, UV detector (----) at 195 nm, position 2; **b** conductivity detector (———) 30 mg l^{-1} ClO$_3$, UV detector (----) at 195 nm respectively (From [11.14])

The UV detector can also be used in some cases to resolve overlapping peaks. Determination of the nitrite peak by using the conductivity detector is complicated by both the ion exclusion effect and the incomplete resolution between the large chloride peak and the much smaller nitrite peak.

The ion exclusion interference can be eliminated for UV active anions by placing the UV detector between the separator and suppressor columns (position 1). In addition, the problem of overlapping peaks can sometimes be resolved spectrophotometrically by proper choice of wavelength.

Sulphide cannot be detected under normal ion chromatographic conditions. It is converted into hydrogen sulphide in the suppressor column. Hydrogen sulphide acid is a very weak acid that does not ionize sufficiently to be detected with the conductivity detector. The UV detector, however, is able to detect sulphide at low levels, as is illustrated in Fig. 11.11a.

Since sulphide is a weak acid anion, the UV detector was placed between the separator and suppressor columns. Figure 11.11a also shows the presence of sulphite and sulphate due to oxidation of the sulphide ion.

Ion exclusion, besides being a source of chromatographic interference, can also be used to separate weak acids. Figure 11.11b shows the separation of sulphide by ion exclusion chromatography using a strong cation exchange column and water as the eluent. Peak 1 is a combined peak resulting from both sulphite and sulphate, as monitored with the conductivity detector. Peaks 2 and

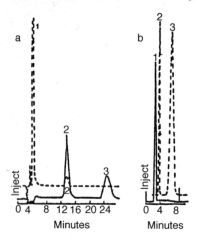

Fig. 11.11a,b. Sulfide separator **a** ion chromatographic conditions conductivity detector (———), UV detector (----) at 215 nm position 1; peak 1, 10 mg l^{-1} S^{2-}; peak 2, SO$_3^{2-}$; peak 3, SO$_4^{2-}$; **b** ion exclusion chromatographic conditions 4 (Table 11.6, column C)-conductivity detector (———), UV detector (----) at 192 nm; position 1; peak 1, SO$_3^{2-}$ and SO$_4^{2-}$; peak 2, SO$_3^{2-}$; peak 3, 15 mg l^{-1} (From [11.14])

3 correspond to sulphite and sulphide respectively, as monitored using the UV detector. Since sulphide can be separated by either ion chromatography or ion exclusion chromatography, the choice of separation mode will depend on the sample matrix.

Figure 11.12 shows the separation of arsenite and arsenate by ion exclusion chromatography using 0.01 mol l^{-1} hydrochloric acid as the eluent. Arsenous acid is a very weak acid and cannot be detected at low levels by the conductivity detector. However, like sulphide, it is easily detected with the UV detector. The simultaneous determination of arsenite and arsenate is possible.

Arsenite can also be separated by ion chromatography using standard conditions (Table 11.6, column A). It elutes early, near the carbonate dip. For this application the UV detector should be placed between the separator and suppressor columns.

11.1.2 Mixtures of Arsenate, Arsenite, Selenate and Selenite

Hoover and Jager [11.15] have described a procedure for the determination of traces of arsenite, selenite and selenate in the presence of major interferences (chloride, nitrate, and sulphate) in potable waters, surface waters and ground waters. By collecting a selected portion of the ion chromatogram, after suppression, on a concentrator column and re-injecting it under the original chromatographic conditions, it was possible to separate selenate, selenite and arsenite from chloride, nitrate and sulphate. Statistical detection limits varied from 0.02 to 1.2 µg of trace element depending on the minor components to be separated and on the water matrix. The maximal reliably separated molar ratio was 2300 for sulphate/selenate.

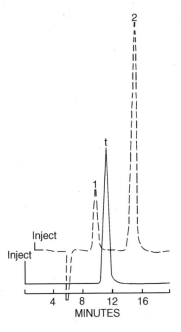

Fig. 11.12. Separation of arsenite and arsenate by ion exclusion chromatography (Table 11.6, column D); conductivity detector (———); UV (----) at 200 nm, position 1; peak 1, 15 mg l^{-1} AsO$_4^{2-}$; peak 2, 30 mg l^{-1} AsO$_3^{2-}$ (From [11.14])

Ion chromatography can readily determine arsenate, selenite, and selenate species in water in the absence of interferences. However, they will usually be completely obscured by the major anions. Arsenic and selenium are ordinarily at their highest oxidation state in surface waters but that is not necessarily the case in ground waters.

Apparatus. Dionex Model 10 ion chromatograph equipped with AG-1 guard column, AS-3 "fast run" separator column and ASC-1 suppressor. AG-3 concentration columns were used for introduction of the sample and for collection of and recycling of selected portions of the chromatogram. For recycle chromatography, modify the plumbing as shown in Fig. 11.13. The original injection valve was connected to two additional dual four-port slider valves. Solid lines in the figure correspond to the "Down" position of the toggle switches. The dummy concentrator in the waste line of the recycle valve was added to maintain approximately constant back pressure on the conductivity detector because the base line conductance of the suppressed eluent is highly pressure sensitive.

Concentrator columns were loaded in place with a Harvard Apparatus infusion pump at 1.5 ml^{-1} using a Hamilton 5-ml glass syringe with a gas tight Teflon tipped plunger and Luer fitting to 0.5 ml i.d. Teflon tubing. The volume of solution loaded was measured by collecting the effluent in a volumetric flask, shown in Fig. 11.13. All recycle runs were made at 15% pump rate (1.51 ml min^{-1}) and at room temperature (25 ± 1 °C).

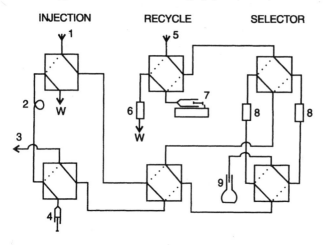

Fig. 11.13. VALVING: (1) eluent supply; (2) sample loop; (3) to separator; (4) injection port; (5) from detector; (6) dummy concentrator column; (7) syringe pump; (8) concentrator columns; (9) volumetric flask 2.5 or 10 ml (From [11.15])

Reagents. As follows.
Stock solution of 1000 mg l^{-1} of the test ions. Prepared from reagent grade sodium salts in water that had been purified by reverse osmosis and mixed bed ion exchange. The stock solutions were adjusted to the composition of the Dionex standard eluent (3.0 m mol l^{-1} NaHCO$_3$, 2.4 m mol l^{-1} Na$_2$CO$_3$) and membrane filtered through 0.45 µm pore size Metricel GA-6. Solutions of 1 or 10 mg l^{-1} were prepared by dilution of the stock with the water under test and further dilutions were made from these shortly before injection. Potable water had to be boiled briefly to remove available chlorine before stable dilute solutions of selenite in it could be obtained.

Results. An eluent composition of 2.0 m mol l^{-1} sodium bicarbonate and 1.5 m mol l^{-1} sodium carbonate was found to elute arsenate about midway between nitrate and sulfate. Selenite eluted between chloride and nitrate in an eluent that was 3.0 m mol l^{-1} in sodium bicarbonate and 2.0 m mol l^{-1} in sodium carbonate.

Suppressed eluent retained in the concentrator columns produced a large carbonate peak on direct reinjection. This peak obscured or interfered with everything eluting before nitrate. In the determination of selenite, the concentrator was washed with deionized water for 5 min after collection and before injection. The wash step was omitted in the determinations of arsenate and selenate where the carbonate did not interfere.

Figures 11.14–11.16 show representative chromatograms for each of the trace anions at close to the statistical detection limit. The initial injection is shown on a logarithmic scale of conductance to include complete peaks of the major constituents. The recycle portion of the chromatogram was obtained at the linear scale, usually 1 µs cm^{-1} full scale sensitivity. Figure 11.15 shows a second collection and recycling of selenite. The process can be repeated indefinitely but there is evidence (below) of 10–15% loss of material each time.

Calibration curves for each trace ion in each matrix are summarized in Table 11.8. Amounts of trace anion injected were run in a randomized order and most calibrations included both 2 ml and 5 ml initial injections. Increasing the sample size allows lower concentrations of the minor constituent to be collected but does not affect the relative proportions that must be separated. Variation of sample sizes between 0.1 and 5 ml did not affect mass detection limits.

The detection limits shown in Table 11.8 are all 3σ (i.e. 3 × standard deviation) limits based on the standard deviation of the ordinates from the least squares linear fit. In all cases this limit is appreciably greater than the detectability of trace amounts in individual runs, as shown in Figs. 11.14–11.16.

One effect of the matrix water on trace determination is the concentration of major ion interference. In the case of selenite, the major interferent is nitrate and the exceptional results in river water may be related to the high nitrate level in that medium. If that were the controlling factor, however, one would expect a greater difference between well and potable water than was observed. Similarly, for selenate the critical separation is from sulphate, which is greatest in drinking water, but river and well water differ more in sulphate content than well and potable water, in contrast to the calibration curves.

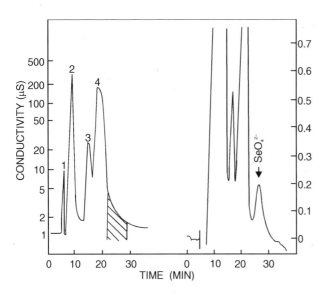

Fig. 11.14. Selenate in drinking water: sample size 5 ml, 10 ng of Se (V) added-major peaks: (1) fluoride; (2) chloride; (3) nitrate; (4) sulfate; *Shaded area* was collected for recycle (From [11.15])

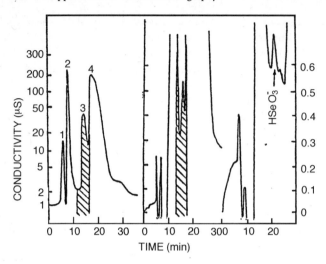

Fig. 11.15. Selenite in potable water, recycled twice: sample size 10 ml, 100 ng of Se (IV) added-major peaks: (1) fluoride; (2) chloride; (3) nitrate; (4) sulfate. *Shaded areas* were collected for recycle (From [11.15])

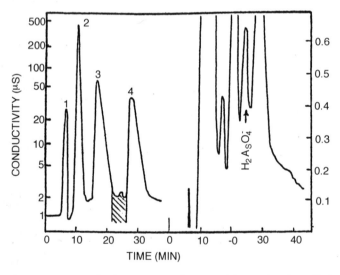

Fig. 11.16. Arsenate in river water:sample size 5 ml, 50 ng of As (V) added-major peaks (1) fluoride; (2) chloride; (3) nitrate; (4) sulfate. *Shaded area* was collected for recycle (From [11.15])

11.1.3 Borate

Hill and Lash [11.16] have described an ion chromatographic method for the determination of down to 0.05 mg l^{-1} of borate in environmental waters., nuclear fuel dissolvent solutions and effluents.

Table 11.8. Calibration data for recycle determination of trace anions in water. (From [11.15]; Copyright 1984, American Chemical Society)

Water	Intercept, S cm^{-1}	Slope, S cm^{-1} g^{-1}	Detection limit, µg	N
Selenate (3.0 mmol l^{-1} NaHCO$_3$, 2.4 mmol l^{-1} Na$_2$CO$_3$, pH 9.72)				
Drinking (1)	0.26 ± 0.15[a]	3.49 ± 0.30[a]	0.21	6
(2)	0.20 ± 0.06	2.37 ± 0.26	0.18	10
Well	0.011 ± 0.03	1.166 ± 0.016	0.020	8
River	0.012 ± 0.004	1.226 ± 0.016	0.018	6
Selenite (3.0 mmol l^{-1} NaHCO$_3$, 2.0 mmol l^{-1} Na$_2$CO$_3$, pH 9.24)				
Drinking	−0.034 ± 0.052	2.23 ± 0.10	0.20	19
Well	−0.089 ± 0.056	2.33 ± 0.16	0.18	11
River	0.77 ± 0.25	1.63 ± 0.25	1.2	13
Arsenate (2.5 mmol l^{-1} NaHCO$_3$, 1.5 mmol l^{-1} Na$_2$CO$_3$, pH 9.56)				
Drinking	0.024 ± 0.10	0.83 ± 0.20	0.47	6
Well	−0.017 ± 0.10	0.781 ± 0.038	0.68	8
River	−0.03 ± 0.09	0.902 ± 0.045	0.83	12

[a]Standard deviations.

Borate is selectively concentrated on Amberlite XE-243 ion exchange resin and converted to tetrafluoroborate using 10% hydrofluoric acid. Tetrafluoroborate is strongly retained by the resin, thus allowing excess fluoride to be eluted without loss of boron. The tetrafluoroborate is eluted with 1 mol l^{-1} sodium hydroxide and is determined in the eluent by ion chromatography. Boron is quantified to a lower limit of 0.05 mg l^{-1}.

Reagents. All reagents analytical reagent grade, except for boric acid which was NBS SRM 951. Water used throughout the study purified with a Milli Q system (Millipore Corporation).

Amberlite XE-243 resin. (Rohm and Haas Company). Ground and wet screened, retain the fraction between 40–80 mesh for column preparation.

Standard boron solutions: standard tetrafluoroborate stock solution. Prepare by mixing approximately 29 mol l^{-1} hydrofluoric acid and approximately 10 mol l^{-1} boric acid in a 100-ml volumetric flask. Allow the solution to react at ambient temperature for 15 min before dilution to volume. Dilute the stock solution daily to 1 to 1000 to yield approximately 8.4 mg BF$_4$ l^{-1}. This dilute solution is not stable for more than two days but the stock solution can be used for at least two weeks.

Apparatus. Model 14 ion chromatograph, Dionex Corporation. Clear polymethylpentane plastic ware (Nagel Company) was used in the preparation of all reagents and standards. Separating columns – standard 3 × 150 and 3 × 500 mm anion separator columns, as well as a 3 × 500-mm brine anion separator were used [11.17]. Suppressor column-standard 6 × 250 mm anion suppressor column, from Dionex Corporation.

Columns of Amberlite XF-243 resin are prepared as described by Carlson and Paul [11.18] except that a disposable polypropylene pipette tip is used for the column. Seal the pipette tip to the polyethylene reservoir with Teflon tape. Column dimensions 4.0 cm × 0.75 cm with a resin bed volume of approximately 0.8 ml. Use a flow rate through the column of about 1 ml min^{-1}. The column has a capacity of approximately 5×10^{-2} m mol boron.

At the lower sensitivity setting (3 μmho full scale) pump pulse noise is encountered in ion chromatography. To alleviate the problem, a pulse damper was constructed. Seal an air filled polyallomer centrifuge tube of 10 ml capacity (available from Scientific Products) and place inside a glass vial (9.7 × 2.5 cm). Then fill the vial with water. Drill two holes in the plastic screw cap to accommodate two 20 gauge syringe needles and cement the needles to the cap and the cap to the glass vial with a fast drying epoxy. Cement TFE tubing (0.5 mm i.d.) to the needles. Use a 3-m section of tubing to connect the damper to the pressure gauge inlet of the manifold. This prevents the eluent from mixing with the contents of the glass vial. Connect the other end of the damper to the pressure gauge. The vial was shatter proofed by wrapping it with tape. The damper improved the signal to noise ratio by a factor of ten on the 3 μmho full scale range.

Procedures. Two procedures are used to prepare samples for injection into the ion chromatograph. The standard procedure using the Amberlite XE 243 column provides the highest degree of specificity and must be used when the boron concentration is less than 10^{-2} mol l^{-1} and/or when large amounts of chloride, nitrate, sulphate, phosphate and chromate are present. The alternate procedure, which circumvents the XE-243 column, is significantly more rapid and involves conversion of the boron to tetrafluoroborate in solution.

Column Regeneration. Prior to passing any samples over a freshly prepared XE-243 column the resin must be regenerated. To do this, pass 2 ml of 10% hydrofluoric acid over the column and allow to stand for 10–15 min. Then pass 2 ml of water over the column followed by 5 ml of sodium hydroxide and 5 ml of water. This elution sequence ensures that any boron introduced onto the resin during column preparation has been completely removed.

Standard Procedure. Pass neutral or alkaline samples containing between 10 and 100 μg of boron (5 μg of boron concentration is at the 0.05 mg l^{-1} level) over the XE-243 column followed by 2 ml of water, 3 ml of 3 mol l^{-1} ammonium hydroxide, 2 ml of water and 2 ml of 10% hydrofluoric acid. The column is allowed to stand for 10–15 min to convert the boron to tetrafluoroborate and then rinse with 2 ml of water followed by 5 ml of 0.3 mol l^{-1} ammonium hydroxide and 2 ml of water. Elute the tetrafluorborate from the column into a 50-ml plastic volumetric flask (a calibrated 10-ml plastic graduated cylinder is used if original boron concentration is at the 0.05 mg l^{-1} level) with 5 ml of 1 mol l^{-1} sodium hydroxide followed by 5 ml of water and dilute to volume with water.

Acidic samples must be neutralized before introduction onto the XE-243 column. This is best performed by passing the sample over a column of weak cation exchange resin (i.e. Bio Rex 70) in the ammonium form.

Alternate Procedure. Samples that are greater than 10^{-2} mol l^{-1} in boron and contain less than a 50 to 1 mole ratio of chloride, nitrate, sulfate, phosphate, or chromate to boron, may be analysed by an alternate procedure, one not requiring the use of the XE-243 column. In this procedure, transfer the sample to a polymethylpentene volumetric flask and add 29 mol l^{-1} hydrofluoric acid to yield approximately a 17 to 1 mole ratio of fluoride to boron. Allow the solution to stand for 15 min then cool to room temperature and dilute to volume. It may be further diluted as necessary for injection into the ion chromatograph.

Ion Chromatographic Measurement. Inject samples and calibration standards containing between 1 and 20 mg l^{-1} tetrafluoroborate into the ion chromatograph using the conditions shown in Table 11.9. Calculate results on a peak height ratio basis.

The effects of the common anions on the standard procedure are shown in Table 11.10. As can be seen, none of these ions interferes at a 100 to 1 mole ratio to boron. In addition, at least a 280 to 1 mole ratio of fluoride to boron, as well as a 100 to 1 mole ratio of chloride to boron, can be tolerated. Figure 11.17 shows an ion chromatogram of a sample which initially contained all of the above anions at a 100 to 1 mole ratio and which was prepared according to the standard procedure. As shown, tetrafluoroborate peaks are well resolved from all the residual amounts of those anions not completely removed by the XE-243 column treatment.

When using the alternate procedure, no interference was encountered from a 100 to 1 mole ratio of fluoride to boron or from a 50 to 1 mole ratio of phosphate, nitrate, sulfate or chromate to boron.

Five water samples prepared to contain boron at levels ranging from 500 to 0.05 mg l^{-1} were analysed according to the standard procedure. The results are shown in Table 11.11. Recoveries ranged from 99.3% to 96.4%.

Table 11.9. Ion chromatographic conditions. (From [11.16]; Copyright 1980, American Chemical Society)

Eluent	0.003 mol l^{-1} NaHCO$_3$/0.004 mol l^{-1} Na$_2$CO$_3$
Flow rate	138 ml/h, 30 %
Analytical column	3 × 500 mm brine anion separator
Suppressor column	6 × 250 mm anion suppressor
Detector sensitivity	3 µmho full scale
Injection volume	100 µl

Table 11.10. Effect of diverse ions. (From [11.16]; Copyright 1980, American Chemical Society)

Ion	Added as	Mole ratio to boron[b]	% recovery
CrO_4^{2-}	K_2CrO_4	100 to 1	101.4
NO_3^-	$NaNO_3$	100 to 1[a]	
SO_4^{2-}	Na_2SO_4	100 to 1[a]	99.1
HPO_4^{2-}	$Na_2HPO_4 \cdot 9H_2O$	100 to 1[a]	

[a] Added as a mixture of NO_3^-, SO_4^{2-}, and HPO_4^{2-}.
[b] 10^{-3} M boron.

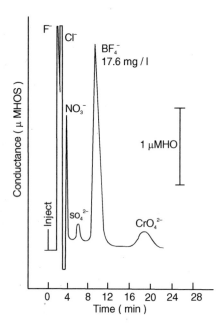

Fig. 11.17. Effects of diverse ions on IC determination of tetrafluoroborate ion (From [11.16])

The method was also applied to a solution prepared to contain 0.01 mg l^{-1} boron. A 100-ml aliquot of the solution was passed over the XE-243 column and treated as described in the standard procedure except that the 1 mol l^{-1} sodium hydroxide eluent was injected directly into the ion chromatograph. Recoveries for two replicates were 79% and 115%. The erratic recoveries at this level are attributed mainly to the large unidentified peak which elutes adjacent to the tetrafluoroborate peak only when the 1 mol l^{-1} sodium hydroxide eluent is injected undiluted. Recoveries may be improved by passing larger volumes of solution over the XE-243 resin so that samples can be diluted before injection.

A nuclear fuel dissolvent solution 0.37 mol l^{-1} in boron and 6.8 mol l^{-1} in hydrofluoric acid was analyzed using the alternate procedure. A relative stan-

Table 11.11. Recovery of boron from water. (From [11.16]; Copyright 1980, American Chemical Society)

Boron content mg l^{-1}	Av. recovery, %[a]	Rel. Std. dev., %
500	99.2	1.8
50	96.4	1.1
5	97.1	0.5
0.5	99.3	2.2
0.05	97.2	2.6

[a]Average of 5 replicates.

dard deviation of 0.3% was obtained for the analysis of six replicate samples. The average recovery was 99.2%.

11.1.4 Miscellaneous Anions

Various other applications to the determination of anions in water are summarised in Table 11.12.

11.2 Rainwater

11.2.1 Mixtures of Chloride, Bromide, Fluoride, Nitrite, Nitrate, Sulphite, Sulphate and Phosphate

Yamamoto et al. [11.31] and Matsuchita [11.32] have described a method for the determination of chloride, phosphate, nitrate, nitrite, sulphate, calcium and magnesium in rainwater. They used 1 m mol l^{-1} EDTA as an eluent and a silica based anion exchange column.

Apparatus. Ion chromatographic system consisting of a Toyo Soda Model HLC-6DL ion chromatograph [11.32]. A Toyo Soda Model CM-8 conductivity detector with a range of 5 µs cm^{-1}. The sample loop volume was 0.1 ml and a flow rate of 1.0 ml min^{-1} was employed. The separation column, 4.6 mm i.d. 50 mm long, made of Teflon, was packed with porous silica based anion exchanger of TSK gel IC-Anion SW (Toyo Soda, particle size 5 ± 1 µm capacity 0.1 mequiv g^{-1}. The column and the conductivity detector was maintained at 27 ± 1 °C.

Table 11.12. Ion chromatography of anions

Anions	Eluent	Column	Type of Sample	Detector	Reference
AsO_3^{3-}/SO_3^{2-}	–	–	Environmental waters	–	11.19
PO_4^{3-}	–	–	Sea	–	11.20
CN^{1-}, S^{1-}, OCl^-	–	–	Ground waters	–	11.21
$SO_3^{2-}/S_2O_3^{2-}$	4.8 mmol l^{-1} NaHCO$_3$ then 6.9 mmol l^{-1} NaHCO$_3$/0.86 m mol l^{-1} Na$_2$CO$_3$	Gradient elution	Not stated	–	11.22
NO_3^{1-}/SO_4^{2-}	–	–	Natural	–	11.23
$F^{1-}/O\text{-}PO_4^{3-}/Br^{1-}$	–	–	River	–	11.24
$Cl^{1-}/NO_3^{1-}/SO_4^{2-}$	–	–	Pure waters	–	11.25
CO_3^{2-}/HCO_3^{1-}	–	Zipax Sax and Westcan	–	Electrical conductivity and uv	11.26
SiO_3^{2-} (22 µg l^{-1})	–	–	Natural	–	11.27
SO_4^{2-} (50 p mole)	5-sulpho-isophthalic acid	Nucleosil SB anion exchange resin	Freshwater sediment	Radio ion chromatography	11.28
Misc. anions	–	Polystyrene divinyl benzene copolymer	Pure water	conductiometric	11.29
NO_2^{1-}/NO_3^{1-}	–	–	Natural	–	11.30
$NO_2, Br, NO_3, SO_4, I, SCN$	–	Na napthalene trisulphonate	Natural	Spectrophotometric	11.75
$F^{1-}, ClO_1^{1-}, Cl^{1-}, NO_2^{1-}, HPO_2^{2-}$ $AsO_4^{3-}, NO_3^{1-}, ClO_3^{1-}, SO_4^{2-}$, CrO_4^{2-}, BF_4^{1-}, Glycolate, acetate, lactate, propionate, aceto acetate, α hydroxy butyrate, chloroacetate, isobutyrate, butyrate (as Ruthenium 11, 1, 10 phenanthroline or ruthenium 11, 2, 2' bipyridy ion interaction complexes)	–	–	Natural	Fluorimetric	11.76
$F^{1-}, BO_3^{3-}, Cl^{1-}, AsO_4^{3-}, NO_3^{1-}$ glycollate, malonate	–	–	Natural	Two dimensional conductiometric detection	11.77

Cl^{-1}, NO_2^{1-}, Br^{1-}, NO_3^{1-}, PO_4^{3-}, SeO_3^{2-} SO_4^{2-}, lactate, acetate, propionate, butyrate, isobutyrate	–	–	Natural	11.78
NO_2^{1-}, NO_3^{1-}, phenylene diamine isomers	–	–	Natural	11.79
F, BO_3, Cl, AsO_4, NO_3, glyoxylate, malonate	–	–	Natural	11.81

EDTA eluent (1 m mol l^{-1}) was prepared by dissolving ethylene diaminetetraacetic acid disodium salt in deionized water and adjusting to pH 6.0 with 0.1 mol l^{-1} sodium hydroxide.

Table 11.13 shows the retention times of chloride, phosphate ($H_2PO_4^-$), nitrite, nitrate, sulphate, calcium and magnesium, All of these ions were eluted within 15 min. The calibration curves of these ions by peak height measurements were linear up to the concentration of 20 mg l^{-1}. The detection limits defined as the concentration corresponding to twice the value of noise of the base line were 0.04 mg l^{-1} for chloride 0.05 mg l^{-1} for nitrite, nitrate, magnesium and calcium and 0.1 mg l^{-1} for phosphate and sulphate.

The retention times observed for calcium and magnesium injected as metal cations and those injected as EDTA chelate anions were not significantly different. Therefore, an alkaline earth metal (M^{2+}) forms a chelate anion ($M(EDTA)^{2-}$) immediately upon contact with EDTA eluent and is then separated according to the following exchange reactions. In this case, the H_2EDTA^{2-} and $HEDTA^{3-}$ anions were assumed to present nearly equimolar concentrations in the eluent at pH 6.0:

$$M(EDTA)^{2-} + RN(H_2EDTA)^{2-} \rightleftharpoons RN(MEDTA)^{2-} + H_2EDTA^{2-}$$

$$M(EDTA)^{2-} + RN(HEDTA)^{3-} \rightleftharpoons RN(MEDTA)^{2-} + HEDTA^{3-}.$$

Using the EDTA eluent, magnesium and calcium chelate anions were eluted between nitrate and sulphate as shown in Table 11.13 and also in Fig. 11.18. Negative peaks of chelate anions might be caused by the lower conductivity of the chelate anion relative to that of the EDTA eluent anion.

The analytical results are given in Table 11.14.

Buchholz et al. [11.33] have described a procedure for the determination of less than 1 µg l^{-1} of chloride, nitrate and sulphate in 2 ml of rainwater sample using non-suppressed ion chromatography. Detection limits are less than

Table 11.13. Adjusted retention times. (From [11.31]; Copyright 1984, American Chemical Society)

Anions	Rt/min^{-1}	Cations	Rt/min^{-1}
Cl$^-$	3.4	Ca^{2+}	7.4
H$_2$PO$_4^-$	3.6	Mg^{2+}	9.0
NO$_2^-$	3.8		
NO$_3^-$	5.0		
SO$_4^{2-}$	14.2		

Table 11.14. Analytical results on rain water. (From [11.31]; Copyright 1984, American Chemical Society)

mg l^{-1}					
	Cl$^-$	NO$_3^-$	SO$_4^{2-}$	Ca^{2+}	Mg^{2+}
	2.5	1.1	0.38	0.9	0.5

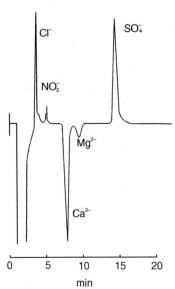

Fig. 11.18. Chromatogram of water (From [11.31])

0.1 mg l^{-1} for chloride and nitrate and 0.25 mg l^{-1} for sulphate. The method can accomplish the simultaneous analysis of chloride, nitrate, nitrite and sulphate in less than 25 min.

Apparatus. Model 110 A high performance liquid chromatograph (Beckman) equipped with a 2-ml constant volume sample loop and a 4.6 × 250 mm Wescan 269–001 anion column with guard column. Detection of anions from column eluent is by measurement of conductivity change as eluent flows at 1.8 ml min^{-1} through the conductivity cell of a conductivity detector (Wescan Model 213 A). The chromatograms, including peak area data for each anion, were obtained on a reporting integrator (Hewlett Packard 3390A). The anion calibration standard was prepared to contain 1.49, 1.64, and 1.52 mg l^{-1} chloride, nitrate and sulphate respectively. The eluent buffer consisted of 2 m mol l^{-1} potassium hydrogen phthalate adjusted to pH 4.5 with potassium hydroxide. All standards, buffers, and other chemicals were certified A.C.S. grade (Fisher). The handling of all samples, standards, and buffers was accomplished with rubber gloves to prevent anion contamination due to contact with hands.

The injection of a 2-ml sample greatly improved detection sensitivity. However, the injection of a sample of this magnitude onto a column with a bed volume of only 4 ml did present a problem. The reduced conductivity of the sample, compared to the eluent buffer, resulted in a large drop in the baseline as the sample front entered the conductivity cell. At a recommended eluent buffer concentration of 4 m mol l^{-1} potassium hydrogen phthalate (pH 4.5) as suggested by the manufacturer, the chloride and nitrate peaks were on an upward sloping baseline, which prevented accurate determination of the peak area with the reporting integrator. The sulphate peak, whose retention time was much longer than that of the other two anions, always exhibited a baseline suitable for accurate quantitation. The chloride and nitrate peaks were shifted from the sloping portion of the baseline by a reduction in the concentration of the eluent buffer. Reduction in buffer concentration increased the retention time of all the anion peaks. All peaks also experienced broadening and, in the case of sulphate, this was significant enough to limit the reduction in buffer concentration to 2 mmol l^{-1} potassium hydrogen phthalate (pH 4.5). At this concentration the chloride and nitrate peaks were shifted from the sloping baseline for more accurate quantitation, while the sulphate peak had not broadened to the point where detection sensitivity was too greatly diminished.

To test the precision of this method, the calibration standard was diluted 1:2, 1:4 and 1:8. The standard and each dilution were then analyzed for anions. The results are summarized in Table 11.15. To verify that detection response was linear for each anion in the concentration range desired for these studies, graphs of peak area vs concentration were prepared.

The accuracy of the method was considered by determination of the concentration of chloride, nitrate and sulphate in an EPA nutrient/mineral standard. The results are shown in Table 11.16. The proximity of the chromatographic values to the EPA values indicates that corrections for matrix effects, due to additional constituents present in this standard, are unnecessary (Table 11.16).

The use of concentration columns in conjunction with ion chromatography is one means of improving the sensitivity of this procedure. Wetzel et al. [11.34] used this procedure to obtain determinations down to µg l^{-1} concentrations of chloride, phosphate, nitrate and sulphate in rainwater samples. Conductivity detector response is linear with concentrations over the mg l^{-1} to µg l^{-1} range for the determinands examined. It was shown that a 3×50 mm pellicular anion exchange resin with tetraethyl ammonium groups attached to a styrene divinyl benzene backbone concentrator column can be loaded remotely and stored for at least 7 days before analysis without significantly affecting ionic determinations.

Sample loading can be performed with the concentrator replacing the injection loop; sample is then loaded by syringe or by pump. Concentration may also be loaded off line, disconnected from the ion chromatograph. In the former case, activation of the injection valve automatically places the concentrator column in line with the separator column. In the latter case, the concentrator must be manually connected to the flow path before the separator. After injection, sample ions are separated and detected in the usual manner. Normal multispecies analysis often requires less than 5 min per ion; trace analysis requires the same time as routine analysis once the concentrator is loaded. Concentrator loading requires 5–15 min per sample.

Reagents. As follows.
Standard solutions of 1000 mg l^{-1}. Prepare for individual species (chloride, phosphate, nitrate and sulphate) from reagent grade chemicals and triply distilled

Table 11.15. Concentrations of calibration standard and various dilutions determined by IC method. (From [11.33]; Copyright 1982, Terra Scientific Publications Co., Japan)

	Concentrations (mg l^{-1})		
	Cl^-	NO_3^-	SO_4^-
A. Laboratory Standard	1.49	1.64	1.52
Concentration by IC Method	1.42	1.57	1.56
	1.48	1.57	1.40
	1.48	1.62	1.53
	1.47	1.62	1.51
Mean	1.46	1.62	1.50
Standard Deviation	0.03	0.03	0.07
B. Laboratory Standard (1:2	0.75	0.82	0.76
dilution) Concentration by	0.76	0.80	0.73
IC Method	0.74	0.80	0.81
	0.74	0.80	0.78
	0.74	0.79	0.81
Mean	0.75	0.80	0.78
Standard Deviation	0.01	0.01	0.04
C. Laboratory Standard (1:4	0.37	0.41	0.38
dilution) Concentration	0.36	0.37	0.51
IC Method	0.36	0.38	0.40
	0.36	0.38	0.45
	0.34	0.40	0.43
Mean	0.36	0.38	0.45
Standard Deviation	0.01	0.01	0.05
D. Laboratory Standard (1:8	0.19	0.21	0.19
Concentration by IC Method	0.19	0.18	0.12
	0.17	0.21	0.23
	0.18	0.20	0.22
	0.17	0.18	0.19
Mean	0.18	0.19	0.19
Standard Deviation	0.01	0.02	0.05

Table 11.16. Comparison of anion concentrations: chromatographic values vs EPA values. (From [11.33]; Copyright 1982, Terra Scientific Publications Co., Japan)

	Concentrations (mg l^{-1})		
	Cl^-	NO_3^-	SO_4^-
1. EPA Value	4.6	0.34	1.8
Chromatography Value			
Run I	5.0	0.32	1.83
Run II	4.84	0.34	1.72
Mean	4.92	0.33	1.78
2. EPA Value	2.3	0.17	0.9
Chromatography Value			
Run I	2.27	0.16	0.88
Run II	2.39	0.14	0.89
Mean	2.33	0.15	0.89

water. Then dilute solutions to 10 or 1 mg l^{-1} before further dilutions to 20 or 2 µg l^{-1}.

Apparatus. Model 14 Ion Chromatograph (Dionex Corporation, Sunnyvale, California, 94086). Linear Instruments strip chart recorder was used to monitor the analyses. Instrumental conditions were: eluent, 0.003 mol l^{-1} NaHCO$_3$/0.0024 mol l^{-1} Na$_2$CO$_3$; flow rate, 30% × 460 ml/h = 138/h; separator column, 3 × 500 mm anion separator; suppressor column, 3 × 250 mm anion suppressor; conductivity meter setting; variable; sample volume, variable; recorder speed, 0.5 cm/min^{-1}.

Pump Loading of Concentrator Columns. To eliminate any possibility of sample contamination, samples to be loaded by pump were placed in a Wheaton flask under a purified nitrogen atmosphere. As supplied, the ion chromatograph has two analytical systems, including two eluent pumps. One of the eluent pumps was used to load the concentrator columns while the other was used to load the chromatographic separation. The concentrators were loaded at a rate of 7 ml/min^{-1} as determined by measuring the effluent from the concentrator wasteline. The sample loading and injection flow system is shown in Fig. 11.19. The two valves constitute one injection valve (upper and lower halves). Solid lines indicate liquid flow connections; dotted lines indicate alternate flow paths used when the valve is in the load position.

During trace analyses, the sample flask always contains water of extremely low ionic content. Attempts to clean the flask often resulted in contamination. All glassware used should be thoroughly rinsed and then soaked for a minimum of 24 h with triple distilled water.

Remote Loading of Concentrator Columns. All concentrators were flushed with eluent for 4 min before loading. A 10-ml glass syringe with a stainless steel Luer

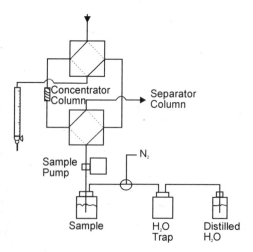

Fig. 11.19. Sample loading and injection flow system (From [11.34])

adaptor was used for remote loading of concentrator columns. A female Luer adaptor was attached to the column and the sample was forced through the column by manual pressure. Excessive force resulted in leakage along the plunger and through the Luer fitting. Remote loading by pump is also possible. Loaded concentrators were then placed in line before the separator column using the flared Teflon tubing connectors provided with each column.

Suppressor Column Use for Trace Determinations. IC suppressor columns are used to remove highly conductive eluent ions from the separator column effluent before entering the conductivity cell. During anion determinations, the suppressor exchanges H^+ for Na^+ thus converting the highly conductive sodium bicarbonate, sodium carbonate eluent to a weakly conductive, dilute carbonic acid solution. One consequence of suppressor column use in that solutions with lower conductance than the carbonic acid exiting the suppressor column cause a negative inflection in the detector output or base line. Because sample water is less conductive than the carbonic acid solution, it causes such a negative deflection or water dip approximately 2 min after injection during anion determinations.

In order to obtain reproducible results, trace sample ions of interest must be separated from the water dip. Using the sodium bicarbonate sodium carbonate eluent listed above a chloride peak will not separate from the dip when using a concentrator column in conjunction with a fully regenerated (100% H^+ form) 6 × 250 mm suppressor column. Acceptable precision is obtained, though, by converting approximately 50% of the H^+ form suppressor resin to the sodium form. This causes the water dip to become narrower and to elute earlier, thus separating the dip from the chloride peak.

Another method which permits chloride trace determination is to decrease the volume of the suppressor column. As with the partially exhausted larger suppressor, the water dip is separated from the chloride peak. Smaller columns naturally have less capacity than larger columns; in order to avoid delays due to frequent regeneration of these smaller columns, a dual suppressor system was used. One suppressor may be regenerated while the other is used for analysis. Figure 11.20 shows the flow system used for the dual suppressor column flow system.

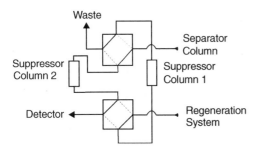

Fig. 11.20. Dual suppressor column flow system (From [11.34])

Adjust the conductivity cell and meter to a reading of 147 μmho cm^{-1} at 25°C using a 0.001 mol l^{-1} potassium chloride solution.

Procedure. Load various sample volumes onto concentrator columns by glass syringe or by pump. Concentrators loaded by syringe were placed before the separator column of the ion chromatograph, those loaded by pump having replaced the injection loop and remained in place during loading and analysis.

Studies on this method have shown the following. (1) Sample ions are quantitatively retained (<7% RSD) in the concentrator, both when large volumes of μg l^{-1} levels or small volumes of mg l^{-1} levels are loaded. Only the total number of milliequivalents is important. (2) Detector response is linear over a large range (2-10^3 μg l^{-1}) of ion concentration. (3) Reproducibility for identical samples in the 5-20 μg l^{-1} range loaded remotely on different concentrators is 18% RSD for phosphate and nitrate and 25% RSD for chloride and sulphate. (4) Concentrators can be loaded and then stored for at least 7 days at room temperature before analysis without significantly affecting results.

Tables 11.17-11.20 show that response for phosphate, nitrate and sulphate was linear over four orders of magnitude (2-10^4 μg l^{-1}). Responses for chloride ion was linear over three orders of magnitude (2-10^3 μg l^{-1}). Correlation coefficients for a linear least squares fit were between 0.9887 and 0.9999.

Table 11.21 lists the results obtained from remotely loaded concentrators analyzed immediately after loading vs seven days after loading. Results show

Table 11.17. Chloride response using a concentrator column. (From [11.34]; Copyright 1979, American Chemical Society)

Concentration (μg l^{-1})	Sample Volume (ml)	Micro-equivalents	Scale μmho cm^{-1}	Response μmho cm^{-1}
1000	5	1.41 × 10^{-1}	100	758.46
900	5	1.27 × 10^{-1}	100	691.13
800	5	1.13 × 10^{-1}	100	609.85
700	5	9.87 × 10^{-2}	100	534.34
600	5	8.46 × 10^{-2}	100	463.16
500	5	7.05 × 10^{-2}	100	387.65
400	5	5.64 × 10^{-2}	30	310.19
300	5	4.23 × 10^{-2}	30	235.04
200	5	2.82 × 10^{-2}	30	150.44
100	5	1.41 × 10^{-2}	30	75.32
70	5	9.87 × 10^{-3}	10	55.12
50	5	7.05 × 10^{-3}	10	41.26
40	5	5.64 × 10^{-3}	10	30.73
30	5	4.23 × 10^{-3}	10	21.99
20	5	2.82 × 10^{-3}	10	14.87
10	5	1.41 × 10^{-3}	3	7.96
5	5	7.05 × 10^{-4}	3	3.76
4	5	5.64 × 10^{-4}	3	3.12
3	5	4.23 × 10^{-4}	3	2.08
2	5	2.82 × 10^{-4}	3	1.50

Table 11.18. Actual and equivalent response of phosphate using a concentrator column.(From [11.34]; Copyright 1979, American Chemical Society)

Concentration µg l^{-1}	Sample Volume ml	Micro-equivalents	Scale µmho cm^{-1}	Actual Response µmho cm^{-1}	Response equivalent to 5 ml
10000	1	.667	30	24.56	122.8
2000	5	.667	30	21.75	21.75
1000	5	.333	30	8.070	8.070
500	5	.1667	30	3.508	3.508
500	5	.1667	3	3.86	3.86
100	5	.0333	3	.702	.702
50	5	.01667	3	.318	.318
20	50	.0667	3	1.648	.1684
15	50	.0500	3	1.228	.1228
10	50	.0333	3	.8070	.0807
8	50	.0267	3	.6140	.0614
5	50	.01667	3	.3509	.0351
2	50	.00667	3	.1228	.0123

Table 11.19. Actual and equivalent response of nitrate using a concentrator column.(From [11.34]; Copyright 1979, American Chemical Society)

Concentration µg l^{-1}	Sample Volume ml	Micro-equivalents	Scale µmho cm^{-1}	Actual Response µmho cm^{-1}	Response equivalent to 5 ml
10000	1	.161	30	18.42	92.10
2000	5	.161	30	17.19	17.19
1000	5	.0806	30	7.368	7.368
500	5	.0403	30	3.508	3.508
500	5	.0403	3	3.930	3.930
100	5	.00806	3	.7018	.7018
50	5	.00403	3	.2807	.2807
20	50	.0161	3	1.912	.1912
15	50	.0121	3	1.386	.1386
10	50	.00806	3	.9474	.0947
8	50	.00645	3	.7719	.0772
5	50	.00403	3	.5965	.0597
2	50	.00161	3	.3684	.0368

that samples can be stored for at least seven days without significantly affecting accuracy. The difference in determined concentrations between the immediate and the postponed analyses is less than the RSD for the remotely loaded concentrators.

Ion chromatography using an amperometric detector has been used to determine down to 10 µg l^{-1} bromide in rainwater and ground waters [11.35]. The equipment features an automated ion chromatograph, including a programme controller, an automatic sampler, an integrator and an amperometric detector. A fixed potential is applied to the cell. Any electroactive species having

Table 11.20. Actual and equivalent response of phosphate using a concentrator column. (From [11.34]; Copyright 1979, American Chemical Society)

Concentration µg l^{-1}	Sample Volume ml	Micro-equivalents	Scale µmho cm^{-1}	Actual Response µmho cm^{-1}	Response equivalent to 5 ml
10000	1	.239	30	19.29	96.49
2000	5	.239	30	15.79	15.79
1000	5	.119	30	6.140	6.14
500	5	.0597	30	2.807	2.81
500	5	.0597	3	3.158	3.16
100	5	.0119	3	.579	.579
50	5	.00597	3	.281	.281
20	50	.0239	3	1.491	.149
15	50	.0179	3	1.105	.111
10	50	.0119	3	.754	.075
8	50	.00955	3	.579	.058
5	50	.00597	3	.351	.035
2	50	.00239	3	.158	.016

Table 11.21. Comparison of trace samples analyzed immediately after loading and after seven days storage. (From [11.34]; Copyright 1979, American Chemical Society)

Ion	Concentration µg l^{-1}	Peak height[a] After 0 days (mm)	Peak height[a] After 7 days (mm)	RSD (%)
Cl$^-$	5	147	165	12.7
PO$_4^{3-}$	20	23	22	0.7
NO$_3^-$	20	24	21	2.0
SO$_4^{2-}$	20	23	27	2.8

[a]Composite standard of 100 ml was loaded on each concentrator.

an oxidation-reduction potential near the applied potential will generate a current that is directly proportional to the concentration of the electroactive species.

The detection limits for bromide is 0.01 mg l^{-1} and the relative standard deviation is less than 5% for bromide concentrations between 0.05 and 0.5 mg l^{-1}. Chloride interferes if the chloride to bromide ratio is greater than 1000:1 for a range of 0.01–0.1 mg l^{-1} bromide; similarly, chloride interferes in the 0.1–1.0 mg l^{-1} range if the ratio is greater than 5000:1. In the latter case, a maximum of 2000 mg l^{-1} of chloride can be tolerated. Recoveries of bromide ranged from 97 to 110%.

Apparatus. Dionex Model 12 ion chromatograph with 200-µl sample loop, an eluent flow rate of 2.3 ml min^{-1} (30% of full capacity) and a sample pump flow rate of 3.8 ml min^{-1} (50% of full capacity). Pulse damper is installed just before the injection valve to reduce flow pulsation. A Dionex amperometric detector consisting of cell and potentiostat. The cell contains a silver working electrode, a platinum counter electrode and an Ag/AgCl reference electrode. This cell is installed between separator and suppressor columns. The working silver electrode should be cleaned monthly by polishing the surface with a small amount of an abrasive toothpaste on a paper tissue to remove the grey tarnish and then rinsing with deionized water. An applied potential of 0.10 V and a detector output range of 100 nA V^{-1} were used.

A Technicon autosampler IV automatically delivered the sample to the ion chromatograph and a Spectra Physics SP-4100 integrator measured the peak heights.

Reagents. All reagent solutions prepared in deionized water. Use reagent grade sodium bromide for the preparation of standards.

Eluent mixture, 0.003 mol l^{-1} sodium bicarbonate and 0.002 mol l^{-1} sodium carbonate.

Dionex 4 × 50 mm fast run anion precolumn and a 4 × 250 mm separator column.

Procedure. Allow eluent to flow for approximately 30 min to reach equilibrium. It takes about 10 min to complete the chromatogram using a 200 µl sample loop. The retention time of bromide was 5.4 min. Use two calibration graphs, one for bromide concentration of 0.10–1.0 mg l^{-1} (with a 10 V input) using a quadratic fit, and another for bromide concentrations of 0.01–0.10 mg l^{-1} (with a 1 V input) using single point calibration with intercept of zero. A reference water sample is used in every twentieth position of the tray.

To determine recoveries, known concentrations of bromide were added to several natural water samples; recoveries ranged from 97 to 110%. Typical results are given in Table 11.22.

Table 11.22. Recovery of bromide in filtered water samples. (From [11.35]; Copyright 1983 Elsevier Science Publisher BV, Netherlands)

Sample No.	Concentration (mg l^{-1})		Bromide added	Bromide found	Recovery (%)
	Chloride	Bromide			
1	31	0.59	0.5	1.13	108
2	263	0.46	0.5	1.03	108
3	63	0.33	0.3	0.65	107
4	56	0.30	0.3	0.61	103
5	2.28	0.22	0.3	0.51	97
6	5.95	0.040	0.05	0.089	98
7	4.59	0.037	0.05	0.086	98
8	4.78	0.031	0.03	0.062	103
9	4.89	0.029	0.01	0.039	100
10	5.05	0.031	0.03	0.064	110

Conductimetric and amperometric detectors are commonly used simultaneously. Seven anions, namely fluoride, chloride, nitrite, phosphate, bromide, nitrate and sulphate, are determined by conductivity detection. At the same time, bromide concentrations less than $0.1\,mg\,l^{-1}$ are determined by amperometric detection. There is excellent agreement between the results obtained with these two detectors. The plot of these comparison data yielded a least squares slope of 0.975 ± 0.021, a y intercept of -0.0015 ± 0.014, and a correlation coefficient of 0.998. If a sample contained a chloride concentration greater than $1000\,mg\,l^{-1}$, the bromide peak would be masked by the chloride peak with conductiometric detection. In contrast, the maximum chloride concentration allowable for amperometric detection is $2000\,mg\,l^{-1}$ as long as the chloride to bromide ratio is less than 5000:1.

Oikawa and Saitoh [11.36] reported studies of the application of ion chromatography to the determination of fluoride, chloride, bromide, nitrite, nitrate, sulphate, sulphite and phosphate ions in 3-ml samples of rainwater. The results show that the most suitable eluent for this purpose is $2\,m\,mol\,l^{-1}$ sodium carbonate/ $5\,m\,mol\,l^{-1}$ sodium hydroxide. The reproducibility of the determination was satisfactory for standard solutions of all the ions except for nitrite. This problem was solved by preparing standard and sample solutions with the same composition as the eluent.

Apparatus. Model 14 ion chromatography (Dionex Corp.). Perkin Elmer Model Sigma 10 Data Processor.

Reagents. As follows.
    Sodium carbonate (anhydrous), Merck ultrapure.
    Sodium nitrite, Merck Analytical grade.
    Water, deionized and distilled.

Procedure. Filter the rainwater sample with a membrane filter having a pore size of 0.45 µm. Use the filtrate as the sample for the test. Inject approximately 2 ml of the sample into a microloop with a syringe. Table 11.23 lists the determination conditions. Obtain the peak areas of the chromatogram by the half value width method (width mm) in half value of peak height x peak height (mm) and prepare a calibration curve.

Table 11.23. Anion determination conditions by ion chromatography. (From [11.36]; Copyright 1982, Pergamon Press Ltd., UK)

Separator colum	3 × 500 mm, filled with low-capacity anion exchange resin
Suppressor colum	6 × 250 nm, filled with high-capacity cation exchange resin
Eluent	Mixture of $2\,m\,mol\,l^{-1}$ Na_2CO_3 + $5\,m\,mol\,l^{-1}$ NaOH
Pumping speed	156 ml h^{-1}
Sample loop	100 µl

In the sodium carbonate – sodium bicarbonate eluent system, four kinds of eluent were tested by Oikawa and Saitoh [11.36] – 2 m mol l^{-1} sodium carbonate, 2 m mol l^{-1} sodium carbonate, 2 m mol l^{-1} sodium bicarbonate, and 3 m mol l^{-1} sodium bicarbonate.

Figure 11.21a shows the results. The higher the concentration of sodium bicarbonate, the closer the peaks of fluoride, chloride and nitrite ions approach one another. This trend was more significant in sulphite and sulphate and the elution time was shorter. When a combination of 2 m mol l^{-1} sodium carbonate/ 1 m mol l^{-1} sodium bicarbonate was used, the ions could not be determined because the peak of phosphate was not separated from that of nitrate and sulphite. The ions also could not be determined when the peak of sulphite overlapped when 2 m mol l^{-1} sodium carbonate/1 m mol l^{-1} sodium bicarbonate and 2 mmol l^{-1} sodium carbonate/3 m mol l^{-1} sodium bicarbonate eluents were used. These results show that the sodium carbonate – sodium bicarbonate eluent system is not suitable when simultaneous determination of fluoride, chloride, nitrite, bromide, phosphate, nitrate, sulphite and sulphate is required in samples containing all of these ions.

Figure 11.21b shows the results of ion determinations using the sodium carbonate sodium hydroxide system. The pH of this eluent system was high, 10–12. When 2 m mol l^{-1} sodium carbonate alone (pH 10.12) and a mixture of 2 m mol l^{-1} sodium carbonate/ 2.5 m mol l^{-1} sodium hydroxide were used, the peak of nitrate overlapped with that of sulphite and that of sulphate overlapped with that of phosphate. Therefore these eluents are not suitable for determination of those ions.

When the 2 m mol l^{-1} sodium carbonate (5 m mol l^{-1} sodium hydroxide (pH 11.63) was used, all the ions were well separated. When the pH was further increased, that is, the 2 m mol l^{-1} sodium carbonate 10 m mol l^{-1} sodium hydroxide eluent was used, the elution time of ions other than phosphate decreased and the peaks approached one another. Conversely, the elution time of phosphate increased; therefore, this eluent is also not suitable.

These results show that the 2 m mol l^{-1} sodium carbonate/ 5 m mol l^{-1} sodium hydroxide eluent is the most suitable when simultaneous determination of fluoride, chloride, nitrite, bromide, sulphite, nitrate, sulphate and phosphate is required or when the sample is considered to contain all of these ions. Further studies were therefore made using this eluent.

Figure 11.21a,b shows that an optimum eluent for the determination of anions can be selected, and that information to determine the interfering ions and the best conditions can be obtained.

Using this procedure, except for nitrite, the coefficient of variation for each ion was less than 5%, and reproducibility was relatively good (Table 11.24). The peak height of nitrite was elevated as the number of sample injections were increased, and after nine tests the increase in peak height was as high as 20%.

The separation of chloride, nitrite, nitrate, sulphite and sulphate ions in an actual rainwater sample was repeatedly carried out using the procedure. Table 11.25 shows the results, which indicate that the peak height of nitrite

510 Applications of Ion Chromatography

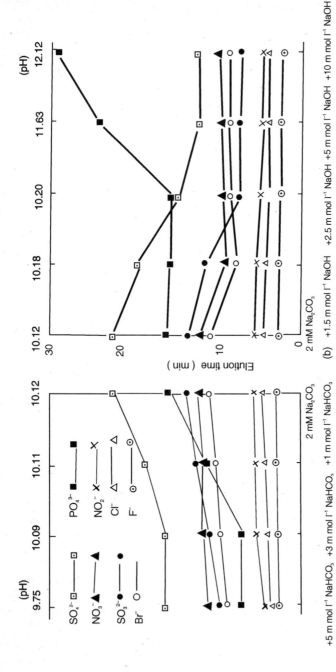

Fig. 11.21a,b. Relationships between concentrations of eluent elements and elution time: **a** mixture of Na_2CO_3 + $NaHCO_3$; **b** mixture of Na_2CO_3 + NaOH (From [11.36])

Table 11.24. Reproducibility of determination of anions in standard mixed solution. (From [11.36]; Copyright 1982, Pergamon Press Ltd., UK). (Peak height unit: cm)

Ion No.	F^- 3 mg l^{-1}	Cl^- 5 mg l^{-1}	NO_2^- 10 mg l^{-1}	NO_3^- 10 mg l^{-1}	SO_3^{2-} 30 mg l^{-1}	SO_4^{2-} 20 mg l^{-1}
1	16.4	11.2	10.0	4.3	3.1	9.4
2	17.0	11.5	10.5	4.4	3.2	9.9
3	16.7	11.3	10.9	4.6	3.2	10.5
4	16.9	11.5	11.0	4.4	3.2	9.7
5	16.1	11.8	11.3	4.5	3.3	10.6
6	17.1	11.9	11.6	4.6	3.3	10.7
7	17.4	12.3	11.8	4.6	3.3	10.3
8	16.8	12.1	11.9	4.5	3.2	10.6
9	16.9	11.9	11.9	4.6	3.2	10.3
$\bar{X}$	16.8	11.7	11.2	4.5	3.2	10.2
CV%	2.3	3.2	6.0	2.5	2.0	4.5

Table 11.25. Reproducibility of anions in actual rainwater samples. (From [11.36]; Copyright 1982, Pergamon Press Ltd., UK)

Ion No.	Cl^-	NO_2^-	NO_3^-	SO_3^{2-}	SO_4^{2-}
1	2.47	0.78	0.43	1.04	0.83
2	2.32	0.79	0.48	1.09	0.84
3	2.24	0.87	0.48	1.31	0.87
4	2.18	0.95	0.42	1.15	0.86
5	2.11	0.95	0.38	0.98	0.81
6	2.15	0.97	0.46	1.09	0.81
7	2.15	0.99	0.49	1.09	0.86
8	2.10	1.13	0.48	1.20	0.83
9	2.12	1.13	0.46	1.20	0.86
$\bar{X}$	2.20	0.95	0.45	1.12	0.84
CV%	2.5	13.3	8.1	8.8	2.7

tended to increase as the number of times the determination was repeated, showing a CV of 13.3%. The CV of other anions was less than 10% as was the reproducibility test of the standards.

The detection limit of each ion was obtained from the peak of the chromatogram and the noise of the base line when 100 µl samples containing fluoride (0.1 mg l^{-1}), chloride (0.2 mg l^{-1}), nitrite (0.5 mg l^{-1}), nitrate (1 mg l^{-1}), sulphite (5 mg l^{-1}) and sulphate (3 mg l^{-1}) were used. When the detection limit was twice the signal to noise ratio of the base line these limits were 0.01 mg l^{-1} for fluoride, 0.02 mg l^{-1} for chloride, 0.05 mg l^{-1} for nitrite, 0.01 mg l^{-1} for nitrate, 0.5 mg l^{-1} for sulphite and 0.1 mg l^{-1} for sulphate.

Little change was found in the concentration of anions during three days storage when the sample was kept at 25 °C in a refrigerator.

Figure 11.22 shows an example of an ion chromatogram of rainwater. Using the following conditions, Schwabe et al. [11.37] determined chloride, fluoride, nitrate, phosphate, and sulphate in 20 min in rainwater and potable water in amounts down to 0.5 mg l^{-1}

Type	Dionex D12
Sensitivity	1 ks
Analytical column	Fast run anion separator
Suppressor column	Anion suppressor
Sample loop	100 µl
Pump flow	115 ml h^{-1}
Pressure	22 bar
Run time	11 min per sample
Eluent	0.0048 mol l^{-1} Na HCO$_3$
	0,0023 mol l^{-1} Na$_2$CO$_3$

Figure 11.23 shows the results for chloride, fluoride, nitrate and sulphate in Berlin rainwater. The month of March shows significantly higher loads than the other months, caused by high rainfall. There is also a high load of sulphate and nitrate for October and November.

Wagner et al. [11.38] and Rowland [11.39] determined chloride, nitrate and sulphate in amounts down to 0.5 mg l^{-1}.

Ion chromatographic methods have been described for the co-determination of anions and cations in rainwater. Thus, Jones and Tarter [11.40] using the conditions given in Table 11.26 reported determinations down to 1 mg l^{-1} of anions (chloride, bromide and sulphate) and cations (sodium, potassium, magnesium and calcium) in rainwater (Fig. 11.24) without converting the cations to anion complexes prior to detection [11.41]. The technique uses a cation separ-

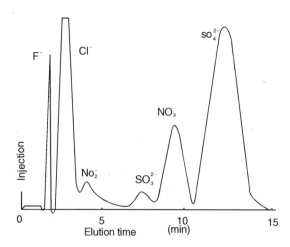

Fig. 11.22. Example of ion-chromatogram of rainwater (From [11.36])

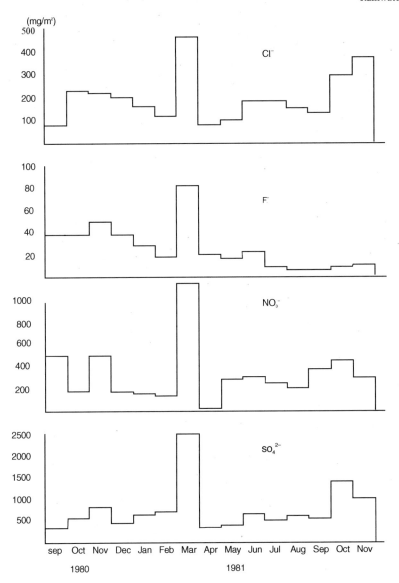

Fig. 11.23. Rainwater (Berlin Marienfelde) (From [11.37])

ator column, a conductivity detector, an anion separator column, an anion suppressor column, and either a second conductivity detector or an electrochemical detector in sequence. The use of different eluants provides a means for the detection of monovalent cations and anions and divalent cations and anions in each of the samples. Using an eluant with a basic pH, it is possible to separate simultaneously and detect the monovalent cations (with the

Table 11.26. Instrumental conditions. (From [11.40]; Copyright 1985, International Scientific Communications Ltd.)

Instuments	Dionex model 2010i ion chromatograph Dionex electrochemical detector Houston Instrument Omniscribe chart recorder
Electrochemical applied potential	0.4 V
Chart recorder speed	0.25 cm min^{-1}
Flow rate	1.5 ml min^{-1}
Suppressor column	Dionex AMS, preproduction prototype, Membrane Suppressor
Injection volume	0.10 ml
Eluant	0.0016 mol l^{-1} Li$_2$CO$_3$ + 0.0024 mol l^{-1} LiCOOH·H$_2$O pH = 10.4 Monovalent cations anions
Separator columns	
Anion separator	150 × 4 mm i.d. Dionex HPIC-AS3 anion
Cation separator	200 × 4 mm i.d. Dionex HPIC-CS1 cation 0.0033 mol l^{-1} Cu phthalate divalent cations and anions
Eluant	
Separator columns	
Anion separator	250 × 4.6 mm i.d. Vydac 3021C4.6 anion
Cation separator	200 × 4 mm i.d. Dionex HPIC-CS1 cation

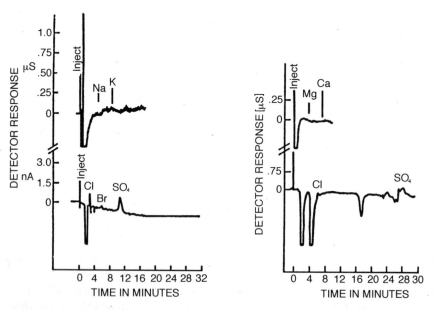

Fig. 11.24. Chromatograms of rainwater (Fort Worth, Texas) Monovalent cations and anions and divalent cations and anions (From [11.40])

exception of the ammonium ion) and anions, while an eluant with an acidic pH allows for the separation and detection of divalent cations and anions.

The equipment and operating conditions used by Jones and Tarter [11.40] are described below. The instrumental set-up consists primarily of two separator columns, two detectors and one suppressor column. The analyte flows through the injection valve into the cation separator, where the cations are separated. The separated cations are then detected using a conductivity detector. (this is, in effect, a single column ion chromatographic analysis at this stage). The anions, which are essentially unretained on the cation column, are separated on the anion separator column. The ions then travel through the anion suppressor column, where the previously separated and detected cations are removed. Finally, the separated anions pass through the second detector, which may be a conductivity detector or an electrochemical detector. The electrochemical detector responds to pH changes in the eluant as the dissociated acids pass through the detector.

Apparatus. Dionex 2010 i ion chromatograph, Dionex electrochemical detector, Houston Instrument Omniscribe chart recorder.

Conditions. As follows.

Electrochemical applied potential	0.4 V
Chart recorder speed	0.25 cm min^{-1}
Flow rate	1.5 ml min^{-1}
Suppressor column	Dionex 4M S$_1$ membrane suppressor
Injection volume	0.1 ml

Reagents. For monovalent ions:

Eluant	0.0016 mol l^{-1} Li$_2$CO$_3$ + 0.0024 mol l^{-1} LiCOOH, H$_2$O pH = 10.4 monovalent cations and anions.
Separator columns	
Anion separator	150 × 4 mm i.d. Dionex HPLC -A 53 anion
Cation separator	200 × 4 mm i.d. Dionex HPLC -CS 2 cation
For divalent ions:	
Eluant	0.0033 mol l^{-1} copper phthalate divalent cations and anions
Separator columns	
Anion separator	250 × 4.6 mm i.d. Vydac 3021 C 4.6 anion
cation separator	200 × 4 mm i.d. Dionex HPLC-CS 1 cation

All standard anion and cation solutions were prepared by the appropriate dilution of standard stock solutions using distilled deionized water. The salts used to prepare these solutions were ACS Certified quality.

Two different eluants were used, lithium carbonate-lithium acetate dihydrate, and copper phthalate. A stock solution of the lithium carbonate lithium acetate dihydrate eluant was prepared from ACS Certified salts using distilled deionized water with the appropriate dilution to obtain the working eluant.

The copper phthalate eluant was prepared by mixing a solution of cupric acetate with an excess of potassium hydroxide and filtering the resulting cupric hydroxide precipitate. The cupric hydroxide precipitate was then mixed with an equimolar amount of phthalic acid and heated gently overnight to produce copper phthalate.

Procedure. All water samples were injected and analysed as received without prior preparative steps. If dirty samples had been analysed filtering is necessary to remove the solid particles prior to injection on the ion chromatograph.

The normal procedure for the use of the ion chromatograph was followed, with the exception of the use of the electrochemical detector after the anion suppressor. The solutions were injected and the detector and chart recorder were adjusted to provide peaks of appropriate height.

Simultaneous analysis of both anions and cations indicates that water samples from various localities contain many of the same ions but in differing amounts.

A comparison of ion chromatography and a wet chemical method, EDTA titration, for calcium and magnesium shows good correlation between the two methods. With ion chromatography, the total calcium and magnesium is approximately 42 mg l^{-1} while the EDTA titration indicates a total hardness of approximately 49 mg l^{-1} with calcium assumed to be the major contributor.

The retention times and detection limits change according to the eluant selected but, under normal daily operating conditions, detection limits will be less than 2 mg l^{-1} and retention times will range from 0.5 to 25 min for most species.

Cheam and Chau [11.74] have discussed the automated simultaneous analysis of anions and mono and divalent cations in water.

11.3 Potable Water

11.3.1 Mixtures of Chloride, Bromide, Nitrate and Sulphate

The techniques described respectively in Sects. 11.1.1 and 11.2.1 for the analysis of natural waters [11.8] and rainwater [11.37] have been applied to potable waters.

In studies of the formation of bromoforms produced during the disinfection of water supplies it is necessary to have a method capable of determining bromide ion at the µg l^{-1} level. Morrow and Minear [11.42] have developed an ion chromatographic procedure using a concentrator column to improve sensitivity. This analysis was carried out on a Dionex model 125 ion chromatograph equipped with concentrator column in place of an injection loop, coupled with an autosampler, which allowed detection at the µg l^{-1} level. A concentrator column is a short separator column used to strip ions from a measured volume of matrix leading to a lowering of detection limits by several orders of magnitude.

To initiate an analysis the system lines were flushed with sample for 2 min prior to loading via the auto sampler. A 50 ml volume of sample was then loaded onto the concentrator, via the auto-sampler, at 100% A (7.67 ml min^{-1}). The concentrated sample was then flushed (injected) into the system using slightly weakened standard eluent (0.003 mol l^{-1} sodium bicarbonate/0.0024 mol l^{-1} sodium carbonate) at 30% Q (2.3 ml min^{-1}). The auto sampler was programmed to allow the ion chromatograph to remain in the inject position for the remainder of each sample elution, after injection of the sample, so that a continual flushing of the concentrator was accomplished prior to loading of the next sample.

When using conventional ion chromatographic separation techniques, it is possible that other matrix anions also common to natural waters may coelute with bromide. For example, bromide and nitrate elute simultaneously using a standard anion separator column (No.30065), standard anion suppressor (No. 30366) and standard eluent (0.003 mol l^{-1} sodium bicarbonate/ 0.0024 mol l^{-1} sodium carbonate at 30% flow (7.67 ml min^{-1}). A representative chromatogram is shown in Fig. 11.25.

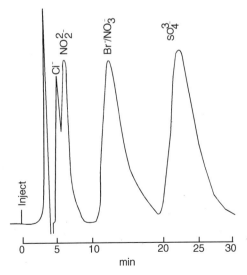

Fig. 11.25. Simultaneous elution of bromide and nitrate using standard eluent (From [11.42])

Separation of bromide from all other matrix anions was shown to be possible using a trace anion separator (No. 30827), a standard anion suppressor, and slightly weakened standard eluent, at 30% flow.

Chloride, nitrate and sulfate were the only ions found to be present in any significant quantities in the representative raw waters analysed. To determine whether response for bromide remained the same regardless of the concentration of these anions, recovery tests were performed on bromide standard solutions of 80 µg l^{-1} spiked with varying concentrations of chloride, nitrate and sulfate, using the concentrator found to have the highest µ-equivalent capacity.

In the presence of 100 mg l^{-1} chloride, there was 100% recovery of 80 µg l^{-1} bromide. Chloride was not expected to interfere with the elution of bromide at any concentration since bromide is held preferentially over chloride by the concentrator column. In the presence of 100 mg l^{-1} nitrate only 16.2% of the bromide was recovered indicating the concentrator's preference for the nitrate ion. There was no response for 80 µg l^{-1} bromide in the presence of 1000 mg l^{-1} sulfate, indicating 0% recovery.

Evaluation of µ-equivalent concentrator capacity yielded 8.5, 10.6, 6.8, 13.9, 11.7 and 14.6 for the 6 columns subjected to analysis. Evaluation of the concentrator found to possess the highest µ-equivalent capacity (14.6) for percent recoveries of 80 µg l^{-1} bromide in the presence of varying amounts of nitrate and sulphate revealed that nitrate in excess of 17 mg l^{-1} was found to interfere with the elution of bromide and that sulphate in excess of 25 mg l^{-1} was found to interfere with the elution of bromide.

In Table 11.27 a comparison is shown of bromide determinations obtained by the ion chromatographic method and a spectrophotometric method [11.43] based on the catalytic effect of bromide on the oxidation of iodide or iodine to iodate by potassium permanganate in acidic solution.

A direct statistical comparison of the two methods reveals that, for standard solutions of 20, 60 and 100 µg l^{-1}, the spectrophotometric method produces standard deviations of 0.37, 0.97 and 0.21 and RSDs of 1.9, 1.6 and 0.2%, while the ion chromatography method produces a standard deviation of 1.5 and a RSD of 3.0% for 50 µg l^{-1} bromide.

Analysis of several raw waters by both methods yielded comparable results on four out of five samples.

Table 11.27. Comparison of IC and spectrophotometric analyses for bromide. (From [11.42]; Copyright 1984, Pergamon Publishing Co., UK)

City I.D. No.	Date	Bromide (µg l^{-1})	
		IC	Spectrophotometer
1	3/27/81	7.5	11.6
2	9/22/81	29.6	30.9
3	12/17/80	104.7	109.8
4	10/16/80	42.6	42.9
5	7/24/81	55.8	55.4

Detection by the ion chromatographic method is subject to interference by excessive levels of nitrate and sulphate but this is readily avoided by altering the concentration volume accordingly.

11.3.2 Mixtures of Sulphide and Cyanide

Bond et al. [11.44] have described a method for the simultaneous determination of down to $1\,mg\,l^{-1}$ free sulphide and cyanide by ion chromatography with electrochemical detection. These workers carried out considerable exploratory work on the development of ion chromatographic conditions for separating sulphide and cyanide in a basic medium (to avoid losses of toxic hydrogen cyanide and hydrogen sulphide) and on the development of a suitable amperometric detector.

Reagents. Distilled deionized water, to prepare all standards and eluents.

Sulfide and cyanide standards. Prepare daily by dissolving sodium sulphide and potassium cyanide respectively in $0.005\,mol\,l^{-1}$ sodium hydroxide. Prior to use filter the eluents through a $0.45\mu m$ filter. Then degas before and during chromatographic runs with high purity nitrogen.

Apparatus. Dionex (Sunnyvale, CA) Model 10 ion chromatograph with the supplied detector or replaced by an electrochemical (amperometric) detector. A waters (Milford, MA) Model 6000A pump in conjunction with a Model U6K injector. Separator column, a Dionex anion exchange column (internal diameter $= 3\,mm$, length $= 250\,mm$). Amperometric detector, Tacussell (Villeurbanne, France) DELC-1 detector cell. Dropping mercury electrode (preferred) and a mercury coated platinum working electrode. Mercury was plated onto the platinum at -1.00 V vs Ag/AgCl for 5 min and then the electrode was dipped into a mercury pool. Glassy carbon auxiliary and Ag/AgCl ($3\,mol\,l^{-1}$ NaCl) reference electrodes. Cathode ray oscilloscope to trace current time curves and hence determine the drop time in the Tacussell cell F.G. & G. Princeton Applied Research Corp., (P.A.R) Princeton, NJ. Model 310 polarographic detector cell with Model 303 static mercury drop electrode with auxiliary and reference electrodes as above.

Polarographic data recovered by use of a PAR Model 174A polarographic analyser, with a P.A.R Model 303 static mercury drop electrode.

Voltammograms in a conventional electrochemical cell recorded by using "mini electrodes" obtained from Metrohm (Herisau, Switzerland) Gold working electrode, glassy carbon auxiliary and Ag/AgCl ($3\,mol\,l^{-1}$) reference electrode for voltammetry.

Procedure. Carry out all experiments in a controlled temperature laboratory at 22 $\pm 1\,°C$.

Despite a higher sensitivity, the mercury film electrode has some disadvantages compared to the dropping mercury electrode. Day to day reproducibility of the mercury film and hence the analyte signal is harder to maintain than at the dropping mercury electrode. If the film layer is too thin, bad tailing occurs (electrode saturates); the electrode can also eventually become poisoned. The length of time it lasts will depend on the nature and content of samples injected. Consequently more frequent calibration is required than at a dropping mercury electrode.

Figure 11.26 shows that the response in this water is not the same as in deionized water, suggesting that only free sulfide and cyanide are being determined. This is not surprising and is characteristic of most methods for determining sulfide or cyanide. Since the drinking water was known to contain a significant amount of copper, some of the cyanide may be bound to the metal. Deliberate addition of copper to the water reduced the cyanide peak, supporting this hypothesis. The nonzero intercept for cyanide calibration curves is believed to be a result of complexation by impurities.

Essentially identical results were obtained by using short drop times at the dropping mercury electrode, d.c. mode (0 V) or using the mercury coated platinum electrode, normal pulse mode (pulse-0.80 to 0 V, pulse delay = 1 s, pulse width = 20 ms.) Examination of industrial effluents known to contain sulphide and cyanide demonstrate that the proposed method works extremely well and that free cyanide rather than total is being determined.

11.3.3 Mixtures of Arsenate, Selenate and Selenite

The procedure described in Sect.11.1.2 for the determination of arsenate, selenate and selenite in natural waters [11.15] has also been applied to potable waters.

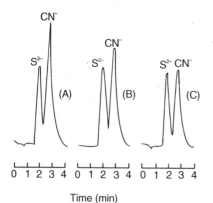

Fig. 11.26A–C. Chromatograms for sulfide and cyanide in deionized water and potable water: **A** deionized water; **B** potable water sample containing 250 ng of sulfide and 850 ng of cyanide; **C** potable water sample as in **B** spiked with copper ions $(2\,mg\,l^{-1})$ and the decrease in the cyanide response can be seen (From [11.44])

11.4 Industrial Effluents and Waste Waters

11.4.1 Mixtures of Chloride, Bromide, Fluoride, Nitrite, Nitrate, Sulphate and Phosphate

The work of Mosko [11.45] is important in that be is one of the few workers who have given serious consideration to the determination of nitrite in water. His paper is concerned with the determination of chloride, sulphate, nitrate, nitrite, orthophosphate, fluoride and bromide in industrial effluents, waste water and cooling water. Two types of analytical columns were evaluated (standard anion and fast run series). Chromatographic conditions, sample pretreatment and the results of interference, sensitivity, linearity, precision, comparative and recovery studies are described. The standard column provided separation capabilities which permitted the determination of all seven anions. The fast run column could not be used for samples containing nitrite or bromide owing to resolution problems.

Reagents. Low conductivity water ($1-1.5$ µmho cm^{-1}) prepared by polishing demineralized water with a strong mixed bed resin exchanger used to prepare all standards and eluents.

Standard eluent (0.003 mol l^{-1} NaHCO$_3$/0.0024 mol^{-1} Na$_2$CO$_3$). (Periodic conditioning and regeneration of the separator columns and suppressor column employed the use of 0.1 mol l^{-1} sodium carbonate and 1 N sulphuric acid solutions, respectively.)

1 + 1 solution of sodium bicarbonate/sodium carbonate (50 g l^{-1} solutions) appropriate volume added to standards and samples to produce a matrix background similar to the eluent. This minimized interference from the "water dip" that is common to ion chromatography analyses.

Apparatus. Dionex Auto Ion System 12 Analyzer (Dionex Corp., Sunnyvale, California) consisting of a Model 12 ion chromatograph (Dionex Corp.) coupled with a Gilson autosampler, a Columbia Scientific Industries Supergrator 3-A integrator and a Gilford Instrument Laboratories dual pen strip chart recorder. A concentrator/precolumn (4×50 mm, Dionex 30825) and separator column (4×250 mm, Dionex 30827) were used for the standard seven anion determination. Two anion concentrator precolumns (4×50 mm Dionex 30825) connected in series (fast run series) were employed for the five anion analyses. A high capacity suppressor column (9×100 mm Dionex 30828) containing a strong acid cation exchange resin in the hydrogen form was used for all chromatography. The ion chromatograph was equipped with a 250-µl sample loop. A conductivity full scale sensitivity setting of 30 µmho was routinely used.

Eluent pump flow rate of 35–40% (161–184 ml h^{-1}) maintained system pressure between 500 and 700 psi. Both the sample and regeneration pumps were operated at 50% flow rate (230 ml h^{-1}).

Procedure. Filter 10 ml of sample through a 0.22 µm membrane and collect the filtrate in a 15-ml capacity polystyrene vial containing 100 µl of 1 + 1 sodium bicarbonate/sodium carbonate solution. Calibration was established initially and after every six sample analyses by using the combination anion solution indicated in Table 11.28. The calibration standard were prepared fresh daily from separate 1000 mg l^{-1} stock solutions, fortified in the same manner as the samples with bicarbonate/carbonate solution.

Analysis time per run (seven anion determination) was about 24 min. Table 11.29 contains information on the calibrated range of the automated method.

A variation in sensitivity of orthophosphate was noted with relation to concentration range. Linear responses were obtained for concentration ranges 0–25 mg l^{-1} phosphate and 30–80 mg l^{-1} phosphate. However, at a concentration between 25 and 30 mg l^{-1} phosphate, the sensitivity changed. To establish a broad linear range, the data were evaluated by using the method of linear regression. Up to a concentration of 55 mg l^{-1} phosphate the actual phosphate values did not vary significantly from the linear regression plot. Attempting to quantitate phosphate at concentrations above the 55 mg l^{-1} level would result in significant positive error. The degree of error would increase as the phosphate concentration increased. Because of the linearity problem the dynamic range of

Table 11.28. Typical retention times and peak heights of the calibration standard. (From [11.45]; Copyright 1984, American Chemical Society)

Concn. mg l^{-1}	Anion	Approx retention time, min	Approx peak height, µV
5.0	fluoride (F)	2.1	54 000
5.0	chloride (Cl)	3.5	30 000
20.0	nitrite (NO_2)	4.5	50 000
25.0	orthophosphate (PO_4)	7.0	24 000
25.0	bromide (Br)	9.0	24 300
5.0	nitrate (NO_3)	11.0	5 000
50.0	sulfate (SO_4)	15.5	50 700

Table 11.29. Detection limits and calibrated range for sequential seven anion determinations. (From [11.45]; Copyright 1984, American Chemical Society)

Anion	Detection limit mg l^{-1}	Calibrated range, mg l^{-1}	Limit of linearity, mg l^{-1}
Fluride (F)	0.02	0.1–15.0	> 20[a]
Chloride (Cl)	0.02	0.1–30.0	> 20[a]
Nitrite (NO_2)	0.25	0.5–30.0	50
Orthophosphate (PO_4)	0.1	0.2–50.0	50–60
Bromide (Br)	0.1	0.5–50.0	> 200[a]
Nitrate (NO_3)	0.1	0.2–30.0	150
Sulfate (SO_4)	0.2	0.5–100.0	120

[a] Maximum concentration level tested.

the method was limited to 0–50 mg l^{-1} phosphate. Calibration of the system was established by using a concentration of phosphate standard (25 mg l^{-1} phosphate) which was mid-range for the method. Also, the phosphate concentration used for calibration fell within the range where the change in sensitivity occurred. Precision and accuracy data confirmed accurate calibration over the 0–55 mg l^{-1} range, using a 25 mg l^{-1} phosphate standard.

A decrease in retention time as concentration increased was noted for nitrate. The average retention times observed during development of precision data are listed in Table 11.30. This phenomenon (peak migration) was a source of concern in the automated method because the integrator identifies peaks on the basis of retention time. The program entered into the integrator permitted only a 10% maximum deviation in retention time between the sample peak and a standard calibration peak. If the sample peak retention time deviated more than 10% from that of the calibration standard, the sample peak would be identified by the integrator as an extraneous peak or would possibly be incorrectly identified as some other anion having a similar retention time. A minimum reporting limit of 0.2 mg l^{-1} nitrate was a pre-established requirement for the automated ion chromatographic method. The concentration of nitrate used to establish calibration had to possess a retention time with a 10% window that encompassed the retention time of a 0.2 mg l^{-1} nitrate peak. Calibration with a 5.0 mg l^{-1} nitrate standard provided a reasonable compromise whereby the widest dynamic range in which identification of the required minimum reporting limit (0.2 mg l^{-1} nitrate) was attainable. With this concentration of nitrate used for calibration the range of the method was 0.2–30 mg l^{-1} nitrate. Concentrations of nitrate > 30 mg l^{-1} nitrate resulted in retention times outside the 10% window and were identified by the integrator as extraneous peaks or sometimes incorrectly as bromide (anion eluting immediately before nitrate).

Reproducibility of nitrite with relation to suppressor column capacity was investigated by continuous analysis of a 20 mg l^{-1} nitrite solution over at 6 h period. During this period a 24% increase in sensitivity was observed. However, it was noted that satisfactory results (within 10% deviation) were obtained if calibration were re-established after every six sample analyses. Partial exhaustion (after 2 h of operation) of the suppressor column was required to achieve sufficient

Table 11.30 Change in retention time with concentration of nitrate. (From [11.45]; Copyright 1984, American Chemical Society)

Nitrate, mg l^{-1}	Retention time, min	Retention time normalized to 5.0 mg l^{-1} NO$_3$, min
0.2	12.4	+1.2
5.0	11.2	0
20.0	10.9	−0.3
25.0	10.9	−0.3
45.0	9.2	−2.0

sensitivity for measurement of the desired $0.5 \, \text{mg} \, l^{-1}$ nitrite minimum reporting level. Since use of a strong eluent to exhaust partially the suppressor column would also decrease operating time, it was decided that any sample requiring nitrite quantitation would be analyzed after the second calibration of the system (calibration performed after every six samples). No nitrite data were reported for samples analyzed by using a freshly regenerated suppressor column.

Table 11.31 contains some examples of comparative data. Most results agreed within experimental limits for the specific concentration range and test method used. Traditional methods of analysis used for comparison purposes were as follows: fluoride, ion selective electrode method; chloride, automated mercuric thiocyanate colorimetric method; nitrite, permanganate titration method; orthophosphate, automated ascorbic acid reduction colorimetric method; nitrate automated copper – cadmium reduction colorimetric method. Sulphate was measured with a barium methylthymol blue colorimetric automated method.

Green and Woods [11.46] have described a method employing a hydrogen form cation exchange column for the analysis of mixtures of chloride, fluoride, phosphate and sulphate in industrial waste waters.

11.4.2 Mixtures of Cyanide, Sulphide, Iodide and Bromide

Rocklin and Johnson [11.47] also used an electrochemical detector in the ion chromatographic determination of cyanide and sulphide. They showed that by placing an ion exchange column in front of an electrochemical detector, using a silver working electrode, they were able to separate cyanide, sulphide, iodide and bromide and detect them in water samples at concentrations of 2, 30, 10 and $10 \, \mu\text{g} \, l^{-1}$ respectively. Cyanide and sulphide could be determined simultaneously. The method has been applied to the analysis of complexed cyanides and it is shown that cadmium and zinc cyanides can be determined as total free cyanide while nickel and copper complexes can only partially be determined in this way. The strongly bound cyanide in gold, iron or cobalt complexes cannot be determined by this method.

This method is based on the work of Pihla and Kosta [11.48, 11.49] who showed that a silver working electrode has the ability to produce a current that is linearly proportional to the concentration of cyanide in an amperometric electrochemical flow through cell. The reaction for cyanide is

$$\text{Ag} + 2\,\text{CN}^- \rightarrow \text{Ag(CN)}_2^- + e^-.$$

Under these conditions sulphides and halides produce insoluble precipitates rather than soluble complexes:

$$2\,\text{Ag} + \text{S}^{2-} \rightarrow \text{Ag}_2\text{S} + 2e^-$$

$$\text{Ag} + \text{X}^- \rightarrow \text{AgX} + e^-$$

Table 11.31. Comparison of results (mg l^{-1}) obtained with automated IC vs. results obtained with traditional methods. (From [11.45]; Copyright 1984, American Chemical Society)

Water type	Fluoride (F)		Chloride (Cl)		Nitrite (NO$_2$)		Orthophosphate (PO$_4$)		Bromide (Br)		Nitrate (NO$_3$)		Sulfate (SO$_4$)	
	IC	Electrode	IC	Wet chem	IC	Wet chem	IC	Wet chem	IC	Wet chem	IC	Wet chem	IC	Wet chem
Boiler feed	0.1		16	16	<0.5		<0.2	<0.05	0.5		2.0	1.9	42	40
Boiler	0.2		1.2	1.5	<0.5		11	9.9	<0.5		<0.2		47	55
Boiler	1.2		140	125	<5		14	14.4	<0.5		1.0	1.0	70	60
Cooling	4.0		20	24	<0.5		2.0	2.5	1.4		14	15	68	75
Cooling	0.4		23	28	<0.5		4.2	4.2	<0.5		6.0		46	45
Zeolite softened	0.1		18	18	<0.5		<0.2	<0.05	<0.5		2.2	2.0	37	35
Softened	0.2	0.2	40	38	<0.5		0.4	0.2	<0.5		<0.2		32	25
Nitrite treated	<0.5		15	16	<2.5	<100	<10		<25		1000	990	88	80
Nitrite treated	5.0		13	16	<2.5	<100	0.6	0.55	<2.5		140	155	38	40
City make-up	1.0	1.0	16	16	<0.5		<0.2	<0.05	<0.5		9.0	8.3	23	25
Well water	0.6	0.6	2.2	3.0	<0.5		<0.2	<0.05	<0.5		<0.2	<0.2	10	5
Waste effluent	<1	0.1	260	280	<5		16	16.1	150		22	19	240	260

526   Applications of Ion Chromatography

Rocklin and Johnson [11.47] overcame the latter problem by placing an ion exchange column in front of the electrochemical detector. Cyanide and sulphide are separated and are thus determined simultaneously. Although bromide and iodide can be determined by ion chromatography with conductivity detection, the use of electrochemical detection results in greater selectivity as well as increased sensitivity.

Reagents. As follows.
Cyanide standard solution. Prepare from a $1000\,mg\,l^{-1}$ sodium cyanide stock solution, standardized by argentometric titration.

Sulphide standards. Prepare by diluting 21% $(NH_4)_2S$ ammonium sulphide.
$K_3Fe(CN)_6$.
$Ni(CN)_4^{2-}$. Prepare from sodium cyanide and nickel acetate.
$Cd(CN)_4^{2-}$. Prepare from $1000\,mg\,l^{-1}$ solutions of cadmium chloride and sodium cyanide.

Copper cyanide solutions. Prepare by adding stoichiometric quantities of sodium cyanide to a $1.0 \times 10^{-5}\,mol\,l^{-1}$ cuprous cyanide solution.

Apparatus. Dionex System 2010 (P/B 3520L) ion chromatograph with sample loop size 50 µl and eluent flow $2.5\,ml\,min^{-1}$. Cyanide and sulphide separated on an HPIC-AS3 (P/N35311) anion exchange column using an eluent consisting of $14.7\,mmol\,l^{-1}$ ethylenediamine, $10\,mmol\,l^{-1}$ NaH_2BO_3 (prepared from H_3BO_3 and NaOH) and $1.0\,mmol\,l^{-1}$ sodium carbonate. The eluent pH is 11.0. Chromatography of iodide was performed by using an HPIC-ASI (P/N 30827) column with an eluent consisting of $2\,mmol\,l^{-1}$ sodium nitrate. Chromatography of bromide was performed using an HPIC-AS3 (P/N 30985) column with an eluent consisting of $2\,mmol\,l^{-1}$ sodium carbonate.

Electrochemical detection performed with an Ion Chrom/Amperometric Detector (P/N 35221). The cell (Fig. 11.27) consists of a silver rod working electrode 1.3 cm long × 0.178 cm in diameter, an Ag/AgCl reference electrode separated from the flowing stream by a Nafion cation exchange membrane and platinum counter electrode. (Nafion is a registered trade mark of E. I. du Pont de Nemours & Co.) The cell geometry is based on one previously reported by Lown et al. [11.50]. The working electrode was occasionally cleaned by mechanical polishing. The applied potential was 0.V for cyanide and sulphide, 0.20 V for iodide, and 0.30 V for bromide.

Figure 11.28 shows the separation achieved on a 12 anion standard by this procedure. Sulphide, cyanide, bromide, and sulphite are detected at the silver electrode while nitrite, nitrate, phosphate and sulphate produce no response. Due to the low dissociation of hydrogen sulphide and hydrogen cyanide following protonation by the suppressor column, they are not detected by the conductivity detector.

The major advantage of chromatography over other analytical methods is its ability to separate interferences. With one exception, the determination of one of the four ions is not affected by the presence of the others. For example, a solution

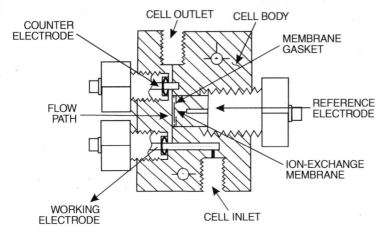

Fig. 11.27. Diagram of amperometric flow through cell (From [11.47])

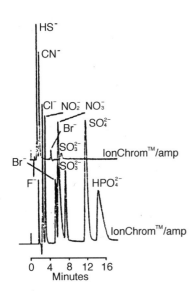

Fig. 11.28. Simultaneous analysis by using electrochemical and conductivity detection. Concentrations are 300 µg l^{-1} sulfide, 500 µg l^{-1} cyanide, 1 mg l^{-1} fluoride, 4 mg l^{-1} chloride, 10 mg l^{-1} nitrites 10 mg l^{-1} bromide, 25 mg l^{-1} nitrate, 30 mg l^{-1} sulphite, 25 mg l^{-1} sulfate and 50 mg l^{-1} phosphate (From [11.47])

containing 2500 times as much chloride as bromide has little effect on the determination of bromide (Table 11.32) even though chloride elutes first. Since E° for the oxidation of silver to silver chloride is with respect to the value of E° for the oxidation of silver to silver bromide, at a potential just on the diffusion controlled plateau for bromide, the current response for bromide will be much greater than that for chloride. This can be exploited in order to determine small quantities of bromide in a large excess of chloride, a difficult process using other methods of analysis due to the similar chemical properties of the two halides. The determina-

Table 11.32. Determination of 1 mg l^{-1} cyanide in the presence of sulfide. (From [11.47]; Copyright 1983, American Chemical Society)

Amt of added S^{2-}, mg l^{-1}	Recovery of CN$^-$, %, ±2%
0	100
0.10	109
1.00	112
10.0	104

tion of 50 µg l^{-1} bromide present in 1000 mg l^{-1} of chloride is shown in Fig. 11.29. The large negative dip following the chloride peak is caused by the reduction of the silver chloride deposited on the electrode. Since there is no longer chloride in the solution next to the electrode (as it has all eluted) and since the applied potential is below the diffusion controlled plateau, silver chloride reduction is favoured in order to satisfy the Nernst equation. A small dip following the bromide peak can also be seen.

The determination of cyanide is affected by the presence of sulphide in the sample as shown in Table 11.33. Since sulphide elutes before cyanide (Fig. 11.28), cyanide does not interfere with the analysis of sulphide. The addition of 0.1 mg l^{-1} sulphide enhances the cyanide peak by 9%. This effect can be minimized or eliminated by using the standard addition method or by matching the cyanide standards as closely as possible to the sample. For example, if the sample is known to contain approximately 100 µg l^{-1} sulphide, then this amount should be added to the standards.

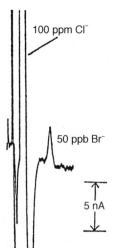

Fig. 11.29. Determination of 50 µg l^{-1} bromide in 1000 mg l^{-1} chloride from reagent grade NaCl. A 100-µl sample loop was used (From [11.47])

Table 11.33. Determination of 400 µg l^{-1} bromide in the presence of chloride. (From [11.47]; Copyright 1983, American Chemical Society)

Amt of added Cl$^-$, mg l^{-1}	Recovery of Br$^-$, %, ±2%
0	100
10	104
100	100
1000	115[a]

[a] The 1000 mg l^{-1} chloride solution contains 50 µg l^{-1} bromide, accounting for 12.5% of the 15% excess.

In general, plots of the log of peak height as a function of the log of concentration are linear at low concentration. At high concentrations an increase in concentration produces a smaller increase in current as shown by the plateau for each ion. This plateau (also observed in FIA studies) is caused by uncompensated resistance in the cell. The upper limit of linearity can be extended by increasing the applied potential or by decreasing the size of the injection loop. With a 100-µl injection loop, the detection limit for cyanide is 2 µg l^{-1}, for sulphide 30 µg l^{-1}, for iodide 10 µg l^{-1} and for bromide 10 µg l^{-1}.

11.4.2.1   Metal Cyanide Complexes

The cyanide ion in inorganic cyanides can be present as both complexed and free cyanide. In order to study the chromatography of metal cyanides, Rocklin and Johnson [11.47] prepared and assayed solutions of cadmium, zinc, copper, nickel, gold, iron and cobalt cyanides. Table 11.34 lists the percentage of total cyanide detected.

The results suggest that the complex cyanides can be grouped into three categories depending on the cumulative formation constant and stability of the complex.

Category 1 includes the weakly complexed and labile cyanides $Cd(CN)_4^{2-}$ ($\log_4 B_4 = 18.78$) and $Zn(CN)_4^{2-}$ ($\log B_4 = 16.7$). These complexes completely dissociate under the chromatographic conditions used, the cyanide being indistinguishable from free cyanide.

Category 2 includes the moderately strong cyanide complexes $Ni(CN)_4^{2-}$ ($\log B_4 = 31.3$) and $Cu(CN)_4^{3-}$ ($\log B_4 = 30.3$). Although the complexes are labile, they are retained on the column and slowly dissociate during the chromatography. This slow dissociation produces tailing which lasts for several minutes as the free cyanide elutes and is detected. As the results presented in Table 11.34 demonstrate, the tailing and the non-quantitative recovery of cyanide preclude the use of direct injection to determine total cyanide in samples containing copper and nickel. These samples may be analyzed after acid distillation and caustic trapping. The cyanide in the caustic solution can then be determined by ion chromatography with electrochemical detection.

Category 3 includes those cyanides which are inert and therefore totally undissociated, such as $Au(CN)_2^-$ (log $B_2 = 38.3$), $Fe(CN)_6^{3-}$ (log $B_6 = 42$) and $Co(CN)_6^{3-}$ (log $B_6 = 64$). No free cyanide was detected for these complexes. Although these complexes do not elute under the chromatographic conditions used, they can be eluted and determined by using different chromatographic conditions and conductivity detection.

Samples containing both free cyanide (or weakly complexed cyanide) and strongly complexed cyanide can be analyzed for free cyanide by direct injection. The determination of total cyanide (both free and strongly complexed) requires distillation of the sample with caustic trapping.

11.4.3 Sulphite and Dithionate

McCormick and Dixon [11.51] and Halcombe and Meserole [11.52], respectively, have described methods for the determination of sulphite in photographic effluents and waste groundwaters.

Petrie et al. [11.80] determined sulphite and dithionate in mineral leachates by ion chromatography.

11.4.4 Borate

The method described in Sect. 11.1.3 for the determination of borate in natural waters [11.16] has also been applied to industrial effluents.

Table 11.34. Percentage of total cyanide in metal complexes determined as "free" cyanide. (From [11.47]; Copyright 1983, American Chemical Society)

Metal complex	log β_r	%
$Cd(CN)_4^{2-}$	18.8	102
$Zn(CN)_4^{2-}$	16.7	102
$Ni(CN)_4^{2-}$	31.3	81
$Cu(CN)_4^{3-}$	30.3	52
$Cu(CN)_3^{2-}$	28.6	42
$Cu(CN)_2^-$	24.0	38
$Au(CN)_2^-$	38.3	0
$Fe(CN)_6^{3-}$	42	0
$Co(CN)_4^{3-}$	64[a]	0

11.5 Boiler Feed Water

11.5.1 Mixtures of Chloride and Sulphate

Roberts et al. [11.53] have described a single column ion chromatographic method for the determination of chloride and sulphate in steam condensate and boiler feed water. This was shown to be a valuable technique for analysing $\mu g\, l^{-1}$ levels of chloride and sulphate in very pure waters. The anions are concentrated on a short precolumn, separated on a low capacity ion exchange column and detected by an electrical conductivity detector. The apparatus is simple and no 'suppressor' column is needed. Adaptation to on-line analysis would be inexpensive and automation would require control of only the load/inject valve.

The preliminary work by Roberts et al. [11.53] was carried out using resin of very low exchange capacity (0.003 m equiv g^{-1}) and a 7.5×10^{-4} mol l^{-1} solution of benzoic acid as the eluent. Unusually good sensitivity can be obtained in single column anion chromatography using this eluent. To increase the sensitivity further, the size of the sample loop was increased to 500 µl from the usual 100 µl. This method adequately separated chloride and sulphate with detection limits of 10 mg l^{-1} chloride and 100 mg l^{-1} sulphate. However, a prolonged dip in the base line occurred sometime after the sulphate peak and made this procedure inefficient for repetitive analyses.

Further experimentation showed that extremely dilute water samples could be concentrated very simply and effectively using a small column containing an anion exchange resin. This column is positioned on a sampling valve in the same way as a sample loop, and the concentrated ions are eluted from the concentrator column to the separator column by placing the concentrator column in the eluent stream.

Chloride can be separated on an XAD-1 anion exchange column (citrate form) using potassium benzoate as an eluent. Sulphate can be separated on an XAD-1 column (nitrate form) using a nitrate eluent.

Equipment used in these analyses is inexpensive and it is feasible to have two complete systems; one for chloride and one for sulphate. Down time for the conversion of the column from one form to another can be avoided in this way. Alternately, two sets of columns could be used.

Chloride must be separated from the water dip before it can be analysed reliably in the same run with a sulphate. This was accomplished by carefully constructing a concentrator column and by careful choice of the column dimensions, resin capacity and eluent strength.

The width of the water dip was reduced by using a small concentrator and low dead-volume fittings. The small column contains only 85 mg of resin, and therefore a high capacity anion exchanger is needed to trap effectively the sample. In this work, a resin of 0.5 mequiv g^{-1} was used. (Fig. 11.30). The sample ions were eluted from the concentrator column with an eluent flow opposite to that of the sample flow. This back flushing mode is needed because the concentrator column may have a column capacity that is higher than the column capacity of the separator column. In this situation, an eluent could not elute sample ions through the added capacity in a reasonable amount of time. A back flush mode does not limit the capacity of the resin used in the concentrator column. It is more difficult to overload or use all the sites of a relatively high capacity concentrator column.

The effect of column and eluent parameters on anion retention times has been discussed previously by Gjerde et al. [11.54, 11.55]. If eluent flow rate and anion selectivity coefficients are taken to be constant then the effects of column resin weight w, resin capacity c, and eluent concentration (E) on the adjusted retention time of anions t', are shown by the following equation where y is the sample anion charge and x is the eluent anion charge:

$$\log t' = \log w + y/x \log c - y/x \log[E] - \text{constant}.$$

It can be seen that the log of column resin weight can be directly proportional to log adjusted retention time regardless of the anionic charges. Increasing the weight of resin in a column will shift both chloride and sulfate to later retention times. However, lowering the resin capacity and/or increasing the eluent concentration will shift the sulphate, faster relative to chloride, to shorter retention times. These parameters were adjusted until chloride eluted away from the water dip and sulphate still eluted in a reasonable amount of time. The determination of chloride and sulphate in one run greatly reduced the time needed to do the analysis. Sample concentrations for which peaks were at least three times the background signal were 3–5 $\mu g l^{-1}$ chloride and 1–2 $\mu g l^{-1}$ sulphate with this procedure.

A chromatogram of the determination of chloride and sulphate in a single run is shown in Fig. 11.31. Data from 20-ml injections of 5–100 $\mu g l^{-1}$ chloride and sulphate in "pure" water were used to plot standard addition calibration graphs. A straight line of the form y = 0.401x + 1.556 and a correlation coefficient of 0.997 were obtained for chloride and y = 0.516x + 3.70 and a correlation coefficient of 0.995 were obtained for sulphate. Extrapolation of these plots showed a 4.0 $\mu g l^{-1}$ chloride concentration and 6.5 $\mu g l^{-1}$ sulphate concentration for this sample of "pure" water. Detection limits under these conditions were 3–5 $\mu g l^{-1}$.

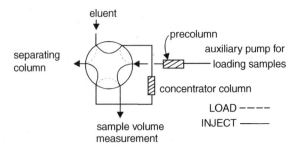

Fig. 11.30. Sampling valve with a concentrator column. Sample loading positions are for a concentrator column containing a low capacity ion exchange resin (From [11.53])

Plots of peak height vs sample volume were not linear for chloride at volumes greater than 20 ml. The reason for this is unknown but perhaps chloride is not taken up quantitatively by the concentrator column under these conditions. The peak heights of chloride from standard 30 $\mu g\,l^{-1}$ chloride solutions were the same with either 30 or 100 $\mu g\,l^{-1}$ sulphate present, and it was shown that calibration curves are linear. The same loading volume should be used for all standard and sample solutions.

11.5.2 Miscellaneous Anions

Other methods for the determination of anions in boiler feed waters are summarized in Table 11.35.

11.6 Plant and Soil Extracts

11.6.1 Mixtures of Chloride, Nitrate, Sulphate and Phosphate

Bradfield and Cooke [11.59] described an ion chromatographic method using a UV detector for the determination of chloride, nitrate, sulphate and phosphate in water extracts of plants and soils. Plant materials are heated for 30 min at 70 °C with water to extract anions. Soils are leached with water and Dowex 50–X4 resin added to the aqueous extract which is then passed through a Sep-Pak C_{18} cartridge and the eluate then passed through the ion chromatographic column. The best separation of these ions was obtained using a 5×10^{-4} mol l^{-1} potassium hydrogen phthalate solution in 2% methanol at pH 4.9. A reverse phase system was employed. Retention times were 5.5, 7.9, 12.6

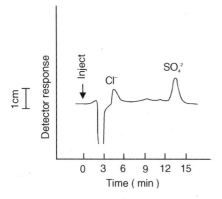

Fig. 11.31. Standard containing 19 $\mu g\,l^{-1}$ chloride and 22 $\mu g\,l^{-1}$ sulfate concentrated from 20.0 ml; eluent, 2.0×10^{-4} potassium phthalate, pH 6.2; concentrator column resin, XAD-4 0.5 mequiv g^{-1}; separator column resin, XAD-1 0.013 mequiv g^{-1} particle size 44-57 um (From [11.53])

Table 11.35. Determination of anions in boiler feed waters

Anions	Eluent	Separation column	Type of sample	Detector	References
SO_3, PO_4 (1 mg l^{-1})	–	–	Bioler blow down water	–	11.56
Cl, NO_2, SO_4 (<1 µg l^{-1})	–	–	High purity water	Electrochemical	11.57
CO_3 (<0.02 mg l^{-1})	Phthalic acid at pH 6.5–7.2	Zipai SAX	High Purity water	Indirect ultra-violet at 321 nm	11.58

and 18 min for chloride, nitrate, phosphate and sulphate respectively. Recoveries ranged from 84 to 108% with a mean of 97%.

Figures 11.32–11.34 show chromatograms obtained in the analysis of plant extracts, soil extracts and fertilizer extracts [11.60].

11.7 Foodstuffs

Capillary electrophoresis combined with suppressed detection of ultraviolet sensitive species has been applied [11.82] to the determination of mixtures of anions and cations in grape juice. Detection limits were of the order of 1–10 µg L^{-1}. The suppression was made from a short ion exchange membrane placed between the separation capillary and the UV detector cell. Retention times were as follows:

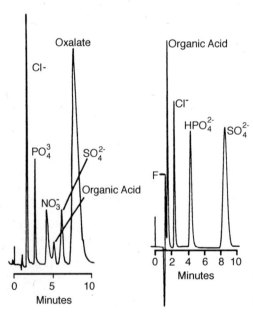

Fig. 11.32 a,b. Determination of anions in plant extracts: a fresh spinach sample; b fresh anion sample. Aqueous extracts of macerated samples

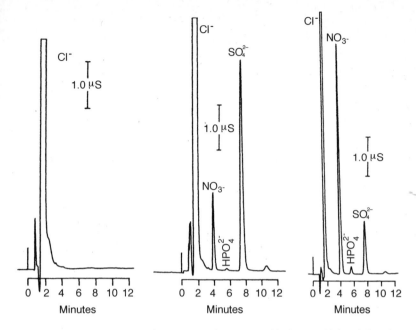

Fig. 11.33a–c. Determination of anions in soil extracts: **a** blank 10 mol l^{-1} KCl; **b** soil sample A 10 mmol l^{-1} KCl extract; **c** soil sample B 10 mmol l^{-1} KCl extract 1:500 dil. AMPIC-NG1 should also be used in series to remove humic acids

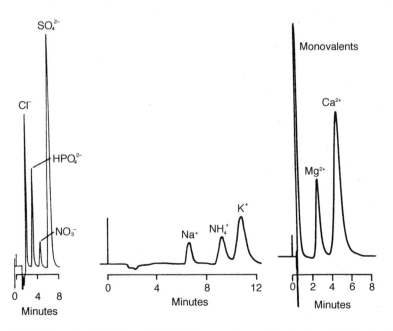

Fig. 11.34a–c. Analysis of granulated fertilizer samples: **a** determination of anions; **b** determination of monovalent cations; **c** determination of divalent cations

HCO_3 (min), ClO_2, F, PO_4, ClO_3, ClO_4, NO_3, NO_2, SO_4, Cl, I, Br, CrO_4 (20 min). A sample of grape juice was found to contain, acetate, succinate, malate, tartrate, citrate and isocitrate.

11.8 Column Coupling Isotachophoresis of Natural Waters

11.8.1 Mixtures of Chloride, Fluoride, Nitrite, Nitrate, Sulphate and Phosphate

This technique offers very similar advantages to ion chromatography in the determination of anions in water, namely multipleion analysis, little or no sample pretreatment, speed, sensitivity and automation.

Very similar advantages are offered by capillary isotachophoresis (ITP) [11.61, 11.62]. However, with the exception of the work by Bocek et al. [11.63] in which the determinations of chloride and sulphate in mineral water were demonstrated, little attention has been paid to this subject.

Zelenski et al. [11.64] applied the technique to the determination of $0.02-0.1$ mg l^{-1} quantities of chloride, nitrate, sulphate, nitrite, fluoride and phosphate in river water. Approximately 25 min was required for a full analysis. These workers used a column coupling technique which, by dividing the analysis into two stages, enables a high load capacity and a low detection limit to be achieved simultaneously without an appreciable increase in the analysis time.

The separation unit of the home made ITP instrument used is shown in Fig. 11.35. A 0.85 mm i.d. capillary tube made of fluorinated ethylene propylene copolymer was used in the pre-separation (first) stage and a capillary tube of 0.30 mm i.d. made of the same material served for the separation in the second stage. Both tubes were provided with conductivity detection cells [11.65] and an a.c. conductivity mode of detection [11.61] was used for making the separations visible. The driving current was supplied by a unit enabling independent currents to be pre-selected for the preseparation and final analytical stages. The run of the analyser was controlled by a programmable timing and control unit. The zone lengths from the conductivity detector, evaluated electronically, were printed on a line printer.

The use of a column coupling configuration of the separation unit provides the possibility of applying a sequence of two leading electrolytes in one analytical run. Therefore, the choice of optimum separation conditions can be advantageously divided into two steps: (1) the choice of a leading electrolyte suitable for the separation and quantitation of the macro-constituents in the first stage (pre-separation column) simultaneously having a retarding effect on the effective mobilities of micro constituents (nitrite, fluoride, phosphate) and (2) the choice of the leading electrolyte for the second stage in which only micro constituents are separated and quantified (macro constituents were removed from the analytical system after their evaluation in the first stage).

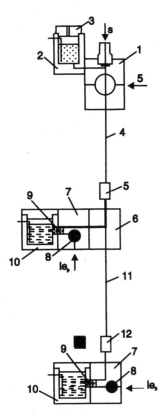

Fig. 11.35. Separation unit in a column coupling configuration as used for the analysis of anions in river water: 1, sampling block with a 30 μl sampling valve; 2, terminating electrolyte compartment with a cap (3); 4, 0.85 mm i.d. capillary tube (pre-separation column); 5 and 12, conductivity detectors; 6, bifurcation block; 7, refilling block with needle valve (8); 9, mechanically supported membranes; 10, leading electrolyte compartments; 11, 0.30 mm i.d. capillary tube; s, positions for the sample introduction (valve or microsyringe); Iep and Ie, positions for refilling of the columns used for the first and second stages respectively (From [11.64])

To satisfy the requirements for the properties of the leading electrolyte applied in the first stage and, consequently, to decide its composition, two facts had to be taken into account, i.e. the pH value of the leading electrolyte needs to be around 4 or less (retardation of nitrite relative to the macro constituents in this stage) and at the same time the separations of the macro constituents need to be optimized by other means than adjusting the pH of the leading electrolyte (anions of strong acids). For the latter reason, complex equilibria and differentiation of anions through the charge number of the counter ions were tested at a low pH as a means of optimizing the separation conditions for chloride, nitrate and sulphate in the presence of nitrite, fluoride and phosphate.

The retardation of chloride and sulphate through complex formation with cadmium enabling the separation of anions of interest [11.62] to be carried out, was found to be unsuitable, as the high concentration of cadmium ions necessary to achieve the desired effect led to the loss of fluoride and phosphate (probably owing to precipitation).

Similarly, the use of calcium and magnesium as complexing co-counter ions [11.66] to decrease the effective mobility of sulphate was found to be ineffective as a very strong retardation of fluoride occurred.

Better results were achieved when a divalent organic cation was used as a co-counter ion in the leading electrolyte [11.67–11.69] employed in the first separation stage when, simultaneously, the pH of the leading electrolyte was 4 or less, and the steady state configuration of the constituents to be separated was chloride, nitrate, sulphate, nitrite, fluoride and phosphate. The detailed composition of the operational system of this type used for quantitative analysis is given in Table 11.36 (system No.1.)

The choice of the leading electrolyte for the second stage, in which the microconstituents were finally separated and quantitatively evaluated, was straightforward: a low concentration of the leading constituent (low detection limit) and a low pH of the leading electrolyte (separation according to pK values). The operational system used throughout this work in the second stage is given in Table 11.36.

An isotachopherogram from the analysis of a model mixture of anions obtained in the first separation stage is shown in Fig. 11.36. This isotachopherogram merely indicates the differences in the effective mobilities of the anions of interest, as the concentration of nitrite, fluoride and phosphate in river water are usually too low to be detected in the first stage.

Both the cross sectional area of the capillary tube and the concentration of the leading constituent used in the second stage were optimized with respect to the determinations of micro constituents. In a search for optimal separating conditions, operational system No.2 (Table 11.36) was used for the determinations of microconstituents.

Table 11.36. Operational systems. (From [11.64]; Copyright 1984, Elsevier Science Publishers BV, Netherlands)

System No.	Parameter	Leading electrolytes		Terminating electrolyte
		1st stage	2nd stage	
1	Anion	Cl^-	Cl^-	$CITR^b$
	Concentration (mmol l^{-1})	8	1	2
	Counter ion	$BALA^a$	$BALA^a$	H^+
	Co-counter ion	BTP^c	–	–
	Concentration (mmol l^{-1})	3	–	–
	Additive to the leading electrolyte	0.1 % HEC^d	0.1% HEC^d	–
	pH	3.55	3.55	ca. 3
2	Anion	Cl^-	Cl^-	$CITR^b$
	Concentration (mmol l^{-1})	1	1	1
	Counter ion	H^+	H^+	H^+
	Additive to the leading electrolyte	0.1% HEC^d	0.1% HEC^d	–
	pH	3.0	3.0	ca. 3

[a] β-Alanine.
[b] Citric acid.
[c] Abbreviations: BALA = β-Alanine; CITR = citric acid; BTP = aminopropane 1, 3-bis-tris(hydroxymethyl)methyl-aminopropane,
[d] HEC = Hydroxyethylcellulose.

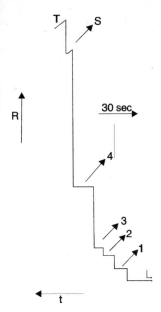

Fig. 11.36. Isotachopherogram for the separation of anions in the first stage using operational system No. 1: 1, nitrate (40 mg l^{-1}); 2, sulphate (20 mg l^{-1}); 3, nitrite (20 mg l^{-1}); 4, fluoride (20 mg l^{-1}); 5, phosphate (20 mg l^{-1}). Driving current 250 μA. L and T = leading and terminating anions, respectively; R = increasing resistence, t = increasing time. The sample was introduced with the aid of a 30 μl valve (From [11.64])

As only anionic constituents of relatively strong acids can achieve the effective mobilities within the chloride phosphate mobility interval at low pH, the number of possible interfering constituents (e.g. organic anions associated with biological processes in water) is also reduced using the proposed operational systems.

Some typical results obtained by applying the isotachophoretic technique to river water samples are shown in Tables 11.37 and 11.38.

These results serve as an illustration of the degree of agreement of the quantitative analyses obtained by this technique and classical methods. In several instances the agreement of the results is good. However, large differences found for some sample constituents (especially macro constituents) must be ascribed to a lower accuracy of the classical methods, as is clear from the analyses of model mixtures summarized in Table 11.39.

11.8.2 Miscellaneous Anions

Keundler and Reich [11.70] have characterized electrolyte systems for the isotachophoresis of anions by cluster analysis. The leading electrolyte system with a pH of 3 and one with a pH between 6 and 10 were found, based on heirarchical and a non-heirarchical clustering procedure, to be the most favourable binary combination for anion identification purposes. Ternary combinations were also selected. Other systems studied do not increase the identification power.

Table 11.37. Determinations of inorganic anionic macro-constituents present in river water. (From [11.64]; Copyright 1984, Elsevier Science Publishers BV, Netherlands)

Sample No.	Sulphate (mg l^{-1})		Nitrate (mg l^{-1})		Chloride (mg l^{-1})	
	ITP	Titrimetry[a]	ITP	Spectrophotometry[b]	ITP	Titrimetry[c]
1	36.6	41.6	13.2	11.7	18.9	31.2
2	40.8	34.4	13.6	12.3	23.0	33.2
3	38.9	42.5	13.3	11.3	20.6	33.2
4	57.6	56.9	13.3	9.0	26.3	39.2
5	50.9	54.4	12.4	10.3	22.2	35.5
6	43.7	55.2	13.3	9.4	22.8	34.6

[a] The procedure described in [11.68].
[b] For details, see [11.69].
[c] Mercurimetric determination (11.69).
The relative standard deviations for ITP determinations were better than 2% whereas for the classical methods they varied within the range 2–15%.

Table 11.38. Determination of inorganic anionic micro-constituents present in river water. (From [11.64]; Copyright 1984, Elsevier Science Publishers BV, Netherlands)

Sample No.[a]	Nitrite (mg l^{-1})		Fluoride by ITP (mg/l)[c] mg l^{-1}	Phosphate (mg l^{-1})	
	ITP	Spectrophotometry[b]		ITP	Spectrotometry[d]
1	0.25	0.13	0.17	0.61	0.55
2	0.29	0.13	0.18	0.56	0.51
3	0.33	0.11	0.18	0.49	0.44
4	0.33	0.16	0.14	0.23	0.16
5	0.33	0.16	0.13	0.27	0.22
6	0.22	0.13	0.10	0.34	0.27

[a] Samples as in Table [11.37].
[b] The procedure described in [11.68].
[c] Not determined by a classical method as only a very labour-consuming distillation method [11.68] was available.
[d] For details, see [11.69].
The relative standard deviations for both ITP and classical were typically 5% or less.

Bidling Meyer et al. [11.71] have studied the ion-pairing chromatographic determination of anions using an ultraviolet absorbing co-ion in the mobile phase.

Bidling Meyer et al. [11.72] have also discussed the ion-pair chromatographic determination of anions using an ultraviolet absorbing co-ion in the molule phase. The separation and spectrophotometric determination of otherwise non-ultraviolet absorbing anions is possible. Spectrophotometric detection was comparable to conductiometric detection at nanogram levels for the eight anions investigated.

It has been reported [11.73] that post suppression membrane-based ion exchange chromatography with fluorescence detection permits detection limits superior to those obtained by conductivity detection in hydroxide eluent suppressed anion chromatography.

Table 11.39. Determinations of anionic macro-constituents present in model mixtures. (From [11.64]; Copyright 1984, Elsevier Science Publishers BV, Netherlands)

Sample	Sulphate mg l^{-1}			Nitrate (mg l^{-1})			Chloride (mg l^{-1})		
	Taken	ITP	Titrimetry	Taken	ITP	Spectrophotmetry	Taken	ITP	Titrimetry
A	50.0	49.7	32.7	15.0	14.4	14.3	25.0	25.4	29.9
B	100.0	100.5	103.7	30.0	30.2	24.2	50.0	49.8	57.5
C	150.0	149.6	–	45.0	a	44.0	75.0	a	83.7

[a] Mixed zone of nitrate and chloride for the column used. For further details see notes in Table 11.38 and the text.

11.9 References

11.1 Small H, Stevens TS, Bauman WC (1975) Anal Chem 47: 1801
11.2 Yamamoto M, Yamamoto H, Yamamoto Y (1984) Anal Chem 56: 832
11.3 Sawicki E, Mullik JD, Wittgenstein E (1978) In: Ion chromatographic analysis of environmental pollutants. Ann Arbor Science, Ann Arbor, Michigan
11.4 Lash RP, Hill CJ (1980) Anal Chem Acta 144: 405
11.5 Pohlandt C (1980) South African J Chem 33: 87
11.6 Cassidy M, Elchuk S (1982) Anal Chem 54: 1558
11.7 Jones VK, Carter G (1985) American Laboratory 17: 48
11.8 Smee BW, Hall GEM, Koop DJ (1978) Journal of Geochemical Exploration 10: 245
11.9 Van Os MJ, Slanina J, De Ligny CL, Hammers WE, Agterdenbos J (1982) Anal Chim Acta 144: 73
11.10 Slanina J, Lingerak WA, Ondelman JE, Borst P, Bakker FD, Mulik JD, Sawicki E (editors) (1979) In: Ion Chromatograpic Analysis and Environmental Pollutants Volume Z. Ann Arbor Science Publishers, Michigan
11.11 Asshauer J, Halazz I (1974) Journal of Chromatographic Science 12: 139
11.12 Marko Varga G, Csiky I, Jonsson JA (1984) Anal Chem 56: 2066
11.13 Cochrane RA, Miller DE (1982) Journal of Chromatography 241: 392
11.14 Williams RJ (1985) Anal Chem 55: 851
11.15 Hoover TB, Jager GD (1984) Analytical Chemistry 56: 221
11.16 Hill CJ, Lash RP (1980) Analytical Chemistry 52: 24
11.17 Small H, Stevens TS, Bauman WC (1975) Analytical Chemistry 47: 1801
11.18 Carlson R, Paul J (1969) J Soil Science 108: 226
11.19 Hansen LD, Richter BE, Rollins DK, Lamb JD, Eatough J (1979) Analytical Chemistry 51: 633
11.20 Peschet JL, Tinet C (1986) Techniques Sciences Methods 81: 351
11.21 Sunden T, Lingron M, Cedergren A (1983) Analytical Chemistry 55: 2
11.22 Marko Varga G, Csiky I, Jonsson JA (1984) Analytical Chemistry 56: 2066
11.23 Dits JS, Smeenk GMM (1985) H20 18: 7
11.24 Wilken RD, Kock HH (1985) Fresenius Zeitschrift fur Analytische Chemie 320: 477
11.25 Brandt G, Kettrip A (1985) Fresenius Zeitschrift fur Analytische Chemie 320: 485
11.26 Tyree SY, Bynum MAO (1984) Limnology and Oceanography 29: 1337
11.27 Okada T, Kuwamoto T (1985) Analytical Chemistry 57: 258
11.28 Hordijk CA, Cappenberg TE (1985) Journal of Microbiological Methods 3: 205
11.29 Sawicki E, Mulik JD, Wittgenstein E (1978) In: Ion Chromatographic Analysis of Environmental Pollutants. Ann Arber Science Publishers Inc Ann Arbor Michigan
11.30 Iskandarani Z, Pietrzyk DJ (1982) Anal Chem 54: 2601
11.31 Yamamoto M, Yamamoto H, Yamamoto Y, Matzushita S, Baba N, Ikeshigi T (1984) Analytical Chemistry 56: 822
11.32 Matsushita S, Tada Y, Baba D, Hosako J (1983) Chromatography 259: 459
11.33 Buchholz AE, Verplough CJ, Smith JL (1982) Journal of Oceanographic Science 20: 499
11.34 Wetzel RA, Anderson CL, Schleicher H, Crook DG (1979) Analytical Chemistry 51: 1532
11.35 Pyen GS, Erdmann DE (1983) Analytica Chimica Acta 149: 355
11.36 Oikawa K, Saitoh H (1982) Chemosphere 11: 933
11.37 Schwabe R, Darimont T, Mohlman T, Pabel E, Sonneborn M (1983) International Journal of Environmental Analytical Chemistry 14: 169
11.38 Wagner F, Valenta P, Nurnberg W (1985) Fresenius Zeitschrift fur Analytische Chemie 320: 470
11.39 Rowland AP (1986) Analytical Proceedings 23: 308
11.40 Jones VK, Tarter JG (1985) International Laboratory 36: September
11.41 Jones VK, Tarter JG (1984) Journal of Chromatography 312: 356
11.42 Morrow CM, Minear RA (1984) Water Research 18: 1165

11.43 Skougstad MW, Fishman MJ, Friedman LC, Erdmann DE, Duncan SS (1978) Techniques of Water Resources investigations in the US.G.S., Book 5, Chapter A1
11.44 Bond AM, Heritage ID, Wallace AG, McCormick MJ (1982) Analytical Chemistry 54: 582
11.45 Mosko J (1984) Analytical Chemistry 56: 629
11.46 Green LW, Woods JR (1981) Analytical Chemistry 53: 2187
11.47 Rocklin RD, Johnson EL (1983) Analytical Chemistry 55: 4
11.48 Pihla B, Kosta L (1980) Analytica Chimica Acta 114: 275
11.49 Pihla B, Kosta L, Hrvstoviski B (1979) Talanta 26: 805
11.50 Lown JA, Koile R, Johnson DC (1980) Analytica Chimica 116: 33
11.51 McCormick MJ, Dixon LM (1985) Journal of Chromatography 322: 478
11.52 Holcombe OJ, Meserole FB (1981) Water Quality Bulletin 6: 37
11.53 Roberts KM, Gjerde DT, Fritz JS (1981) Analytical Chemistry 53: 1691
11.54 Gjerde DT, Schmuckler G, Fritz SS (1980) Journal of Chromatography 35: 187
11.55 Gjerde DT (1980) Ph.D. Dissertation, Iowa State University
11.56 Stevens TS, Turkelson VT (1977) Analytical Chemistry 49: 1176
11.57 Tretter H, Paul G, Blum I, Schreck H (1985) Fresenius Zeitschrift fur Analytische Chemie 321: 650
11.58 Brandt G, Matuschek G, Kettrup A (1985) Fresenius Zeitschrift fur Analytische Chemie 321: 653
11.59 Bradfield G, Cooke DT (1985) Analyst (London) 110: 1409
11.60 Dionex Corporation, Albary Park, Camberley, Surrey G 4 15 2PL, U.K.
11.61 Everaerts DM, Beckers JL, Verheggen Th PEM (1976) In: Isotachophoresis–Theory Instrumentation and applications Elsevier Amsterdam Oxford New York
11.62 Hjalmarsson SG, Baldesten A (1981) CRC Crit Rev Anal Chem 11: 261
11.63 Bocek P, Miedziak I, Demi M, Janak J (1987) J Chromatography 137: 83
11.64 Zelenski I, Zelenska V, Kanmiansky D, Havasi P, Lednarova V (1984) Journal of Chromatography 294: 317
11.65 Kanianski D, Koval M, Stankoviansky S (1983) Journal of Chromatography 267: 67
11.66 Kaniansky D, Evereerts FM (1978) Journal of Chromatography 148: 441
11.67 Kaniansky D, Madajova V, Zelensky I, Stankoviansky S (1980) Journal of Chromatography 194: 11
11.68 The Testing of water (1980) E Merck Darmstdt Germany
11.69 Williams MJ (1979) In: Handbook of Anion Determination Butterworth London
11.70 Keundler E, Reich G (1988) Analytical Chemistry 60: 120
11.71 Bidling Meyer BA, Sanastasia CT, Warren FV (1988) Analytical Chemistry 60: 192
11.72 Bidling Meyer BA, Sanastasia CT, Warren FV (1987) Analytical Chemistry 59: 1843
11.73 Shintani H, Dasgupta PK (1987) Analytical Chemistry 59: 1963
11.74 Cheam V, Chau ASY (1987) Analyst (London) 112: 993.
11.75 Maki SA, Danielson ND (1991) Analytical Chemistry 63: 699
11.76 Rigas PG, Pietrzyk DJ (1988) Analytical Chemistry 60: 1650
11.77 Berglund I, Dasgupta PT, Lopez JL, Nara O (1993) Analytical Chemistry 65: 1192
11.78 Saari Nordhaus R, Anderson JM (1992) Analytical Chemistry 64: 2283
11.79 Marengo E, Gennaro MC, Abrigo C (1992) Analytical Chemistry 64: 1885
11.80 Petrie LM, Jakely ME, Brandvig RL, Kroening JG (1993) Analytical Chemistry 65: 952
11.81 Berglund I, Dasgupta PK, Lopez JL, Nara O (1993) Analytical Chemistry 65: 1192
11.82 Avdalovic N, Pohl CA, Rocklin RD, Stillman JR (1993) Analytical Chemistry 65: 1470.

12 On-Site Measurement of Anions

This chapter discusses available portable instrumentation that can be carried around by inspectors to carry out measurements for anions on rivers, reservoirs, effluents, etc.

Merck supply a range of test kits covering many different determinands (Table 12.1). Test strips and test kits, both visual and photometric are available and all are suitable for use in the field.

Merckoquant Test Strips. These consist of a reactive test zone firmly bonded to a plastic backing. The test zone is impregnated with reagents, buffers and other substances. The strips are used for rapid exploratory testing of substance concentrations as low as 1 mg l^{-1}. To test a sample the test zone is immersed in the water sample for 1–2 s and matched against a colour scale.

Aquamerck Test Kits. These consist of titrimetric and colorimetric test kits supplied in boxes or blister packs. The simplest titration tests contain a dropping bottle or precision dropping pipette. The number of drops of reagent required to change the colour of the indicator is a measure of the concentration of the substance being tested for. Aquamerck colorimetric tests incorporate a waterproof scale with precise directions, or alternatively a testing vessel consisting of a 10-mm cell with coloured reference blocks to the right and left of it so that a colour comparison can be made.

Aquaquant Test Kits. The aquaquant system for rapid water analysis fully meets the stringent requirements relating to detection sensitivities, ease of use and economy. Each Aquaquant test kit consists of a plastic case containing a sliding colour comparator, which is of a moulded construction designed to hold test solutions, reagents and accessories as a test is being performed. Short-tube and long-tube test kits are available. The different path lengths allow for lower and higher sensitivities.

Microquant Test Kits. There is often a need for a rapid colorimetric test in a medium sensitivity range roughly equivalent to that covered by the Aquamerck and Aquaquant short tube tests, but which will also permit coloured or turbid

Table 12.1. Test kits supplied by Merck

Determination	Measuring range or graduation mg l^{-1}	Merckoquant test strips	Aquamerck titrimetric test with dropping bottle or titrating pipette	Aquamerck colorimetric test in a blister pack	Aquamerck colorimetric test with colour scale	Aquamerck colorimetric test with testing vessel	Exchange pack	Aquaquant test with sliding colour comparator	Microquant test with comparator disc	Spectroquant test for rapid photometric determinations
Carbonate	1–100		8048				8041			
Chloride	25–2500/25		11132							
	2–800/2		11106							
	5–300									
	3–300							14401		
	0.4–40								14753	14755
Chromate	3–100	10012								
	0.005–0.1							14402		
	0.1–1.6							14441		
	0.1–10								14756	14758
	0.025–2.5									
Nitrite	0.1–10								14774	14776
	0.03–3									
Phosphate (vanadate-molybdate method)	2.2–29(P_2O_5) 5.67(Na_3PO_4)					8016				
	1–40						11125			
	1.5–100						11125	14449	14840	14842
	0.25–25									
(ammonium hepta-molybdate method)	0.1–2.0									
	1–10(P_2O_5)					11138	8046			
							8046			
	0.01–0.16							14409		
	0.1–3.0								14786	14788
	0.024–2.4									

Table 12.1 (Contd.)

Determination	Measuring range or graduation mg l^{-1}	Merckoquant test strips	Aquamerck titrimetric test with dropping bottle or titrating pipette	Aquamerck colorimetric test in a blister pack	Aquamerck colorimetric test with colour scale	Aquamerck colorimetric test with testing vessel	Exchange pack	Aquaquant test with sliding colour comparator	Microquant test with comparator disc	Spectroquant test for rapid photometric determinations
Phosphate silicate	1–10(P_2O_5) 0.3–3 (SiO_2)					11119	11123			
Sulphate	200–1600	10019								
	25–300									
	25–300							14411		
	10–600								14789	14791
Sulphide (lead acetate paper)	10–500	9511								
Sulphite	2.5–200/2.5	10013	11148							

samples to be measured. The Microquant tests, which work with transmitted light, are eminently suitable for this type of analysis.

Spectroquant Analysis System. This comprises an SQ 115 digital photometer, a TR 205 Thermoreactor for elevated temperature test, 100 and 148 °C, and the Spectroquant test kits. The SQ 115 photometer measures absorbance and concentration in the range 370–1000 nm. The photometer is conveniently calibrated with calibration cells, which incorporate transparent coloured windows to simulate standard solutions of precisely defined concentrations. The inconvenience of calibrating against standard solutions and the unreliability of using calibration factors are thus avoided.

Merck also supply compact laboratories consisting of packs of reagent bottles and accessories. These include the determination of carbonate, nitrite and nitrate.

Palintest Rapid Test Kits. These are available in a visual test form (tablet count, colour match and turbidity tests) or as a spectrophotometric version using the Palintest Photometer 5000 (Table 12.2)

Palintest also supply swimming pool and spa test kits for the determination of alkalinity and calcium hardness (0–1000 mg l^{-1}), chloride (0–5000 mg l^{-1}), and sulphate (0–200 mg l^{-1}). Comparator 2000 and photometer 5000 versions are available.

Table 12.2. Test kits available from Palintest

Order code	Test	Range	Test method
PS 440	Nitrate	0.2–1.0 mg l^{-1} N	CM
		0.88–4.4 mg l^{-1} NO_3	CM
PS 445	Nitrite	0.05–0.5 mg l^{-1} N	CM
		0.15–1.6 mg l^{-1} NO_2	CM
PS 450	Phosphate	0–80 mg l^{-1} PO_4	CM
CS 112	Total alkalinity	0–500 mg l^{-1}	TC
CS113	Chloride	0–1000 mg l^{-1}	TC
CS120	Nitrite	0–1500 mg l^{-1}	TC
CS 131	Phosphate limit	30$70 mg l^{-1}	CM
CS 132	Sulphate	0–3000 mg l^{-1}	TU
CS 123	Sulphite LR	0–50 mg l^{-1}	TC
CS 124	Sulphite HR	0–500 mg l^{-1}	TC
PM 179	Fluoride	0–1.5 mg l^{-1}	
PM 175	Molybdate HR	0–100 mg l^{-1} MoO_4	
PM 163	Nitrate	0–1.0 mg l^{-1} N	
PM 109	Nitrite	0–0.5 mg l^{-1} N	
PM 177	Phosphate LR	0–4.0 mg l^{-1}	
PM 114	Phosphate	0–100 mg l^{-1}	
PM 154	Sulphate	0–200 mg l^{-1}	
PM 168	Sulphide	0–0.5 mg l^{-1}	

LR = low range, HR = high range

Table 12.3. De Lange test kits

	Measuring range mg l^{-1}	Method	LASA Aqua[a]	LPIW[a]	CADAS[b] 100	Reaction temperature
LCK 311 chloride	1–70	Iron-III-thiocyanate	x	x	x	Ambient
LCK 313 chromium total and hexavalent	0.05–1.0 0.01–0.25	Diphenylcarbazide in analogy to DEVE 10	x	x x	x x	Ambient Ambient and 100°C
LCK 323 fluoride	0.1–1.5	Spadns		x	x	Ambient
LCK 315 cyanide	0.01–0.5	Barbituric pyridine in analogy to DIN 38 405 D 13	x	x	x	Ambient
LCK 339 nitrate	3–80	2, 6-Dimethyl phenol in analogy to DIN 38 405 D9	x	x	x	Ambient
LCK 341 nitrite	0.1–3.0 0.01–0.1	Sulphanilic acid/naphtylamine in analogy to DIN 38 405 D 10	x	x x	x x	Ambient
LCK 049 o-phosphate	5–100	Vanadate molbydate (VMR)	x	x	x	Ambient
LCK 349 total and o-phosphate	0.1–5.0 0.01–0.5	Phosphomolybdic blue in analogy to DIN 38 405 D 10	x	x x	x x	Ambient and 100 °C
LCK 153 sulphate	10–150	Barium sulphate	x	x	x	Ambient
pipette tests						
LCW 053 sulphide	0.05–1.0	Dimethyl-p-phenylenediamine in analogy DEV 07	–	x	x	Ambient
LCW 054 sulphite	0.1–5.0	Iodide/iodate	–	x	x	Ambient

De Lange Cuvette and Pipette Tests. These are all based on the use of portable spectrophotometers, the LASA Aqua filter photometer for small numbers of water analyses, the LPIW filter photometer for water and sewage analysis and the top-of-the-range Cadas 100 with print-out of results for water and sewage analysis (Table 12.3). The latter instrument covers the spectral range 200–900 nm amongst many others.

13 On-Line Measurement of Anions

On-line process analysers are used throughout the water industry in applications such as the monitoring of rainwater, water in distribution systems, potable water and effluent and sewage treatment processes. A wide range of instrumentation is available from various suppliers. Some instruments determine single parameters and some are multi-parameter instruments. It is convenient to discuss these under separate headings.

Industrial scale on-line instrumentation supplied by Kent Instrumentation Ltd. for the determination of anions (chloride, nitrate, phosphate, ammonia, fluoride and silica) in various types of water are reviewed in Table 13.1. All of these instruments are microcomputer-based and, with the exception of silica, have autocalibration built-in. Generally the sensitivities are adequate for the requirements of the water industry.

In addition Iskra produce an industrial-scale cyanide analyser (model MA 540). This is a sensitive instrument based on the flow electrode measuring cyanide at concentrations down to $1\ \mu g\, l^{-1}$ and, as such, might be of interest in the water industry for cyanide monitoring of effluents or reservoir water. PHOX also supply a digital ammonia monitoring module with auto-clearing.

Some multi-parameter instruments are reviewed in Table 13.2. The Kent System 19 has been specifically designed for use in the water industry. In these systems continuous single-point and multi-stream analytical instruments gather data for data logging or transmission by telemetry to a control station. Control features of this component are the model 7975 multi-parameter water quality monitor which can analyse up to six streams simultaneously and the 8080 series microprocessor. In abstraction from rivers, reservoirs or boreholes, the instrumentation and systems allow raw water to be monitored continuously before and during pumping to the water treatment plant.

Additionally, mobile or permanent water-quality monitoring stations can be supplied to check environmental conditions and either log the data locally or transmit it by telemetry to a central station. The data thus gathered can be used for control intervention and the generation of reports on, for example, the effects of long-term abstraction.

Both before and during raw water treatment, the instruments and systems play an important role in helping to ensure the quality of the drinking water supply.

Table 13.1. Single-parameter microprocessor-controlled on-line industrial anion and cation analysers from Kent Instrumentation

Parameter	Samples type	Model No.	Ranges Concentration $\mu g\,l^{-1}$	Flow	Temp. (°C)	Pressure (lb in^{-2})	Accuracy	Description	Standardization
Anions									
Chloride	Boiler water water treatment plants	Cabinet version 8024 Panel version 8034	0–100 0–200 0–1000 0–2000 0–5000	2.5–750 ml min^{-1}	5–45	2–20	$\pm 5\,\mu g\,l^{-1}$ or $\pm 5\%$	Solid-state chloride ion-selective electrode and sulphate reference electrode for potentiometric determination of chloride	Auto
Nitrate	Potable water Surface waters	8026	0–1000 0–500 500–1000	0.5 l h^{-1}	5–45	2–20	$\pm 5\%$	Liquid membrane nitrate ion selective electrode	Auto
Phosphate	Water quality monitor reservoirs	8086 Wall mounting 8063 Panel mounting 8064	0–5000 mg l^{-1} 0–20 mg l^{-1} 0–10 mg l^{-1}	0.5 l h^{-1} 0.5 l h^{-1} 0.5 l h^{-1}	5–55 5–50 5–50	2–20 2–20 2–20	$\pm 5\%$ $\pm 5\%$ $\pm 5\%$	Liquid membrane nitrate ion selective electrode. Spectrometric method. Up to 40 $\mu g\,l^{-1}$ silica results in positive error of $\pm 2\,\mu g\,l^{-1}$ phosphate error increasing at higher levels of silica	Auto Auto Auto
Fluoride	Potable water	8081	0.1–1000 mg l^{-1}	0.5 l h^{-1}	5–55	2–20	$\pm 5\%$	Fluoride-lanthanum chloride ion-selective electrode with calomel reference electrode	Auto
Silica	Power generation industry	Cabinet mounting 8061 Panel mounting 8062	0–50 0–200 0–500 0–1000 0–2000	6 ml min^{-1}	5–40	220	$\pm 2\,\mu g\,l^{-1}$ or $\pm 2\%$	sodium hexametaphosphate-sodium chloride reagents added to suppress aluminium and iron. Spectrophotometric ascorbic acid-H$_2$SO$_4$-ammonia citric acid. Reduction up to 5 mg l^{-1} Phosphate causes positive error of $<2\,\mu g\,l^{-1}$ silica 30 mg l^{-1} phosphate results in positive error of 4 $\mu g\,l^{-1}$ silica	Manual monthly

Table 13.2. Multi-parameter industrial instrumentation

Type of sample	Supplier	Model No
Raw water ex rivers, boreholes and reservoris	Kent Industrial Measurements	System 19 instrumentation package for determining nitrate, ammonia, chloride, silica, fluoride and phosphate
Potable water treatment	Kent Industrial measurements	Nitrate fluoride, ammonia
Effluent and sewage treatment	Kent Industrial Measurements	nitrate,
Distribution water, water towers, etc.	Kent Industrial Measurements	nitrate,
Surface water, water treatment, potable water, sewage effluent, sea water	Skalar	SA9000 on-line process analyser, alkalinity, chloride, total cyanide, nitrate, nitrite, silica, sulphate, sulphide,
Waste water treatment processes	Skalar	Series 8100 process analyser, chloride, fluoride, nitrite, phosphate, bromide, nitrate, sulphate, silica,

A wide range of standard discrete monitoring and control products are offered, many of which are also available in System 19 rack-mounting form for compact and economical installation.

These are ably complemented by other, more specialized equipment and systems for both analytical and water treatment applications, to meet the exacting requirements of the industry.

Kent instruments and systems and water meters are widely used throughout the water distribution network in pipelines, water towers, reservoirs and so on.

In recent years a variety of microprocessor-based and computer-compatible add-on systems have been developed to provide hard data on leakage control, water distribution and other useful analyses.

The System 19 instrumentation and systems can be found in all areas monitoring the outfalls from industrial processes, in rivers and at the treatment plant itself.

From individual monitors to complete on-line systems, these analysers are deployed throughout many treatment processes to improve efficiency and ensure effective treatment as well as to protect the environment and ultimately the drinking water supply.

The Skalar SA 9000 on-line process analyser is available in two models – as a free-standing unit complete with its own storage section for reagents or as a unit which may be bench or wall mounted or built into an even more comprehensive control system. The main unit consists of the on-line analyser, in which the analysis of the process stream is fully automated. The analysis process is based on the automated addition of sample to reagents which form a coloured complex under controlled conditions. The complex is measured photometrically. The signal is fed into the microprocessor which offers the flexibility for the specific application. Sampling is carried out directly from the process stream.

Table 13.3. Applications of Skalar SA-9000 on-line process analyser

Process stream component	Surface water	Water treatment	Potable water	Boiler water	Cooling water	Condensate	Sewage effluent	Sea water
Chloride	●		●		●	●	●	●
Chromium	●		●		●			
Cyanide (free)	●		●				●	
Cyanide (total)	●		●				●	
Fluoride			●					●
Nitrate	●		●				●	●
Nitrite	●		●				●	●
Phosphate	●		●	●			●	●
Silica	●			●	●	●		
Sulphate	●		●					●
Sulphide							●	

For analysing the same process stream component from several sources, a multi-stream attachment can be supplied. For samples carrying suspended particles, a special continuous in-line filter system is available. Calibration valves allow the introduction of standards and blanks for automatic calibration. Some applications of the SA-9000 analyser are shown in Table 13.3.

The Dionex series 8100 process analyser employs ion chromatography, high-performance liquid chromatography, or flow-injection techniques for the on-line determination of a variety of constituents in process streams. This is a modular instrument which can be modified should processing needs change, (Fig. 13.1)

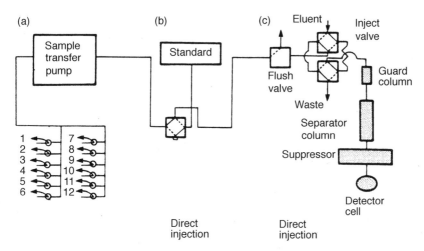

Fig. 13.1. Flowchart: Dionex series 8100 process analyser: **a** sample selection module-up to 18 sample points, continuous or sample on demand, non-metallic, NEMA 4 enclosure; **b** sample preparation module: dilution, reagent addition, standard preparation, preconcentration, matrix elimination; **c** chromatography hydraulics module: ion chromatography, HPLC, flow injection analysis

Due to the modularity, the series 8100 can be used in a batch-sampling mode where an operator manually collects the sample, or in an on-line mode where the analyser controls all sample selection and sample pretreatment. A series 8100 is easily utilized in an automated batch mode and can be completely upgraded to on-line operation at any time in the future.

A sample preparation module provides automatic dilutions of samples in the range of 1/10 and 1/5000 with precisions normally better than 2%. Up to six multi-level calibration standards may be automatically prepared from a stock standard for calibration of the analyser.

Kent supply the series 1800 continuously operating ion-selective electrode monitor which has been used for monitoring concentrations of fluoride and nitrate amongst other determinands. These instruments have a microprocessor-based logging system and are equipped with alarm facilities.

14 Preconcentration of Anions

Despite the great strides forward in analytical instrumentation that have been made in the last decade, the analyst working in the fields of potable water analysis and environmental analysis of non-saline and saline waters finds that, frequently, the equipment has insufficient sensitivity to be able to detect the low concentrations of substances present in his samples with the consequence that he has to report less than the detection limit of the method. Consequently, trends upwards or downwards in the levels of background concentrations of these substances in the environment cannot be followed. This is a very unsatisfactory situation which is being made worse by the extremely low detection limits being set in new directives on levels of pollution, issued by the European Community and other international bodies. To overcome the problem, there has been a move in recent years, to apply preconcentration to the sample prior to analysis so that, effectively, the detection limit of the method is considerably reduced to the point that actual results can be reported and trends followed.

The principle of preconcentration is quite simple. Suppose that we need to determine $5 \, ng \, l^{-1}$ of a substance in a sample and that the best technique has a detection limit of $1 \, \mu g \, l^{-1}$ ($1000 \, ng \, l^{-1}$). To reduce the detection limit to $5 \, ng \, l^{-1}$, we might, for example, pass 1l of the sample down a small column of a substance that absorbs the substance with 100% efficiency. We would then pass down the column 5 ml of a solvent or reagent which completely dissolves the substance from the column thereby achieving a preconcentration of $1000/5 = 200$. Thus, if the detection limit of the analytical method without preconcentration were $1000 \, ng \, l^{-1}$, then with preconcentration it would be reduced to approximately $5 \, ng \, l^{-1}$.

The use of a column is but one of many possible methods of achieving preconcentration.

A combination of preconcentration with the newest, most sensitive, and, by definition, most expensive analytical techniques now becoming available is achieving previously undreamt of detection limits at the very time when the requirements for such sensitive analysis is increasing at a rapid pace. Thus the combination of preconcentration with graphite furnace, Zeeman or inductively coupled plasma atomic absorption spectrometry and, particularly, the combination of the latter technique with mass spectrometry is enabling exceedingly low concentrations of metals in the $ng \, l^{-1}$ or lower range to be determined. Preconcentration prior to gas or high performance liquid chromatography is achieving similar results in the analysis of organics.

Another aspect of preconcentration is, however, worthy of mention, particularly in the case of smaller laboratories which cannot afford to purchase the full range of modern analytical instrumentation. Using older, less sensitive instrumentation, preconcentration will still achieve very useful reductions in detection limits which will be adequate in many but not all instances. Thus, if conventional atomic absorption spectrometry achieves detection limits of 1 and 5 mg l^{-1} for cadmium and lead in water, then a 200-fold preconcentration will reduce these limits to approximately 5 and 25 µg l^{-1} and a 1000-fold preconcentration will achieve 1 and 5 µg l^{-1}.

14.1 Iodide

Palagni [14.1] used the pulsed column bed technique to selectively separate and preconcentrate low levels of iodide-127 in natural water. The pulsed bed consisted of a syringe containing a scintillation detector and a gamma-ray spectrometer. The adsorbent consisted of open cell polyurethane foam impregnated with long chain tri-n-alkylamine which formed a complex with iodide.

14.2 Phosphate

Khan and Chow [14.2] have described a method in which the phosphate is converted into phosphomolybdate which is then extracted into polyurethane foam and analysed for molybdenum by X-ray fluorescence spectrometry directly on the foam. A polyether type polyurethane foam disc is squeezed for an hour in a mixture of phosphate, sodium molybdate and hydrochloric acid, spiked with phosphorus-32. After washing and drying, the foam disc is placed on plastic foam and stretched across the X-ray source holder. The method is simple and rapid and the precision is 5% for 0.25 mg l^{-1} and 2% for 2.5 mg l^{-1} phosphate. Equimolar amounts of silicon, germanium and arsenic IV appear to interfere with the determination.

A preconcentration technique based on membranes has been described for phosphate [14.3]. Phosphorus was collected as phospho-molybdenum on a nitrocellulose or acetylcellulose membrane in the presence of n-dodecyltrimethylammonium bromide. The membrane was dissolved in dimethyl-sulphoxide and the absorbance of dimethyl sulphoxide solution measured. Moderate concentrations of silicate, anionic and non-ionic surfactants and high concentrations of sodium-chloride did not interfere. Arsenate interference could be eliminated by reducing arsenate to arsenite. Determination of condensed and organic phosphates was possible following their conversion to orthophosphoric acid. The limit of determination was 0.02 µg l^{-1} phosphate.

In a method developed to preconcentrate orthophosphate from natural water, Hashitani et al. [14.4] used activated carbon loaded with zirconium. A 0.1–1l water sample at pH 1.5 was passed down the column of this material on which phosphate was adsorbed quantitatively and instantly at a pH below 8 and was desorbed at a pH above 13.5. Pyrophosphate, tripolyphosphate and metaphosphate behaved as orthophosphate.

Hori et al. [14.5] examined the adsorption behaviour of various phosphorus containing anions on ferric hydroxide as a function of solution pH. Adsorbed phosphorus compounds were determined spectrophotometrically and percent adsorption calculated from adsorbed and initial amounts. This was plotted against solution pH. Orthophosphate was adsorbed quantitatively at pH 4.0–8.0; triphosphate and pyrophosphate at pH 4.0–9.3 and monomethylphosphate, phosphite and α and β glycerophosphates at pH 4.0–6.8. Dimethylphosphate and hypophosphite were only slightly adsorbed at all pH values examined. The adsorbed ions were desorbed readily into a small volume of alkaline solution.

A solution of iron containing ferric chloride and acidified with hydrochloric acid is treated with aqueous ammonia to precipitate ferric hydroxide co-precipitated with ferric phosphate produced by any phosphate ions in the sample [14.6]. The precipitate is filtered off and dissolved in dilute hydrochloric acid and ammonia and uranyl acetate added to produce uranyl phosphate which is estimated polarographically. To determine phosphate plus arsenate the ferric hydroxide precipitate, containing coprecipitated iron phosphate and arsenite, is made acid with hydrochloric acid and potassium iodide added. The arsenic trichloride produced is extracted with carbon tetrachloride and this solution back extracted into hydrochloric acid. Application of the ammonia ferric chloride precipitation technique to this gives a preconcentrate containing arsenic only.

In one procedure [14.7, 14.8] phosphate is adsorbed onto anion exchange resin. Orthophosphate is quantitatively adsorbed by a A,G Dowex 1 – X8 anion exchange resin, then eluted and reacted with an acid molybdate reagent for estimation. Arsenic and organic phosphorus compounds did not interfere with the estimation of orthophosphate while polyphosphates did interfere if present in equal or greater amounts than orthophosphate. It is concluded that the use of the anion exchange technique results in a more valid estimate than direct action with the acid molybdate reagent.

14.3 Sulphide

De Salvo and Street [14.9] preconcentrated sulphide from well water on a column of cadmium exchanged zeolite sorbent prior to determination by the methylene blue spectrophotometric method. This method had an appreciably lower detection limit than conventional sulphide methods.

A flow injection gas diffusion method has been described [14.10] for the preconcentration and determination of traces of sulphide. Accumulation and preconcentration of the analyte is accomplished using the acceptor stream of a diffusion unit in a closed-loop recirculating mode. The detection limit from a 5 ml sample is 0.15 µg l^{-1} level is 4%.

Sulphide has been preconcentrated on a column of Amberlite IRA 400 ion exchange resin. The sulphide is removed from the column with 4 mol l^{-1} [14.11] sodium hydroxide and determined spectrophotometrically by the N, N'-dimethyl-p-phenylene diamine method. Down to 0.1 µg l^{-1} sulphide can be determined by this procedure.

14.4 Borate

Yoshimura et al. [14.12] have described a highly sensitive method for preconcentrating borate based on adsorption on Sephadex G-25 gel in alkaline medium and reversible desorption acid medium. The borate is then determined spectrophotometrically by the azomethine-4-method.

Jun et al. [14.13] preconcentrated boron from natural waters as its complex with chromotropic acid and octyltrimethyl ammonium chloride on an anion exchange column (TSK gel, I.C. Anion PW).

The eluted concentrate was analysed by high performance liquid chromatography.

14.5 Arsenite

Howard et al. [14.14] electively preconcentrated arsenite onto mercapto modified silica gel. Arsenate, monomethylarsonate, and dimethyl arsenite, which commonly occur in arsenic contaminated natural water samples, do not interfere in this procedure.

14.6 Selenate / Selenite

2,2′-Diethylaminocellulose filters enabled selenate (SeO_4^{2-}) and selenite (SeO_3^{2-}) to be preconcentrated from natural water [14.15, 14.16] at pH 3–6 with a detection limit of 0.05 µg l^{-1}.

Selenite has been preconcentrated on a bismuthiol 11 modified anion exchange resin (Amberlite 1 RA–400) [14.17] followed by fluorimetric estimation using diamino-napthalene. Selenite adsorbed on the column as selenotrisulphate was desorbed with a small volume of 0.1 mol l^{-1} penicillamine or 0.1 mol l^{-1} cysteine prior to fluorimetric determination of selenium [14.18].

Selenate and selenite can be determined by energy dispersive X-ray spectroscopy after preconcentration of elementary selenium on activated carbon, ascorbic acid being used to reduce selenium to its elemental form [14.19–14.21].

14.7 Chromate

The preconcentration of chromium(IV) (chromate) in natural waters by coprecipitation with barium sulphate has been shown to be a selective and accurate procedure [14.22]. Analysis of the preconcentrate is carried out by determining down to 0.02 µg l^{-1} chromium spectrophotometrically. The method is based on the similarity between the solubility products of barium sulphate and barium chromate. Interference from ferric iron, aluminium, and trivalent chromium was overcome by using salicylic acid as a masking agent. The method was used on river water samples and the results obtained were in good agreement with those obtained by an aluminium hydroxide coprecipitation method. In order to prevent contamination, the membrane filter, the sample bottle, and other vessels were carefully precleaned with concentrated hydrochloric acid and redistilled water. The coprecipitation step for chromium (VI) was carried out as soon as possible after sampling to minimize the sample storage problem.

Mullins [14.23] has described a procedure for determining the concentrations of dissolved chromium species in sea water. Chromium(III) and chromium (VI) separated by coprecipitation with hydrated iron(III) oxide and total dissolved chromium are determined separately by conversion to chromium(VI), extraction with ammonium pyrrolidinediethyl-dithiocarbamate into methyl isobutyl ketone and determination by atomic absorption spectroscopy. The detection limit is 40 ng l^{-1} Cr. The dissolved chromium not amenable to separation and direct extraction is calculated by difference. In the waters investigated, total concentrations were relatively high (1–5µg l^{-1}) with chromium (VI) the predominant species in all areas sampled with one exception, where organically bound chromium was the major species. A standard contact time of 4h was found to be necessary for the quantitative coprecipitation of chromium on ferric oxide. The r.s.d values for the determination of chromium(III), chromium(VI), and total dissolved chromium were generally 10.0, 5.0, and 5.0% respectively. From these results, the r.s.d for the calculated concentration of the bound species was 20%.

The sorption of hexavalent chromium (chromate) onto hydrous ferric oxide can be explained in terms of ligand exchange, hydroxyl ions being released from the surface of the hydrated oxide particles. For magnetic particles, however, the sorption effect was explained by the reduction of traces of hexavalent chromium at the magnetite/water interface, as a result of which a very fine ferric hydroxide/chromium hydroxide coating was produced which exhibited further reduction of hexavalent chromium.

Nakayama et al. [14.24] have described a method for the determination of chromium (III), chromium (VI), and organically bound chromium in sea water. They found that sea water in the sea of Japan contained about 9×10^{-9} mol l^{-1} dissolved chromium. This is shown to be divided as about 15% inorganic Cr^{III}, about 25% inorganic Cr^{IV}, and about 60% organically bound chromium. These workers studied the coprecipitation behaviours of chromium species with hydrated iron(III) and bismuth oxides.

The collection behaviour of chromium species was examined as follows. Sea water (400 ml), spiked with 10^{-8} mol l^{-1} Cr^{III}, Cr^{VI}, and Cr^{III} organic complexes labelled with ^{51}Cr, was adjusted to the desired pH with hydrochloric acid or sodium hydroxide. An appropriate amount of hydrated iron(III) or bismuth oxide was added; the oxide precipitates were prepared separately and washed thoroughly with distilled water before use [14.25]. After about 24 h, the samples were filtered on 0.4 µm Nuclepore filters. The separated precipitates were dissolved with hydrochloric acid and the solutions thus obtained were used for γ-activity measurements. In the examination of solvent extraction, chromium was measured by using ^{51}Cr, while iron and bismuth were measured by electrothermal atomic absorption spectrometry. The decomposition of organic complexes and other procedures were also examined by electrothermal atomic absorption spectrometry.

The percentage collection of Cr^{III} with hydrated iron(III) oxide may decrease considerably in the neutral pH range when organic materials capable of combining with Cr^{III}, such as citric acid and certain amino acids, are added to the sea water [14.26]. Moreover, synthesized organic Cr^{III} complexes are scarcely collected with hydrated iron(III) oxide over a wide pH range [14.26]. As it was not known what kind of organic matter acts as the major ligand for chromium in sea water Nakayama et al. [14.24] used EDTA and 8-quinolinol-5-sulphonic acid to examine the collection and decomposition of organic chromium species, because these ligands form quite stable water-soluble complexes with Cr^{III} although they are not actually present in sea water. Both these Cr^{III} chelates are stable in sea water at pH 8.1 and are hardly collected with either of the hydrated oxides. The organic chromium species were then decomposed to inorganic chromium(III) and chromium(VI) species by boiling with 1 g ammonium persulphate per 400 ml sea water and acidified to 0.1 mol l^{-1} of acid with hydrochloric acid. Iron and bismuth, which would interfere in atomic absorption spectrometry, were 99.9% removed by extraction from 2 mol l^{-1} hydrochloric acid solution with a p-xylene solution of 5% tri-iso-octylamine. Cr^{III} remained almost quantitatively in the aqueous phase in the concentration range $10^{-9} - 10^{-6}$ mol l^{-1} whether or not iron or bismuth was present. However, as about 95% of Cr^{VI} was extracted by the same method, samples which may contain Cr^{VI} should be treated with ascorbic acid before extraction so as to reduce Cr^{VI} to Cr^{III}.

When the residue obtained by the evaporation of the aqueous phase after the extraction was dissolved in 0.1 mol l^{-1} nitric acid and the resulting solution was used for electrothermal atomic absorption spectrometry, a negative interference, which was seemingly due to residual organic matter, was observed. This interference was successfully removed by digesting the residue on a hot plate with 1ml of concentrated hydrochloric acid and 3 ml of concentrated nitric acid. This process had the advantage that the interference of chloride in the atomic absorption spectroscopy was eliminated during the heating with nitric acid.

Cranston and Murray [14.27, 14.28] took the samples in polyethylene bottles that had been precleaned at 20 °C for 4 days with 1% hydrochloric acid. This acid had been previously distilled to reduce metal impurity levels. Total chromium, $Cr^{IV} + Cr^{III} + Cr_p$ (particulate chromium) was coprecipitated with iron(II) hydroxide and reduced chromium ($Cr^{III} + Cr_p$) was coprecipitated with iron(III) hydroxide. These coprecipitation steps were completed within minutes of sample collection to minimize storage problems. The iron hydroxide precipitates were filtered through o.4 μm Nuclepore filters and stored in polyethylene vials for later analyses in the laboratory. Particulate chromium, was also obtained by filtering unaltered samples through o.4 μm filters. In the laboratory the iron hydroxide coprecipitates were dissolved in 6 mol l^{-1} hydrochloric acid and analysed by flameless atomic absorption. The limit of detection of this method is about $0.1 - 0.2$ nmol l^{-1}. Precision is about 5%.

Alumina has been used to preconcentrate chromate before determination by inductively coupled plasma atomic emission spectrometry [14.29]. Trivalent chromium is not adsorbed under these conditions. A detection limit of 0.2 μg l^{-1} was achieved by this procedure.

Use of immobilized chelating agents for sequestering trace metals from aqueous and saline media presents several significant advantages over chelation–solvent extraction approaches to this problem. [14.30, 14.31] With little sample manipulation, large preconcentration factors can generally be realized in relatively short times with low analytical blanks. As a consequence of these considerations Willie et al. [14.32] developed a new approach to the determination of total chromium. This involves preliminary concentration of dissolved chromium from sea water by means of an immobilized diphenyl-carbazone chelating agent, prior to determination by atomic absorption spectrometry. A Perkin-Elmer Model 500 atomic absorption spectrometer fitted with a HGA-500 furnace with Zeeman background correction capability was used for chromium determinations. Chromium was first reduced to Cr^{III} by addition of 0.5 ml aqueous sulphur dioxide and allowing the sample to stand for several minutes. Aliquots of sea water were then adjusted to pH 9.0 ± 0.2 using high-purity ammonium hydroxide and gravity fed through a column of silica at a nominal flow rate of 10 ml min^{-1}.

The sequestered chromium was then eluted from the column with 10.0 ml 0.2 mol l^{-1} nitric acid. More than 93% of chromium was recovered in the first 5 ml of eluate by this method. Extraction of 80 ng spikes of Cr^{III} from 200 ml aliquots of sea water was semiquantitative.

Boniforti et al. [14.33] compared several preconcentration methods in the determination of metals in sea water. A comparison was made of ammonium pyrrolidinedithiocarbamate-8-quinolinol complexation followed by extraction with methyl isobutyl ketone or Freon-113, coprecipitation with magnesium hydroxide or iron(II) hydroxide or chelating by batch treatment with Chelex-100 for the determination of chromium. Atomic absorption spectrometry and inductively coupled plasma atomic emission spectrometry were used for analysis. Interferences, recovery, precision, accuracy, and detection limits were compared. The Chelex-100 resin method was most suitable for the preconcentration of all determinands except chromium, whereas preconcentration of Cr^{III} and Cr^V was achieved only by coprecipitation with iron(II) hydroxide.

Schaller and Neeb [14.34] extracted di(trifluoroethyl) dithiocarbamate chelates of hexavalent and trivalent chromium and cobalt from aqueous solution at pH3 using a carbon-18 column. Adsorbed chelates were eluted with toluene before gas chromatographic analysis with electron capture detection. Broadening of the peaks of cobalt and chromium was reduced by arranging that the end of the capillary was 1 cm within the detector body, the base of which was isolated with glass wool and aluminium foil. The last 0.2 cm of the column was also protected with a sleeve of braided wire. The detection limit for chromium(VI) was 0.05 µg l^{-1}. Various ion exchange resins have been used to preconcentrate chromate ions including Dowex AG 1—X4 anion exchange resin [14.35].

Parkow [14.36] acidified water samples containing chromate to pH 5 and passed them upwards through an anion exchange resin, so that the chromate is adsorbed in a narrow zone at the lower end of the resin bed. The chromate is eluted rapidly with small volumes of an acidic reductant solution which reacts with chromate on the column to form trivalent chromium during elution, thus producing very high concentration factors.

Parkow and Januer [14.37] acidified water samples containing chromate to pH5 and then passed them upwards through a Dowex Ag 1 — X4 anion exchange resin, so that the chromate was adsorbed in a narrow zone at the lower end of the resin bed. The chromate was eluted rapidly with small volumes of an acidic reductant solution which reacts with chromate on the column to form trivalent chromium during elution, thus producing very high concentration factors. Amberlite LA liquid anion exchange resin has also been used for this purpose[14.38].

Farag et al. [14.39] determined chromium(VI) (chromate) in water using 1,5-diphenylcarbazide-loaded polyurethane foam.

A particularly recent innovation that has been used for metals and also for chromate preconcentration consists of a modification of the flow injection analysis technique whereby the samples pass through a microcolumn containing an adsorbant for the ion of interest, thereby achieving a concentration factor. Following an automatic switch to an acidic reagent, the adsorbed anion is desorbed in a sharp pulse of the flowing reagent and then passes on to the detection system. Systems such as this, therefore, combine preconcentration and automation. Syty et al. [14.40] separated chromium (III) from chromium (VI) on a microcolumn of alumina which preconcentrated chromium(VI). Chromium(VI) was then flushed from the column with a small volume of acid before determination by inductively coupled plasma atomic emission spectrometry in amounts down to 0.2 µg l^{-1}.

14.8 Molybdate

Hidalgo et al. [14.41] preconcentrated molybdenum(VI) in potable water and sea water by co-flotation on iron(III) hydroxide. Co-flotation was achieved by means of surfactants; octadecylamine for tap water samples, hexadecyltrimethylammonium bromide for natural waters, and both hexadecyltrimethylammonium bromide and octadecylamine for sea water samples. Differential pulse polarography using the catalytic wave caused by molybdenum(VI) in nitrate medium was applied to the preconcentrate. Good reproducibility was obtained with mean values of 0.7 ng ml^{-1} and 5.7 ng ml^{-1} for molybdenum (VI) in tap water and sea water, respectively.

Murthey and Ryan [14.42] used colloid flotation as a means of preconcentration prior to neutron activation analysis for molybdenum. Hydrous iron (III) is floated in the presence of sodium dodecyl sulphate with small nitrogen bubbles from 1l of sea water at pH 5.7. Recoveries of molybdenum were better than 95%. This method has been used [14.43] to precipitate from water samples traces of molybdenum. The trace element was concentrated by coprecipitation with thionalide at pH 9.1. Coprecipitation with thionalide allowed the concentration of both ions and colloids.

Molybdenum has been determined after preconcentration on Sephadex G. 25 gel at pH 3.5. Ethylenediaminetetra acetic acid desorbs molybdenum from the gel. This preconcentration method brings the detection limit for molybdenum in fresh and sea waters down to 1 µg l^{-1} in 250-ml samples. The procedure was applied to 100-ml solutions containing 1.25 µmol of molybdenum and 1- to 10000-fold amounts of various ions. Interference from vanadium (V) and tungsten (VI) is probably due to complex formation with molybdenum. Iron(III) at a 100-fold level lowered the recovery, probably because its hydroxides or hydroxo complexes adsorb molybdenum, but the interference could be overcome in the presence of 0.5 mol l^{-1} acetate. The method is suitable for the determination of molybdenum at the levels normally encountered in river water (0.2–0.6 µg l^{-1}) and is particularly effective for sea water.

Shriadal et al. [14.44] determined molybdenum VI in sea water spectrophotometrically after enrichment as the Tiron complex on a thin layer of anion exchange resin. There were no interferences from trace elements or major constituents of sea water except for chromium and vanadium. These were reduced by the addition of ascorbic acid. The concentration of dissolved molybdenum (VI) determined in Japanese sea water was 11.5 µg l^{-1} with a relative standard deviation of 1.1%.

Kuroda et al. [14.45] preconcentrated trace amounts of molybdenum from acidified sea water an a strongly basic anion exchange resin (Bio-Rad AgI, X8 in the chloride form) by treating the water with sodium azide. Molybdenum VI complexes with azide were stripped from the resin by elution with ammonium chloride/ammonium hydroxide solution (2 mol l^{-1}/mol l^{-1}). Relative standard deviations of better than 8% of levels of 10 µg l^{-1} were attained for sea water using graphite furnace atomic absorption spectrometry.

Vasquez-Gonzalez et al. [14.46] have described a method for preconcentrating and determining molybdate by electrothermal atomization atomic absorption spectrometry after preconcentration by means of anion-exchange using Amberlite IRA-400 resin in citrate form. The optimal analytical parameters were established by drying, carbonization, charring, atomization and cleaning in a graphite furnace. The precision and accuracy of the method were investigated. Less than 0.2 µg l^{-1} molybdate could be determined by this procedure.

14.9 Rhenate

Matthews and Riley [14.47] have described the following procedure for determining down to 0.06 µg l^{-1} rhenium in sea water. From 6 to 8 µg l^{-1} rhenium was found in Atlantic sea water. The rhenium in a 15-l sample of sea water, acidified with hydrochloric acid, is concentrated by adsorption on a column of De-Acidite FF anion-exchange resin (Cl$^-$ form), followed by elution with 4 mol l^{-1} nitric acid and evaporation of the eluate. The residue (0.2 ml), together with standard and blanks, is irradiated in a thermal neutron flux of at least 3×10^{12} neutrons cm^{-2} s^{-1} for at least 50h. After a decay period of 2 days, the sample solution and blank are treated with potassium perrhenate as carrier and evaporated to dryness with a slight excess of sodium hydroxide. Each residue is dissolved in 5 mol l^{-1} sodium hydroxide. Hydroxylammonium chloride is added (to reduce Tc^{VII}) which arises as ^{99m}Tc from activation of molybdenum present in the samples, and the Re^{VII} is extracted selectively with ethyl methyl ketone. The extracts are evaporated, the residue is dissolved in formic acid:hydrochloric acid (19:1), the rhenium is adsorbed on a column of Dowex 1 and the column is washed with the same acid mixture followed by water and 0.5 mol l^{-1} hydrochloric acid; the rhenium is eluted at 0°C with acetone:hydrochloric acid (19:1) and is finally isolated by precipitation as tetraphenylarsonium perrhenate. The precipitate is weighed to determine the chemical yield and the ^{186}Re activity is counted with an end-window Geiger-Muller tube. The irradiated standards are dissolved in water together with potassium perrhenate. At a level of 0.057 µg l^{-1} rhenium the coefficient of variation was ± 7%.

14.10 Complex Anions

Simple cationic metal species do not react with anion exchange resins. However, if the metal ion M$^+$ is first reacted with a reagent with which it forms a

Table 14.1. Preconcentration of anionic metal species from natural water

Metal	Type of water	Resin	Medium	Eluting agent	Detection limit	Analytical finish	Ref
	Potable	Anion exchange	Pyro catechol violet	–	–	–	14.48
d	Natural	Dowex 1-X8	HCl or HBr	HNO_3	–	Spectrophotometric	14.49
d	Natural	Amberlite IRA-400	Cyanide	CH_3COOH	–	Spectrophotometric	14.50
Cd	Natural	Biorad 140–1242 A91-X2	8-hydroxy quinoline 5-sulphonic acid	2 mol l^{-1} HNO_3	< 0.2 µg l^{-1}	AAS	14.51, 14.52
Cd	Sea water	Anion exchange	1 mol l^{-1} thiocynate 0.1 mol l^{-1} HCl	2 mol l^{-1} HNO_3	–	AAS	14.53
Cd	Sea water	Polyacrylamidoxime Resin	–	1:1 HCl H_2O	–	AAS	14.54
Cd	Natural and potable	Dowex 1-X8	MeOH:HBr	aq HNO_3	–	AAS	14.55
Cd	Sea Water	Single bead anion exchange	–	–	–	Mass spectroscopy	14.56
Co	Natural	Biorad 140-1242, A91-X2	8-hydroxy quinoline 5-sulphonic acid	12 mol l^{-1} HCl	< 0.2 µg l^{-1}	AAS	14.51, 14.52
Co	Natural	Dowex A91-X8	HCl-thiocyanate	HCl	–	Spectrophotometric	14.57
Co	Sea water	Anion exchange	0.1 mol l^{-1} cyanate 0.1 mol l^{-1} HCl	2 mol l^{-1} HNO_3	–	AAS	14.52
Cu	Sea water	Anion exchange	1 mol l^{-1} cyanate 0.1 mol l^{-1} HCL	2 mol l^{-1} HNO_3	–	AAS	14.53
Cu	Sea water	Polyacryl amidoxime resin	–	1:1 HCl H_2O	–	AAS	14.54
Cu	Natural and potable	Dowex 1-X8	MeOH:HBr	aq NH_3	–	AAS	14.55
Au	Natural	Anion exchange	–	–	–	NAA	14.58
Au	Sea water	Single bead anion exchange	–	–	–	Mass spectrometry	14.56
In	Sea water	Single bead anion exchange	–	–	–	Mass spectrometry	14.56

Table 14.1 (Contd.)

Metal	Type of water	Resin	Medium	Eluting agent	Detection limit	Analytical finish	Ref
Ir	Sea water	Single bead anion exchange	–	–	–	Mass spectrometry	14.59
Ir	Sea water	Single bead Anion exchange	–	–	–	AAS	14.60
Fe	Sea water	Polyacryl amidoxine	1:1 H_2O :HCl	–	–	AAS	14.54
Fe	Sea water	Amberlyst A27 (thin layer chromatography)	Bathophenanthroline	HCl NH_2OH NH_4OH pH 4–5	–	Photodensitometry	14.61
Fe	Natural	Anion exchange	–	–	1–2 ng	Spectrophotometric	14.62
Pb	Natural	Biorad 140–1242 AG1-X2	8 Hydroxy quinoline-5-sulphonic acid	2 mol l^{-1} HNO_3	< 0.2 µg l^{-1}	AAS	14.51, 14.52
Pb	Sea water	Polyacryl amidoxime	–	1:1 HCl:-H_2O	–	AAS	14.54
Pb	Sea water and natural	Dowex 1-X8	MeOH: HBR	aq HNO_3	–	AAS	14.55
Pb	Sea water	Single bead anion exchange	–	–	–	Mass spectrometry	14.56
Pb	Natural	Dowex AG X-8	NH_2OH NaCl	4N HNO_3	2 ng l^{-1}	CV AAS	14.63
Ni	Mineral water	Dowex A-1	NaOO CCH_3	HCl	0.5 µg	Spectro photometric	14.64
Mo	Mineral and potable	Dowex 1-X8	Acid thiocyanate ascorbic acid	–	–	Spectro photometric as Tiron complex	14.65–14.69
Mo	Sea water	Amberlite (GC–400)	Ascorbic acid	4N HNO_3	10 µg l^{-1}	Spectro photometric as thiocyanate complex	14.70
Pt	Sea water	Polyacryl amidoxime	–	1:1 H_2O :HCl	–	AAS	14.60
Pu-239 Pu-240	Natural	Anion exchange	–	HF-HCl	–	Deposited on Pt γ-ray spectrometry	14.71
Pu-240	Sea water	Single bead anion exchange	–	–	–	Mass sapectrometry	14.56

Table 14.1 (Contd.)

Metal	Type of water	Resin	Medium	Eluting agent	Detection limit	Analytical finish	Ref
	Natural	Anion exchange	HCl	0.4 mol l^{-1} Thiourea	–	NAA	14.72
	Natural	Dowex Ag 1 × 8	–	Acetone: HNO$_3$:H$_2$O (20:1:1)	–	Ammonium pyrolidone dithiocarbamate extraction -AAS	14.73
	Sea water	Single bead anion exchange	–	–	–	Mass spectrometry	14.56
	Natural	Strongly basic anion exchange	Conversion to Tl Cl$_4^{3-}$ ion	H$_2$SO$_3$	–	GFAAS	14.74
	Natural	Anion exchange	Conversion Tl Cl$_4^{3-}$ ion with Cr IV Sulphate	(NH$_4$)SO$_4$	3.3 ng l^{-1}	AAS	14.75
Th	Natural	Dowex AG1-X8	HNO$_3$	HCl	–	Spectrophotometric	14.76
Sn	Sea water	Dowex 1X(Cl from)	HCl	2 mol l^{-1}	–	Spectrophotometric with catechol violet	14.77
U	Sea water	Basic anion exchange	azide ions	1 mol l^{-1}	–	Spectrometric	14.78
U	Natural	Amberlite LA-1	–	–	0.04 µg l^{-1}	Fluorimetric	14.79–14.81
U	Natural	Dowex AG1-X8	H$_2$SO$_4$	H$_2$SO$_4$	0.3 µg l^{-1}	Fluorimetric	14.82
V	Sea water	Anion exchange	1 M thiocyanate 0.1 MHCl	2 mol l^{-1} HNO$_3$	–	AAS	14.53
V	Sea water	Dowex 1-X8 (SCN form)	Thiocyanate -HCl	cone HCl	1.65 µg l^{-1}	Spectrophotometric as 2-pyridyl azoresorcinol	14.83
V	Sea water	Dowex 1-X8	Thiocyanate -ascorbic acid	–	–	AAS	14.84, 14.85
Zn	Sea water	EDE-10P	HCl	HCl	–	Spectrophotometric	14.86
Zn	Natural	Biorad 140-1242	8-hydroxy quinoline 5-sulphonic acid	2 mol l^{-1} HNO$_3$	< 0.2 µg l^{-1}	AAS	14.51, 14.52

Table 14.1 (Contd.)

Metal	Type of water	Resin	Medium	Eluting agent	Detection limit	Analytical finish	Ref
Zn	Sea water	Anion exchange	1M thiocyanate 0.1M HCl	2 mol l^{-1} HNO_3	–	AAS	14.53
Zn	Sea water	Polyacryl amidoxime resin	–	1:1 $HCl:H_2O$	–	AAS	14.54
Various radio nucleides	Natural	Single bead anion exchange	–	–	–	Point source mass spectrometry	14.87
22 elements	Natural	Anion exchange	–	–	–	NAA	14.88

negatively charged anionic complex, then the resulting negatively charged metal-containing ions are retained on an anion exchange resin. Thus, cadmium II forms a soluble anionic complex upon reaction with potassium cyanide:

$$CdCl_2 + 4KCN \rightarrow K_2[Cd(CN)_4]^{2-} + 2KCl.$$

This complex is retained in a column of strong base anion exchange resin which, for convenience, is represented as follows:

$$2\,\text{resin}\,CH_2R_3N^+Cl^- + [Cd(CN)_4]^{2-} \rightarrow [\text{resin}\,CH_2R_3N^+]^{2-}_2$$
$$[Cd(CN)_4]^{2-} + 2Cl^-.$$

The cadmium complex can then be dissolved off the column with a small volume of aqueous acetic acid to produce the acetate form of the resin and cadmium acetate:

$$[\text{resin-}CH_2R_3N^+]_2[Cd(CN)_4]^{2-} + 4CH_3COOH \rightarrow$$
$$2[\text{resin}-CH_2R_3N^+][OOCH_3]^- + Cd(OOCH_3)_2 + 4HCN.$$

Other examples include the formation of the following.

(a) Complexes of the type $H_2[MBr_4]^{2-}$ upon reaction of cadmium, copper or lead with hydrobromic acid:

$$MCl_2 + 4HBR = H_2[MBR_4]^{2-} + 2HCl.$$

(b) Complexes of the type $H_2M(CN_5)_4^{2-}$ upon reaction of cadmium, cobalt or zinc with acidic ammonium thiocyanate:

$$MCl_2 + 4KCNS + 2HCl = H_2[M(CNS)_4]^{2-} + 4KCl.$$

(c) Formation of $TlCl_4^{3-}$ by reaction of thallous ion with hydrochloric acid.

High preconcentration factors can be achieved by such techniques, some further examples of which are reviewed in Table 14.1.

14.11 References

14.1 Palagni S (1983) International J Applied Radiation and Isotopes 34: 755
14.2 Khan AS, Chow A (1983) Analytical Letters (London) 16: 265
14.3 Taguchi S, Ito-oko E, Matsuyama K, Kashara I (1985) Gotok Talanta 32: 391
14.4 Hashitani H, Okumura M, Fujinaga K (1987) Fresenins Zeitschrift fur Analytische Chemie. 326: 540
14.5 Hori T, Moriguchi M, Sassaki K, Kitagawa S, Munakato M (1985) Analytica Chimica Acta 173: 299
14.6 Rozaiski L (1972) Chemica Analit 17: 55
14.7 Westland AD, Bouchair I (1974) Water Research 8: 467
14.8 Blanchar W, Riego D (1975) Journal of Environmental Quality 4: 45
14.9 De Salvo DD, Street KW (1986) Analyst (London) 111: 1307
14.10 Milosavljevic EB, Solujieh L, Hendrix JL, Nelson JH (1988) Analytical Chemistry 60: 2791
14.11 Paez DM, Guagnini OA (1971) Mikrochimica Acta 2: 220
14.12 Yoshimura K, Kariya R, Torntani T (1979) Analytica Chimica Acta 109: 115
14.13 Jun Z, Oshima M, Motomizu S (1988) Analyst (London) 113: 1631
14.14 Howard AG, Volkan M, Ataman Y (1987) Analyst (London) 112: 159
14.15 Smits J, Van Grieken R (1981) Analytica Chimica Acta 123: 9
14.16 Smits J, Van Grieken R (1981) International Journal of Environmental Analytical Chemistry 9: 81
14.17 Wu TL, Lambert L, Hastings D, Banning D (1980) Bulletin of Environmental Contamination and Toxicology 24: 411
14.18 Smith DH, Christie WH (1980) International Journal of Environmental Analytical Chemistry 8: 241
14.19 Rubberecht H, Van Grieken R (1982) Talanta 29: 823
14.20 Massee R, Van der Sloot HA, Das HA (1977) Journal of Radioanalytical Chemistry 38: 157
14.21 Orvini E, Ladola L, Gallorini M, Zerlia T (1981) In Heavy Metals in the Environment. p657. Elsevier Amsterdam
14.22 Yamazaki H (1980) Analytica Chimica Acta 113: 131
14.23 Mullins TL (1984) Analytica Chimica Acta 165: 97
14.24 Nakayama E, Kuwamoto T, Takoro H, Fujinaka F (1981) Analytica Chimica Acta 131: 247
14.25 Nakayama E, Kuwamoto T, Tokoro H, Fujinaka (1981) Analytica Chimica Acta 130: 401
14.26 Takayama E, Kuwamoto T, Tokoro H, Fujinaka (1981) Analytica Chimica Acta 130: 289
14.27 Cranston RE, Murray JW (1978) Analytica Chimica Acta 99: 275
14.28 Cranston RE, Murray JW (1980) Limnology and Oceanography 25: 1104
14.29 Cox AG, Cook TG, McLeod CW (1985) Analyst (London) 110: 331
14.30 Myasoedova GV, Savvin SG, Zhu R (1982) Analytica Chim 37: 499
14.31 Leyden DE, Wegschneider W (1981) Analytical Chemistry 63: 1059A
14.32 Willie SN, Sturgeon RE, Berman SS (1983) Analytical Chemistry 55: 981
14.33 Boniforti R, Ferraroli R, Frigieri P, Heltai D, Queirezza G (1984) Analytica Chimica Acta 162: 33
14.34 Schaller H, Neeb R (1987) Fresenius Zeitschrift fur Analytische Chemie 327: 170
14.35 Aoyama M, Habo T, Suzuki S (1981) Analytica Chimica Acta 129: 237
14.36 Parkow JF, Lieta DP, Lin JW, Ohl SE, Shum WP, Januer GE (1977) Science of the Total Environment 7: 17
14.37 Parkow JF, Januer GE (1974) Analytica Chimica Acta 69: 97
14.38 Muzzuetelli A, Minoia C, Pozzoli L, Ariati L (1982) Applied Spectroscopy 4: 182
14.39 Farag AG, El-Wakl AM, El-Shahawi MS (1981) Analyst (London) 106: 809
14.40 Syty A, Christenson RG, Raius TC (1986) Atomic Spectroscopy 7: 89
14.41 Hidalgo JL, Gomez MA, Caballero M, Cela R, Perez-Bustamonte A (1988) Talanta 35: 301
14.42 Murthey RSS, Ryan DE (1983) Analytical Chemistry 55: 682

14.43 Zmijewska W, Polkowska Motrenko H, Stakowska H (1984) Journal of Radioanalytical and Nuclear Chemistry Articles 84: 319
14.44 Shriadal HMA, Katoaka M, Ohzeki K (1985) Analyst (London) 110: 125
14.45 Kuroda R, Matsumoto N, Ogmura K (1988) Fresenius Zeutschrift fur Analytische Chemie 330: 111
14.46 Vasquez-Gonzalez JF, Bermejo-Barrera P, Bermejo-Martinez F (1987) Atomic Spectroscopy 8: 159
14.47 Matthews AD, Riley JP (1970) Analytica Chimica Acta 51: 455
14.48 Sarzanini C, Mentasti E, Porta V, Gennaro MC (1987) Analytical Chemistry 59: 484
14.49 Korkische J, Dimitriades D (1973) Talanta 20: 1295
14.50 Ashizawa T, Hosoya K (1971) Japan Analyst 20: 1416
14.51 Berge DG, Going JE (1981) Analytica Chimica Acta 123: 19
14.52 Parkow JF, Januer JE (1974) Analytica Chimica Acta 69: 97
14.53 Kiriyama T, Kuroda R (1985) Mikrochimica Acta 1: 405
14.54 Colleta MB, Siggia S, Barnes RM (1980) Analytical Chemistry 52: 2347
14.55 Korkische J, Sario A (1977) Analytica Chimica Acta 76: 393
14.56 Koide M, Lee DS, Stallard MO (1984) Analytical Chemistry 56: 1956
14.57 Korkische J, Dimitriades D (1973) Talanta 20: 1287
14.58 Asamov KA, Abdullah AA, Zakhidov AS (1969) Doklady Acad Nauk Uzbek SSR 31: 26
14.59 Kingston H, Pella PA (1986) Analytical Chemistry 58: 616
14.60 Hodge V, Stallard M, Koide M, Goldberg ED (1986) Analytical Chemistry 58: 616
14.61 Shriadah MMA, Ohseki K (1986) Analyst (London) 111: 555
14.62 Shi Yu L, Wei Ping G (1984) Talanta 31: 844
14.63 Mandel S, Das AK (1982) Atomic Spectroscopy 3: 56
14.64 Nevoral V, Okae A (1988) Czlka Form 17: 478
14.65 Shriadah HMA, Katoeka M, Ohzebi K (1985) Analyst (London) 110: 125
14.66 Riley JP and Taylor D (1968) Analytica Chimica Acta 40: 479
14.67 Kawabuchi K and Kuroda K (1969) Analytica Chimica Acta 46: 23
14.68 Kawabuchi K and Kuroda R (1969) Analytica Chimica Acta 46: 23
14.69 Kuroda R and Tarui T (1974) Fresenius Zeutschrift fur Analytische Chemie 269: 22
14.70 Kiriyama T and Kuroda R (1984) Talanta 31: 472
14.71 Golchert NW, Sedlet J (1972) Radiochemical and Radioanalytical letters 12: 215
14.72 Kawabuchi K, Riley JP (1973) Analytica Chimica Acta 65: 271
14.73 Chau TT, Fishmann MJ, Ball TW (1969) Analytica Chimica Acta 43: 189
14.74 Riley JP, Siddique SA (1986) Analytica Chimica Acta 181: 177
14.75 De Ruck A, Van decasteele C, Daws R (1987) Mikrochim Acta No 4/6: 187
14.76 Korkishe J and Dimitraiades D (1973) Talalnta 20: 1303
14.77 Kodama Y, Tsubota H (1971) Japan Analyst 20: 1554
14.78 Kuroda R, Oguma K, Mukai N, Imamoto M (1987) Talanta 34: 433
14.79 Brits RJS, Smit MCB (1977) Analytical Chemistry 49: 67
14.80 Gladney ES, Owens JW, Starner JW (1976) Analytical Chemistry 48: 973
14.81 Zielinski RA, Mc Kown M (1984) Journal of Radioanalytical and Nuclear Chemistry Articles 84: 207
14.82 Danielsson A, Roennholm B, Kiellstoem LE, Ingman F (1973) Talanta 20: 185
14.83 Kirkyama T, Kuroda R (1972) Analytica Chimica Acta 62: 464
14.84 Weiss HV, Guttman HA, Korkische J, Steffan I (1977) Talanta 24: 509
14.85 Korkische J, Gross H (1973) Talanta 20: 1153
14.86 Kurochkina NI, Lyakh VI, Pervelyaeva GL (1972) Nauch Trudy irkutsh gos Nauchno issled Inst Redk tsvet Metall (24) 149 (1972). Ref: Zhur Khim 19 GD (13) Abstract No 13 G 148
14.87 Carter JA, Walker RL, Smith DH (1980) International Journal of Environmental Analytical Chemistry 8: 241
14.88 Bergerioux C, Blanc PC, Haerdi W (1977) Journal of Radio-analytical Chemistry 39: 823

Subject Index

Acetate in
  natural water   453, 466, 467, 496, 497
  sea water   467
Acetoacetate in
  body fluid   466
  natural water   466, 467, 496, 497
Acrylate in seawater   467
Air
  free cyanide in   17–19, 262
  sulphide in   17–19, 336
Aminoacetate in natural water   2–20
Amperometry of
  ascorbate   458
  nitrite   210
Animal feeds
  free cyanide in   17–19, 261
  iodide in   103
Anisates in natural water   466, 467
Anodic stripping voltammetry of
  phosphate   17–19
  selenite   424
  thiosulphate   318
Arsenate in
  natural water   419–421, 466, 467, 486–490
  potable water   420, 520
Arsenite in
  natural water   419–421, 425, 486–490
  potable water   486
Arsenite, preconcentration   558
Ascorbate in
  crustacea   459
  foodstuffs   458
  natural water   457, 458
  pharmaceuticals   458
Atomic absorption spectrometry of
  arsenate   2–20, 30, 31, 420
  arsenite   2–20, 30, 31, 419
  borate   2–20, 30, 31
  chloride   2–20, 30, 31, 64, 65
  chromate   2–20, 30, 31, 436–439, 444, 445
  dichromate   2–20, 30, 31, 444, 445
  free cyanide   2–20, 30, 31, 246
  iodide   2–20, 30, 31, 103
  molybdate   2–20, 30, 31, 445, 446
  nitrate   2–20, 30, 31, 159
  phosphate   2–20, 30, 31, 356, 365, 369
  polyphosphate   2–20, 30, 31, 369
  selenate   2–20, 30, 31, 426–429
  selenite   2–20, 30, 31
  silicate   2–20, 30, 31, 385, 386, 426–429
  sulphate   2–20, 30, 31, 296, 310
  sulphide   2–20, 30, 31, 334, 335
  sulphite   2–20, 30, 31, 296
  thiosulphate   2–20, 30, 31, 319
  total cyanide   2–20, 30, 31, 262
  tungstate   2–20, 30, 31, 446

Benzoate in
  natural water   466, 467
  seawater   467
Beverages
  citrate in   454
  lactate in   456
  malate in   455
  oxalate in   455
Bicarbonate in
  foodstuffs   534, 536
  natural water   411, 415–417, 469–486, 496, 497
  potable water   411, 414, 416, 417, 469–486
Biological materials
  flouride in   17–20, 114
  free cyanide in   17–20, 262
  nitrate in   17–20, 240, 241
  nitrite in   17–20, 240, 241
  selenite in   17–20, 424
Blood
  phosphate in   17–20, 374
  sulphide in   17–20, 335
Body fluids, acetoacetate in   466
Boiler feed water
  carbonate in   534
  chloride in   530–534
  nitrite in   216, 534
  phosphate in   373, 534
  silicate in   386, 387

Boiler feed water (*contd.*)
  sulphate in 309, 530–534
  sulphite in 315, 316, 534
Borate in
  industrial effluents 396, 490–494, 496, 497
  natural water 388–396, 467, 490–494, 496, 497, 530
  plant extracts 398
  plants 397
  sewage 397
  soil 397
Borate, preconcentration 558
Borofluoride in natural water 467, 496, 497
Bromate in
  flour 129
  natural water 128, 133, 480
  potable water 129, 133
Bromide in
  foodstuffs 91, 92, 534–536
  industrial effluents 91, 521–529
  natural water 78–88, 466, 467, 470–486, 496, 497
  potable water 89–90, 124, 125, 127, 128, 470–486, 516–519
  rain 89, 119–124, 129, 495–516
  soil 91, 92
Bromide, process analyser 550–554
Bromite in rain 128, 129
Butyrate in
  natural water 466, 467
  seawater 467

Carbonate in
  boiler feed water 534
  high purity water 414
  natural water 414, 415, 496, 497
  soil 415
  waste water 414
Carbonate, test kit 544–549
Carboxylates in
  natural water 463, 464, 466, 467
  potable water 466
Cement
  chromate in 445
  dichromate in 445
Chlorate in
  foodstuffs 534–536
  natural water 126, 127, 133, 467, 496, 497
  potable water 125, 128, 133
  soil extracts 128
Chloride
  process analyser 550–554
  test kit 544–549
Chloride in
  boiler feed water 72–77, 531–534
  foodstuffs 78, 534, 536
  industrial effluents 70, 71, 521–524
  natural water 58–68, 116–119, 127–128, 133, 466, 467, 469, 496, 497
  plants 77
  plant extracts 533, 534
  potable water 69, 70, 124, 125, 133, 516–519
  rain 119–124, 495–516
  sewage 72
  soil 77, 78
  soil extracts 533, 534
  waste water 71, 72, 125
Chlorite in
  natural water 126, 127
  potable water 128
  waste water 128
Chloroacetate in natural water 467, 496, 497
Chlorobenzoate in natural water 466
Chromate in
  cement 445
  foodstuffs 534, 536
  glass 445
  natural water 434–442, 466, 467
  oxides 445
  potable water 445
  silicates 445
  waste water 442–444
Chromate
  preconcentration 559–563
  sample preservation 442
  test kit 544–549
Chronopotentiometry analysis of free cyanide 261
Citrate in
  beverages 454
  foodstuffs 454
  natural water 466, 467
Coal, fluoride in 20, 115, 116
Cobalt cyanide complexes in industrial effluent 449, 524–530
Colometric analysis of chloride 68
Column coupling capillary isotachophoresis of
  chloride 2–20, 53, 54, 68, 123, 298, 360
  fluoride 2–20, 53, 54, 106, 107, 123, 124, 360, 536–539
  nitrate 2–20, 53, 54, 232, 360, 536–539
  nitrite 2–20, 53, 54, 232, 536–539
  organic anions 539
  phosphate 2–20, 53, 54, 360, 536–539
  sulphate 2–20, 53, 54, 360, 536–539
Condensed polyphosphates in natural water 374–378
Continuous flow analysis of
  chloride 69, 301
  free cyanide 248–250

nitrate   170, 174–178, 301
nitrite   209, 301
phosphate   301, 302, 366, 367
sulphate   301
sulphite   314
Cooling water, iodide in   102
Copper cyanide complexes in industrial effluents   449, 450, 524–530
Crustacea ascorbate in   458, 459
Cyanate in
  natural water   271
  water   2–20, 44–48
Cyanide in
  industrial effluent   524–530
  natural water   496, 497
  potable water   519, 520
Cyanide, process analyser   459, 460

Dehydroascorbate in foodstuffs   459, 460
Dichromate in
  cements   445
  glass   445
  natural water   434–442
  oxides   445
  potable water   445
  silicates   445
  waste water   442–444
Dichromate
  preconcentration   559–563
  sample preservation   442
Dimethylbenzoate in water   446, 447
Dithionate in industrial effluents   529, 530
Drager tubes, determination of sulphide   331–334

Electrochemical analysis of nitrate   171, 172
Electrophoresis of amino acids   461
Enzymic assay of
  nitrate   167, 168
  phosphate   364, 365
Estuary water, phosphate in   366, 367
Ethylene diamine tetraacetate in
  natural water   462
  waste water   462

Ferrocyanide in
  industrial effluents   448
  water   2–20
Fertilizers, free cyanide in   17–20, 261
Flame emission spectrometry of sulphate   304
Flour, bromate in   129
Flow injection analysis of
  arsenate   2–20, 26–30, 419

arsenite   2–20, 26–30
borate   2–20, 26–30, 396
bromate   2–20, 26–30, 129
bromide   2–20, 26–30, 91
chlorate   2–20, 26–30, 127
chloride   2–20, 26–30, 69, 73, 367
chlorite   2–20, 26–30
chromate   2–20, 26–30
free cyanide   2–20, 26–30, 245
germanate   2–20, 26–30, 445
hypochlorite   2–20, 26–30
nitrate   2–20, 26–30, 159, 160, 170, 225, 228, 301–304, 367
nitrite   2–20, 26–30, 202–205, 217, 225, 228, 301–304, 367
phosphate   2–20, 26–30, 301–304, 354, 355, 367, 374
polysulphide   296, 314, 318, 319, 328
pyruvate   2–20, 26–30, 460
silicate   2–20, 26–30, 383
sulphate   2–20, 26–30, 289–296, 301–304, 306, 310, 318, 319, 328, 367
sulphide   2–20, 26–30, 296, 314, 318, 328
sulphite   2–20, 26–30, 296, 314, 316–318, 328
thiocyanate   2–20, 26–30, 323, 324
thiosulphate   2–20, 26–30, 296, 313, 314, 318, 319, 328, 329
total alkalinity   2–20, 26–30, 400–413
Fluoride in
  biological materials   114
  coal   115, 116
  foodstuffs   534, 536
  industrial effluents   113, 521–524
  milk   115
  natural water   104–107, 116–119, 466, 467, 469–477, 479–486, 496, 497, 536–541, 544–549
  plant extracts   114, 115
  potable water   107–113, 125, 126, 469–477, 479–486, 508–512
  rain   107, 117, 119–124, 129, 508–512
  sewage   114, 115
  vegetative matter   114, 115
  waste water   113, 114, 125
Fluoride, process analyser   550, 554
Foodstuffs
  ascorbate in   458
  bicarbonate in   534, 536
  bromate in   17–20
  bromide in   17–20, 91, 92, 534, 536
  chlorate in   534, 536
  chloride in   17–20, 78, 534, 536
  chromate in   534, 536
  citrate in   17–20, 454

Foodstuffs (*contd.*)
  dihydroascorbate in   459, 460
  fluoride in   534, 536
  glutamate in   17-20, 460, 461
  isocitrate in   17-20, 454, 455
  iodide in   17-20, 103, 534, 536
  lactate in   17-20, 456, 457
  malate in   17-20, 456
  nitrate in   17-20, 534, 536
  nitrite in   534, 536
  oxalate in   17-20, 534, 536
  perchlorate in   17-20, 534, 536
  phosphate in   534, 536
  sulphate in   534-536
  sulphite in   17-20, 317
Formate in natural water   453, 466, 467
Free cyanide in
  air   262
  animal feeds   261, 262
  biological fluids   261, 262
  fertilizers   261
  industrial effluents   247-251
  natural water   241-247
  plants   261
  potable water   247
  sewage   262
  waste water   251-261
Fumarate in natural water   466, 467

Gallate in natural water   466, 467
Gas chromatography of
  amino acids   2-20, 54-56, 461
  bromide   2-20, 54-56, 83-86, 90-92, 124, 125
  carboxylates   2-20, 54-56
  chloride   2-20, 54-56, 66, 70, 124, 125
  cyanide   2-20, 54-56, 320, 322
  fluoride   2-20, 54-56,
  free cyanide   2-20, 54-56, 246, 247, 255-260
  iodide   2-20, 54-56, 94, 102, 124
  nitrate   2-20, 54-56, 305
  nitrite   2-20, 54-56, 171, 206-208, 236-238
  phosphate   2-20, 54-56, 305, 367
  polysulphide   2-20, 54-56, 336
  selenate   2-20, 54-56, 429-433
  selenite   2-20, 54-56, 423, 424, 429-431
  sulphate   2-20, 54-56, 305
  sulphide   2-20, 54-56, 328, 330
  sulphonate   2-20, 54-56
  thiocyanate   2-20, 54-56, 320, 322
  total cyanide   2-20, 54-56, 263
Gas chromatography-atmospheric pressure helinum microwave induced plasma emission spectrometry of fluoride   106

Gas phase luminescence analysis of silicate   383
Germanate in natural water   445
Glass
  chromate in   445
  dichromate in   445
Glutamate in foodstuffs   460, 461
Glycollate in natural water   466, 467
Glyoxylate in natural water   496, 497
Gold cyanide complexes in industrial effluents   449, 524-530
Grain, sulphate in   17-20, 311

Hexabenzene carboxylate in natural water   467
Hexathionate in industrial effluents   320
High performance liquid chromatography of
  acetate   453
  arsenic heteropolyacids   356-360
  ascorbate   458
  bromide   2-20, 47-52, 86, 90, 117, 119, 120, 126-129, 133, 320, 321
  bromite   2-20, 47-52
  carboxylates   463
  chlorate   2-20, 47-52
  chloride   2-20, 47-52, 66, 70, 117-220, 125, 127-129, 130, 320, 321
  chlorite   2-20, 47-52, 117-125, 127-129, 133
  chromate   2-20, 47-52, 439, 440
  dichromate   2-20, 47-52, 439-440
  ethylene diamine tetraacetate   2-20, 47-52, 462
  fluoride   2-20, 47-52, 113
  formate   2-20, 47-52, 453
  germanium heteropoly acids   359, 360
  hexathionate   320
  hypochlorite   2-20, 47-52
  iodate   2-20, 47-52, 117, 125, 127
  iodide   2-20, 47-52, 94, 102, 104, 117-120, 127-128, 133, 320
  monofluorophosphate   378
  nitriloacetate   2-20, 47, 52, 462
  nitrate   2-20, 47-52, 119, 120, 228-231, 234, 320, 321
  nitrite   2-20, 47-52, 228-231, 234
  pentathionate   320
  phosphate   2-20, 47-52, 356-360, 369, 373, 375-378
  phosphorus heteropoly acids   395, 360
  pyrophosphate   375-378
  selenate   2-20, 47-52, 431-433
  selenite   2-20, 47-52, 431-433
  silicon heteropoly acids   358-360
  sulphate   2-20, 47-52
  sulphite   314
  sulphonates   2-20, 47-52, 463

tetrathionate 320
thiocyanate 2-20, 47-52, 119, 120, 320, 321
thiosulphate 2-20, 47-52, 320
tripolyphosphate 375-378
trithionate 320
High purity water, carbonate in 414, 415
Hydroxybenzoate in natural water 466, 467
α-Hydroxybutyrate in natural water 466, 467, 496, 497
Hypochlorite in
  ground water 127, 128
  natural water 126, 127, 496, 497

Inductively coupled plasma atomic emission spectrometry of
  bromide 2-20, 31, 34, 35, 83
  chloride 2-20, 31, 34, 35, 64, 65
  fluoride 2-20, 31, 34, 35, 105
  iodide 2-20, 31, 34, 35, 93
  molybdate 2-20, 31, 34, 35, 446, 447
  sulphate 2-20, 31, 34, 35, 296, 297
  tungstate 2-20, 31, 34, 35, 447
Industrial effluents
  borate in 396, 490-494, 530, 496, 497
  bromide in 91, 521-530
  chloride in 70, 71, 521-524
  cobalt cyanide complexes in 524-530
  cyanide in 524-530
  dithionate in 529-530
  ferrocyanide in 448-450
  fluoride in 113, 521-524
  gold cyanide complexes in 524-530
  hexathionate in 319, 320
  iodide in 102, 524-530
  iron cyanide complexes in 524-530
  nickel cyanide complexes in 524-530
  nitrate in 183, 239, 521-524
  nitrite in 210, 211, 239, 521-524
  pentathionate in 319, 320
  phosphate in 368, 521-524
  polysulphide in 336
  polythionite in 319, 320
  silicate in 385
  sulphate in 307, 308, 521-524
  sulphide in 329, 330, 524-530
  sulphite in 315, 530
  tetrathionate in 319, 320
  trithionate in 319, 320
  thiocyanate in 321, 322
  thiosulphate in 319-320
Iodate in
  natural water 128, 130-133, 480, 483
  potable water 124, 125, 133
  rain 123
Iodide, preconcentration 556

Iodide in
  animal foods 103
  cooling water 102
  foodstuffs 103, 534-536
  industrial effluents 102, 524-529
  milk 104
  natural waters 92-95, 127-133, 496, 497
  pharmaceuticals 104
  potable water 96-102, 124, 125, 128
  rain 96, 117-120, 123, 124
  seaweed 103
  table salt 103, 104
Iodine in natural water 130-133
Ion chromatography of
  acetate 463, 467, 496, 497
  acetoacetate 466, 467, 496, 497
  aminoacetate 2-20
  anisate 466, 467
  arsenate 2-20, 44-47, 315, 420, 421, 467, 486-490, 496, 497, 520
  arsenite 2-20, 44-47, 421, 422, 425, 433, 490
  ascorbate 2-20, 44-47
  benzoate 466, 467
  bicarbonate 2-20, 44-47, 415-417, 496, 497, 534-536
  borate 2-20, 44-47, 396, 466, 467, 489-497
  borofluoride 466, 467, 496, 497
  bromate 482-486
  bromide 2-20, 44-47, 89-91, 123-125, 330, 360, 367, 368, 466, 467, 473-477, 479-486, 496, 497, 504-529, 534, 536
  bromite 2-20, 44-47
  butyrate 466, 467
  carbonate 2-20, 44-47, 414, 496, 497
  carboxylate 463
  chloroacetate 466, 467, 496, 497
  chlorobenzoate 466, 467
  chlorate 2-20, 44-47, 466, 467, 496, 497, 534, 536
  chloride 2-20, 44-47, 66, 68, 70-72, 123-125, 129, 306-311, 315, 360, 370, 423, 466-486, 495-524, 531-536
  chlorite 2-20, 44-47, 128
  chromate 466, 467, 496, 497, 534, 536
  citrate 463, 467
  cobalt cyanide complexes 449, 524-530
  copper cyanide complexes 449, 524-530
  cyanate 2-20, 44-47
  cyanide 329-330, 496, 497, 519, 520, 524-529
  dimethyl benzoate 466, 467
  dithionate 315, 319, 530
  fluoride 2-20, 44-47, 107, 113, 115, 116, 125, 129, 306, 308, 309, 360, 367, 370, 466, 467, 469-516, 521-524, 534, 536
  formate 2-20, 44-47, 463-467

Ion chromatography of (contd.)
    free cyanide 2–20, 44–47, 247, 250
    fumarate 466, 467
    gallate 466, 467
    glycollate 466, 467, 496, 497
    glyoxylate 496, 497
    gold cyanide complexes 449, 524–530
    hydroxybenzoate 466, 467
    hydroxybutyrate 466, 467, 496, 497
    hypochlorite 2–20, 44–47, 127, 128, 329, 496, 497
    iodate 482–486
    iodide 2–20, 44–47, 102, 123–125, 330, 496, 497, 524–530, 534, 536
    isobutyrate 466, 467, 496, 497
    iron cyanide complexes 448, 449, 524–530
    lactate 466, 467, 496, 497
    malate 466, 467
    malonate 466, 467, 496, 497
    metal cyanide complexes 2–20, 44–47, 448, 449
    nickel cyanide complexes 448, 449
    nitrate 2–20, 44–47, 167, 171, 180, 183, 185–187, 216, 232, 239, 306, 307, 310, 311, 360, 367, 368, 374, 466, 467, 469–486, 496, 497, 533–536
    nitrite 2–20, 44–47, 129, 208–210, 232, 239, 309, 310, 360, 367, 368, 423, 466, 467, 469–486, 495–516, 521–524, 534, 536
    perchlorate 466, 467, 496, 497, 534, 536
    phosphate 2–20, 44–47, 308, 309, 311, 315, 316, 360, 367, 368, 370, 373, 423, 466, 467, 473–486, 495–516, 521–524, 533–536
    polythionate 319
    propionate 466, 467, 496, 497
    protocatechuate 466, 467
    pyruvate 463–466
    resorcylate 466–467
    salicylate 466–467
    selenate 2–20, 44–47, 422, 425, 433, 486–490, 520
    selenite 2–20, 44–47, 315, 422, 423, 425, 433, 466, 467, 486–490, 496, 497
    silicate 2–20, 44–47, 383, 496, 497
    sulphate 2–20, 44–47, 129, 299, 305, 306, 308–311, 360, 370, 466, 467, 469–486, 495–519, 521–524, 531–536
    sulphide 2–20, 44–47, 329, 330, 485, 486, 496, 497, 519, 520, 524–529
    sulphite 2–20, 44–47, 315, 316, 319, 496, 497, 508–512, 534
    sulphonate 2–20, 44–47
    tartrate 466, 467
    thiocyanate 2–20, 44–47, 496, 497
    thiosulphate 2–20, 44–47, 319, 496, 497
    total cyanide 2–20, 44–47, 263
    zinc cyanide complexes 448, 449, 524–530
Ion exchange chromatography of
    bromide 2–20, 52, 53, 85, 120–123
    carbonate 2–20, 52, 53, 414, 416
    chloride 2–20, 52, 53, 68, 120–123, 298, 299
    fluoride 2–20, 52, 53, 106
    nitrate 2–20, 52, 53, 167, 238, 239
    nitrite 2–20, 52, 53, 238, 297
    phosphate 2–20, 52, 53, 414
    salicylate 2–20, 52, 53
    selenite 2–20, 52, 53, 424
    sulphate 2–20, 52, 53, 298, 299, 414
Ion exclusion chromatography of
    bicarbonate 416, 417
    carbonate 309
    carboxylates 463
    chloride 71, 309
    phosphate 309, 370, 375
    pyrophosphate 375
    sulphate 309
    tripolyphosphate 375
Ion pair chromatography of
    inorganic anions 540
    organic anions 540
Ion selective electrodes in
    acetate 2–20, 38–41, 453
    borate 2–20, 38–41, 396
    bromide 2–20, 38–41, 83
    chloride 2–20, 38–41, 71–78, 116–119
    fluoride 2–20, 38–41, 107–113
    free cyanide 2–20, 38–41, 246, 250, 254, 255
    iodate 2–20, 38–41, 132, 133
    iodide 2–20, 38–41, 93, 94, 103, 132, 133
    nitrate 2–20, 38–41, 116, 117, 161–166, 184, 186, 240, 241, 246
    nitrite 2–20, 38–41, 240, 241, 246
    perchlorate 2–20, 38–41, 127
    sulphate 2–20, 38–41, 297, 298, 304, 305
    sulphide 2–20, 38–41, 328, 331, 335
    sulphite 2–20, 38–41, 314
    thiocyanate 2–20, 38–41, 320, 323
    thiosulphate 2–20, 38–41
    total cyanide 2–20, 38–41, 263, 265–268
Iron cyanide complexes in industrial effluents 448–450, 524–530
Iso ascobate in water 2–20
Isobutyrate in natural water 466, 467, 496, 497
Isocitrate in foodstuffs 454, 455
Isotope dilution analysis of
    bromide 89
    chloride 66, 67

Lactate in
  beverages   456, 457
  foodstuffs   456, 457
  natural waters   466–467, 496, 497
  silage   456

Malate in
  beverages   456
  foodstuffs   456
  natural waters   466, 467
Malonate in water   466, 467, 496, 497
Mass spectrometry of nitrate   167
Meat
  nitrate in   17–20
  nitrite in   17–20
Metal cyanide complexes
  preconcentration   564, 568
Metal cyanide complexes in natural waters   448, 449
Micelle ion exclusion chromatography of
  bromide   86, 119, 123, 133
  iodate   123, 133
  iodide   94, 117, 123, 133
  iodite   119, 123, 133
  nitrate   119, 123, 133, 232
  nitrite   119, 232
Milk
  fluoride in   17–20, 115
  iodide in   17–20, 104
  nitrate in   17–20
  nitrite in   17–20, 217
  pyruvate in   17–20, 460
  selenite in   17–20, 423, 424
Molybdate in
  natural waters   445–447
  seawater   446
Molybdate, preconcentration   563, 564
Monofluorophosphate in natural water   378

Natural waters
  acetate in   453, 466, 467
  acetoacetate in   466, 467, 496, 497
  amino acetates in   461
  anisate in   466, 467
  arsenate in   419–422, 466, 467, 486–490, 496, 497
  arsenite in   419–421, 425, 486–490
  ascorbate in   457, 458
  benzoate in   466, 467
  bicarbonate in 415, 496, 497
  borate in   388–396, 466, 467, 490–494, 496, 497
  borofluoride in   466, 467, 496, 497

bromate in   128, 133, 482–485
bromide in   78–88, 126, 127, 466, 467, 474–486, 496, 497
butyrate in   466, 467, 496, 497
carbonate in   414, 463–467, 496, 497
chlorate in   126, 127, 133, 466, 467, 496, 497
chloride in   58–68, 116–119, 127, 133, 466, 467, 469–486, 496, 497, 536–539
chlorite in   126, 127
chromate in   439–442, 466, 467, 496, 497
citrate in   466, 467
condensed polyphosphates in   374–378
cyanate in   271
cyanide in   496, 497
dichromate in   434–432
dimethyl benzoate in   466, 467
ethylene diamine tetraacetate in   462
fluoride in   104–107, 116–119, 466, 467, 474–486, 496, 497, 536–539
formate in   453, 466, 467
free cyanide in   241–247
fumarate in   466, 467
gallate in   466, 467
germanate in   445
glycollate in   466, 467, 496, 497
glyoxylate in   496, 497
hexabenzene carboxylate in   466, 467
hydroxybenzoate in   466, 467
α-hydroxybutyrate in   466, 467, 496, 497
hypochlorite in   126, 127, 496, 497
iodate in   128, 130–133, 482–485
iodide in   92–95, 496, 497
iodine in   130–133
isobutyrate in   466, 467, 496, 497
lactate in   466, 467, 496, 497
malate in   466, 467
malonate in   466, 467, 496, 497
metal cyanide complexes in   448–450
molybdate in   445, 446
monofluorophosphate in   378
nitrate in   116–119, 133, 140–168, 217–234, 466, 467, 469–486, 496, 497, 536–539
nitrite in   133, 187–209, 217–234, 466, 467, 474–485, 496, 497, 536–539
nitriloacetate in   462
octane dioate in   467
octanoiate in   467
perchlorate in   466, 467, 496, 497
polysulphide in   336
polythionate in   318, 319
phosphate in   342–365, 375–378, 419, 420, 466, 467, 469–486, 496, 497, 536–539
phosphite in   466, 467,
propionate in   466, 467, 496, 497

Natural waters (*contd.*)
  protocatchuate in   466, 467
  pyrophosphate in   374–378
  resorcylate in   466, 467
  salicylate in   466, 467
  selenate in   425, 429–433, 486–490
  selenite in   421–423, 429–433, 466, 467, 486–490, 496, 497
  silicate in   381–383, 496, 497
  sulphate in   278–300, 466, 467, 469–486, 496, 497, 536–539
  sulphide in   324–329, 485, 486, 496, 497
  sulphite in   311–315, 321, 496, 497
  sulphonate in   463, 464, 466
  tartrate in   466, 467
  tellurate in   434
  thiocyanate in   320–321, 496, 497
  thiosulphate in   318, 319, 496, 497
  total alkalinity in   400–411
  total cyanide in   262, 263
  tripolyphosphate in   374–378
  tungstate in   446, 447
  uranate in   447
  vanadate in   447
Nephelometry of sulphate   288
Neutron activation analysis of
  bromide   88
  iodide   95
  phosphate   365
Nickel cyanide complexes in industrial effluents   448, 449, 524–530
Nitrate
  process analyser   550, 554
  test kit   544–549
Nitrate in
  biological fluids   240
  industrial effluents   183, 239, 521–524
  natural waters   116, 117, 119, 133, 140–168, 466, 467, 469–486, 496, 497, 536–539
  plants   186
  plant extracts   185, 533, 534
  potable waters   172–183, 234–239, 469–486, 516–519
  rain   119–124, 129, 168–172, 234, 495–516
  sewage   184, 239, 240
  snow   168–172
  soil   185, 186, 240
  soil extracts   185, 533, 534
  vegetables   241
  waste water   128, 183, 184
Nitriloacetic acid in
  natural waters   462
  waste waters   462
Nitrite
  process analyser   550–554
  test kit   544–549
Nitrate in
  biological fluids   240
  boiler feed water   216, 534
  foodstuffs   534, 536
  high purity water   216
  industrial effluents   210, 238, 521–524
  milk   217
  natural water   217–234, 466, 467, 469–486, 496, 497, 536–539
  potable water   234–239, 516–520
  rain   123, 129, 209, 234, 495–516
  sewage   239, 240
  soil   216, 217, 240
  vegetables   241
  waste water   127, 128, 211–216, 271

Octane dioate in natural water   467
Octanoiate in natural water   467
Oxalate in
  beverages   455
  foodstuffs   455
Oxides
  chromate in   445
  dichromate in   445

Paper chromatography of
  cyanate   271, 321
  cyanide   271, 321, 323
  thiocyanate   271, 321, 323
Pentathionate in industrial effluents   319, 320
Perchlorate in
  foodstuffs   534, 536
  natural water   466, 467, 496, 497
Pharmaceuticals
  ascorbate in   17–20, 458
  iodide in   17–20, 103
  salicylate in   17–20, 457
  sulphite in   17–20, 318
Phosphate
  adsorption on glass   365
  preconcentration   566, 567
  process analyser   550–554
  sample preservation   370–371
  test kit   544–549
Phosphate in
  blood   374
  boiler feed water   373, 534
  estuary water   366, 367
  industrial effluents   368, 521–524
  natural water   342–366, 374–378, 419, 420, 469–486, 496, 497, 536–539
  plant extracts   373

potable water   17–20, 368
rain   129, 367, 368, 495–516
seawater   366, 367
serum   374
sewage   374
soil extracts   373, 533, 534
waste water   368–372
Phosphite in natural water   466, 467
Plant extracts
  borate in   17–20, 398
  bromide in   17–20
  chloride in   17–20, 533, 534
  fluoride in   17–20, 115
  nitrate in   17–20, 185, 533, 534
  phosphate in   17–20, 373, 533, 534
  sulphate in   17–20, 533, 534
  vanadate in   17–20, 447, 448
Plants
  borate in   17–20, 397
  chloride in   17–20, 77
  free cyanide in   17–20, 261
  nitrate in   17–20, 186, 187
  sulphate in   17–20, 310, 311
  vanadate in   447, 448
Polarography of
  arsenate   2–20, 41–44, 421
  arsenite   2–20, 41–44, 421
  bromide   2–20, 41–44, 87, 88
  chromate   2–20, 41–44, 440, 442, 444
  dichromate   2–20, 41–44, 440, 444
  ethylene diamine tetra acetate   2–20, 41–44
  free cyanide   2–20, 41–44, 250
  iodide   2–20, 41–44, 95, 103
  nitrate   2–20, 41, 44, 183, 187, 232, 233
  nitrite   2–20, 41–44, 208, 209, 232, 233
  nitriloacetate   2–20, 41–44, 462
  phosphate   2–20, 41–44, 360–364
  selenate   2–20, 41–44, 433, 434
  selenite   2–20, 41–44, 422, 423, 433, 434
  silicate   2–20, 41–44, 283, 384
  sulphide   2–20, 41–44, 328–330
Polysulphide in
  industrial effluents   336
  natural water   296, 336
Polythionate in
  industrial effluents   319, 320
  natural waters   318, 319
Post suppression membrane based ion exchange chromatography of
  inorganic anions   540
  organic anions   540
Potable water
  arsenate in   420, 520
  arsenite in   486–489
  bicarbonate in   415–517
  bromate in   129, 133
  bromide in   88–90, 124, 125, 516–519
  carboxylate in   463, 464, 466
  chlorate in   128
  chloride in   69, 70, 125, 494–505, 516–519
  chlorite in   128
  chromate in   445
  cyanide in   519, 520
  dichromate in   445
  fluoride in   107–113, 125, 508–512
  free cyanide in   247
  iodate in   129, 133
  iodide in   96–102, 124, 125, 133
  nitrate in   171–183, 234–239, 516–519
  nitrite in   234–239
  phosphate in   368
  selenate in   433, 486–489
  selenite in   433, 520, 486–489
  silicate in   383, 384
  sulphate in   305–307, 516–519
  sulphide in   329, 485, 486, 519, 520
  sulphite in   519, 520
  sulphonate in   464–466
  total alkalinity in   411–414
Potentiometric analysis of
  glutamate   461
  iodide   95
  polysulphide   336
Preconcentration of
  arsenite   558
  borate   558
  chromate   559–562
  dichromate   559–562
  iodide   556
  metal cyanide complexes   564, 568
  molybdate   563, 564
  phosphate   556, 557
  rhenate   564
  selenate   558, 559
  selenite   558, 559
  sulphide   557, 558
Process analysers
  bromide   550–554
  chloride   550–554
  cyanide   550–554
  fluoride   550–554
  nitrate   550–554
  nitrite   550–554
  phosphate   550–554
  silicate   550–554
  sulphate   550–554
  sulphide   550–554
  total alkalinity   550–554

Propionate in
  natural water 466, 467, 496, 497
  sea water 467
Protocatechuate in natural water 466, 467
Pyrolysis gas chromatography of
    chlorolignosulphonates 464–466
Pyrophosphate in natural waters 374–378
Pyruvate in
  milk 460
  sea water 467

Radioactivation analysis of phosphate 365
Radiochemical analysis of
  chloride 68
  iodide 95
Rain
  bromate in 129
  bromide in 89, 119–124, 129, 315, 495–516
  chloride in 69, 119–124, 129, 209, 315, 367, 368, 495–516
  fluoride in 107, 117, 119–124, 129, 315, 495–516
  iodide in 95, 119–124, 129
  nitrate in 119–124, 129, 168–172, 209, 234, 315, 367, 368, 495–516
  nitrite in 119–124, 129, 209, 234, 315, 495–516
  phosphate in 129, 209, 315, 367, 368, 495–516
  sulphate in 119–124, 129, 209, 300–305, 367, 368, 495–516
  sulphite in 315, 495–516
  thiocyanate in 117–121
Raman spectroscopy of
  nitrate 183, 184
  nitrite 564
Recorcylate in natural water 466, 467
Rhenate, preconcentration 564

Salicylate in
  natural water 466, 467
  pharmaceuticals 457
  urine 457
Sample preservation
  chromate 442
  dichromate 442
  phosphate 370, 371
  selenate 433
  selenite 433
Seawater
  acetate in 467
  acrylate in 467
  benzoate in 467
  butyrate in 467
  molybdate in 446

  phosphate in 366, 367
  propionate in 467
  pyruvate in 467
  selenate in 425
  selenite in 425
  valerate in 467
Seaweed, iodide in 103
Segmented flow analysis of
  bicarbonate 2–20, 24–26, 415
  borate 2–20, 24–26, 396
  chromate 2–20, 24–26, 435
  fluoride 2–20, 24–26, 105
  free cyanide 2–20, 24–26, 245
  nitrate 2–20, 24–26, 225
  nitrite 2–20, 24–26, 201, 225
  phosphate 2–20, 24–26, 354
  silicate 2–20, 24–26, 383
  sulphite 2–20, 24–26, 314
  total alkalinity 2–20, 24–26, 402
Selenate
  preconcentration 558, 559
  sample preservation 433
Selenate in
  natural water 419, 422, 425, 429–433
  potable water 425, 433, 520
  seawater 425–429
  treated water 433, 434
Selenite
  preconcentration 558, 559
  sample preservation 433
Selenite in
  biological samples 424
  milk 423, 424
  natural waters 421–423, 429–433, 466, 467
  potable water 423, 433
  seawater 425–429
  soil extracts 423
  treated water 433
  urine 424
Serum
  Phosphate in 17–20, 374
  sulphide in 17–20
Sewage
  chloride in 72
  fluoride in 114
  free cyanide in 262
  nitrate in 184, 185
  nitrite in 239–241
  phosphate in 372, 373
  sulphide in 331–334
Silage, lactate in 456
Silicate
  process analyser 550–554
  test kit 544–549

Silicate in
  boiler feed water   386, 387
  industrial effluents   385
  natural waters   381–383, 496, 497
  potable waters   383–386
  waste waters   386
Silicates
  chromate in   445
  dichromate in   445
Snow, nitrate in   168–172
Soil
  borate in   17–20, 397
  bromide in   17–20, 91
  carbonate in   17–20, 415
  chloride in   17–20, 77
  free cyanide in   17–20
  nitrate in   17–20, 185, 240
  nitrite in   17–20, 216, 217, 240
  sulphate in   17–20, 309, 310
  sulphide in   17–20, 334, 335
  vanadate in   17–20, 447, 448
Soil extract
  chlorate in   17–20, 128, 129
  chloride in   17–20, 533, 534
  chlorite in   17–20
  hypochlorite in   17–20
  nitrate in   17–20, 185, 533, 534
  phosphate in   17–20, 533, 534
  selenite in   17–20, 423
  sulphate in   17–20, 310, 533, 534
  vanadate in   17–20, 447, 448
Spectrofluorometry of
  arsenate   422
  arsenite   422
  ascorbate   458
  borate   388–394
  chloride   2–24, 73
  free cyanide   2–24, 250
  isoascorbate   2–24
  nitrate   2–24, 155–158
  nitrite   2–24
  salycylate   2–24
  selenate   2–24, 425–429, 433, 434
  selenite   2–24, 425–429, 433, 434
  sulphate   2–24, 288, 289
  sulphide   2–24, 327
Spectrophotometry of
  acetoacetate   466
  arsenate   2–22, 419–421
  arsenite   2–22, 420, 421
  ascorbate   2–22, 457, 458
  borate   2–22, 388–396
  bromide   2–22, 79–82, 89, 90
  chlorate   2–22, 128

  chloride   2–22, 59–62
  chlorite   2–22, 128
  chromate   2–22, 434, 435, 442–444
  citrate   2–22, 454
  dichromate   459, 460
  dihydroascorbate
  ferrocyanide   2–22, 448
  fluoride   2–22, 104, 105, 107, 108, 114
  free cyanide   2–22, 241–245, 248, 251–254, 262
  glutamate   2–22, 460, 461
  hypochlorite   2–22
  iodate   2–22, 130–132
  iodide   2–22, 96–102, 130–133
  isoascorbate   2–22
  isocitrate   2–22
  lactate   2–22
  malate   2–22
  nitrate   2–22, 140–155, 168–170, 172–174, 183, 185, 186, 217–225, 240, 241
  nitrite   2–22, 140–155, 210–213, 216–225, 240, 241
  oxalate   2–22
  phosphate   2–22, 342–365, 372, 373
  selenite   2–22, 421, 422
  silicate   2–22, 381–383, 385–387
  sulphate   2–22, 281–284, 300, 301, 305, 308, 311, 313, 315
  sulphide   2–22, 307, 308, 315, 325–327, 330, 331, 336
  sulphite   2–22, 307, 308, 313, 315, 317, 330, 331
  tellurate   2–22, 434
  thiocyanate   2–22, 320, 323
  thiosulphate   2–22
  titanate   2–22, 448
  total alkalinity   2–22, 400–402, 411
  total cyanide   2–22, 264, 265
  uranate   2–22, 447
  vanadate   2–22, 447, 448
Sulphate
  process analyser   550–554
  test kit   544–549
Sulphate in
  boiler feed water   309, 530–534
  foodstuffs   534, 536
  grain   311
  industrial effluents   307, 308, 521–524
  natural water   278–300, 466, 467, 469–486, 496, 497
  plants   310, 311
  plant extracts   533, 534
  potable water   305–307
  rain   119–124, 129, 300–305, 495–516

Sulphate in (contd.)
    soil   309, 310
    soil extracts   310, 311, 533, 534
    waste water   308, 309
Sulphide
    preconcentration   557, 558
    process analyser   550-554
    test kit   544-549
Sulphide in
    air   336
    ground water   127
    industrial effluents   329, 330, 524-530
    natural water   324-329, 482, 496, 497
    potable water   329, 485, 519, 520
    sewage   331-334
    soil   334, 335
    waste water   330, 331
Sulphite
    test kit   544-549
Sulphite in
    boiler feed water   315, 316
    foodstuffs   316
    industrial effluents   314, 526-530
    natural water   296, 311-315, 321, 420, 496, 497
    pharmaceuticals   318
    rain   315, 508-512
    waste water   315
    wine   316

Table salt, iodide in   17-20, 103, 104
Tartrate in natural water   466, 467
Tellurate in natural water   434
Test kits
    carbonate   544-549
    chloride   544-549
    chromate   544-549
    fluoride   544-549
    nitrate   544-549
    nitrite   544-549
    phosphate   544-549
    silicate   544-549
    sulphide   544-549
    sulphate   544-549
    sulphite   544-549
    total alkalinity   544-549
Tetrathionate in industrial effluents   320
Thiocyanate in
    industrial effluents   321, 322
    natural waters   320-331, 496, 497
    rain   117-124
    urine   323
    waste water   323
Thiosulphate in
    industrial effluent   319, 320
    natural waters   296, 318, 319, 496, 497
Tissue, tunstate in   17-20, 447
Titanate in natural waters   448
Titration of
    bicarbonate   2-20, 35-38, 415
    bromide   2-20, 35-38, 78
    chlorate   2-20, 35-38, 126, 127
    chloride   2-20, 35-38, 58, 59, 70, 71, 72, 77, 78
    chlorite   2-20, 35-38, 127
    free cyanide   2-20, 35-38, 247, 248, 321, 322
    hypochlorite   2-20, 35-38
    iodide   2-20, 35-38, 92, 102
    polysulphide   2-20, 35-38, 336
    polythionate   2-20, 35-38, 319
    suphate   2-20, 35-38, 278-287, 306-309, 318
    sulphide   2-20, 35-38, 324, 329, 330, 336
    sulphite   2-20, 35-38, 311-313, 317, 336
    thiocyanate   2-20, 35-38, 321, 322
    thiosulphate   2-20, 35-38, 318-320, 326
    total alkalinity   2-20, 35-38, 400, 401
    total cyanide   2-20, 35-38, 263
Total alkalinity, process analyser   550-554
Total alkalinity in
    natural waters   400-402
    potable water   411
Total cyanide in water   2-20
Treated water
    selenate in   433, 434
    selenite in   433, 434
Tripolyphosphate in natural waters   374-378
Trithionate in industrial effluents   320
Tungstate in
    natural water   446, 447
    tissue   447
Turbidimetry of sulphate   287, 288

Ultraviolet spectroscopy of
    ascorbate   458
    citrate   2-20, 454
    dihydroascorbate   2-20
    ethylene diamine tetraacetate   2-20
    isocitrate   2-20, 454, 455
    lactate   2-20, 456, 457
    malate   2-20, 456
    nitriloacetate   2-20, 462
    nitrate   2-20, 119, 158, 159, 178-180, 225
    nitrite   2-20, 201, 202, 225
    oxalate   455
    phosphate   2-20, 369
    pyruvate   2-20, 460
    sulphite   2-20, 316
    total phosphate   369
Uranate in natural water   447

Urine
  salicylate in 17–20, 457
  selenite in 424
  thiocyanate in 323, 324
Valerate in seawater 467
Vanadate in
  natural waters 447
  plant extracts 447, 448
  plants 447, 448
  soil 447, 448
  soil extracts 447, 448
Vegetative matter
  fluoride in 115
  nitrate in 241
  nitrite in 241

Waste water
  carbonate in 414
  chromate in 442–444
  chloride in 71, 72, 125
  chlorite in 128
  cyanide in 251–261, 323
  dichromate in 442–444
  ethyl diamine tetraacetate in 462
  fluoride in 125
  nitrate in 128, 183, 184
  nitriloacetate in 462
  nitrite in 210–216
  phosphate in 368–372
  silicate in 386
  sulphate in 308, 309
  sulphide in 330, 331
  sulphite in 315
  thiocyanate in 323
  total cyanide in 263–271
Wine, sulphite in 17–20, 315

X-ray fluorescence spectroscopy
    of phosphate 365
X-ray spectrometry of bromide 88

Zinc cyanide complexes in
  industrial effluents 524–530
  water 488–450

Springer Verlag and the environment

We at the Springer-Verlag firmly believe that an international science publisher has a special obligation to the environment, and our corporate policies consistently reflect this conviction. We also expect our business partners – paper mills, printers, packaging manufacturers, etc. – to commit themselves to using environmentally friendly materials and production processes. The paper in this book is made from low- or no-chlorine pulp and is acid free, in conformance with international standards for paper permanency.